PROBLEMS AND SOLUTIONS ON ELECTROMAGNETISM

Major American Universities Ph.D.
Qualifying Questions and Solutions

PROBLEMS AND SOLUTIONS ON ELECTROMAGNETISM

Compiled by:

**The Physics Coaching Class
University of Science and
Technology of China**

Refereed by:

Zhao Shu-ping, You Jun-han & Zhu Jun-jie

Edited by:

Lim Yung-kuo

World Scientific

NEW JERSEY · LONDON · SINGAPORE · BEIJING · SHANGHAI · HONG KONG · TAIPEI · CHENNAI

Published by

World Scientific Publishing Co. Pte. Ltd.

5 Toh Tuck Link, Singapore 596224

USA office: 27 Warren Street, Suite 401-402, Hackensack, NJ 07601

UK office: 57 Shelton Street, Covent Garden, London WC2H 9HE

British Library Cataloguing-in-Publication Data
A catalogue record for this book is available from the British Library.

First published 1993
Reprinted 1994, 1996, 1998, 2000, 2002, 2003, 2005, 2007

Major American Universities Ph.D. Qualifying Questions and Solutions
PROBLEMS AND SOLUTIONS ON ELECTROMAGNETISM

ISBN-13 978-981-02-0625-3
ISBN-10 981-02-0625-9
ISBN-13 978-981-02-0626-0 (pbk)
ISBN-10 981-02-0626-7 (pbk)

PREFACE

This series of physics problems and solutions, which consists of seven volumes — Mechanics, Electromagnetism, Optics, Atomic, Nuclear and Particle Physics, Thermodynamics and Statistical Physics, Quantum Mechanics, Solid State Physics and Relativity, contains a selection of 2550 problems from the graduate school entrance and qualifying examination papers of seven U.S. universities — California University Berkeley Campus, Columbia University, Chicago University, Massachusetts Institute of Technology, New York State University Buffalo Campus, Princeton University, Wisconsin University — as well as the CUSPEA and C. C. Ting's papers for selection of Chinese students for further studies in U.S.A. and their solutions which represent the effort of more than 70 Chinese physicists plus some 20 more who checked the solutions.

The series is remarkable for its comprehensive coverage. In each area the problems span a wide spectrum of topics while many problems overlap several areas. The problems themselves are remarkable for their versatility in applying the physical laws and principles, their uptodate realistic situations, and their scanty demand on mathematical skills. Many of the problems involve order of magnitude calculations which one often requires in an experimental situation for estimating a quantity from a simple model. In short, the exercises blend together the objectives of enhancement of one's understanding of the physical principles and ability of practical application.

The solutions as presented generally just provide a guidance to solving the problems, rather than step by step manipulation, and leave much to the students to work out for themselves, of whom much is demanded of the basic knowledge in physics. Thus the series would provide an invaluable complement to the textbooks.

The present volume for Electromagnetism consists of five parts: electrostatics, magnetostatic and quasi-stationary electromagnetic fields, circuit analysis, electromagnetic waves, relativity and particle-field interactions, and contains 440 problems. 34 Chinese physicists were involved in the task of preparing and checking the solutions.

In editing, no attempt has been made to unify the physical terms, units and symbols. Rather, they are left to the setters' and solvers' own preference so as to reflect the realistic situation of the usage today. Great

v

pains has been taken to trace the logical steps from the first principles to the final solutions, frequently even to the extent of rewriting the entire solution. In addition, a subject index has been included to facilitate the location of topics. These editorial efforts hopefully will enhance the value of the volume to the students and teachers alike.

Yung-Kuo Lim
Editor

INTRODUCTION

Solving problems in school work is the exercise of the mind and examination questions are usually picked from the problems in school work. Working out problems is an essential and important aspect of the study of Physics.

Major American University Ph.D. Qualifying Questions and Solutions is a series of books which consists of seven volumes. The subjects of each volume and the respective referees (in parentheses) are as follows:

1. Mechanics (Qiang Yan-qi, Gu En-pu, Cheng Jia-fu, Li Ze-hua, Yang De-tian)
2. Electromagnetism (Zhao Shu-ping, You Jun-han, Zhu Jun-jie)
3. Optics (Bai Gui-ru, Guo Guang-can)
4. Atomic, Nuclear and Particle Physics (Jin Huai-cheng, Yang Bao-zhong, Fan Yang-mei)
5. Thermodynamics and Statistical Physics (Zheng Jiu-ren)
6. Quantum Mechanics (Zhang Yong-de, Zhu Dong-pei, Fan Hong-yi)
7. Solid State Physics, Relativity and Miscellaneous Topics (Zhang Jia-lu, Zhou You-yuan, Zhang Shi-ling)

This series covers almost all aspects of University Physics and contains 2550 problems, most of which are solved in detail.

The problems have been carefully chosen from 3100 problems, of which some came from the China–U.S. Physics Examination and Application Program, some were selected from the Ph.D. Qualifying Examination on Experimental High Energy Physics sponsored by Chao Chong Ting. The rest came from the graduate school entrance examination questions of seven world-renowned American universities: Columbia University, University of California at Berkeley, Massachusetts Institute of Technology, University of Wisconsin, University of Chicago, Princeton University and State University of New York, Buffalo.

In general, examination problems in physics in American universities do not involve too much mathematics; however, they are to a large extent characterized by the following three aspects: some problems involving various frontier subjects and overlapping domains of science are selected by professors directly from their own research work and show a "modern style". Some problems involve broad fields and require a quick mind to analyse, while the methods needed for solving the other problems are simple and practical but requires a full "touch of physics". Indeed, we venture to opine that the problems, as a whole, embody to some extent the characteristics of American science and culture, as well as the philosophy underlying American education.

Therefore, we considered it worthwhile to collect and solve these problems and introduce them to students and teachers, even though the effort involved was extremely strenuous. As many as a hundred teachers and graduate students took part in this time-consuming task.

A total of 440 problems makes up this volume of five parts: electrostatics (108), magnetostatic and quasi-stationary electromagnetic fields (119), circuit analysis (90), electromagnetic waves (67), and relativity, particle-field interactions (56).

In scope and depth, most of the problems conform to the undergraduate physics syllabi for electromagnetism, circuit analysis and electrodynamics in most universities. However, many of them are rather profound, sophisticated and broad-based. In particular, problems from American universities often fuse fundamental principles with the latest research activities. Thus the problems may help the reader not only to enhance understanding in the basic principles, but also to cultivate the ability of solving practical problems in a realistic environment.

International units are used whenever possible, but in order to conform to some of the problems, Gaussian units are also used. This in fact would give the student broader training and wider experience.

This volume is the result of collective efforts of 34 physicists involved in working out and checking of the solutions, among them Zheng Dao-chen, Hu You-qiu, Ning Bo, Zhu Xue-liang, and Zhao Shu-ping.

CONTENTS

Part IV   Electromagnetic Waves

Part V   Relativity, Particle-Field Interactions

PART 1

ELECTROSTATICS

1. BASIC LAWS OF ELECTROSTATICS (1001–1023)

1001

A static charge distribution produces a radial electric field

$$\mathbf{E} = A\frac{e^{-br}}{r}\mathbf{e}_r\,,$$

where A and b are constants.

(a) What is the charge density? Sketch it.

(b) What is the total charge Q?

(MIT)

Solution:

(a) The charge density is given by Maxwell's equation

$$\rho = \nabla \cdot \mathbf{D} = \epsilon_0 \nabla \cdot \mathbf{E}.$$

As $\nabla \cdot u\mathbf{v} = \nabla u \cdot \mathbf{v} + u\nabla \cdot \mathbf{v}$,

$$\nabla \cdot \mathbf{E} = A\left[\nabla(e^{-br}) \cdot \frac{\mathbf{e}_r}{r^2} + e^{-br}\nabla \cdot \left(\frac{\mathbf{e}_r}{r^2}\right)\right].$$

Making use of Dirac's delta function $\delta(\mathbf{r})$ with properties

$$\delta(\mathbf{r}) = 0 \quad \text{for} \quad \mathbf{r} \neq 0,$$
$$= \infty \quad \text{for} \quad \mathbf{r} = 0,$$

$$\int_V \delta(\mathbf{r})dV = 1 \quad \text{if } V \text{ encloses} \quad \mathbf{r} = 0,$$
$$= 0 \quad \text{if otherwise},$$

we have

$$\nabla^2\left(\frac{1}{r}\right) = \nabla \cdot \nabla\left(\frac{1}{r}\right) = \nabla \cdot \left(-\frac{\mathbf{e}_r}{r^2}\right) = -4\pi\delta(\mathbf{r}).$$

Thus

$$\rho = \epsilon_0 A\left[-\frac{be^{-br}}{r^2}\mathbf{e}_r \cdot \mathbf{e}_r + 4\pi e^{-br}\delta(\mathbf{r})\right]$$

$$= -\frac{\epsilon_0 Ab}{r^2}e^{-br} + 4\pi\epsilon_0 A\,\delta(\mathbf{r}).$$

3

Hence the charge distribution consists of a positive charge $4\pi\varepsilon_0 A$ at the origin and a spherically symmetric negative charge distribution in the surrounding space, as shown in Fig. 1.1.

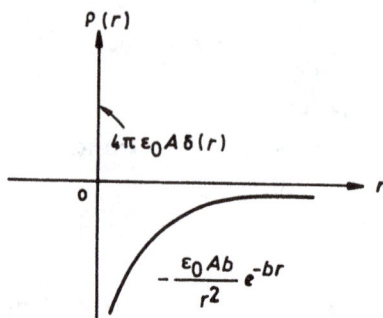

Fig. 1.1

(b) The total charge is

$$Q = \int_{\text{all space}} \rho\, dV$$

$$= -\int_0^\infty \frac{\varepsilon_0 Abe^{-br}}{r^2} \cdot 4\pi r^2 dr + \int_{\text{all space}} 4\pi\varepsilon_0 A\, \delta(\mathbf{r}) dV$$

$$= 4\pi\varepsilon_0 A[e^{-br}]_0^\infty + 4\pi\varepsilon_0 A$$

$$= -4\pi\varepsilon_0 A + 4\pi\varepsilon_0 A = 0\,.$$

It can also be obtained from Gauss' flux theorem:

$$Q = \lim_{r\to\infty} \oint_S \varepsilon_0 \mathbf{E} \cdot d\mathbf{S}$$

$$= \lim_{r\to\infty} \frac{\varepsilon_0 Ae^{-br}}{r^2} \cdot 4\pi r^2$$

$$= \lim_{r\to\infty} 4\pi\varepsilon_0 Ae^{-br} = 0\,,$$

in agreement with the above.

1002

Suppose that, instead of the Coulomb force law, one found experimentally that the force between any two charges q_1 and q_2 was

$$\mathbf{F}_{12} = \frac{q_1 q_2}{4\pi\varepsilon_0} \cdot \frac{(1 - \sqrt{\alpha r_{12}})}{r_{12}^2}\mathbf{e}_r \,,$$

where α is a constant.

(a) Write down the appropriate electric field $\mathbf{E}$ surrounding a point charge q.

(b) Choose a path around this point charge and calculate the line integral $\oint \mathbf{E} \cdot d\mathbf{l}$. Compare with the Coulomb result.

(c) Find $\oint \mathbf{E} \cdot d\mathbf{S}$ over a spherical surface of radius r_1 with the point charge at this center. Compare with the Coulomb result.

(d) Repeat (c) at radius $r_1 + \Delta$ and find $\nabla \cdot \mathbf{E}$ at a distance r_1 from the point charge. Compare with the Coulomb result. Note that Δ is a small quantity.

(*Wisconsin*)

Solution:

(a) The electric field surrounding the point charge q is

$$\mathbf{E}(r) = \frac{q}{4\pi\varepsilon_0}\frac{1}{r^2}(1 - \sqrt{\alpha r})\mathbf{e}_r \,,$$

where r is the distance between a space point and the point charge q, and $\mathbf{e}_r$ is a unit vector directed from q to the space point.

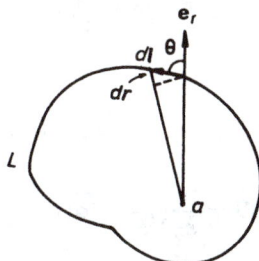

Fig. 1.2

(b) As in Fig. 1.2, for the closed path L we find

$$d\mathbf{l} \cdot \mathbf{e}_r = dl \cos\theta = dr$$

and

$$\oint_L \mathbf{E} \cdot d\mathbf{l} = \oint \frac{q}{4\pi\varepsilon_0} \frac{1}{r^2}(1 - \sqrt{\alpha r})dr$$

$$= \frac{q}{4\pi\varepsilon_0}\left[-\oint_L d\left(\frac{1}{r}\right) + 2\sqrt{\alpha}\oint_L d\left(\frac{1}{\sqrt{r}}\right)\right] = 0.$$

From Coulomb's law $\mathbf{F}_{12} = \frac{q_1 q_2}{4\pi\varepsilon_0 r_{12}^2}\mathbf{e}_{r_{12}}$, we can obtain the electric field of the point charge

$$\mathbf{E}(r) = \frac{q}{4\pi\varepsilon_0 r^2}\mathbf{e}_r.$$

Clearly, one has

$$\oint_L \mathbf{E} \cdot d\mathbf{l} = 0.$$

So the Coulomb result is the same as that of this problem.

(c) Let S be a spherical surface of radius r_1 with the charge q at its center. Defining the surface element $d\mathbf{S} = dS\mathbf{e}_r$, we have

$$\oint_s \mathbf{E} \cdot d\mathbf{S} = \oint_s \frac{q}{4\pi\varepsilon_0} \frac{1}{r_1^2}(1 - \sqrt{\alpha r_1})dS$$

$$= \frac{q}{\varepsilon_0}(1 - \sqrt{\alpha r_1}).$$

From Coulomb's law and Gauss' law, we get

$$\oint_s \mathbf{E} \cdot d\mathbf{S} = \frac{q}{\varepsilon_0}.$$

The two results differ by $\frac{q}{\varepsilon_0}\sqrt{\alpha r_1}$.

(d) Using the result of (c), the surface integral at $r_1 + \Delta$ is

$$\oint_s \mathbf{E} \cdot d\mathbf{S} = \frac{q}{\varepsilon_0}(1 - \sqrt{\alpha(r_1 + \Delta)}).$$

Consider a volume V' bounded by two spherical shells S_1 and S_2 with radii $r = r_1$ and $r = r_1 + \Delta$ respectively. Gauss' divergence theorem gives

$$\oint_{S_1+S_2} \mathbf{E} \cdot d\mathbf{S} = \int_{V'} \nabla \cdot \mathbf{E}\, dV.$$

As the directions of $d\mathbf{S}$ on S_1 and S_2 are outwards from V', we have for small Δ

$$\frac{q}{\varepsilon_0}\left[-\sqrt{\alpha(r_1+\Delta)}+\sqrt{\alpha r_1}\right]=\frac{4\pi}{3}[(r_1+\Delta)^3-r_1^3](\Delta\cdot\mathbf{E})|_{r=r_1}\,.$$

As $\frac{\Delta}{r_1}\ll1$, we can approximately set

$$\left(1+\frac{\Delta}{r_1}\right)^n\approx1+n\frac{\Delta}{r_1}\,.$$

Thus one gets

$$\nabla\cdot\mathbf{E}(r=r_1)=-\frac{\sqrt{\alpha}\,q}{8\pi\varepsilon_0 r_1^{5/2}}\,.$$

On the other hand, Coulomb's law would give the divergence of the electric field produced by a point charge q as

$$\nabla\cdot\mathbf{E}(r)=\frac{q}{\varepsilon_0}\delta(r)\,.$$

1003

Static charges are distributed along the x-axis (one-dimensional) in the interval $-a\le x'\le a$. The charge density is

$$\begin{array}{ll}\rho(x') & \text{for}\quad|x'|\le a\\ 0 & \text{for}\quad|x'|>a\,.\end{array}$$

(a) Write down an expression for the electrostatic potential $\Phi(x)$ at a point x on the axis in terms of $\rho(x')$.

(b) Derive a multipole expansion for the potential valid for $x>a$.

(c) For each charge configuration given in Fig. 1.3, find
    (i) the total charge $Q=\int\rho dx'$,
    (ii) the dipole moment $P=\int x'\rho dx'$,
    (iii) the quadrupole moment $Q_{xx}=2\int x'^2\rho dx'$,
    (iv) the leading term (in powers of $1/x$) in the potential Φ at a point $x>a$.

(Wisconsin)

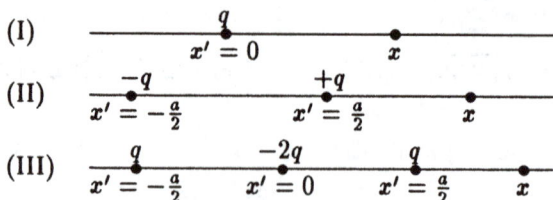

Fig. 1.3

Solution:

(a) The electrostatic potential at a point on x-axis is

$$\Phi(x) = \frac{1}{4\pi\varepsilon_0} \int_{-a}^{a} \frac{\rho(x')}{|x - x'|} dx' \ .$$

(b) For $x > a, a > x' > -a$, we have

$$\frac{1}{|x - x'|} = \frac{1}{x} + \frac{x'}{x^2} + \frac{x'^2}{x^3} + \dots \ .$$

Hence the multipole expansion of $\Phi(x)$ is

$$\Phi(x) = \frac{1}{4\pi\varepsilon_0} \left[\int_{-a}^{a} \frac{\rho(x')}{x} dx' + \int_{-a}^{a} \frac{\rho(x')x'}{x^2} dx' + \int_{-a}^{a} \frac{\rho(x')x'^2}{x^3} dx' + \dots \right] .$$

(c) The charge configuration (I) can be represented by

$$\rho(x') = q\delta(x') \,,$$

for which

(i) $Q = q$; (ii) $P = 0$; (iii) $Q_{xx} = 0$; (iv) $\Phi(x) = \dfrac{q}{4\pi\varepsilon_0 x}$.

The charge configuration (II) can be represented by

$$\rho(x') = -q\delta\left(x' + \frac{a}{2}\right) + q\delta\left(x' - \frac{a}{2}\right),$$

for which

(i) $Q = 0$; (ii) $P = qa$; (iii) $Q_{xx} = 0$; (iv) $\Phi(x) = -\dfrac{qa}{4\pi\varepsilon_0 x^2}$.

The charge configuration (III) can be represented by

$$\rho(x') = q\delta\left(x' + \frac{a}{2}\right) + q\delta\left(x' - \frac{a}{2}\right) - 2q\delta(x'),$$

for which

(i) $Q = 0$; (ii) $P = 0$; (iii) $Q_{xx} = qa^2$; (iv) $\Phi(x) = \dfrac{qa^2}{8\pi\varepsilon_0 x^3}$.

1004

Two uniform infinite sheets of electric charge densities $+\sigma$ and $-\sigma$ intersect at right angles. Find the magnitude and direction of the electric field everywhere and sketch the lines of **E**.

(*Wisconsin*)

Solution:

First let us consider the infinite sheet of charge density $+\sigma$. The magnitude of the electric field caused by it at any space point is

$$E = \frac{\sigma}{2\varepsilon_0}.$$

The direction of the electric field is perpendicular to the surface of the sheet. For the two orthogonal sheets of charge densities $\pm\sigma$, superposition of their electric fields yields

$$E = \frac{\sqrt{2}\,\sigma}{2\varepsilon_0}.$$

The direction of **E** is as shown in Fig. 1.4.

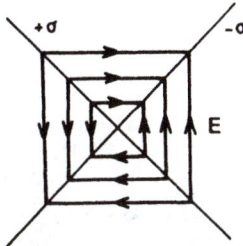

Fig. 1.4

1005

Gauss' law would be invalid if

(a) there were magnetic monopoles,

(b) the inverse-square law were not exactly true,

(c) the velocity of light were not a universal constant.

(*CCT*)

Solution:

The answer is (b).

1006

An electric charge can be held in a position of stable equilibrium:

(a) by a purely electrostatic field,

(b) by a mechanical force,

(c) neither of the above.

(*CCT*)

Solution:

The answer is (c).

1007

If **P** is the polarization vector and **E** is the electric field, then in the equation $\mathbf{P} = \alpha\mathbf{E}$, α in general is:

(a) scalar,   (b) vector,   (c) tensor.

(*CCT*)

Solution:

The answer is (c).

1008

(a) A ring of radius R has a total charge $+Q$ uniformly distributed on it. Calculate the electric field and potential at the center of the ring.

(b) Consider a charge $-Q$ constrained to slide along the axis of the ring. Show that the charge will execute simple harmonic motion for small displacements perpendicular to the plane of the ring.

(*Wisconsin*)

Solution:

As in Fig. 1.5, take the z-axis along the axis of the ring. The electric field and the potential at the center of the ring are given by

$$\mathbf{E} = 0, \quad \varphi = \frac{Q}{4\pi\varepsilon_0 R}.$$

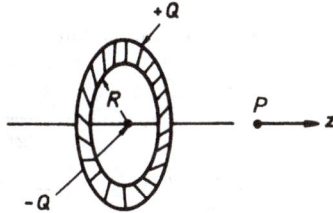

Fig. 1.5

The electric field at a point P on the z-axis is given by

$$\mathbf{E}(z) = \frac{Qz}{4\pi\varepsilon_0(R^2 + z^2)^{3/2}}\mathbf{e}_z.$$

Thus a negative charge $-Q$ at point p is acted upon by a force

$$\mathbf{F}(z) = -\frac{Q^2 z}{4\pi\varepsilon_0(R^2 + z^2)^{3/2}}\mathbf{e}_z.$$

As $z \ll R, F(z) \propto z$ and $-Q$ will execute simple harmonic motion.

1009

An amount of charge q is uniformly spread out in a layer on the surface of a disc of radius a.

(a) Use elementary methods based on the azimuthal symmetry of the charge distribution to find the potential at any point on the axis of symmetry.

(b) With the aid of (a) find an expression for the potential at any point $\mathbf{r}\,(|\mathbf{r}| > a)$ as an expansion in angular harmonics.

(*Wisconsin*)

Solution:

(a) Take coordinate axes as in Fig. 1.6 and consider a ring formed by circles with radii ρ and $\rho + d\rho$ on the disc. The electrical potential at a point $(0,0,z)$ produced by the ring is given by

$$d\varphi = \frac{1}{4\pi\varepsilon_0} \cdot \frac{q}{\pi a^2} \cdot \frac{2\pi\rho d\rho}{\sqrt{\rho^2 + z^2}} \,.$$

Integrating, we obtain the potential due to the whole ring:

$$\varphi(z) = \int_0^a \frac{q}{2\pi\varepsilon_0 a^2} \cdot \frac{\rho d\rho}{\sqrt{\rho^2 + z^2}}$$

$$= \frac{q}{2\pi\varepsilon_0 a^2}\left(\sqrt{a^2 + z^2} - |z|\right) .$$

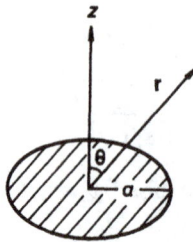

Fig. 1.6

(b) At a point $|\mathbf{r}| > a$, Laplace's equation $\nabla^2\varphi = 0$ applies, with solution

$$\varphi(r,\theta) = \sum_{n=0}^{\infty}\left(a_n r^n + \frac{b_n}{r^{n+1}}\right)P_n(\cos\theta) \,.$$

As $\varphi \to 0$ for $r \to \infty$, we have $a_n = 0$.

In the upper half-space, $z > 0$, the potential on the axis is $\varphi = \varphi(r,0)$. As $P_n(1) = 1$, we have

$$\varphi(r,0) = \sum_{n=0}^{\infty}\frac{b_n}{r^{n+1}} \,.$$

In the lower half-space, $z < 0$, the potential on the axis is $\varphi = \varphi(r, \pi)$. As $P_n(-1) = (-1)^n$, we have

$$\varphi(r, \pi) = \sum_{n=0}^{\infty} (-1)^n \frac{b_n}{r^{n+1}} .$$

Using the results of (a) and noting that for a point on the axis $|\mathbf{r}| = z$, we have for $z > 0$

$$\sum_{n=0}^{\infty} \frac{b_n}{r^{n+1}} = \frac{2q}{4\pi\varepsilon_0 a^2} (\sqrt{a^2 + r^2} - r)$$

$$= \frac{qr}{2\pi\varepsilon_0 a^2} \left(\sqrt{1 + \frac{a^2}{r^2}} - 1 \right) .$$

However, as

$$\left(1 + \frac{a^2}{r^2}\right)^{1/2} = 1 + \frac{1}{2}\left(\frac{a^2}{r^2}\right) + \frac{\frac{1}{2}(\frac{1}{2} - 1)}{2!}\left(\frac{a^2}{r^2}\right)^2 + \cdots$$

$$+ \frac{\frac{1}{2}(\frac{1}{2} - 1)\ldots\ldots(\frac{1}{2} - n + 1)}{n!}\left(\frac{a^2}{r^2}\right)^n + \cdots ,$$

the equation becomes

$$\sum_{n=0}^{\infty} \frac{b_n}{r^{n+1}} = \frac{qr}{2\pi\varepsilon_0 a^2} \sum_{n=1}^{\infty} \frac{\frac{1}{2}(\frac{1}{2} - 1)\ldots(\frac{1}{2} - n + 1)}{n!}\left(\frac{a^2}{r^2}\right)^n .$$

Comparing the coefficients of powers of r gives

$$b_{2n-1} = 0, \quad b_{2n-2} = \frac{q}{2\pi\varepsilon_0 a^2} \frac{\frac{1}{2}(\frac{1}{2} - 1)\ldots(\frac{1}{2} - n + 1)}{n!} a^{2n} .$$

Hence, the potential at any point $\mathbf{r}$ of the half-plane $z > 0$ is given by

$$\varphi(\mathbf{r}) = \frac{q}{2\pi\varepsilon_0 a} \sum_{n=1}^{\infty} \frac{\frac{1}{2}(\frac{1}{2} - 1)\ldots(\frac{1}{2} - n + 1)}{n!}$$

$$\times \left(\frac{a}{r}\right)^{2n-1} P_{2n-2}(\cos\theta), \quad (z > 0) .$$

Similarly for the half-plane $z < 0$, as $(-1)^{2n-2} = 1$ we have

$$\varphi(\mathbf{r}) = \frac{q}{2\pi\varepsilon_0 a} \sum_{n=1}^{\infty} \frac{\frac{1}{2}(\frac{1}{2}-1)\ldots(\frac{1}{2}-n+1)}{n!}$$

$$\times \left(\frac{a}{r}\right)^{2n-1} P_{2n-2}(\cos\theta) \quad (z < 0).$$

Thus the same expression for the potential applies to all points of space, which is a series in Legendre polynomials.

1010

A thin but very massive disc of insulator has surface charge density σ and radius R. A point charge $+Q$ is on the axis of symmetry. Derive an expression for the force on the charge.

(Wisconsin)

Solution:

Refer to Problem **1009** and Fig. 1.6. Let Q be at a point $(0,0,z)$ on the axis of symmetry. The electric field produced by the disc at this point is

$$E = -\frac{\sigma}{2\varepsilon_0}\left(\frac{z}{\sqrt{a^2+z^2}} - 1\right),$$

whence the force on the point charge is

$$F = QE = \frac{\sigma Q}{2\varepsilon_0}\left(1 - \frac{z}{\sqrt{a^2+z^2}}\right).$$

By symmetry the direction of this force is along the axis of the disc.

1011

The cube in Fig. 1.7 has 5 sides grounded. The sixth side, insulated from the others, is held at a potential ϕ_0. What is the potential at the center of the cube and why?

(MIT)

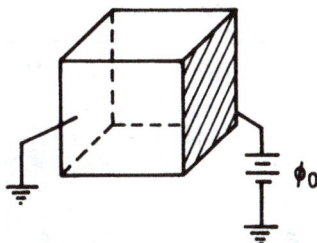

Fig. 1.7

Solution:
The electric potential ϕ_c at the center of the cube can be expressed as a linear function of the potentials of the six sides, i.e.,

$$\phi_c = \sum_i C_i \phi_i \, ,$$

where the C_i's are constants. As the six sides of the cube are in the same relative geometrical position with respect to the center, the C_i's must have the same value, say C. Thus

$$\phi_c = C \sum_i \phi_i \, .$$

If each of the six sides has potential ϕ_0, the potential at the center will obviously be ϕ_0 too. Hence $C = \frac{1}{6}$. Now as the potential of one side only is ϕ_0 while all other sides have potential zero, the potential at the center is $\phi_0/6$.

1012

A sphere of radius R carries a charge Q, the charge being uniformly distributed throughout the volume of the sphere. What is the electric field, both outside and inside the sphere?

(*Wisconsin*)

Solution:
The volume charge density of the sphere is

$$\rho = \frac{Q}{\frac{4}{3}\pi R^3} \, .$$

Take as the Gaussian surface a spherical surface of radius r concentric with the charge sphere. By symmetry the magnitude of the electric field at all points of the surface is the same and the direction is radial. From Gauss' law

$$\oint \mathbf{E} \cdot d\mathbf{S} = \frac{1}{\varepsilon_0} \int \rho dV$$

we immediately obtain

$$\mathbf{E} = \frac{Q\mathbf{r}}{4\pi\varepsilon_0 r^3} \qquad (r \geq R),$$

$$\mathbf{E} = \frac{Q\mathbf{r}}{4\pi\varepsilon_0 R^3} \qquad (r \leq R).$$

1013

Consider a uniformly charged spherical volume of radius R which contains a total charge Q. Find the electric field and the electrostatic potential at all points in the space.

(Wisconsin)

Solution:

Using the results of Problem **1012**

$$\mathbf{E}_1 = \frac{Q\mathbf{r}}{4\pi\varepsilon_0 R^3}, \qquad (r \leq R)$$

$$\mathbf{E}_2 = \frac{Q\mathbf{r}}{4\pi\varepsilon_0 r^3}, \qquad (r \geq R)$$

and the relation between electrostatic field intensity and potential

$$\varphi(p) = \int_p^\infty \mathbf{E} \cdot d\mathbf{l},$$

we obtain

$$
\begin{aligned}
\varphi_1(r) &= \int_r^R \mathbf{E}_1 \cdot d\mathbf{r} + \int_R^\infty \mathbf{E}_2 \cdot d\mathbf{r} \\
&= \int_r^R \frac{Qr dr}{4\pi\varepsilon_0 R^3} + \int_R^\infty \frac{Q dr}{4\pi\varepsilon_0 r^2} \\
&= \frac{Q}{8\pi\varepsilon_0 R} \left(3 - \frac{r^2}{R^2} \right) \qquad (r \leq R),
\end{aligned}
$$

$$\varphi_2(r) = \int_r^\infty \mathbf{E}_2 \cdot d\mathbf{r} = \frac{Q}{4\pi\varepsilon_0 r} \qquad (r \geq R).$$

1014

For a uniformly charged sphere of radius R and charge density ρ,

(a) find the form of the electric field vector $\mathbf{E}$ both outside and inside the sphere using Gauss' law;

(b) from $\mathbf{E}$ find the electric potential ϕ using the fact that $\phi \to 0$ as $r \to \infty$.

(Wisconsin)

Solution:

(a) Same as for Problem **1013**.

(b) Referring to Problem **1013**, we have

$$\text{for} \quad r > R, \ \phi = \frac{R^3 \rho}{3\varepsilon_0 r},$$

$$\text{for} \quad r < R, \ \phi = \frac{\rho R^3}{6\varepsilon_0} \left(3 - \frac{r^2}{R^2} \right).$$

1015

In the equilibrium configuration, a spherical conducting shell of inner radius a and outer radius b has a charge q fixed at the center and a charge density σ uniformly distributed on the outer surface. Find the electric field for all r, and the charge on the inner surface.

(Wisconsin)

Solution:

Electrostatic equilibrium requires that the total charge on inner surface of the conducting shell be $-q$. Using Gauss' law we then readily obtain

$$\mathbf{E}(r) = \frac{q}{4\pi\varepsilon_0 r^2}\mathbf{e}_r \qquad\qquad \text{for } r < a,$$

$$\mathbf{E} = 0 \qquad\qquad\qquad\qquad \text{for } a < r < b,$$

$$\mathbf{E}(r) = \frac{1}{4\pi\varepsilon_0}\frac{4\pi b^2 \sigma}{r^2}\mathbf{e}_r = \frac{\sigma b^2}{\varepsilon_0 r^2}\mathbf{e}_r \quad \text{for } r > b.$$

1016

A solid conducting sphere of radius r_1 has a charge of $+Q$. It is surrounded by a concentric hollow conducting sphere of inside radius r_2 and outside radius r_3. Use the Gaussian theorem to get expressions for

(a) the field outside the outer sphere,

(b) the field between the spheres.

(c) Set up an expression for the potential of the inner sphere. It is not necessary to perform the integrations.

(*Wisconsin*)

Solution:

Because of electrostatic equilibrium the inner surface of the hollow conducting sphere carries a total charge $-Q$, while the outer surface carries a total charge $+Q$. Using Gauss' law

$$\oint_s \mathbf{E} \cdot d\mathbf{S} = \frac{Q_{\text{tot}}}{\varepsilon_0},$$

where Q_{tot} is the algebraic sum of all charges surrounded by a closed surface s, we obtain

(a) $\mathbf{E}(\mathbf{r}) = \frac{Q}{4\pi\varepsilon_0 r^2}\mathbf{e}_r$ $(r > r_3)$

(b) $\mathbf{E}(\mathbf{r}) = \frac{Q}{4\pi\varepsilon_0 r^2}\mathbf{e}_r$ $(r_2 > r > r_1)$

(c) Using the expression for the potential $\varphi(p) = \int_p^\infty \mathbf{E} \cdot d\mathbf{l}$, we find the potential of the inner sphere:

$$\varphi(r_1) = \int_{r_1}^{r_2} \frac{Q}{4\pi\varepsilon_0 r^2} dr + \int_{r_3}^\infty \frac{Q}{4\pi\varepsilon_0 r^2} dr.$$

1017

The inside of a grounded spherical metal shell (inner radius R_1 and outer radius R_2) is filled with space charge of uniform charge density ρ. Find the electrostatic energy of the system. Find the potential at the center.

(*Wisconsin*)

Solution:

Consider a concentric spherical surface of radius $r(r < R_1)$. Using Gauss' law we get

$$\mathbf{E} = \frac{r}{3}\frac{\rho}{\varepsilon_0}\mathbf{e}_r.$$

As the shell is grounded, $\varphi(R_1) = 0, E = 0\,(r > R_2)$. Thus

$$\varphi(r) = \int_r^{R_1} E\,dr = \frac{\rho}{6\varepsilon_0}(R_1^2 - r^2).$$

The potential at the center is

$$\varphi(0) = \frac{1}{6\varepsilon_0}\rho R_1^2.$$

The electrostatic energy is

$$W = \int \frac{1}{2}\rho\varphi dV = \frac{1}{2}\int_0^{R_1} \frac{\rho}{6\varepsilon_0}(R_1^2 - r^2)\cdot\rho\cdot4\pi r^2 dr = \frac{2\rho^2 R_1^5}{45\varepsilon_0}.$$

1018

A metal sphere of radius a is surrounded by a concentric metal sphere of inner radius b, where $b > a$. The space between the spheres is filled with a material whose electrical conductivity σ varies with the electric field strength E according to the relation $\sigma = KE$, where K is a constant. A potential difference V is maintained between the two spheres. What is the current between the spheres?

(*Wisconsin*)

Solution:

Since the current is

$$I = j\cdot S = \sigma E\cdot S = KE^2\cdot S = KE^2\cdot 4\pi r^2,$$

the electric field is

$$E = \frac{1}{r}\sqrt{\frac{I}{4\pi K}}$$

and the potential is

$$V = -\int_b^a E\cdot dr = -\int_b^a \sqrt{\frac{I}{4\pi K}}\frac{1}{r}dr = \sqrt{\frac{I}{4\pi K}}\ln\left(\frac{b}{a}\right).$$

Hence the current between the spheres is given by

$$I = 4\pi KV^2/\ln(b/a).$$

1019

An isolated soap bubble of radius 1 cm is at a potential of 100 volts. If it collapses to a drop of radius 1 mm, what is the change of its electrostatic energy?

(*Wisconsin*)

Solution:

If the soap bubble carries a charge Q, its potential is

$$V = \frac{Q}{4\pi\varepsilon_0 r} \cdot$$

For $r = r_1 = 1$ cm, $V = V_1 = 100$ V, we have $Q = 4\pi\varepsilon_0 r_1 V_1$. As the radius changes from r_1 to $r_2 = 1$ mm, the change of electrostatic energy is

$$\Delta W = \frac{Q^2}{8\pi\varepsilon_0 r_2} - \frac{Q^2}{8\pi\varepsilon_0 r_1} = 2\pi\varepsilon_0 (r_1 V_1)^2 \left(\frac{1}{r_2} - \frac{1}{r_1} \right)$$

$$= 2\pi \times 8.85 \times 10^{-12} \times (10^{-12} \times 100)^2 \times \left(\frac{1}{10^{-3}} - \frac{1}{10^{-2}} \right)$$

$$= 5 \times 10^{-8} \text{ J}.$$

1020

A static electric charge is distributed in a spherical shell of inner radius R_1 and outer radius R_2. The electric charge density is given by $\rho = a + br$, where r is the distance from the center, and zero everywhere else.

(a) Find an expression for the electric field everywhere in terms of r.

(b) Find expressions for the electric potential and energy density for $r < R_1$. Take the potential to be zero at $r \to \infty$.

(*SUNY, Buffalo*)

Solution:

Noting that ρ is a function of only the radius r, we can take a concentric spherical surface of radius r as the Gaussian surface in accordance with the symmetry requirement. Using Gauss' law

$$\oint_s \mathbf{E} \cdot d\mathbf{S} = \frac{1}{\varepsilon_0} \int \rho(r) dr,$$

we can get the following results:

(a) Electric field strength.

For $r < R_1$, $\mathbf{E}_1 = 0$.

For $R_1 < r < R_2$, using the relation $4\pi r^2 E_2 = \frac{4\pi}{\varepsilon_0} \int_{R_1}^{r} (a + br') r'^2 dr'$ we find

$$\mathbf{E}_2 = \frac{1}{\varepsilon_0 r^3} \left[\frac{a}{3} (r^3 - R_1^3) + \frac{b}{4} (r^4 - R_1^4) \right] \mathbf{r} \,.$$

For $R_2 > r$, from $4\pi r^2 E_3 = \frac{4\pi}{\varepsilon_0} \int_{R_1}^{R_2} (a + br') r'^2 dr'$ we get

$$\mathbf{E}_3 = \frac{1}{\varepsilon_0 r^3} \left[\frac{a}{3} (R_2^3 - R_1^3) + \frac{b}{4} (R_2^4 - R_1^4) \right] \mathbf{r} \,.$$

(b) Potential and the energy density for $r < R_1$.

Noting that $\varphi(\infty) = 0$, the potential is

$$\varphi(r) = \int_r^\infty \mathbf{E} \cdot d\mathbf{l} = \left(\int_r^{R_1} + \int_{R_1}^{R_2} + \int_{R_2}^\infty \right) \mathbf{E} \cdot d\mathbf{r}$$

$$= \frac{1}{\varepsilon_0} \left[\frac{a}{3} (R_2^2 - R_1^2) + \frac{b}{4} (R_2^3 - R_1^3) \right] \,.$$

Also, as $\mathbf{E}_1 = 0 \, (r < R_1)$, the energy density for $r < R_1$ is

$$W = \frac{\varepsilon_0}{2} E_1^2 = 0 \,.$$

1021

An electric charge Q is uniformly distributed over the surface of a sphere of radius r. Show that the force on a small charge element dq is radial and outward and is given by

$$dF = \frac{1}{2} E dq \,,$$

where $E = \frac{1}{4\pi\varepsilon_0} \frac{Q}{r^2}$ is the electric field at the surface of the sphere.

(*Wisconsin*)

Solution:

The surface charge density is given by

$$\sigma = \frac{Q}{4\pi r^2} \,.$$

As shown in Fig. 1.8, we consider a point P inside the sphere close to an area element ds. The charge dq on this area element will produce at the point P an electric field which is approximately that due to a uniformly charged infinite plate, namely,

$$\mathbf{E}_{1P} = -\frac{\sigma}{2\varepsilon_0}\mathbf{n}\,,$$

where $\mathbf{n}$ is a unit vector normal to ds in the outward direction.

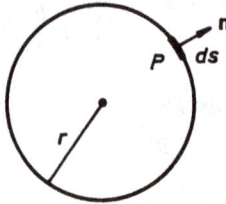

Fig. 1.8

The electric field is zero inside the sphere. Hence, if we take $\mathbf{E}_{2P}$ as the electric field at P due to all the charges on the spherical surface except the element ds, we must have

$$\mathbf{E}_P = \mathbf{E}_{1P} + \mathbf{E}_{2P} = 0\,.$$

Therefore,

$$\mathbf{E}_{2P} = \frac{\sigma}{2\varepsilon_0}\mathbf{n} = \frac{Q}{8\pi\varepsilon_0 r^2}\mathbf{n}\,.$$

As P is close to ds, $\mathbf{E}_{2P}$ may be considered as the field strength at ds due to the charges of the spherical surface. Hence, the force acting on ds is

$$d\mathbf{F} = dq\mathbf{E}_{2P} = \frac{1}{2}Edq\mathbf{n}\,,$$

where $E = Q/4\pi\varepsilon_0 r^2$ is just the field strength on the spherical surface.

1022

A sphere of radius R_1 has charge density ρ uniform within its volume, except for a small spherical hollow region of radius R_2 located a distance a from the center.

(a) Find the field **E** at the center of the hollow sphere.

(b) Find the potential ϕ at the same point.

(*UC, Berkeley*)

Solution:

(a) Consider an arbitrary point P of the hollow region (see Fig. 1.9) and let

$$OP = \mathbf{r}, \quad Q'P = \mathbf{r}', \quad OO' = \mathbf{a}, \quad \mathbf{r}' = \mathbf{r} - \mathbf{a}.$$

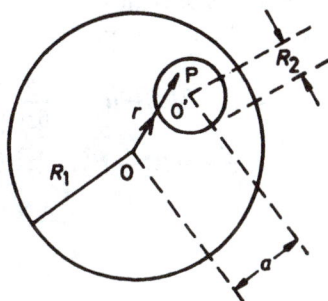

Fig. 1.9

If there were no hollow region inside the sphere, the electric field at the point P would be

$$\mathbf{E}_1 = \frac{\rho}{3\varepsilon_0}\mathbf{r}.$$

If only the spherical hollow region has charge density ρ the electric field at P would be

$$\mathbf{E}_2 = \frac{\rho}{3\varepsilon_0}\mathbf{r}'.$$

Hence the superposition theorem gives the electric field at P as

$$\mathbf{E} = \mathbf{E}_1 - \mathbf{E}_2 = \frac{\rho}{3\varepsilon_0}\mathbf{a}.$$

Thus the field inside the hollow region is uniform. This of course includes the center of the hollow.

(b) Suppose the potential is taken to be zero at an infinite point. Consider an arbitrary sphere of radius R with a uniform charge density ρ. We can find the electric fields inside and outside the sphere as

$$\mathbf{E}(r) = \begin{cases} \frac{\rho r}{3\varepsilon_0}, & r < R, \\ \frac{\rho R^3}{3\varepsilon_0 r^3}\mathbf{r}, & r > R. \end{cases}$$

Then the potential at an arbitrary point inside the sphere is

$$\phi = \left(\int_r^R + \int_R^\infty \right) \mathbf{E} \cdot d\mathbf{r} = \frac{\rho}{6\varepsilon_0} (3R^2 - r^2), \tag{1}$$

where r is the distance between this point and the spherical center.

Now consider the problem in hand. If the charges are distributed throughout the sphere of radius R_1, let ϕ_1 be the potential at the center O' of the hollow region. If the charge distribution is replaced by a small sphere of uniform charge density ρ of radius R_2 in the hollow region, let the potential at O' be ϕ_2. Using (1) and the superposition theorem, we obtain

$$\phi_{O'} = \phi_1 - \phi_2 = \frac{\rho}{6\varepsilon_0}(3R_1^2 - a^2) - \frac{\rho}{6\varepsilon_0}(3R_2^2 - 0)$$

$$= \frac{\rho}{6\varepsilon_0}[3(R_1^2 - R_2^2) - a^2].$$

1023

The electrostatic potential at a point P due to an idealized dipole layer of moment per unit area τ on surface S is

$$\phi_P = \frac{1}{4\pi\varepsilon_0} \int \frac{\tau \cdot \mathbf{r}}{r^3} dS,$$

where $\mathbf{r}$ is the vector from the surface element to the point P.

(a) Consider a dipole layer of infinite extent lying in the x-y plane of uniform moment density $\tau = \tau e_z$. Determine whether ϕ or some derivative of it is discontinuous across the layer and find the discontinuity.

(b) Consider a positive point charge q located at the center of a spherical surface of radius a. On this surface there is a uniform dipole layer τ and a uniform surface charge density σ. Find τ and σ so that the potential inside the surface will be just that of the charge q, while the potential outside will be zero. (You may make use of whatever you know about the potential of a surface charge.)

(*SUNY, Buffalo*)

Solution:

(a) By symmetry the electrostatic potential at point P is only dependent on the z coordinate. We choose cylindrical coordinates (R, θ, z) such that P is on the z-axis. Then the potential at point P is

$$\phi_P = \frac{1}{4\pi\varepsilon_0} \int \frac{\boldsymbol{\tau}\cdot\mathbf{r}}{r^3}\,dS = \frac{1}{4\pi\varepsilon_0} \int \frac{\tau z}{r^3}\,dS\,.$$

As $r^2 = R^2 + z^2, dS = 2\pi R dR$, we get

$$\phi_P = \frac{2\pi\tau z}{4\pi\varepsilon_0} \int_0^\infty \frac{RdR}{\sqrt{(R^2+z^2)^3}} = \begin{cases} \frac{\tau}{2\varepsilon_0}, & z>0, \\ -\frac{\tau}{2\varepsilon_0}, & z<0. \end{cases}$$

Hence, the electrostaic potential is discontinous across the x-y plane (for which $z = 0$). The discontinuity is given by

$$\Delta\phi = \frac{\tau}{2\varepsilon_0} - \left(-\frac{\tau}{2\varepsilon_0}\right) = \frac{\tau}{\varepsilon_0}\,.$$

(b) It is given that $\phi = 0$ for $r > a$. Consequently $\mathbf{E} = 0$ for $r > a$. Using Gauss' law

$$\oint \mathbf{E}\cdot d\mathbf{S} = \frac{Q}{\varepsilon_0}\,,$$

we find that $\sigma \cdot 4\pi a^2 + q = 0$. Thus

$$\sigma = -\frac{q}{4\pi a^2}\,.$$

If the potential at infinity is zero, then the potential outside the spherical surface will be zero everywhere. But the potential inside the sphere is $\varphi = \frac{q}{4\pi\varepsilon_0 r}$. For $r = a, \varphi = \frac{q}{4\pi\varepsilon_0 a}$, so that the discontinuity at the spherical surface is

$$\Delta\phi = -\frac{q}{4\pi\varepsilon_0 a}\,.$$

We then have $\frac{\tau}{\varepsilon_0} = -\frac{q}{4\pi\varepsilon_0 a}$, giving

$$\boldsymbol{\tau} = -\frac{q}{4\pi a}\mathbf{e}_r\,.$$

2. ELECTROSTATIC FIELD IN A CONDUCTOR (1024–1042)

1024

A charge Q is placed on a capacitor of capacitance C_1. One terminal is connected to ground and the other terminal is insulated and not connected to anything. The separation between the plates is now increased and the capacitance becomes C_2 ($C_2 < C_1$). What happens to the potential on the free plate when the plates are separated? Express the potential V_2 in terms of the potential V_1.

(*Wisconsin*)

Solution:

In the process of separation the charge on the insulated plate is kept constant. Since $Q = CV$, the potential of the insulated plate increases as C has decreased. If V_1 and V_2 are the potentials of the insulated plate before and after the separation respectively, we have

$$V_2 = \frac{C_1}{C_2}V_1 .$$

1025

Figure 1.10 shows two capacitors in series, the rigid center section of length b being movable vertically. The area of each plate is A. Show that the capacitance of the series combination is independent of the position of the center section and is given by $C = \frac{A\varepsilon_0}{a-b}$. If the voltage difference between the outside plates is kept constant at V_0, what is the change in the energy stored in the capacitors if the center section is removed?

(*Wisconsin*)

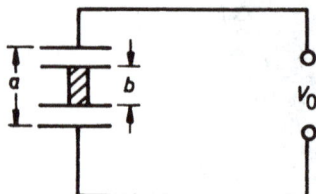

Fig. 1.10

Solution:

Let d_1 be the distance between the two upper plates and d_2 be the distance between the two lower plates. From Fig. 1.10 we see that

$$d_1 + d_2 = a - b,$$

$$C_1 = \frac{\varepsilon_0 A}{d_1}, \quad C_2 = \frac{\varepsilon_0 A}{d_2}.$$

For the two capacitors in series, the total capacitance is

$$C = \frac{C_1 C_2}{C_1 + C_2} = \frac{A\varepsilon_0}{d_1 + d_2} = \frac{A\varepsilon_0}{a - b}.$$

As C is independent of d_1 and d_2, the total capacitance is independent of the position of the center section. The total energy stored in the capacitor is

$$W = \frac{1}{2}CV_0^2 = \frac{A\varepsilon_0 V_0^2}{2(a - b)}.$$

The energy stored if the center section is removed is

$$W' = \frac{A\varepsilon_0 V_0^2}{2a},$$

and we have

$$W - W' = \frac{A\varepsilon_0 V_0^2}{2(a - b)} \frac{b}{a}.$$

1026

A parallel-plate capacitor is charged to a potential V and then disconnected from the charging circuit. How much work is done by slowly changing the separation of the plates from d to $d' \neq d$? (The plates are circular with radius $r \gg d$.)

(Wisconsin)

Solution:

Neglecting edge effects, the capacitance of the parallel-plate capacitor is $C = \frac{\varepsilon_0 \pi r^2}{d}$ and the stored energy is $W = \frac{1}{2}CV^2$. As the charges on the plates, $Q = \pm CV$, do not vary with the separation, we have

$$V' = \frac{C}{C'}V.$$

The energy stored when separation is d' is

$$W' = \frac{1}{2}C'\left(\frac{C}{C'}V\right)^2 = \frac{1}{2}\cdot\frac{C^2}{C'}V^2.$$

Thus the change of the energy stored in the capacitor is

$$\Delta W = W' - W = \frac{1}{2}CV^2\left(\frac{C}{C'} - 1\right) = \frac{1}{2}CV^2\left(\frac{d'}{d} - 1\right).$$

Therefore, the work done in changing the separation from d to d' is

$$\frac{\varepsilon_0\pi r^2(d' - d)V^2}{2d^2}.$$

1027

A parallel-plate capacitor of plate area 0.2 m² and plate spacing 1 cm is charged to 1000 V and is then disconnected from the battery. How much work is required if the plates are pulled apart to double the plate spacing? What will be the final voltage on the capacitor?

$$(\varepsilon_0 = 8.9 \times 10^{-12} \text{ C}^2/(\text{N}\cdot\text{m}^2))$$

(*Wisconsin*)

Solution:

When the plates are pulled apart to double the plate spacing, the capacitance of the capacitor becomes $C' = \frac{C}{2}$, where $C = \frac{\varepsilon_0 A}{d}$ is the capacitance before the spacing was increased. If a capacitor is charged to a voltage U, the charge of the capacitor is $Q = CU$. As the magnitude of the charge Q is constant in the process, the change of the energy stored in the capacitor is

$$\Delta W = \frac{1}{2}\frac{Q^2}{C'} - \frac{1}{2}\frac{Q^2}{C} = \frac{1}{2}\frac{Q^2}{C} = \frac{1}{2}CU^2$$
$$= \frac{\varepsilon_0 A U^2}{2d} = \frac{8.9 \times 10^{-12} \times 0.2 \times (10^3)^2}{2 \times 0.01}$$
$$= 8.9 \times 10^{-5} \text{ J}.$$

ΔW is just the work required to pull the plates apart to double the plate spacing. As the charge Q is kept constant, the final voltage is given by

$$CU = C'U', \text{ or } U' = 2U = 2000 \text{ V}.$$

1028

Given two plane-parallel electrodes, space d, at voltages 0 and V_0, find the current density if an unlimited supply of electrons at rest is supplied to the lower potential electrode. Neglect collisions.

(Wisconsin; UC Berkeley)

Solution:

Choose x-axis perpendicular to the plates as shown in Fig. 1.11. Both the charge and current density are functions of x. In the steady state

$$\frac{dj(x)}{dx} = 0.$$

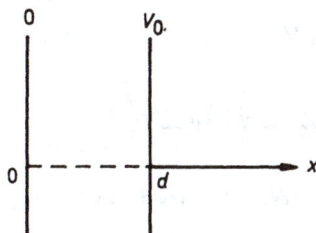

Fig. 1.11

Hence $\mathbf{j} = -j_0\mathbf{e}_x$, where j_0 is a constant. Let $v(x)$ be the velocity of the electrons. Then the charge density is

$$\rho(x) = -\frac{j_0}{v(x)}.$$

The potential satisfies the Poisson equation

$$\frac{d^2V(x)}{dx^2} = -\frac{\rho(x)}{\varepsilon_0} = \frac{j_0}{\varepsilon_0 v(x)}.$$

Using the energy relation $\frac{1}{2}mv^2(x) = eV$, we get

$$\frac{d^2V(x)}{dx^2} = \frac{j_0}{\varepsilon_0}\sqrt{\frac{m}{2eV(x)}}.$$

To solve this differential equation, let $u = \frac{dV}{dx}$. We then have

$$\frac{d^2V}{dx^2} = \frac{du}{dx} = \frac{du}{dV}\frac{dV}{dx} = u\frac{du}{dV},$$

and this equation becomes

$$udu = AV^{-\frac{1}{2}}dV,$$

where $A = \frac{j_0}{\varepsilon_0}\sqrt{\frac{m}{2e}}$. Note that $\frac{dV}{dx} = 0$ at $x = 0$, as the electrons are at rest there. Integrating the above gives

$$\frac{1}{2}u^2 = 2AV^{\frac{1}{2}},$$

or

$$V^{-\frac{1}{4}}dV = 2A^{\frac{1}{2}}dx.$$

As $V = 0$ for $x = 0$ and $V = V_0$ for $x = d$, integrating the above leads to

$$\frac{4}{3}V_0^{\frac{3}{4}} = 2A^{\frac{1}{2}}d = 2\left(\frac{j_0}{\varepsilon_0}\sqrt{\frac{m}{2e}}\right)^{\frac{1}{2}}d.$$

Finally we obtain the current density from the last equation:

$$\mathbf{j} = -j_0\mathbf{e}_x = -\frac{4\varepsilon_0 V_0}{9d^2}\sqrt{\frac{2eV_0}{m}}\mathbf{e}_x.$$

1029

As can be seen from Fig. 1.12, a cylindrical conducting rod of diameter d and length l ($l \gg d$) is uniformly charged in vacuum such that the electric field near its surface and far from its ends is E_0. What is the electric field at $r \gg l$ on the axis of the cylinder?

(*UC, Berkeley*)

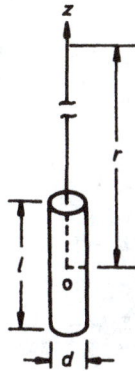

Fig. 1.12

Solution:

We choose cylindrical coordinates with the z-axis along the axis of the cylinder and the origin at the center of the rod. Noting $l \gg d$ and using Gauss' theorem, we can find the electric field near the cylindrical surface and far from its ends as

$$\mathbf{E}_0 = \frac{\lambda}{\pi \varepsilon_0 d} \mathbf{e}_\rho \,,$$

where λ is the charge per unit length of the cylinder and $\mathbf{e}_\rho$ is a unit vector in the radial direction. For $r \gg l$, we can regard the conducting rod as a point charge with $Q = \lambda l$. So the electric field intensity at a distant point on the axis is approximately

$$E = \frac{Q}{4\pi \varepsilon_0 r^2} = \frac{E_0 dl}{4r^2} \,.$$

The direction of $\mathbf{E}$ is along the axis away from the cylinder.

1030

An air-spaced coaxial cable has an inner conductor 0.5 cm in diameter and an outer conductor 1.5 cm in diameter. When the inner conductor is at a potential of +8000 V with respect to the grounded outer conductor,

(a) what is the charge per meter on the inner conductor, and

(b) what is the electric field intensity at $r = 1$ cm?

(*Wisconsin*)

Solution:

(a) Let the linear charge density for the inner conductor be λ. By symmetry we see that the field intensity at a point distance r from the axis in the cable between the conductors is radial and its magnitude is given by Gauss' theorem as

$$E = \frac{\lambda}{2\pi\varepsilon_0 r}.$$

Then the potential difference between the inner and outer conductors is

$$V = \int_a^b E dr = \frac{\lambda}{2\pi\varepsilon_0} \ln(b/a)$$

with $a = 1.5$ cm, $b = 0.5$ cm, which gives

$$\lambda = \frac{2\pi\varepsilon_0 V}{\ln(b/a)} = \frac{2\pi \times 8.9 \times 10^{-12} \times 8000}{\ln(1.5/0.5)}$$
$$= 4.05 \times 10^{-7}\,\text{C/m}.$$

(b) The point $r = 1$ cm is outside the cable. Gauss' law gives that its electric intensity is zero.

1031

A cylindrical capacitor has an inner conductor of radius r_1 and an outer conductor of radius r_2. The outer conductor is grounded and the inner conductor is charged so as to have a positive potential V_0. In terms of V_0, r_1, and r_2,

(a) what is the electric field at r? $(r_1 < r < r_2)$

(b) what is the potential at r?

(c) If a small negative charge Q which is initially at r drifts to r_1, by how much does the charge on the inner conductor change?

(*Wisconsin*)

Solution:

(a) From Problem **1030**, we have

$$\mathbf{E}(r) = \frac{V_0}{\ln(r_2/r_1)} \frac{\mathbf{r}}{r^2} \qquad (r_1 < r < r_2).$$

(b)

$$V(r) = V_0 - \int_{r_1}^{r} \mathbf{E} \cdot d\mathbf{r} = \frac{V_0 \ln(r_2/r)}{\ln(r_2/r_1)} \ .$$

(c) Let the change of the charge on the inner conductor be $\Delta Q = Q_1 - Q_2$ with $Q_1 = CV_0$. When a negative charge Q moves from r to r_1, the work done by electrostatic force is $Q(V_0 - V)$. This is equal to a decrease of the electrostatic energy in the capacitor of

$$\frac{Q_1^2}{2C} - \frac{Q_2^2}{2C} = Q(V_0 - V) \ .$$

As Q is a small quantity, we have approximately

$$Q_1 + Q_2 \approx 2Q_1 \ .$$

Hence

$$\frac{2Q_1}{2C} \Delta Q = Q(V_0 - V) \ ,$$

or

$$\Delta Q = \frac{Q}{V_0}(V_0 - V) = \frac{Q \ln(r/r_1)}{\ln(r_2/r_1)} \ .$$

1032

A very long hollow metallic cylinder of inner radius r_0 and outer radius $r_0 + \Delta r \, (\Delta r \ll r_0)$ is uniformly filled with space charge of density ρ_0. What are the electric fields for $r < r_0, r > r_0 + \Delta r$, and $r_0 + \Delta r > r > r_0$? What are the surface charge densities on the inner and outer surfaces of the cylinder? The net charge on the cylinder is assumed to be zero. What are the fields and surface charges if the cylinder is grounded?

(*Wisconsin*)

Solution:

Use cylindrical coordinates (r, φ, z) with the z-axis along the axis of the cylinder. Gauss' law gives the field intensity as

$$\mathbf{E}_1(r) = \frac{\rho_0 r}{2\varepsilon_0} \mathbf{e}_r \qquad \text{for} \quad r < r_0 \ ,$$

$$\mathbf{E}_2(r) = \frac{\rho_0 r_0^2}{2r\varepsilon_0} \mathbf{e}_r \qquad \text{for} \quad r > r_0 + \Delta r \ ,$$

$$\mathbf{E}_3(r) = 0 \ . \qquad \text{for} \quad r_0 < r < r_0 + \Delta r \ .$$

The surface charge density σ on a conductor is related to the surface electric intensity E by $E = \frac{\sigma}{\varepsilon_0}$ with E in the direction of an *outward* normal to the conductor. Thus the surface charge densities at $r = r_0$ and $r = r_0 + \Delta r$ are respectively

$$\sigma(r_0) = -\varepsilon_0 E_1(r_0)$$
$$= -\frac{\rho_0 r_0}{2},$$
$$\sigma(r_0 + \Delta r) = \varepsilon_0 E_2(r_0 + \Delta r)$$
$$= \frac{\rho_0 r_0^2}{2(r_0 + \Delta r)}.$$

If the cylinder is grounded, then one has

$$E = 0 \qquad \text{for} \quad r > r_0 + \Delta r,$$
$$\sigma(r_0 + \Delta r) = 0 \quad \text{for} \quad r = r_0 + \Delta r,$$

E and σ in other regions remaining the same.

1033

An air-filled capacitor is made from two concentric metal cylinders. The outer cylinder has a radius of 1 cm.

(a) What choice of radius for the inner conductor will allow a maximum potential difference between the conductors before breakdown of the air dielectric?

(b) What choice of radius for the inner conductor will allow a maximum energy to be stored in the capacitor before breakdown of the dielectric?

(c) Calculate the maximum potentials for cases (a) and (b) for a breakdown field in air of 3×10^6 V/m.

(*UC, Berkeley*)

Solution:

(a) Let E_b be the breakdown field intensity in air and let R_1 and R_2 be the radii of the inner and outer conductors respectively. Letting τ be the charge per unit length on each conductor and using Gauss' theorem, we obtain the electric fiield intensity in the capacitor and the potential difference between the two conductors respectively as

$$\mathbf{E}_r = \frac{\tau}{2\pi\varepsilon_0 r}\mathbf{e}_r, \qquad V = \int_{R_1}^{R_2} \frac{\tau}{2\pi\varepsilon_0 r}dr = \frac{\tau}{2\pi\varepsilon_0}\ln\frac{R_2}{R_1}.$$

As the electric field close to the surface of the inner conductor is strongest we have

$$E_b = \frac{\tau}{2\pi\varepsilon_0 R_1} \,.$$

Accordingly, we have

$$V_b = E_b R_1 \ln \frac{R_2}{R_1} \,,$$

$$\frac{dV_b}{dR_1} = E_b \left[\ln \frac{R_2}{R_1} + R_1 \frac{R_1}{R_2} \left(-\frac{R_2}{R_1^2} \right) \right] = E_b \left(\ln \frac{R_2}{R_1} - 1 \right).$$

In order to obtain the maximum potential difference, R_1 should be such that $\frac{dV_b}{dR_1} = 0$, i.e., $\ln \frac{R_2}{R_1} = 1$ or $R_1 = \frac{R_2}{e}$. The maximum potential difference is then

$$V_{\max} = \frac{R_2}{e} E_b \,.$$

(b) The energy stored per unit length of the capacitor is

$$W = \frac{1}{2}\tau V = \pi\varepsilon_0 E_b^2 R_1^2 \ln \frac{R_2}{R_1}$$

and

$$\frac{dW}{dR_1} = \pi\varepsilon_0 E_b^2 \left[2R_1 \ln \frac{R_2}{R_1} + R_1^2 \frac{R_1}{R_2} \left(-\frac{R_2}{R_1^2} \right) \right]$$

$$= \pi\varepsilon_0 E_b^2 R_1 \left(2\ln \frac{R_2}{R_1} - 1 \right).$$

For maximum energy storage, we require $\frac{dW}{dR_1} = 0$, i.e., $2\ln \frac{R_2}{R_1} = 1$ or $R_1 = \frac{R_2}{\sqrt{e}}$. In this case the potential difference is

$$V = \frac{1}{2\sqrt{e}} R_2 E_b \,.$$

(c) For (a),

$$V_{\max} = \frac{R_2}{e} E_b = \frac{0.01}{e} \times 3 \times 10^6 = 1.1 \times 10^4 \text{ V} \,.$$

For (b),

$$V_{\max} = \frac{1}{2\sqrt{e}} R_2 E_b = \frac{0.01 \times 3 \times 10^6}{2\sqrt{e}} = 9.2 \times 10^3 \text{ V} \,.$$

1034

In Fig. 1.13 a very long coaxial cable consists of an inner cylinder of radius a and electrical conductivity σ and a coaxial outer cylinder of radius b. The outer shell has infinite conductivity. The space between the cylinders is empty. A uniform constant current density $\mathbf{j}$, directed along the axial coordinate z, is maintained in the inner cylinder. Return current flows uniformly in the outer shell. Compute the surface charge density on the inner cylinder as a function of the axial coordinate z, with the origin $z = 0$ chosen to be on the plane half-way between the two ends of the cable.

(*Princeton*)

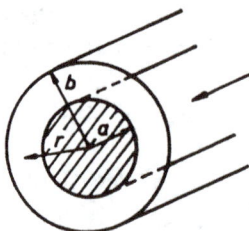

Fig. 1.13

Solution:

Assume that the length of the cable is $2l$ and that the inner and outer cylinders are connected at the end surface $z = -l$. (The surface $z = l$ may be connected to a battery.) The outer cylindrical shell is an ideal conductor, whose potential is the same everywhere, taken to be zero. The inner cylinder has a current density $\mathbf{j} = \sigma \mathbf{E}$, i.e., $\mathbf{E} = \frac{\mathbf{j}}{\sigma} = \frac{j}{\sigma}\mathbf{e}_z$, so that its cross section $z = \text{const.}$ is an equipotential surface with potential

$$V(z) = -\frac{j}{\sigma}(z + l).$$

In cylindrical coordinates the electric field intensity at a point (r, φ, z) inside the cable can be expressed as

$$\mathbf{E}(r, \varphi, z) = E_r(r, z)\mathbf{e}_r + E_z(r, z)\mathbf{e}_z.$$

As the current does not change with z, $E_z(r, z)$ is independent of z also. Take for the Gaussian surface a cylindrical surface of radius r and length dz with z-axis as the axis. We note that the electric fluxes through its two end surfaces have the same magnitude and direction so that their contributions cancel out. Gauss' law then becomes

$$E_r(r, z) \cdot 2\pi r dz = \lambda(z)dz/\varepsilon_0,$$

where $\lambda(z)$ is the charge per unit length of the inner cylinder, and gives

$$E_r(r,z) = \frac{\lambda(z)}{2\pi r \varepsilon_0}.$$

Hence, we obtain the potential difference between the inner and outer conductors as

$$V(z) = \int_a^b E_r(r,z)dr = \frac{\lambda(z)}{2\pi\varepsilon_0}\ln\frac{b}{a}.$$

As $V(z) = -\frac{i}{\sigma}(z+l)$, the above gives

$$\lambda(z) = \frac{2\pi\varepsilon_0 V(z)}{\ln(b/a)} = -\frac{2\pi\varepsilon_0 j(z+l)}{\sigma\ln(b/a)}.$$

The surface charge density at z is then

$$\sigma_s(z) = \frac{\lambda(z)}{2\pi a} = -\frac{\varepsilon_0 j(z+l)}{a\sigma\ln(b/a)}.$$

Choosing the origin at the end surface with $z = -l$, we can write

$$\sigma_s(z) = -\frac{\varepsilon_0 j z}{a\sigma\ln(b/a)}.$$

1035

A finite conductor of uniform conductivity σ has a uniform volume charge density ρ. Describe in detail the subsequent evolution of the system in the two cases:

(a) the conductor is a sphere,

(b) the conductor is not a sphere.

What happens to the energy of the system in the two cases?

(UC, Berkeley)

Solution:

Let the permittivity of the conductor be ε. From $\nabla\cdot\mathbf{E} = \rho/\varepsilon, \nabla\cdot\mathbf{J} + \frac{\partial\rho}{\partial t} = 0$ and $\mathbf{J} = \sigma\mathbf{E}$, we get

$$\frac{\partial\rho}{\partial t} = -\frac{\sigma}{\varepsilon}\rho, \quad\text{or}\quad \rho = \rho_0 e^{-\frac{\sigma}{\varepsilon}t}, \quad\text{and}\quad \nabla\cdot\mathbf{E} = \frac{\rho_0}{\varepsilon_0}e^{-\frac{\sigma}{\varepsilon}t}.$$

(a) If the conductor is a sphere, spherical symmetry requires that $\mathbf{E} = E_r \mathbf{e}_r$. Hence

$$\nabla \cdot \mathbf{E} = \frac{1}{r^2} \frac{\partial}{\partial r} (r^2 E_r) = \frac{\rho_0}{\varepsilon} e^{-\frac{\sigma}{\varepsilon} t},$$

giving

$$\mathbf{E}(r,t) = \frac{\rho_0 \mathbf{r}}{3\varepsilon} e^{-\frac{\sigma}{\varepsilon} t} + \mathbf{E}(0,t) = \frac{\rho_0 \mathbf{r}}{3\varepsilon} e^{-\frac{\sigma}{\varepsilon} t},$$

$$\mathbf{J} = \sigma \mathbf{E} = \frac{\sigma \rho_0 r}{3\varepsilon} e^{-\frac{\sigma}{\varepsilon} t} \mathbf{e}_r.$$

Note that $\mathbf{E}(0,t) = 0$ for symmetry. It is evident that for $t \to \infty$, $\mathbf{E} = 0, \rho = 0$, and $\mathbf{J} = 0$ inside the conductor. Thus the charge is uniformly distributed on the spherical surface after a sufficiently large time.

(b) If the conductor is not a sphere, the solution is more complicated. However we still have that

$$|\mathbf{E}| \propto e^{-\frac{\sigma}{\varepsilon} t}, \quad |\mathbf{J}| \propto e^{-\frac{\sigma}{\varepsilon} t}, \quad \rho \propto e^{-\frac{\sigma}{\varepsilon} t}.$$

This means that $\mathbf{E}$, $\mathbf{J}$ and ρ inside the conductor each decays exponentially to zero with the time constant $\frac{\varepsilon}{\sigma}$. Eventually the charge will be distributed only on the conductor's surface. As for the energy change let us first consider the case (a). The electric field outside the conductor is always the same, while the field inside will change from a finite value to zero. The net result is that the electric energy decreases on account of loss arising from conversion of electric energy into heat. For case (b) the field outside the conductor will depend also on θ and φ but the qualitative result is still the same, namely, the electric energy decreases with time being transformed into heat. In short, the final surface charge distribution is such that the electric energy of the system becomes a minimum. In other words, the conductor will become an equipotential volume.

1036

A spherical conductor A contains two spherical cavities as shown in Fig. 1.14. The total charge on the conductor itself is zero. However, there is a point charge $+q_b$ at the center of one cavity and $+q_c$ at the center of the other. A large distance r away is another charge $+q_d$. What forces act on each of the four objects A, q_b, q_c, and q_d? Which answers, if any, are only approximate and depend on r being very large. Comment on the uniformities of the charge distributions on the cavity walls and on A if r is not large.

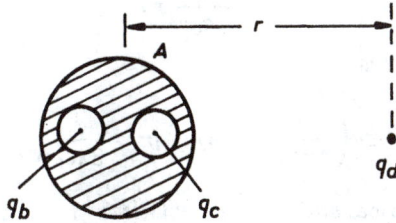

Fig. 1.14

Solution:

Charges outside a cavity have no influence on the field inside because of the electrostatic shielding by the conductor. On account of spherical symmetry the forces acting on the point charges q_b and q_c at the center of the cavities are equal to zero. By electrostatic equilibrium we see that the surfaces of the two spherical cavities carry a total charge $-(q_b + q_c)$, and, since the sphere A was not charged originally, its spherical surface must carry induced charges $q_b + q_c$. As r is very large, we can approximate the interaction between sphere A and point charge q_d by an electrostatic force between point charges $q_b + q_c$ at the center and q_d, namely

$$F = \frac{q_d(q_b + q_c)}{4\pi\varepsilon_0 r^2} .$$

This equation, however, will not hold for r not sufficiently large.

The charge distribution over the surface of each cavity is always uniform and independent of the magnitude of r. However, because of the effect of q_d, the charge distribution over the surface of sphere A will not be uniform, and this nonuniformity will become more and more evident as r decreases.

1037

A spherical condenser consists of two concentric conducting spheres of radii a and b $(a > b)$. The outer sphere is grounded and a charge Q is placed on the inner sphere. The outer conductor then contracts from radius a to radius a'. Find the work done by the electric force.

(*UC, Berkeley*)

Solution:

The electric fields at $r < b$ and $r > a$ are both zero. At $b < r < a$ the electric field is

$$\mathbf{E} = \frac{Q}{4\pi\varepsilon_0 r^2}\mathbf{e}_r .$$

Hence the field energy is

$$W = \int_b^a \frac{1}{2}\varepsilon_0 \left(\frac{Q}{4\pi\varepsilon_0 r^2}\right)^2 4\pi r^2 dr = \frac{Q^2}{8\pi\varepsilon_0}\left(\frac{1}{b} - \frac{1}{a}\right).$$

When the outer spherical surface contracts from $r = a$ to $r = a'$, the work done by the electric force is equal to the decrease of the electric field energy

$$W_a - W_{a'} = \frac{Q^2}{8\pi\varepsilon_0}\left(-\frac{1}{a} + \frac{1}{a'}\right) = \frac{Q^2(a - a')}{8\pi\varepsilon_0 aa'} .$$

1038

A thin metal sphere of radius b has charge Q.

(a) What is the capacitance?

(b) What is the energy density of the electric field at a distance r from the sphere's center?

(c) What is the total energy of the field?

(d) Compute the work expended in charging the sphere by carrying infinitesimal charges from infinity.

(e) A potential V is established between inner (radius a) and outer (radius b) concentric thin metal spheres. What is the radius of the inner sphere such that the electric field near its surface is a minimum?

(*Wisconsin*)

Solution:

(a) Use spherical coordinates (r, θ, φ). The electric field outside the sphere is

$$\mathbf{E}(r) = \frac{Q}{4\pi\varepsilon_0 r^2}\mathbf{e}_r .$$

Let the potential at infinity be zero, then the potential at r is

$$V(r) = \int_r^\infty \frac{Q}{4\pi\varepsilon_0 r'^2}dr' = \frac{Q}{4\pi\varepsilon_0}\cdot\frac{1}{r} .$$

Hence the capacitance is

$$C = \frac{Q}{V(b)} = 4\pi\varepsilon_0 b .$$

(b) $w_e(r) = \frac{1}{2}\mathbf{D} \cdot \mathbf{E} = \frac{1}{2}\varepsilon_0 \mathbf{E}^2 = \frac{Q^2}{32\pi^2\varepsilon_0 r^4}$.

(c) $W_e = \frac{1}{2}V(b)Q = \frac{Q^2}{8\pi\varepsilon_0 b}$.

It may also be calculated from the field energy density $w_e(r)$:

$$W_e = \int_{r>b} w_e(r')dV' = \int_b^\infty \frac{Q^2}{2 \times 16\pi^2\varepsilon_0}\frac{1}{r'^4} \cdot 4\pi r'^2 dr' = \frac{Q^2}{8\pi\varepsilon_0 b} .$$

(d) The work expended in charging the sphere by carrying infinitesimal charges in from infinity is

$$W = \int_0^Q V(Q')dQ' = \int_0^Q \frac{Q'}{4\pi\varepsilon_0}\frac{dQ'}{b} = \frac{Q^2}{8\pi\varepsilon_0 b} = W_e$$

as expected.

(e) Suppose that the inner sphere carries a charge Q. For $a < r < b$ the field intensity is

$$\mathbf{E}(r) = \frac{Q}{4\pi\varepsilon_0 r^2}\mathbf{e}_r .$$

The potential difference between the concentric spheres is

$$V = \int_a^b \mathbf{E}(r) \cdot d\mathbf{r} = \int_a^b \frac{Q}{4\pi\varepsilon_0}\frac{1}{r^2}dr = \frac{Q}{4\pi\varepsilon_0}\left(\frac{1}{a} - \frac{1}{b}\right) .$$

In terms of V we have

$$Q = \frac{4\pi\varepsilon_0 V}{(\frac{1}{a} - \frac{1}{b})}$$

and

$$E(r) = \frac{4\pi\varepsilon_0 V}{4\pi\varepsilon_0 r^2(\frac{1}{a} - \frac{1}{b})} = \frac{V}{r^2(\frac{1}{a} - \frac{1}{b})} .$$

In particular, we have

$$E(a) = \frac{V}{a^2(\frac{1}{a} - \frac{1}{b})} = \frac{Vb}{ab - a^2} .$$

From $\frac{dE(a)}{da} = 0$, we see that $E(a)$ is a minimum at $a = \frac{b}{2}$, and the minimum value is

$$E_{min}(a) = \frac{4V}{b}.$$

1039

A conducting sphere with total charge Q is cut into half. What force must be used to hold the halves together?

(*MIT*)

Solution:

The charge is entirely distributed over the surface with a surface charge density of $\sigma = Q/4\pi R^2$, where R is the radius of the sphere. We know from Problem **1021** that the force exerting on a surface element dS of the conducting sphere is

$$d\mathbf{F} = \frac{\sigma^2}{2\varepsilon_0} dS.$$

Use the coordinate system shown in Fig. 1.15. The plane where the sphere is cut in half is taken to be the xoz plane. The repulsive force between the two half-spheres is perpendicular to the cut plane, so that the resultant force on the right-half must be along the y-axis. The magnitude of the resultant force is

$$F = \int dF \sin\theta \sin\varphi = \frac{\sigma^2}{2\varepsilon_0} R^2 \int_0^\pi \sin\varphi d\varphi \int_0^\pi \sin^2\theta d\theta$$
$$= \frac{\pi\sigma^2 R^2}{2\varepsilon_0} = \frac{Q^2}{32\pi\varepsilon_0 R^2}.$$

This is the force needed to hold the two halves together.

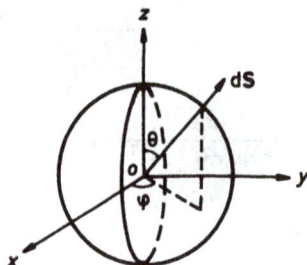

Fig. 1.15

1040

A particle of charge q is moved from infinity to the center of a hollow conducting spherical shell of radius R, thickness t, through a very tiny hole in the shell. How much work is required?

(*Princeton*)

Solution:

The work done by the external force is equal to the increase of the electric field energy of the whole system.

The electric field intensity at a point distance r from the point charge q is $E = \frac{q}{4\pi\varepsilon_0 r^2}$. When q is at infinity the electric energy of the whole system is

$$W = \int_\infty \frac{\varepsilon_0}{2} E^2 dV \, ,$$

integrating over all space, since, as the distance between the spherical shell and q is infinite, the field due to q at the conducting sphere can be taken to be zero.

After q is moved to the center of the conducting spherical shell, as the shell has no effect on the field inside, the electric intensity at a point inside the shell is still $\frac{q}{4\pi\varepsilon_0 r^2}$, r being the distance of the point from q. At a point outside the shell, Gauss' law shows that the electric intensity is still $\frac{q}{4\pi\varepsilon_0 r^2}$. Hence the electric energy of the system remains the same as W but minus the contribution of the shell itself, inside whose thickness the field is zero. Thus there is a decrease of energy

$$-\Delta W = \int_R^{R+t} \frac{\varepsilon_0}{2} \left(\frac{Q}{4\pi\varepsilon_0 r^2} \right)^2 4\pi r^2 dr$$

$$= \frac{q^2}{8\pi\varepsilon_0} \left(\frac{1}{R} - \frac{1}{R+t} \right) ,$$

which is equal to the negative work done by the external force.

1041

A capacitor is made of three conducting concentric thin spherical shells of radii a, b and $d \, (a < b < d)$. The inner and outer spheres are connected by a fine insulated wire passing through a tiny hole in the intermediate sphere. Neglecting the effects of the hole,

(a) find the capacitance of the system,

(b) determine how any net charge Q_B placed on the middle sphere distributes itself between the two surfaces of the sphere.

(Columbia)

Solution:

(a) Suppose that the charge of the inner spherical shell is Q_1 and the charge of the outer shell is $-Q_2$. Then the charges on the inner and outer surfaces of the middle spherical shell are $-Q_1$ and $+Q_2$ ($Q_1, Q_2 > 0$) as shown in Fig. 1.16. The electric field intensities are as follows:

$$\mathbf{E} = \frac{Q_1 \mathbf{r}}{4\pi\varepsilon_0 r^3}, \quad (a < r < b),$$

$$\mathbf{E} = \frac{Q_2 \mathbf{r}}{4\pi\varepsilon_0 r^3}, \quad (b < r < d),$$

$$\mathbf{E} = 0, \quad (r < a, r > d).$$

The potential at a point P is given by

$$\varphi(p) = \int_P^\infty \mathbf{E} \cdot d\mathbf{r}$$

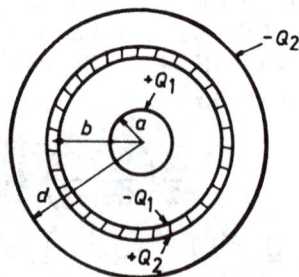

Fig. 1.16

with $\varphi(\infty) = 0$. Thus we have

$$\varphi(d) = 0, \quad \varphi(b) = \frac{Q_2}{4\pi\varepsilon_0}\left(\frac{1}{b} - \frac{1}{d}\right).$$

As the inner and outer spherical shells are connected their potentials should be equal. Hence

$$\varphi(a) = \frac{Q_1}{4\pi\varepsilon_0}\left(\frac{1}{a} - \frac{1}{b}\right) + \frac{Q_2}{4\pi\varepsilon_0}\left(\frac{1}{b} - \frac{1}{d}\right) = 0,$$

whence

$$Q_1\left(\frac{1}{a}-\frac{1}{b}\right)=-Q_2\left(\frac{1}{b}-\frac{1}{d}\right).$$

The potential differences of the spherical shells are

$$V_{ab}=\varphi(a)-\varphi(b)=-\varphi(b),$$
$$V_{db}=\varphi(d)-\varphi(b)=-\varphi(b).$$

Thus the capacitance between the inner sphere and the inner surface of the middle spherical shell is

$$C_{ab}=\frac{Q_1}{V_{ab}}=-\frac{Q_1}{\varphi(b)},$$

and the capacitance between the outer surface of the middle shell and the outer shell is

$$C_{bd}=\frac{Q_2}{V_{bd}}=\frac{Q_2}{\varphi(b)}.$$

The capacitance of the whole system can be considered as C_{ab} and C_{bd} in series, namely

$$C=\left(\frac{1}{C_{ab}}+\frac{1}{C_{bd}}\right)^{-1}=\frac{1}{\varphi(b)}\left(\frac{1}{Q_2}-\frac{1}{Q_1}\right)^{-1}=\frac{4\pi\varepsilon_0 ad}{d-a}.$$

(b) The net charge Q_B carried by the middle shell must be equal to Q_2-Q_1, so that

$$Q_1=-\frac{a(d-b)}{b(d-a)}Q_B,\quad Q_2=\frac{d(b-a)}{b(d-a)}Q_B.$$

This is to say, the inner surface of the middle shell will carry a total charge $\frac{a(d-b)}{b(d-a)}Q_B$ while the outer surface, $\frac{d(b-a)}{b(d-a)}Q_B$.

1042

A long conducting cylinder is split into two halves parallel to its axis. The two halves are held at V_0 and 0, as in Fig. 1.17(a). There is no net charge on the system.

(a) Calculate the electric potential distribution throughout space.

(b) Calculate the electric field for $r \gg a$.

(c) Calculate the electric field for $r \ll a$.

(d) Sketch the electric field lines throughout space.

(MIT)

Solution:

(a) Use conformal mapping to map the interior of the circle $|z| = a$ onto the upper half of the w-plane by the transformation Fig. 1.17(b)

$$w = i\left(\frac{z-a}{z+a}\right).$$

The upper and lower arcs of the circle are mapped onto the negative and positive axes (u-axis) of the w-plane respectively.

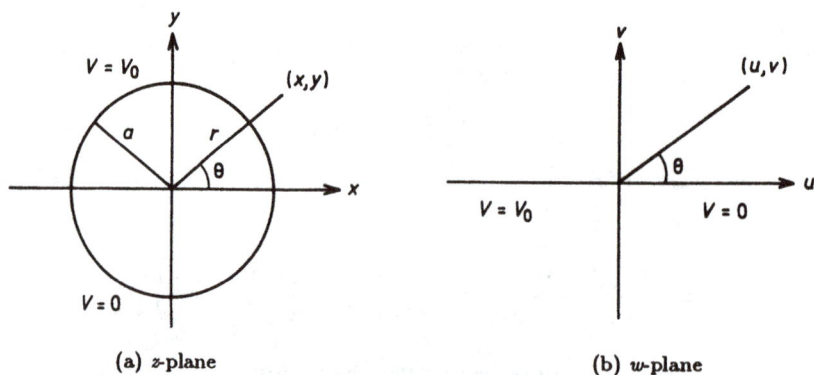

(a) z-plane   (b) w-plane

Fig. 1.17

The problem is now reduced to finding a function V harmonic in the upper half of w-plane and taking the values 0 for $u > 0$ and V_0 for $u < 0$. Use the function $V = A\theta + B$, where A, B are real constants, as $\theta = \text{Im}\{\ln w\}$ is harmonic. The boundary conditions give $B = 0, A = V_0/\pi$. Hence

$$V = \frac{V_0}{\pi}\,\text{Im}\left\{\ln\left[i\frac{z-a}{z+a}\right]\right\}$$

$$= \frac{V_0}{\pi}\,\text{Im}\left\{\ln\left[i\frac{r\cos\theta - a + ir\sin\theta}{r\cos\theta + a + ir\sin\theta}\right]\right\}$$

$$= \frac{V_0}{\pi}\,\text{Im}\left\{\ln\left[i\frac{r^2 - a^2 + 2iar\sin\theta}{(r\cos\theta + a)^2 + r^2\sin^2\theta}\right]\right\}$$

$$= \frac{V_0}{\pi}\left[\frac{\pi}{2} + \arctan\frac{2ar\sin\theta}{|r^2 - a^2|}\right].$$

(b) For $r \gg a$, we have

$$V \approx \frac{V_0}{\pi} \left[\frac{\pi}{2} + \frac{2a \sin \theta}{r} \right] = \frac{V_0}{2} + \frac{2V_0 a \sin \theta}{\pi r},$$

and hence

$$E_r = -\frac{\partial V}{\partial r} = \frac{2V_0 a \sin \theta}{\pi r^2},$$

$$E_\theta = -\frac{1}{r} \frac{\partial V}{\partial \theta} = -\frac{2Va}{\pi r^2} \cos \theta.$$

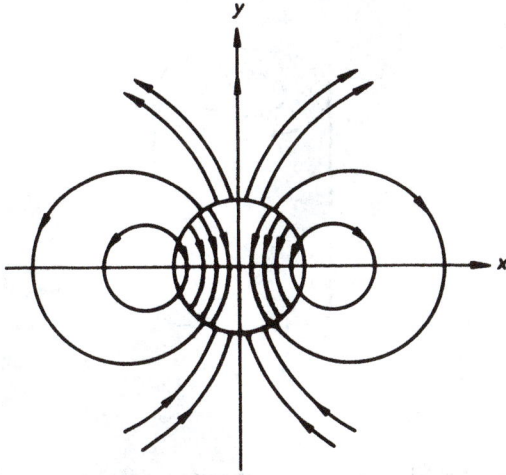

Fig. 1.18

(c) For $r \ll a$, we have

$$V \approx \frac{V_0}{\pi} \left[\frac{\pi}{2} + \frac{2r \sin \theta}{a} \right] = \frac{V_0}{2} + \frac{2V_0 r \sin \theta}{\pi a},$$

and hence

$$E_r = -\frac{\partial V}{\partial r} = -\frac{2V_0 \sin \theta}{\pi a},$$

$$E_\theta = -\frac{1}{r} \frac{\partial V}{\partial \theta} = -\frac{2V_0}{\pi a} \cos \theta.$$

(d) The electric field lines are shown in Fig. 1.18.

3. ELECTROSTATIC FIELD IN A DIELECTRIC MEDIUM (1043–1061)

1043

The space between two long thin metal cylinders is filled with a material with dielectric constant ε. The cylinders have radii a and b, as shown in Fig. 1.19.

(a) What is the charge per unit length on the cylinders when the potential between them is V with the outer cylinder at the higher potential?

(b) What is the electric field between the cylinders?

(*Wisconsin*)

Fig. 1.19

Solution:

This is a cylindrical coaxial capacitor with a capacitance per unit length of

$$C = \frac{2\pi\varepsilon}{\ln(\frac{a}{b})} .$$

As the outer cylinder is at the higher potential, we have from $Q = CV$ the charges per unit length on the inner and outer cylinders:

$$\lambda_i = -\frac{2\pi\varepsilon V}{\ln(\frac{a}{b})} , \quad \lambda_o = \frac{2\pi\varepsilon V}{\ln(\frac{a}{b})} .$$

Gauss' law then gives the electric field intensity in the capacitor:

$$\mathbf{E} = \frac{\lambda_i}{2\pi\varepsilon r}\mathbf{e}_r = -\frac{V}{r\ln(\frac{a}{b})}\mathbf{e}_r .$$

1044

Calculate the resistance between the center conductor of radius a and the coaxial conductor of radius b for a cylinder of length $l \gg b$, which is filled with a dielectric of permittivity ε and conductivity σ. Also calculate the capacitance between the inner and outer conductors.

(*Wisconsin*)

Solution:

Letting V be the voltage difference between the inner and outer conductors, we can express the electric field intensity between the two conductors as

$$\mathbf{E}(r) = \frac{V}{r \ln(\frac{b}{a})} \mathbf{e}_r \ .$$

Ohm's law $\mathbf{J} = \sigma \mathbf{E}$ then gives the current between the two conductors as

$$I = 2\pi r l J = \frac{2\pi \sigma l V}{\ln(\frac{b}{a})} \ .$$

The resistance between the inner and outer conductors is thus

$$R = \frac{V}{I} = \frac{\ln(\frac{b}{a})}{2\pi l \sigma} \ .$$

Since the field is zero inside a conductor, we find the surface charge density ω of the inner conductor from the boundary relation $E = \frac{\omega}{\varepsilon}$, i.e.,

$$\omega = \varepsilon \frac{V}{a \ln(\frac{b}{a})} \ .$$

Thus the inner conductor carries a total charge $Q = 2\pi a l \omega$. Hence the capacitance between the two conductors is

$$C = \frac{Q}{V} = \frac{2\pi \varepsilon l}{\ln(\frac{b}{a})} \ .$$

1045

Two conductors are embedded in a material of conductivity $10^{-4} \Omega/\text{m}$ and dielectric constant $\varepsilon = 80\varepsilon_0$. The resistance between the two conductors is measured to be $10^5 \Omega$. Derive an equation for the capacitance between the two conductors and calculate its value.

(*UC, Berkeley*)

Solution:

Suppose that the two conductors carry free charges Q and $-Q$. Consider a closed surface enclosing the conductor with the charge Q (but not the other conductor). We have, using Ohm's and Gauss' laws,

$$I = \oint \mathbf{j} \cdot d\mathbf{S} = \oint \sigma \mathbf{E} \cdot d\mathbf{S} = \sigma \oint \mathbf{E} \cdot d\mathbf{S} = \sigma \frac{Q}{\varepsilon} .$$

If the potential difference between the two conductors is V, we have $V = IR = \frac{\sigma Q}{\varepsilon} R$, whence

$$C = \frac{Q}{V} = \frac{\varepsilon}{\sigma R} .$$

Numerically the capacitance between the conductors is

$$C = \frac{80 \times 8.85 \times 10^{-12}}{10^{-4} \times 10^5} = 7.08 \times 10^{-11} \text{ F} .$$

1046

Consider a long cylindrical coaxial capacitor with an inner conductor of radius a, an outer conductor of radius b, and a dielectric with a dielectric constant $K(r)$, varying with cylindrical radius r. The capacitor is charged to voltage V. Calculate the radial dependence of $K(r)$ such that the energy density in the capacitor is constant (under this condition the dielectric has no internal stresses). Calculate the electric field $E(r)$ for these conditions.

(*Wisconsin*)

Solution:

Let λ be the charge per unit length carried by the inner conductor. Gauss' law gives

$$D(r) = \frac{\lambda}{2\pi r} ,$$

as D is along the radial direction on account of symmetry.

The energy density at r is

$$U(r) = \frac{1}{2} E D = \frac{D^2}{2\varepsilon_0 K(r)} = \frac{\lambda^2}{8\pi^2 \varepsilon_0 r^2 K(r)} .$$

If this is to be independent of r, we require $r^2 K(r) = \text{constant} = k$, say, i.e., $K(r) = kr^{-2}$.

The voltage across the two conductors is

$$V = -\int_a^b E\,dr = -\frac{\lambda}{2\pi\varepsilon_0 k}\int_a^b r\,dr$$
$$= -\frac{\lambda}{4\pi\varepsilon_0 k}(b^2 - a^2).$$

Hence

$$\lambda = -\frac{4\pi\varepsilon_0 kV}{b^2 - a^2},$$

giving

$$E(r) = -\frac{2rV}{b^2 - a^2}.$$

1047

Find the potential energy of a point charge in vacuum a distance x away from a semi-infinite dielectric medium whose dielectric constant is K.

(*UC, Berkeley*)

Solution:

Use cylindrical coordinates (r, φ, z) with the surface of the semi-infinite medium as the $z = 0$ plane and the z-axis passing through the point charge q, which is located at $z = x$. Let $\sigma_p(r)$ be the bound surface charge density of the dielectric medium on the $z = 0$ plane, assuming the medium to carry no free charge.

The normal component of the electric intensity at a point (r, φ, o) is

$$E_{z1}(r) = -\frac{qx}{4\pi\varepsilon_0(r^2 + x^2)^{3/2}} + \frac{\sigma_p(r)}{2\varepsilon_0}$$

on the upper side of the interface $(z = 0_+)$. However, the normal component of the electeric field is given by

$$E_{z2}(r) = -\frac{qx}{4\pi\varepsilon_0(r^2 + x^2)^{3/2}} - \frac{\sigma_p(r)}{2\varepsilon_0}$$

on the lower side of the interface $(z = 0_-)$. The boundary condition of the displacement vector at $z = 0$ yields

$$\varepsilon_0 E_{z1}(r) = \varepsilon_0 K E_{z2}(r).$$

Hence

$$\sigma_p(r) = \frac{(1-K)qx}{2\pi(1+K)(r^2+x^2)^{3/2}} \, .$$

The electric field at the point $(0,0,x)$, the location of q, produced by the distribution of the bound charges has only the normal component because of symmetry, whose value is obtained by

$$E = \int \frac{\sigma_p(r)x dS}{4\pi\varepsilon_0(r^2+x^2)^{3/2}} = \frac{(1-K)qx^2}{4\pi(1+K)\varepsilon_0} \int_0^\infty \frac{rdr}{(r^2+x^2)^3} = \frac{(1-K)q}{16\pi(1+K)\varepsilon_0 x^2},$$

where the surface element dS has been taken to be $2\pi rdr$. Hence the force acted on the point charge is

$$F = qE = \frac{(1-K)q^2}{16\pi(1+K)\varepsilon_0 x^2} \, .$$

The potential energy W of the point charge q equals the work done by an external force in moving q from infinity to the position x, i.e.,

$$W = -\int_\infty^x F dx' = -\int_\infty^x \frac{(1-K)q^2}{16\pi(1+K)\varepsilon_0 x'^2} dx' = \frac{(1-K)q^2}{16\pi(1+K)\varepsilon_0 x} \, .$$

1048

The mutual capacitance of two thin metallic wires lying in the plane $z = 0$ is C. Imagine now that the half space $z < 0$ is filled with a dielectric material with dielectric constant ε. What is the new capacitance?

(*MIT*)

Solution:

As shown in Fig. 1.20, before filling in the dielectric material, one of the thin conductors carries charge $+Q$, while the other carries charge $-Q$. The potential difference between the two conductors is V and the capacitance of the system is $C = Q/V$. The electric field intensity in space is $\mathbf{E}$. After the half space is filled with the dielectric, let $\mathbf{E'}$ be the field intensity in space. This field is related to the original one by the equation $\mathbf{E'} = K\mathbf{E}$, where K is a constant to be determined below.

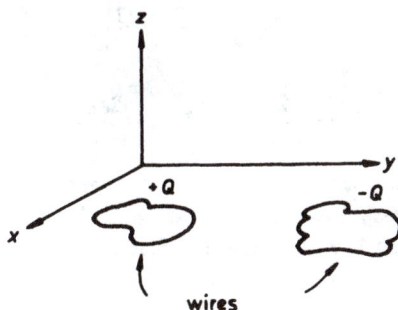

Fig. 1.20

We consider a short right cylinder across the interface $z = 0$ with its cross-section at $z = 0$ just contains the area enclosed by the wire carrying charge $+Q$ and the wire itself. The upper end surface S_1 of this cylinder is in the space $z > 0$ and the lower end surface S_2 is in the space $z < 0$. Apply Gauss' law to this cylinder. The contribution from the curved surface may be neglected if we make the cylinder sufficiently short. Thus we have, before the introduction of the dielectric,

$$\oint_s \mathbf{D} \cdot d\mathbf{S} = \varepsilon_0 \int_{s_1} \mathbf{E} \cdot d\mathbf{S} + \varepsilon_0 \int_{s_2} \mathbf{E} \cdot d\mathbf{S} = Q, \tag{1}$$

and, after introducing the dielectric,

$$\oint_s \mathbf{D}' \cdot d\mathbf{S} = \varepsilon_0 \int_{s_1} \mathbf{E}' \cdot d\mathbf{S} + \varepsilon \int_{s_2} \mathbf{E}' \cdot d\mathbf{S} = Q. \tag{2}$$

Note that the vector areas $\mathbf{S}_1$ and $\mathbf{S}_2$ are equal in magnitude and opposite in direction. In Eq. (1) as the designation of 1 and 2 is interchangeable the two contributions must be equal. We therefore have

$$\int_{s_1} \mathbf{E} \cdot d\mathbf{S} = \int_{s_2} \mathbf{E} \cdot d\mathbf{S} = \frac{Q}{2\varepsilon_0}.$$

Equation (2) can be written as

$$K\left(\varepsilon_0 \int_{s_1} \mathbf{E} \cdot d\mathbf{S} + \varepsilon \int_{s_2} \mathbf{E} \cdot d\mathbf{S}\right) = Q,$$

or

$$\frac{(\varepsilon_0 + \varepsilon)K}{2\varepsilon_0} Q = Q,$$

whence we get

$$K = \frac{2\varepsilon_0}{\varepsilon_0 + \varepsilon}, \quad \mathbf{E}' = \frac{2\varepsilon_0 \mathbf{E}}{\varepsilon + \varepsilon_0}.$$

To calculate the potential difference between the two conductors, we may select an arbitrary path of integration L from one conductor to the other. Before filling in the dielectric material, the potential is

$$V = -\int_L \mathbf{E} \cdot d\mathbf{l},$$

while after filling in the dielectric the potential will become

$$V' = -\int_L \mathbf{E}' \cdot d\mathbf{l} = -K \int_L \mathbf{E} \cdot d\mathbf{l} = KV.$$

Hence, the capacitance after introducing the dielectric is

$$C' = \frac{Q}{V'} = \frac{Q}{KV} = \frac{\varepsilon + \varepsilon_0}{2\varepsilon_0} C.$$

1049

A parallel plate capacitor (having perfectly conducting plates) with plate separation d is filled with two layers of material (1) and (2). The first layer has dielectric constant ε_1, conductivity σ_1, the second, ε_2, σ_2, and their thicknesses are d_1 and d_2, respectively. A potential V is placed across the capacitor (see Fig. 1.21). Neglect edge effects.

(a) What is the electric field in material (1) and (2)?

(b) What is the current flowing through the capacitor?

(c) What is the total surface charge density on the interface between (1) and (2)?

(d) What is the free surface charge density on the interface between (1) and (2)?

(*CUSPEA*)

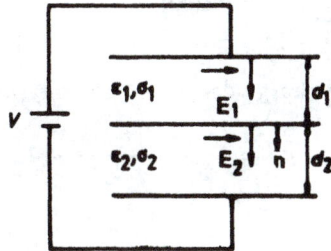

Fig. 1.21

Solution:

(a) Neglecting edge effects, the electric fields E_1 and E_2 in material (1) an (2) are both uniform fields and their directions are perpendicular to the parallel plates. Thus we have

$$V = E_1 d_1 + E_2 d_2 . \qquad (1)$$

As the currents flowing through material (1) and (2) must be equal, we have

$$\sigma_1 E_1 = \sigma_2 E_2 . \qquad (2)$$

Combination of Eqs. (1) and (2) gives

$$E_1 = \frac{V\sigma_2}{d_1\sigma_2 + d_2\sigma_1}, \quad E_2 = \frac{V\sigma_1}{d_1\sigma_2 + d_2\sigma_1} .$$

(b) The current density flowing through the capacitor is

$$J = \sigma_1 E_1 = \frac{\sigma_1\sigma_2 V}{d_1\sigma_2 + d_2\sigma_1} .$$

Its direction is perpendicular to the plates.

(c) By using the boundary condition (see Fig. 1.21)

$$\mathbf{n} \cdot (\mathbf{E}_2 - \mathbf{E}_1) = \sigma_t/\varepsilon_0 ,$$

we find the total surface charge density on the interface between material (1) and (2)

$$\sigma_t = \varepsilon_0(E_2 - E_1) = \frac{\varepsilon_0(\sigma_1 - \sigma_2)V}{d_1\sigma_2 + d_2\sigma_1} .$$

(d) From the boundary condition

$$\mathbf{n} \cdot (\mathbf{D}_2 - \mathbf{D}_1) = \mathbf{n} \cdot (\varepsilon_2 \mathbf{E}_2 - \varepsilon_1 \mathbf{E}_1) = \sigma_f,$$

we find the free surface charge density on the interface

$$\sigma_f = \frac{(\sigma_1 \varepsilon_2 - \sigma_2 \varepsilon_1)V}{d_1 \sigma_2 + d_2 \sigma_1}.$$

1050

In Fig. 1.22, a parallel-plate air-spaced condenser of capacitance C and a resistor of resistance R are in series with an ac source of frequency ω. The voltage-drop across R is V_R. Half the condenser is now filled with a material of dielectric constant ε but the remainder of the circuit remains unchanged. The voltage-drop across R is now $2V_R$. Neglecting edge effects, calculate the dielectric constant ε in terms of R, C and ω.

(*Columbia*)

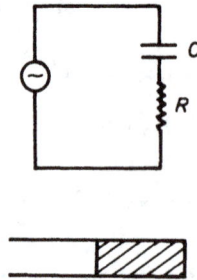

Fig. 1.22

Solution:

When half the condenser is filled with the material, the capacitance of the condenser (two condensers in parallel) becomes

$$C' = \frac{C}{2} + \frac{\varepsilon C}{2\varepsilon_0} = \frac{1}{2}\left(1 + \frac{\varepsilon}{\varepsilon_0}\right)C.$$

The voltage across R is VR/Z, where V is the voltage of the ac source and Z is the total impedance of the circuit. Thus

$$\left|\frac{R}{R + \frac{1}{j\omega C'}}\right| = 2\left|\frac{R}{R + \frac{1}{j\omega C}}\right|,$$

where $j = \sqrt{-1}$. Therefore we get

$$4R^2 + \frac{16}{\omega^2 C^2 (1 + \frac{\varepsilon}{\varepsilon_0})^2} = R^2 + \frac{1}{\omega^2 C^2} .$$

Solving this equation, we obtain

$$\varepsilon = \left(\frac{4}{\sqrt{1 - 3R^2 C^2 \omega^2}} - 1 \right) \varepsilon_0 .$$

1051

A capacitor is made of two plane parallel plates of width a and length b separated by a distance d $(d \ll a, b)$, as in Fig. 1.23. The capacitor has a dielectric slab of relative dielectric constant K between the two plates.

(a) The capacitor is connected to a battery of emf V. The dielectric slab is partially pulled out of the plates such that only a length x remains between the plates. Calculate the force on the dielectric slab which tends to pull it back into the plates.

(b) With the dielectric slab fully inside, the capacitor plates are charged to a potential difference V and the battery is disconnected. Again, the dielectric slab is pulled out such that only a length x remains inside the plates. Calculate the force on the dielectric slab which tends to pull it back into the plates. Neglect edge effects in both parts (a) and (b).

(Columbia)

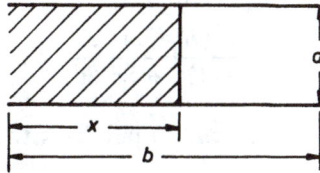

Fig. 1.23

Solution:

Treating the capacitor in Fig. 1.23 as two capacitors in parallel, we obtain the total capacitance as

$$C = \varepsilon_0 \frac{Kxa}{d} + \frac{\varepsilon_0 (b - x)a}{d} = \frac{\varepsilon_0 (K - 1)ax}{d} + \frac{\varepsilon_0 ba}{d} = \frac{\varepsilon_0 [(K - 1)x + b]a}{d} .$$

Consider the charging of the capacitor. The energy principle gives

$$V dQ = d\left(\frac{1}{2}V^2 C\right) + F dx .$$

(a) As $V = $ constant, $Q = CV$ gives

$$V dQ = V^2 dC .$$

Hence

$$F = \frac{1}{2}V^2 \frac{dC}{dx} = \frac{\varepsilon_0(K-1)aV^2}{2d} .$$

Since $K > 1, F > 0$. This means that F tends to increase x, i.e., to pull the slab back into the plates.

(b) Since the plates are isolated electrically, $dQ = 0$. Let the initial voltage be V_0. As initially $x = b, C_0 = \varepsilon_0 \frac{Kba}{d}$ and $Q = C_0 V_0$. The energy principle now gives

$$F = -\frac{d}{dx}\left(\frac{1}{2}V^2 C\right) = -VC\frac{dV}{dx} - \frac{V^2}{2}\frac{dC}{dx} .$$

As

$$\frac{dV}{dx} = \frac{d}{dx}\left(\frac{Q}{C}\right) = -\frac{Q}{C^2}\frac{dC}{dx} ,$$

the above becomes

$$F = Q\frac{dV}{dx} - \frac{Q^2}{2C^2}\frac{dC}{dx} = \frac{Q^2}{2C^2}\frac{dC}{dx}$$
$$= \frac{\varepsilon_0 K^2(K-1)}{[(K-1)x + b]^2}\frac{ab^2}{2d}V_0^2 .$$

Again, as $F > 0$ the force will tend to pull back the slab into the plates.

1052

A dielectric is placed partly into a parallel plate capacitor which is charged but isolated. It feels a force:

(a) of zero   (b) pushing it out   (c) pulling it in .

(CCT)

Solution:

The answer is (c).

1053

A cylindrical capacitor of length L consists of an inner conductor wire of radius a, a thin outer conducting shell of radius b. The space in between is filled with nonconducting material of dielectric constant ε.

(a) Find the electric field as a function of radial position when the capacitor is charged with Q. Neglect end effects.

(b) Find the capacitance.

(c) Suppose that the dielectric is pulled partly out of the capacitor while the latter is connected to a battery of potential V. Find the force necessary to hold the dielectric in this position. Neglect fringing fields. In which direction must the force be applied?

(*CUSPEA*)

Solution:

(a) Supposing that the charge per unit length of the inner wire is $-\lambda$ and using cylindrical coordinates (r, φ, z), we find the electric field intensity in the capacitor by Gauss' theorem to be

$$\mathbf{E} = -\frac{\lambda}{2\pi\varepsilon r}\mathbf{e}_r = \frac{-Q}{2\pi\varepsilon L r}\mathbf{e}_r \; .$$

(b) The potential difference between the inner and outer capacitors is

$$V = -\int_a^b \mathbf{E} \cdot d\mathbf{r} = \frac{\lambda}{2\pi\varepsilon}\ln\left(\frac{b}{a}\right) \; .$$

Hence the capacitance is

$$C = \frac{\lambda L}{V} = \frac{2\pi\varepsilon L}{\ln(\frac{b}{a})} \; .$$

(c) When the capacitor is connected to a battery, the potential difference between the inner and outer conductors remains a constant. The dielectric is now pulled a length x out of the capacitor, so that a length

$L - x$ of the material remains inside the capacitor, as shown schematically in Fig. 1.24. The total capacitance of the capacitor becomes

$$C = \frac{2\pi\varepsilon_0 x}{\ln(\frac{b}{a})} + \frac{2\pi\varepsilon(L - x)}{\ln(\frac{b}{a})}$$

$$= \frac{2\pi\varepsilon_0}{\ln(\frac{b}{a})} \left[\frac{\varepsilon}{\varepsilon_0 L} + \left(1 - \frac{\varepsilon}{\varepsilon_0}\right)x \right].$$

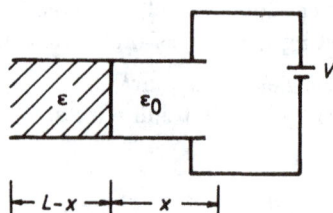

Fig. 1.24

Pulling out the material changes the energy stored in the capacitor and thus a force must be exerted on the material. Consider the energy equation

$$F\,dx = V\,dQ - \frac{1}{2}V^2 dC.$$

As V is kept constant, $dQ = V\,dC$ and we have

$$F = \frac{1}{2}V^2 \frac{dC}{dx} = \frac{\pi\varepsilon_0 V^2}{\ln(\frac{b}{a})} \left(1 - \frac{\varepsilon}{\varepsilon_0}\right)$$

as the force acting on the material.

As $\varepsilon > \varepsilon_0$, $F < 0$. Hence F will tend to decrease x, i.e., F is attractive. Then to hold the dielectric in this position, a force must be applied with magnitude F and a direction away from the capacitor.

1054

As in Fig. 2.15, you are given the not-so-parallel plate capacitor.

(a) Neglecting edge effects, when a voltage difference V is placed across the two conductors, find the potential everywhere between the plates.

(b) When this wedge is filled with a medium of dielectric constant ε, calculate the capacitance of the system in terms of the constants given.

(Princeton)

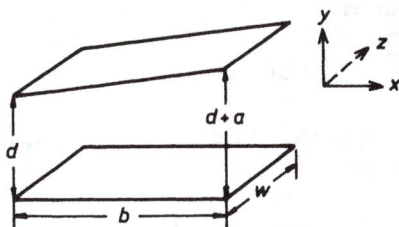

Fig. 1.25

Solution:

(a) Neglecting edge effects, this problem becomes a 2-dimensional one. Take the z-axis normal to the diagram and pointing into the page as shown in Fig. 1.25. The electric field is parallel to the xy plane, and independent of z.

Suppose that the intersection line of the planes of the two plates crosses the x-axis at O', using the coordinate system shown in Fig. 1.26. Then

$$\overline{OO'} = \frac{bd}{a}, \quad \theta_0 = \arctan\frac{a}{b},$$

where θ_0 is the angle between the two plates. Now use cylindrical coordinates (r, θ, z') with the z'-axis passing through point O' and parallel to the z-axis. Any plane through the z'-axis is an equipotential surface according to the symmetry of this problem. So the potential inside the capacitor will depend only on θ:

$$\varphi(r, \theta, z') = \varphi(\theta).$$

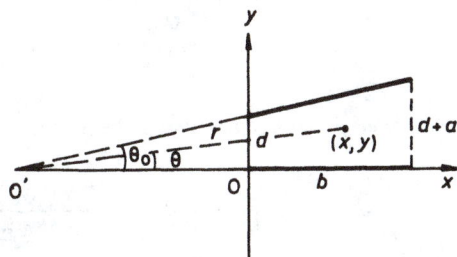

Fig. 1.26

The potential φ satisfies the Laplace equation

$$\nabla^2\varphi = \frac{1}{r^2}\frac{d^2\varphi}{d\theta^2} = 0,$$

whose general solution is

$$\varphi(\theta) = A + B\theta .$$

Since both the upper and lower plates are equipotential surfaces, the boundary conditions are

$$\varphi(0) = 0, \quad \varphi(\theta_0) = V ,$$

whence $A = 0, B = V/\theta_0$. For a point (x, y) inside the capacitor,

$$\theta = \arctan\left[y/\left(x + \frac{bd}{a}\right)\right] .$$

Hence

$$\varphi(x, y) = \frac{V\theta}{\theta_0} = \frac{V \arctan\left[y/(x + \frac{bd}{a})\right]}{\arctan(\frac{a}{b})} .$$

(b) Let Q be the total charge on the lower plate. The electric field inside the capacitor is

$$\mathbf{E} = -\nabla\varphi = -\frac{\partial\varphi}{\partial\theta}\frac{\mathbf{e}_\theta}{r} = -\frac{V}{\theta_0 r}\mathbf{e}_\theta .$$

For a point $(x, 0)$ on the lower plate, $\theta = 0, r = \frac{bd}{a} + x$ and $\mathbf{E}$ is normal to the plate. The surface charge density σ on the lower plate is obtained from the boundary condition for the displacement vector:

$$\sigma = \varepsilon E = -\frac{V\varepsilon}{\theta_0(\frac{bd}{a} + x)} .$$

Integrating over the lower plate surface, we obtain

$$Q = \int \sigma dS = -\int_0^w dz \int_0^b \frac{V\varepsilon}{\theta_0(\frac{bd}{a} + x)} dx = -\frac{\varepsilon V w}{\arctan\frac{a}{b}} \ln\left(\frac{d+a}{d}\right) .$$

Hence, the capacitance of the capacitor is

$$C = \frac{|Q|}{V} = \frac{\varepsilon w}{\arctan\frac{a}{b}} \ln\left(\frac{d+a}{d}\right) .$$

1055

Two large parallel conducting plates, each of area A, are separated by distance d. A homogeneous anisotropic dielectric fills the space between the plates. The dielectric permittivity tensor ε_{ij} relates the electric displacement $\mathbf{D}$ and the electric field $\mathbf{E}$ according to $D_i = \sum_{j=1}^{3} \varepsilon_{ij} E_j$. The principal axes of this permittivity tensor are (see Fig. 1.27): Axis 1 (with eigenvalue ε_1) is in the plane of the paper at an angle θ with respect to the horizontal. Axis 2 (with eigenvalue ε_2) is in the paper at an angle $\frac{\pi}{2} - \theta$ with respect to the horizontal. Axis 3 (with eigenvalue ε_3) is perpendicular to the plane of the paper. Assume that the conducting plates are sufficiently large so that all edge effects are negligible.

(a) Free charges $+Q_F$ and $-Q_F$ are uniformly distributed on the left and right conducting plates, respectively. Find the horizontal and vertical components of $\mathbf{E}$ and $\mathbf{D}$ within the dielectric.

(b) Calculate the capacitance of this system in terms of A, d, ε_i and θ.

(*Columbia*)

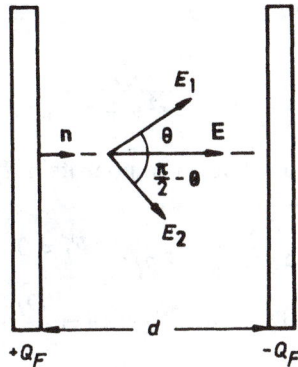

Fig. 1.27

Solution:

(a) Let $\mathbf{n}$ be a unit normal vector to the left plate. As $\mathbf{E} = 0$ inside the plates, the tangential component of the electric field inside the dielectric is also zero because of the continuity of the tangential component of $\mathbf{E}$. Hence, the electric field intensity inside the dielectric can be expressed as

$$\mathbf{E} = E\mathbf{n}.$$

Resolving $\mathbf{E}$ along the principal axes we have

$$E_1 = E\cos\theta, \quad E_2 = E\sin\theta, \quad E_3 = 0.$$

In the coordinates $(\hat{e}_1, \hat{e}_2, \hat{e}_3)$ based on the principal axes, tensor ε_{ij} is a diagonal matrix

$$(\varepsilon_{ij}) = \begin{pmatrix} \varepsilon_1 & 0 & 0 \\ 0 & \varepsilon_2 & 0 \\ 0 & 0 & \varepsilon_3 \end{pmatrix}$$

and along these axes the electric displacement in the capacitor has components

$$D_1 = \varepsilon_1 E_1 = \varepsilon_1 E \cos\theta, \quad D_2 = \varepsilon_2 E \sin\theta, \quad D_3 = 0. \tag{1}$$

The boundary condition of $\mathbf{D}$ on the surface of the left plate yields

$$D_n = \sigma_f = Q_F/A.$$

That is, the normal component of the electric displacement is a constant. Thus

$$D_1 \cos\theta + D_2 \sin\theta = D_n = \frac{Q_F}{A}. \tag{2}$$

Combining Eqs. (1) and (2), we get

$$E = \frac{Q_F}{A(\varepsilon_1 \cos^2\theta + \varepsilon_2 \sin^2\theta)}.$$

Hence the horizontal and vertical components of $\mathbf{E}$ and $\mathbf{D}$ are

$$E_n = E = \frac{Q_F}{A(\varepsilon_1 \cos^2\theta + \varepsilon_2 \sin^2\theta)}, \quad E_t = 0,$$

$$D_n = \frac{Q_F}{A}, \quad D_t = D_1 \sin\theta - D_2 \cos\theta = \frac{Q_F(\varepsilon_1 - \varepsilon_2)\sin\theta\cos\theta}{A(\varepsilon_1 \cos^2\theta + \varepsilon_2 \sin^2\theta)},$$

where the subscript t denotes components tangential to the plates.

(b) The potential difference between the left and right plates is

$$V = \int E_n dx = \frac{Q_F d}{A(\varepsilon_1 \cos^2\theta + \varepsilon_2 \sin^2\theta)}.$$

Therefore, the capacitance of the system is

$$C = \frac{Q_F}{V} = \frac{A(\varepsilon_1 \cos^2\theta + \varepsilon_2 \sin^2\theta)}{d}.$$

1056

It can be shown that the electric field inside a dielectric sphere which is placed inside a large parallel-plate capacitor is uniform (the magnitude and direction of $\mathbf{E}_0$ are constant). If the sphere has radius R and relative dielectric constant $K_e = \varepsilon/\varepsilon_0$, find $\mathbf{E}$ at point p on the outer surface of the sphere (use polar coordinates R, θ). Determine the bound surface charge density at point p.

(*Wisconsin*)

Solution:

The electric field inside the sphere is a uniform field $\mathbf{E}_0$, as shown in Fig. 1.28. The field at point p of the outer surface of the sphere is $\mathbf{E} = E_r \mathbf{e}_r + E_t \mathbf{e}_\theta$, using polar coordinates. Similarly $\mathbf{E}_0$ may be expressed as

$$\mathbf{E}_0 = E_0 \cos\theta \mathbf{e}_r - E_0 \sin\theta \mathbf{e}_\theta.$$

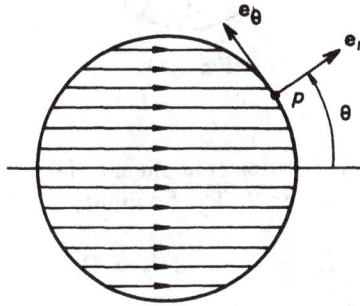

Fig. 1.28

From the boundary conditions for the electric vectors at p we obtain

$$\varepsilon E_0 \cos\theta = \varepsilon_0 E_r, \quad -E_0 \sin\theta = E_t.$$

Hence

$$\mathbf{E} = K_e E_0 \cos\theta \mathbf{e}_r - E_0 \sin\theta \mathbf{e}_\theta.$$

The bound surface charge density at point p is $\sigma_b = \mathbf{P} \cdot \mathbf{e}_r$, where $\mathbf{P}$ is the polarization vector. As $\mathbf{P} = (\varepsilon - \varepsilon_0)\mathbf{E}_0$, we find

$$\sigma_p = (\varepsilon - \varepsilon_0)E_0 \cos\theta = \varepsilon_0(K_e - 1)E_0 \cos\theta.$$

1057

One half of the region between the plates of a spherical capacitor of inner and outer radii a and b is filled with a linear isotropic dielectric of permittivity ε_1 and the other half has permittivity ε_2, as shown in Fig. 1.29. If the inner plate has total charge Q and the outer plate has total charge $-Q$, find:

(a) the electric displacements $\mathbf{D}_1$ and $\mathbf{D}_2$ in the region of ε_1 and ε_2;

(b) the electric fields in ε_1 and ε_2;

(c) the total capacitance of this system.

$(SUNY, \ Buffalo)$

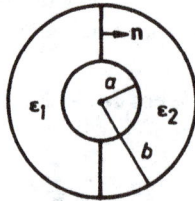

Fig. 1.29

Solution:

We take the normal direction $\mathbf{n}$ at the interface between the dielectrics ε_1 and ε_2 as pointing from 1 to 2. The boundary conditions at the interface are

$$E_{1t} = E_{2t}, \quad D_{1n} = D_{2n}.$$

If we assume that the field $\mathbf{E}$ still has spherical symmetry, i.e.,

$$\mathbf{E}_1 = \mathbf{E}_2 = A\mathbf{r}/r^3,$$

then the above boundary conditions may be satisfied. Take as Gaussian surface a concentric spherical surface of radius $r \, (a < r < b)$. From

$$\oint \mathbf{D} \cdot d\mathbf{S} = Q,$$

we obtain

$$2\pi(\varepsilon_1 + \varepsilon_2)A = Q,$$

or

$$A = \frac{Q}{2\pi(\varepsilon_1 + \varepsilon_2)}.$$

We further find the electric intensity and displacement in regions 1 and 2:

$$\mathbf{E}_1 = \mathbf{E}_2 = \frac{Q\mathbf{r}}{2\pi(\varepsilon_1 + \varepsilon_2)r^3},$$

$$\mathbf{D}_1 = \frac{\varepsilon_1 Q\mathbf{r}}{2\pi(\varepsilon_1 + \varepsilon_2)r^3}, \quad \mathbf{D}_2 = \frac{\varepsilon_2 Q\mathbf{r}}{2\pi(\varepsilon_1 + \varepsilon_2)r^3}.$$

Consider the semispherical capacitor 1. We have

$$V_{ab} = -\int_b^a \frac{A}{r^2} dr = \frac{A(b-a)}{ab}$$

and

$$C_1 = \frac{Q_1}{V_{ab}} = \frac{2\pi\varepsilon_1 ab}{b-a}.$$

A similar expression is obtained for C_2. Treating the capacitor as a combination of two semispherical capacitors in parallel, we obtain the total capacitance as

$$C = \frac{2\pi(\varepsilon_1 + \varepsilon_2)ab}{b-a}.$$

1058

Two concentric metal spheres of radii a and $b\,(a < b)$ are separated by a medium that has dielectric constant ε and conductivity σ. At time $t = 0$ an electric charge q is suddenly placed on the inner sphere.

(a) Calculate the total current through the medium as a function of time.

(b) Calculate the Joule heat produced by this current and show that it is equal to the decrease in electrostatic energy that occurs as a consequence of the rearrangement of the charge.

(*Chicago*)

Solution:

(a) At $t = 0$, when the inner sphere carries electric charge q, the field intensity inside the medium is

$$E_0 = \frac{q}{4\pi\varepsilon r^2}$$

and directs radially outwords. At time t when the inner sphere has charge $q(t)$, the field intensity is

$$E(t) = \frac{q(t)}{4\pi\varepsilon r^2}.$$

Ohm's law gives the current density $\mathbf{j} = \sigma\mathbf{E}$. Considering a concentric spherical surface of radius r enclosing the inner sphere, we have from charge conservation

$$-\frac{d}{dt}q(t) = 4\pi r^2 j(t) = 4\pi r^2 \sigma E(t) = \frac{\sigma}{\varepsilon}q(t).$$

The differential equation has solution

$$q(t) = qe^{-\frac{\sigma}{\varepsilon}t}.$$

Hence

$$E(t,r) = \frac{q}{4\pi\varepsilon r^2}e^{-\frac{\sigma}{\varepsilon}t},$$

$$j(t,r) = \frac{\sigma q}{4\pi\varepsilon r^2}e^{-\frac{\sigma}{\varepsilon}t}.$$

The total current flowing through the medium at time t is

$$I(t) = 4\pi r^2 j(t,r) = \frac{\sigma q}{\varepsilon}e^{-\frac{\sigma}{\varepsilon}t}.$$

(b) The Joule heat loss per unit volume per unit time in the medium is

$$w(t,r) = \mathbf{j}\cdot\mathbf{E} = \sigma E^2 = \frac{\sigma q^2}{(4\pi\varepsilon)^2 r^4}e^{-\frac{2\sigma}{\varepsilon}t},$$

and the total Joule heat produced is

$$W = \int_0^{+\infty} dt \int_a^b dr \cdot 4\pi r^2 w(t,r) = \frac{q^2}{8\pi\varepsilon}\left(\frac{1}{a} - \frac{1}{b}\right).$$

The electrostatic energy in the medium before discharging is

$$W_0 = \int_a^b dr \cdot 4\pi r^2 \cdot \frac{\varepsilon E_0^2}{2} = \frac{q^2}{8\pi\varepsilon}\left(\frac{1}{a} - \frac{1}{b}\right).$$

Hence $W = W_0$.

1059

A condenser comprises two concentric metal spheres, an inner one of radius a, and an outer one of inner radius d. The region $a < r < b$ is filled with material of relative dielectric constant K_1, the region $b < r < c$ is vacuum $(K = 1)$, and the outermost region $c < r < d$ is filled with material of dielectric constant K_2. The inner sphere is charged to a potential V with respect to the outer one, which is grounded $(V = 0)$. Find:

(a) The free charges on the inner and outer spheres.

(b) The electric field, as a function of the distance r from the center, for the regions: $a < r < b$, $b < r < c$, $c < r < d$.

(c) The polarization charges at $r = a$, $r = b$, $r = c$ and $r = d$.

(d) The capacitance of this condenser.

(*Columbia*)

Solution:

(a) Suppose the inner sphere carries total free charge Q. Then the outer sphere will carry total free charge $-Q$ as it is grounded.

(b) Using Gauss' law and the spherical symmetry, we find the following results:

$$\mathbf{E} = \frac{Q}{4\pi K_1 \varepsilon_0 r^2} \mathbf{e}_r, \quad (a < r < b),$$

$$\mathbf{E} = \frac{Q}{4\pi \varepsilon_0 r^2} \mathbf{e}_r, \quad (b < r < c),$$

$$\mathbf{E} = \frac{Q}{4\pi \varepsilon_0 K_2 r^2} \mathbf{e}_r, \quad (c < r < d).$$

(c) Using the equations

$$\sigma_P = \mathbf{n} \cdot (\mathbf{P}_1 - \mathbf{P}_2), \quad \mathbf{P} = \varepsilon_0 (K - 1)\mathbf{E},$$

we obtain the polarization charge densities

$$\sigma_P = \frac{Q}{4\pi a^2} \frac{1 - K_1}{K_1} \quad \text{at} \quad r = a,$$

$$\sigma_P = \frac{Q}{4\pi b^2} \frac{K_1 - 1}{K_1} \quad \text{at} \quad r = b$$

$$\sigma_P = \frac{Q}{4\pi c^2} \frac{1 - K_2}{K_2} \quad \text{at} \quad r = c$$

$$\sigma_P = \frac{Q}{4\pi d^2} \frac{K_2 - 1}{K_2} \quad \text{at} \quad r = d.$$

(d) The potential is

$$V = -\int_d^a \mathbf{E} \cdot d\mathbf{r} = \frac{Q}{4\pi\varepsilon_0}\left[\left(\frac{1}{a}-\frac{1}{b}\right)\frac{1}{K_1} + \left(\frac{1}{b}-\frac{1}{c}\right) + \left(\frac{1}{c}-\frac{1}{d}\right)\frac{1}{K_2}\right].$$

Therefore, the charge in the inner sphere is

$$Q = \frac{4\pi\varepsilon_0 K_1 K_2 abcd V}{K_1 ab(d-c) + K_1 K_2 ad(c-b) + K_2 cd(b-a)},$$

and the capacitance is

$$C = \frac{Q}{V} = \frac{4\pi\varepsilon_0 K_1 K_2 abcd}{K_1 ab(d-c) + K_1 K_2 ad(c-b) + K_2 cd(b-a)}.$$

1060

The volume between two concentric conducting spherical surfaces of radii a and $b\,(a < b)$ is filled with an inhomogeneous dielectric constant

$$\varepsilon = \frac{\varepsilon_0}{1+Kr},$$

where ε_0 and K are constants and r is the radial coordinate. Thus $\mathbf{D}(r) = \varepsilon\mathbf{E}(r)$. A charge Q is placed on the inner surface, while the outer surface is grounded. Find:

(a) The displacement in the region $a < r < b$.

(b) The capacitance of the device.

(c) The polarization charge density in $a < r < b$.

(d) The surface polarization charge density at $r = a$ and $r = b$.

(*Columbia*)

Solution:

(a) Gauss' law and spherical symmetry give

$$\mathbf{D} = \frac{Q}{4\pi r^2}\mathbf{e}_r, \quad (a < r < b).$$

(b) The electric field intensity is

$$\mathbf{E} = \frac{Q}{4\pi\varepsilon_0 r^2}(1+Kr)\mathbf{e}_r, \quad (a < r < b).$$

Hence, the potential difference between the inner and outer spheres is

$$V = \int_a^b \mathbf{E} \cdot d\mathbf{r} = \frac{Q}{4\pi\varepsilon_0} \left(\frac{1}{a} - \frac{1}{b} + K \ln \frac{b}{a} \right).$$

The capacitance of the device is then

$$C = \frac{Q}{V} = \frac{4\pi\varepsilon_0 ab}{(b-a) + abK \ln(b/a)}.$$

(c) The polarization is

$$\mathbf{P} = (\varepsilon - \varepsilon_0)\mathbf{E} = -\frac{QK}{4\pi r}\mathbf{e}_r.$$

Therefore, the volume polarization charge density at $a < r < b$ is given by

$$\rho_P = -\nabla \cdot \mathbf{P} = \frac{1}{r^2} \frac{\partial}{\partial r} \left(\frac{QKr}{4\pi} \right) = \frac{QK}{4\pi r^2}.$$

(d) The surface polarization charge densities at $r = a, b$ are

$$\sigma_P = \frac{QK}{4\pi\sigma} \quad \text{at} \quad r = a; \quad \sigma_P = -\frac{QK}{4\pi b} \quad \text{at} \quad r = b.$$

1061

For steady current flow obeying Ohm's law find the resistance between two concentric spherical conductors of radii $a < b$ filled with a material of conductivity σ. Clearly state each assumption.

(*Wisconsin*)

Solution:

Suppose the conductors and the material are homogeneous so that the total charge Q carried by the inner sphere is uniformly distributed over its surface. Gauss' law and spherical symmetry give

$$\mathbf{E}(r) = \frac{Q}{4\pi\varepsilon r^2}\mathbf{e}_r,$$

where ε is the dielectric constant of the material. From Ohm's law $\mathbf{j} = \sigma\mathbf{E}$, one has

$$\mathbf{j} = \frac{\sigma Q}{4\pi\varepsilon r^2}\mathbf{e}_r.$$

Then the total current is

$$I = \oint \mathbf{j} \cdot d\mathbf{S} = \frac{\sigma}{\epsilon} Q.$$

The potential difference between the two conductors is

$$V = -\int_b^a \mathbf{E} \cdot d\mathbf{r} = -\int_b^a \frac{I}{4\pi\sigma r^2} dr = \frac{I}{4\pi\sigma} \left(\frac{1}{a} - \frac{1}{b} \right),$$

giving the resistance as

$$R = \frac{V}{I} = \frac{1}{4\pi\sigma} \left(\frac{1}{a} - \frac{1}{b} \right).$$

4. TYPICAL METHODS FOR SOLUTION OF ELECTROSTATIC PROBLEMS — SEPARATION OF VARIABLES, METHODS OF IMAGES, GREEN'S FUNCTION AND MULTIPOLE EXPANSION (1062–1095)

1062

A dielectric sphere of radius a and dielectric constant ϵ_1 is placed in a dielectric liquid of infinite extent and dielectric constant ϵ_2. A uiform electric field $\mathbf{E}$ was originally present in the liquid. Find the resultant electric field inside and outside the sphere.

(SUNY, Buffalo)

Solution:

Let the origin be at the spherical center and take the direction of the original field $\mathbf{E}$ to define the polar axis z, as shown in Fig. 1.30. Let the electrostatic potential at a point inside the sphere be Φ_1, and the potential at a point outside the sphere be Φ_2. By symmetry we can write Φ_1 and Φ_2 as

$$\Phi_1 = \sum_{n=0} \left(A_n r^n + \frac{B_n}{r^{n+1}} \right) P_n(\cos\theta),$$

$$\Phi_2 = \sum_{n=0} \left(C_n r^n + \frac{D_n}{r^{n+1}} \right) P_n(\cos\theta),$$

where A_n, B_n, C_n, D_n are constants, and P_n are Legendre polynomials. The boundary conditions are as follows:

(1) Φ_1 is finite at $r = 0$.

(2) $\Phi_2|_{r\to\infty} = -Er\cos\theta = -ErP_1(\cos\theta)$.

(3) $\Phi_1 = \Phi_2|_{r=a}, \varepsilon_1\frac{\partial\Phi_1}{\partial r} = \varepsilon_2\frac{\partial\Phi_2}{\partial r}|_{r=a}$.

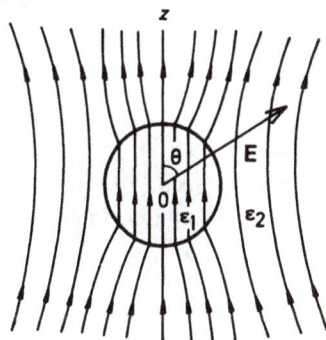

Fig. 1.30

From conditions (1) and (2), we obtain

$$B_n = 0, \quad C_1 = -E, \quad C_n = 0 \ (n \neq 1).$$

Then from condition (3), we obtain

$$-EaP_1(\cos\theta) + \sum_n \frac{D_n}{a^{n+1}} P_n(\cos\theta) = \sum_n A_n a^n P_n(\cos\theta),$$

$$-\varepsilon_2\left[EP_1(\cos\theta) + \sum_n (n+1)\frac{D^n}{a^{n+2}} P_n(\cos\theta)\right] = \varepsilon_1 \sum_n A_n a^{n-1} P_n(\cos\theta).$$

These equations are to be satisfied for each of the possible angles θ. That is, the coefficients of $P_n(\cos\theta)$ on the two sides of each equation must be equal for every n. This gives

$$A_1 = -\frac{3\varepsilon_2}{\varepsilon_1 + 2\varepsilon_2}E, \quad D_1 = \frac{\varepsilon_1 - \varepsilon_2}{\varepsilon_1 + 2\varepsilon_2}Ea^3,$$

$$A_n = D_n = 0, \quad (n \neq 1).$$

Hence, the electric potentials inside and outside the sphere can be expressed as

$$\Phi_1 = -\frac{3\varepsilon_2}{\varepsilon_1 + 2\varepsilon_2}Er\cos\theta,$$

$$\Phi_2 = -\left[1 - \frac{\varepsilon_1 - \varepsilon_2}{\varepsilon_1 + 2\varepsilon_2}\left(\frac{a}{r}\right)^3\right]Er\cos\theta,$$

and the electric fields inside and outside the sphere by

$$\mathbf{E}_1 = -\nabla\Phi_1 = \frac{3\varepsilon_2}{\varepsilon_1 + 2\varepsilon_2}\mathbf{E}, \quad (r < a),$$

$$\mathbf{E}_2 = -\nabla\Phi_2 = \mathbf{E} + \frac{\varepsilon_1 - \varepsilon_2}{\varepsilon_1 + 2\varepsilon_2}a^3\left[\frac{3(\mathbf{E}\cdot\mathbf{r})\mathbf{r}}{r^5} - \frac{\mathbf{E}}{r^3}\right], \quad (r > a).$$

1063

Determine the electric field inside and outside a sphere of radius R and dielectric cosntant ε placed in a uniform electric field of magnitude E_0 directed along the z-axis.

(*Columbia*)

Solution:

Using the solution of Problem **1062**, we have

$$\mathbf{E}_1 = \frac{3\varepsilon_0}{\varepsilon + 2\varepsilon_0}\mathbf{E}_0, \quad (r < R),$$

$$\mathbf{E}_2 = \mathbf{E}_0 + \frac{\varepsilon - \varepsilon_0}{\varepsilon + 2\varepsilon_0}R^3\left[\frac{3(\mathbf{E}_0\cdot\mathbf{r})\mathbf{r}}{r^5} - \frac{\mathbf{E}_0}{r^3}\right], \quad (r > R).$$

1064

A sphere of dielectric constant ε is placed in a uniform electric field $\mathbf{E}_0$. Show that the induced surface charge density is

$$\sigma(\theta) = \frac{\varepsilon - \varepsilon_0}{\varepsilon + 2\varepsilon_0}3\varepsilon_0 E_0 \cos\theta,$$

where θ is measured from the $\mathbf{E}_0$ direction. If the sphere is rotated at an angular velocity ω about the direction of $\mathbf{E}_0$, will a magnetic field be produced? If not, explain why no magnetic field is produced. If so, sketch the magnetic field lines.

(*Wisconsin*)

Solution:

The solution of Problem **1063** gives the electric field inside the sphere as

$$E = \frac{3\varepsilon_0}{\varepsilon + 2\varepsilon_0} E_0 ,$$

which gives the polarization of the dielectric as

$$p = (\varepsilon - \varepsilon_0)E = \frac{3\varepsilon_0(\varepsilon - \varepsilon_0)}{\varepsilon + 2\varepsilon_0} E_0 .$$

The bound charge density on the surface of the dielectric sphere is

$$\sigma(\theta) = n \cdot p = \frac{3\varepsilon_0(\varepsilon - \varepsilon_0)}{\varepsilon + 2\varepsilon_0} E_0 \cos\theta ,$$

n being the unit vector normal to the surface. The total electric dipole moment is then

$$P = \frac{4}{3}\pi R_0^3 p = \frac{4\pi\varepsilon_0(\varepsilon - \varepsilon_0)}{\varepsilon + 2\varepsilon_0} R_0^3 E_0 .$$

Note that P has the same direction as E_0. Then when the sphere is rotated about the direction of E_0, P will not change. This implies that the rotation will not give rise to a polarization current and, therefore, will not produce a magnetic field.

1065

A perfectly conducting sphere is placed in a uniform electric field pointing in the z-direction.

(a) What is the surface charge density on the sphere?

(b) What is the induced dipole moment of the sphere?

(*Columbia*)

Solution:

(a) The boundary conditions on the conductor surface are

$$\Phi = \text{constant} = \Phi_s , \quad \text{say,}$$

$$\varepsilon_0 \frac{\partial \Phi}{\partial r} = -\sigma ,$$

where Φ_s is the potential of the conducting sphere and σ is its surface charge density. On account of symmetry, the potential at a point (r, θ, φ) outside the sphere is, in spherical coordinates with origin at the center of the sphere,

$$\Phi = \sum_{n=0} \left(C_n r^n + \frac{D_n}{r^{n+1}} \right) P_n(\cos \theta). \tag{1}$$

Let E_0 be the original uniform electric intensity. As $r \to \infty$,

$$\Phi = -E_0 r \cos \theta = -E_0 r P_1(\cos \theta).$$

By equating the coefficients of $P_n(\cos \theta)$ on the two sides of Eq. (1), we have

$$C_0 = 0, \quad C_1 = -E_0, \quad D_1 = E_0 a^3, \quad C_n = D_n = 0 \quad \text{for} \quad n > 1.$$

Hence

$$\Phi = -E_0 r \cos \theta + \frac{E_0 a^3}{r^2} \cos \theta, \tag{2}$$

where a is the radius of the sphere. The second boundary condition and Eq. (2) give

$$\sigma = 3\varepsilon_0 E_0 \cos \theta.$$

(b) Suppose that an electric dipole $\mathbf{P} = P\mathbf{e}_z$ is placed at the origin, instead of the sphere. The potential at r produced by the dipole is

$$\Phi_P = -\frac{1}{4\pi\varepsilon_0} \mathbf{P} \cdot \nabla\left(\frac{1}{r}\right) = \frac{P \cos \theta}{4\pi\varepsilon_0 r^2}.$$

Comparing this with the second term of Eq. (2) shows that the latter corresponds to the contribution of a dipole having a moment

$$\mathbf{P} = 4\pi\varepsilon_0 a^3 \mathbf{E}_0,$$

which can be considered as the induced dipole moment of the sphere.

1066

A surface charge density $\sigma(\theta) = \sigma_0 \cos \theta$ is glued to the surface of a spherical shell of radius R (σ_0 is a constant and θ is the polar angle). There is a vacuum, with no charges, both inside and outside of the shell. Calculate the electrostatic potential and the electric field both inside and outside of the spherical shell.

(*Columbia*)

Solution:

Let Φ_+, Φ_- be respectively the potentials outside and inside the shell. Both Φ_+ and Φ_- satisfy Laplace's equation and, on account of cylindrical symmetry, they have the expressions

$$\Phi_+ = \sum_{n=0} b_n r^{-n-1} P_n(\cos\theta), \quad (r > R);$$

$$\Phi_- = \sum_{n=0} a_n r^n P_n(\cos\theta), \quad (r < R).$$

The boundary conditions at $r = R$ for the potential and displacement vector are

$$\Phi_- = \Phi_+ ,$$

$$\sigma(\theta) = \sigma_0 P_1(\cos\theta) = \varepsilon_0 \left(\frac{\partial \Phi_-}{\partial r} - \frac{\partial \Phi_+}{\partial r} \right).$$

Substituting in the above the expressions for the potentials and equating the coefficients of $P_n(\cos\theta)$ on the two sides of the equations, we obtain

$$a_n = b_n = 0 \qquad \text{for} \quad n \neq 1,$$

$$a_1 = \frac{\sigma_0}{3\varepsilon_0}, \quad b_1 = \frac{\sigma_0}{3\varepsilon_0} R^3 \quad \text{for} \quad n = 1.$$

Hence

$$\Phi_+ = \frac{\sigma_0 R^3}{3\varepsilon_0 r^2} \cos\theta, \quad r > R,$$

$$\Phi_- = \frac{\sigma_0 r}{3\varepsilon_0} \cos\theta, \quad r < R.$$

From $\mathbf{E} = -\nabla\Phi$ we obtain

$$\mathbf{E}_+ = \frac{2\sigma_0 R^3}{3\varepsilon_0 r^3} \cos\theta\, \mathbf{e}_r + \frac{\sigma_0 R^3}{3\varepsilon_0 r^3} \sin\theta\, \mathbf{e}_\theta, \quad r > R,$$

$$\mathbf{E}_- = -\frac{\sigma_0}{3\varepsilon_0} \mathbf{e}_z, \quad r < R.$$

1067

Consider a sphere of radius R centered at the origin. Suppose a point charge q is put at the origin and that this is the only charge inside or outside the sphere. Furthermore, the potential is $\Phi = V_0 \cos\theta$ on the surface of the sphere. What is the electric potential both inside and outside the sphere?

(*Columbia*)

Solution:

The potential is given by either Poisson's or Laplace's equation:

$$\nabla^2 \Phi_- = -\frac{q}{\varepsilon_0}\delta(r), \quad r < R;$$

$$\nabla^2 \Phi_+ = 0, \qquad r > R.$$

The general solutions finite in the respective regions, taking account of the symmetry, are

$$\Phi_- = \frac{q}{4\pi\varepsilon_0 r} + \sum_{n=0}^{\infty} A_n r^n P_n(\cos\theta), \quad r < R,$$

$$\Phi_+ = \sum_{n=0}^{\infty} \frac{B_n}{r^{n+1}} P_n(\cos\theta), \quad r > R.$$

Then from the condition $\Phi_- = \Phi_+ = V_0 \cos\theta$ at $r = R$ we obtain $A_0 = -\frac{q}{4\pi\varepsilon_0 R}, A_1 = \frac{V_0}{R}, B_1 = V_0 R^2, B_0 = 0, A_n = B_n = 0$ for $n \neq 0, 1$, and hence

$$\Phi_- = \frac{q}{4\pi\varepsilon_0 r} - \frac{q}{4\pi\varepsilon_0 R} + \frac{V_0 \cos\theta}{R} r, \quad r < R;$$

$$\Phi_+ = \frac{V_0 R^2}{r^2}\cos\theta, \qquad r > R.$$

1068

If the potential of a spherical shell of zero thickness depends only on the polar angle θ and is given by $V(\theta)$, inside and outside the sphere there being empty space,

(a) show how to obtain expressions for the potential $V(r, \theta)$ inside and outside the sphere and how to obtain an expression for the electric sources on the sphere.

(b) Solve with $V(\theta) = V_0 \cos^2 \theta$.

The first few of the Legendre polynomials are given as follows:

$$P_0(\cos\theta) = 1, \quad P_1(\cos\theta) = \cos\theta, \quad P_2(\cos\theta) = \frac{3\cos^2\theta - 1}{2}.$$

We also have

$$\int_0^\pi P_n(\cos\theta)P_m(\cos\theta)\sin\theta d\theta = 0 \quad \text{if} \quad n \neq m$$

and

$$\int_0^\pi P_n^2(\cos\theta)\sin\theta d\theta = \frac{2}{2n+1}.$$

(*Wisconsin*)

Solution:

(a) Since both the outside and inside of the spherical shell are empty space, the potential in the whole space satisfies Laplace's equation. Thus the potential inside the sphere has the form

$$\Phi_1 = \sum_{n=0}^\infty a_n r^n P_n(\cos\theta),$$

while that outside the sphere is

$$\Phi_2 = \sum_{n=0}^\infty \frac{b_n}{r^{n+1}} P_n(\cos\theta).$$

Letting the radius of the shell be R, we have

$$V(\theta) = \sum_{n=0}^\infty a_n R^n P_n(\cos\theta).$$

Multiplying both sides by $P_n(\cos\theta)\sin\theta d\theta$ and integrating from 0 to π, we obtain

$$a_n R^n = \frac{2n+1}{2} \int_0^\pi V(\theta)P_n(\cos\theta)\sin\theta d\theta.$$

Hence

$$\Phi_1 = \sum_{n=0}^\infty \left[\frac{2n+1}{2R^n} \int_0^\pi V(\theta)P_n(\cos\theta)\sin\theta d\theta \right] P_n(\cos\theta)r^n,$$

and similarly

$$\Phi_2 = \sum_{n=0}^{\infty} \left[\frac{(2n+1)R^{n+1}}{2} \int_0^{\pi} V(\theta) P_n(\cos\theta) \sin\theta d\theta \right] \frac{P_n(\cos\theta)}{r^{n+1}} .$$

The charge distribution on the spherical shell is given by the boundary condition for the displacement vector:

$$\sigma(\theta) = \varepsilon_0 \frac{\partial \Phi_1}{\partial r}\Big|_{r=R} - \varepsilon_0 \frac{\partial \Phi_2}{\partial r}\Big|_{r=R}$$

$$= \varepsilon_0 \sum_{n=0}^{\infty} \left[\frac{(2n+1)^2}{2R} \int_0^{\pi} V(\theta) P_n(\cos\theta) \sin\theta d\theta \right] P_n(\cos\theta) .$$

(b) From

$$V(\theta) = V_0 \cos^2\theta = \frac{2V_0}{3} P_2(\cos\theta) + \frac{V_0}{3} P_0(\cos\theta) ,$$

we obtain

$$\int_0^{\pi} V(\theta) P_n(\cos\theta) \sin\theta d\theta = \frac{2}{3} V_0 , \qquad \text{for} \quad n = 0 ;$$

$$\int_0^{\pi} V(\theta) P_n(\cos\theta) \sin\theta d\theta = \frac{4}{15} V_0 , \qquad \text{for} \quad n = 2 ;$$

$$\int_0^{\pi} V(\theta) P_n(\cos\theta) \sin\theta d\theta = 0 , \qquad \text{for} \quad n \neq 0, 2 .$$

Hence

$$\Phi_1 = \frac{V_0}{3} + \frac{2V_0}{3} P_2(\cos\theta) \frac{r^2}{R^2} , \quad (r < R)$$

$$\Phi_2 = \frac{V_0 R}{3r} + \frac{2V_0}{3} P_2(\cos\theta) \frac{R^3}{r^3} , \quad (r > R)$$

$$\sigma(\theta) = \frac{\varepsilon_0 V_0}{3R} \left[1 + 5 P_2(\cos\theta) \right] .$$

1069

A conducting sphere of radius a carrying a charge q is placed in a uniform electric field $\mathbf{E}_0$. Find the potential at all points inside and outside of the sphere. What is the dipole moment of the induced charge on the sphere? The three electric fields in this problem give rise to six energy terms. Identify these six terms; state which are finite or zero, and which are infinite or unbounded.

(*Columbia*)

Solution:

The field in this problem is the superposition of three fields: a uniform field $\mathbf{E}_0$, a dipole field due to the induced charges of the conducting sphere, and a field due to a charge q uniformly distributed over the conducting sphere.

Let Φ_1 and Φ_2 be the total potentials inside and outside the sphere respectively. Then we have

$$\nabla^2 \Phi_1 = \nabla^2 \Phi_2 = 0, \quad \Phi_1 = \Phi_0,$$

where Φ_0 is a constant. The boundary conditions are

$$\Phi_1 = \Phi_2, \quad \text{for} \quad r = a,$$
$$\Phi_2 = -E_0 r P_1(\cos\theta) \quad \text{for} \quad r \to \infty.$$

On account of cylindrical symmetry the general solution of Laplace's equation is

$$\Phi_2 = \sum_n \left(a_n r^n + \frac{b_n}{r^{n+1}} \right) P_n(\cos\theta).$$

Inserting the above boundary conditions, we find

$$a_1 = -E_0, \quad b_0 = a\Phi_0, \quad b_1 = E_0 a^3,$$

while all other coefficients are zero. As $\sigma = -\varepsilon_0 \left(\frac{\partial \Phi_2}{\partial r} \right)_{r=a}$, we have

$$q = \int_0^\pi \left(3\varepsilon_0 E_0 \cos\theta + \frac{\varepsilon_0 \Phi_0}{a} \right) 2\pi a^2 \sin\theta \, d\theta = 4\pi a \varepsilon_0 \Phi_0,$$

or

$$\Phi_0 = \frac{q}{4\pi\varepsilon_0 a}.$$

So the potentials inside and outside the sphere are

$$\Phi_1 = \frac{q}{4\pi\varepsilon_0 a}, \quad (r < a),$$

$$\Phi_2 = -E_0 r \cos\theta + \frac{q}{4\pi\varepsilon_0 r} + \frac{E_0 a^3}{r^2} \cos\theta, \quad (r > a).$$

The field outside the sphere may be considered as the superposition of three fields with contributions to the potential Φ_2 equal to the three terms on the right-hand side of the last expression: the uniform field $\mathbf{E}_0$, a field due to the charge q uniformly distributed over the sphere, and a dipole field due to charges induced on the surface of the sphere. The last is that which would be produced by a dipole of moment $\mathbf{P} = 4\pi\varepsilon_0 a^3 \mathbf{E}_0$ located at the spherical center.

The energies of these three fields may be divided into two kinds: electrostatic energy produced by each field alone, interaction energies among the fields.

The energy density of the uniform external field $\mathbf{E}_0$ is $\frac{\varepsilon_0}{2} E_0^2$. Its total energy $\int \frac{\varepsilon_0}{2} E_0^2 dV$ is infinite, i.e. $W_1 \to \infty$, since $\mathbf{E}_0$ extends over the entire space.

The total electrostatic energy of an isolated conducting sphere with charge q is

$$W_2 = \int_a^\infty \frac{\varepsilon_0}{2} \left(\frac{q}{4\pi\varepsilon_0 r^2} \right)^2 4\pi r^2 dr = \frac{q^2}{8\pi\varepsilon_0 a} \,,$$

which is finite.

The electric intensity outside the sphere due to the dipole $\mathbf{P}$ is

$$\mathbf{E}_3 = -\nabla \left(\frac{E_0 a^3}{r^2} \cos\theta \right) = \frac{2a^3 E_0 \cos\theta}{r^3} \mathbf{e}_r + \frac{a^3 E_0 \sin\theta}{r^3} \mathbf{e}_\theta \,.$$

The corresponding energy density is

$$w_3 = \frac{1}{2}\varepsilon_0 E_3^2 = \frac{\varepsilon_0}{2} \frac{a^6 E_0^2}{r^6} (4\cos^2\theta + \sin^2\theta)$$

$$= \frac{\varepsilon_0}{2} \frac{a^6 E_0^2}{r^6} (1 + 3\cos^2\theta) \,.$$

As the dipole does not give rise to a field inside the sphere the total electrostatic energy of $\mathbf{P}$ is

$$W_3 = \int w_3 dV = \frac{\varepsilon_0 a^6 E_0^2}{2} \int_a^\infty \frac{1}{r^4} dr \int_0^{2\pi} d\varphi \int_{-1}^1 (1 + 3\cos^2\theta) d\cos\theta$$

$$= \frac{4\pi\varepsilon_0 a^3}{3} E_0^2 \,,$$

which is also finite.

For the conducting sphere with total charge q, its suface charge density is $\sigma = q/4\pi a^2$. The interaction energy between the sphere and the external field $\mathbf{E}_0$ is then

$$W_{12} = \int \sigma \cdot (-E_0 a \cos\theta) 2\pi a^2 \sin\theta d\theta$$

$$= \frac{qaE_0}{2} \int_0^\pi \cos\theta d\cos\theta = 0 .$$

Similarly, the interaction energy of the conducting sphere with the field of dipole $\mathbf{P}$ is

$$W_{23} = \int_0^\pi \sigma \cdot \left(\frac{E_0 a^3}{r^2} \cos\theta\right) 2\pi a^2 \sin\theta d\theta = 0 .$$

The interaction energy between dipole $\mathbf{P}$ and external field $\mathbf{E}_0$ is

$$W_{13} = -\frac{1}{2}\mathbf{P} \cdot \mathbf{E}_0 = -2\pi\varepsilon_0 a^3 E_0^2 ,$$

which is finite. The appearance of the factor $\frac{1}{2}$ in the expression is due to the fact that the dipole $\mathbf{P}$ is just an equivalent dipole induced by the external field $\mathbf{E}_0$.

1070

A conducting spherical shell of radius R is cut in half. The two hemispherical pieces are electrically separated from each other but are left close together as shown in Fig. 1.31, so that the distance separating the two halves can be neglected. The upper half is maintained at a potential $\phi = \phi_0$, and the lower half is maintained at a potential $\phi = 0$. Calculate the electrostatic potential ϕ at all points in space outside of the surface of the conductors. Neglect terms falling faster than $1/r^4$ (i.e. keep terms up to and including those with $1/r^4$ dependence), where r is the distance from the center of the conductor. (Hints: Start with the solution of Laplace's equation in the appropriate coordinate system. The boundary condition of the surface of the conductor will have to be expanded in a series of Legendre polynomials: $P_0(x) = 1$, $P_1(x) = x$, $P_2(x) = \frac{3}{2}x^2 - \frac{1}{2}$, $P_3(x) = \frac{5}{2}x^3 - \frac{3}{2}x$.

(*Columbia*)

Solution:

Use spherical coordinates (r, θ, ϕ) with the origin at the spherical center. The z-axis is taken perpendicular to the cutting seam of the two hemi-spheres (see Fig. 1.31). It is readily seen that the potential ϕ is a function of r and θ only and satisfies the following 2-dimensional Laplace's equation,

$$\frac{1}{r^2}\frac{\partial}{\partial r}\left(r^2\frac{\partial\phi}{\partial r}\right) + \frac{1}{r^2\sin\theta}\frac{\partial}{\partial\theta}\left(\sin\theta\frac{\partial\phi}{\partial\theta}\right) = 0, \quad (r \geq R).$$

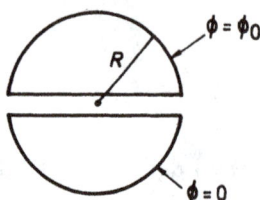

Fig. 1.31

The general solution of this equation is

$$\phi = \sum_{l=0}^{\infty}\frac{A_l}{r^{l+1}}P_l(\cos\theta), \quad (r \geq R).$$

Keeping only terms up to $l = 3$ as required, we have

$$\phi \approx \sum_{l=0}^{3}\frac{A_l}{r^{l+1}}P_l(\cos\theta).$$

The boundary condition at $r = R$ is

$$\phi|_{r=R} = f(\theta) = \begin{cases} \phi_0, & \text{for } 0 \leq \theta < \frac{\pi}{2}, \\ 0, & \text{for } \frac{\pi}{2} < \theta \leq \pi. \end{cases}$$

$f(\theta)$ can be expanded as a series of Legendre polynomials, retaining terms up to $l = 3$:

$$f(\theta) \approx \sum_{l=0}^{3}B_l P_l(\cos\theta),$$

where, making use of the orthogonality of the Legendre polynomials,

$$B_l = \frac{2l+1}{2} \int_0^\pi f(\theta) P_l(\cos\theta) \sin\theta d\theta$$

$$= \frac{(2l+1)\phi_0}{2} \int_0^1 P_l(x) dx \ .$$

Integrating the first few Legendre polynomials as given, we obtain

$$\int_0^1 P_0(x) dx = \int_0^1 dx = 1 \ ,$$

$$\int_0^1 P_2(x) dx = \int_0^1 x dx = \frac{1}{2} \ ,$$

$$\int_0^1 P_2(x) dx = \int_0^1 \frac{1}{2}(3x^2 - 1) dx = 0 \ ,$$

$$\int_0^1 P_3(x) dx = \int_0^1 \frac{1}{2}(5x^3 - 3x) dx = -\frac{1}{8} \ ,$$

which in turn give

$$B_0 = \frac{1}{2}\phi_0 \ , \quad B_1 = \frac{3}{4}\phi_0 \ , \quad B_2 = 0 \ , \quad B_3 = -\frac{7}{16}\phi_0 \ .$$

From

$$\phi|_{r=R} \approx \sum_{l=0}^3 \frac{A_l}{R^{l+1}} P_l = \sum_{l=0}^3 B_l P_l \ ,$$

we further get

$$A_l = R^{l+1} B_l \ .$$

Hence

$$\phi \approx \sum_{l=0}^3 B_l \left(\frac{R}{r}\right)^{l+1} P_l(\cos\theta) = \phi_0 \left\{ \frac{R}{2r} + \frac{3}{4}\left(\frac{R}{r}\right)^2 \cos\theta \right.$$

$$\left. - \frac{7}{32}\left(\frac{R}{r}\right)^4 (5\cos^2\theta - 3) \cos\theta \right\} \ .$$

1071

As can be seen in Fig. 1.32, the inner conducting sphere of radius a carries charge Q, and the outer sphere of radius b is grounded. The distance between their centers is c, which is a small quantity.

(a) Show that to first order in c, the equation describing the outer sphere, using the center of the inner sphere as origin, is

$$r(\theta) = b + c\cos\theta.$$

(b) If the potential between the two spheres contains only $l = 0$ and $l = 1$ angular components, determine it to first order in c.

(*Wisconsin*)

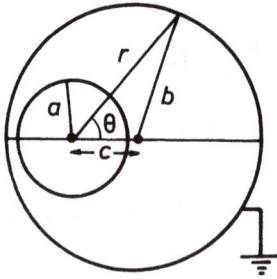

Fig. 1.32

Solution:

(a) Applying the cosine theorem to the triangle of Fig. 1.32 we have to first order in c

$$b^2 = c^2 + r^2 - 2cr\cos\theta \simeq r^2 - 2cr\cos\theta,$$

or

$$r \approx \frac{1}{2}\left(2c\cos\theta + \sqrt{4c^2\cos^2\theta + b^2}\right) \approx b + c\cos\theta.$$

(b) Using Laplace's equation $\nabla^2\Phi = 0$ and the axial symmetry, we can express the potential at a point between the two spheres as

$$\Phi = \sum_{l=0}^{\infty}\left(A_r r^l + \frac{B_l}{r^{l+1}}\right)P_l(\cos\theta).$$

Then retaining only the $l = 0, 1$ angular components, we have,

$$\Phi = A_0 + \frac{B_0}{r} + \left(A_1 r + \frac{B_1}{r^2}\right)\cos\theta.$$

As the surface of the inner conductor is an equipotential, Φ for $r = a$ should not depend on θ. Hence

$$A_1 a + \frac{B_1}{a^2} = 0. \tag{1}$$

The charge density on the surface of the inner sphere is

$$\sigma = -\varepsilon_0 \left(\frac{\partial \Phi}{\partial r} \right)_{r=a}$$

and we have

$$\int_0^\pi \sigma 2\pi a^2 \sin \theta d\theta = Q.$$

This gives

$$B_0 = \frac{Q}{4\pi\varepsilon_0}. \tag{2}$$

Then as the outer sphere is grounded, $\Phi = 0$ for $r \approx b + c \cos \theta$. This gives

$$A_0 + \frac{B_0}{b + c \cos \theta} + \left[A_1(b + c \cos \theta) + \frac{B_1}{(b + c \cos \theta)^2} \right] \cos \theta = 0. \tag{3}$$

To first order in c, we have the approximations

$$(b + c \cos \theta)^{-1} = b^{-1} \left(1 + \frac{c}{b} \cos \theta \right)^{-1} \approx \frac{1}{b} \left(1 - \frac{c}{b} \cos \theta \right),$$

$$(b + c \cos \theta)^{-2} = b^{-2} \left(1 + \frac{c}{b} \cos \theta \right)^{-2} \approx \frac{1}{b^2} \left(1 - \frac{2c}{b} \cos \theta \right).$$

Substituting these expressions in Eq. (3) gives

$$A_0 + \frac{B_0}{b} + \left(-\frac{B_0 c}{b^2} + A_1 b + \frac{B_1}{b^2} \right) \cos \theta \approx 0, \tag{4}$$

neglecting $c \cos^2 \theta$ and higher order terms. As (4) is valid for whatever value of θ, we require

$$A_0 + \frac{B_0}{b} = 0,$$

$$-\frac{B_0 c}{b^2} + A_1 b + \frac{B_1}{b^2} = 0.$$

The last two equations, (1) and (2) together give

$$A_0 = -\frac{Q}{4\pi\varepsilon_0 b},$$

$$A_1 = \frac{Qc}{4\pi\varepsilon_0(b^3 - a^3)}, \quad B_1 = -\frac{Qca^3}{4\pi\varepsilon_0(b^3 - a^3)}.$$

Hence the potential between the two spherical shells is

$$\Phi = \frac{Q}{4\pi\varepsilon_0}\left\{\frac{1}{r} - \frac{1}{b} + \frac{cr}{b^3 - a^3}\left[1 - \left(\frac{a}{r}\right)^3\right]\cos\theta\right\}.$$

1072

Take a very long cylinder of radius r made of insulating material. Spray a cloud of electrons on the surface. They are free to move about the surface so that, at first, they spread out evenly with a charge per unit area σ_0. Then put the cylinder in a uniform applied electric field perpendicular to the axis of the cylinder. You are asked to think about the charge density on the surface of the cylinder, $\sigma(\theta)$, as a function of the applied electric field E_a. In doing this you may neglect the electric polarizability of the insulating cylinder.

(a) In what way is this problem different from a standard electrostatic problem in which we have a charged conducting cylinder? When are the solutions to the two problems the same? (Answer in words.)

(b) Calculate the solution for $\sigma(\theta)$ in the case of a conducting cylinder and state the range of value of E_a for which this solution is applicable to the case described here.

(*Chicago*)

Solution:

Use cylindrical coordinates (ρ, θ, z) with the z-axis along the cylindrical axis and the direction of the applied field given by $\theta = 0$. Let the potentials inside and outside the cylinder be φ_I and φ_{II} respectively. As a long cylinder is assumed, φ_I and φ_{II} are independent of z. As there is no charge inside and outside the cylinder, Laplace's equation applies: $\nabla^2\varphi_I = \nabla^2\varphi_{II} = 0$. The boundary conditions are

$$\varphi_I = \varphi_{II}|_{\rho=r}, \quad \left(\frac{\partial\varphi_I}{\partial\theta}\right) = \left(\frac{\partial\varphi_{II}}{\partial\theta}\right)\bigg|_{\rho=r},$$

$$\varphi_I|_{\rho\to 0} \quad \text{is finite},$$

$$\varphi_{II}|_{\rho\to\infty} \to -E_a\rho\cos\theta.$$

Note the first two conditions arise from the continuity of the tangential component of the electric intensity vector. Furthermore as the electrons are free to move about the cylindrical surface, $E_\theta = 0$ on the surface at equilibrium. As z is not involved, try solutions of the form

$$\varphi_I = A_1 + B_1 \ln \rho + C_1 \rho \cos \theta + \frac{D_1}{\rho} \cos \theta ,$$

$$\varphi_{II} = A_2 + B_2 \ln \rho + C_2 \rho \cos \theta + \frac{D_2}{\rho} \cos \theta .$$

Then the above boundary conditions require that

$$B_1 = D_1 = 0 , \quad C_2 = -E_a ,$$

and

$$A_1 + C_1 r \cos \theta = A_2 + B_2 \ln r - E_a r \cos \theta + \frac{D_2}{r} \cos \theta ,$$

$$- C_1 \sin \theta = E_a \sin \theta - \frac{D_2}{r^2} \sin \theta = 0 .$$

The last equation gives

$$C_1 = 0 , \quad D_2 = E_a r^2 .$$

Applying Gauss' law to unit length of the cylinder:

$$\oint \mathbf{E}_{II} \cdot d\mathbf{S} = \frac{q}{\varepsilon_0} ,$$

i.e.,

$$-\frac{B_2}{r} \cdot 2\pi r = \frac{1}{\varepsilon_0} \sigma_0 \cdot 2\pi r ,$$

we obtain

$$B_2 = -\frac{\sigma_0 r}{\varepsilon_0} .$$

Neglecting any possible constant potential, we take $A_2 = 0$. Then

$$A_1 = B_2 \ln r = -\frac{\sigma_0 a \ln r}{\varepsilon_0} .$$

We ultimately obtain the following expressions

$$\varphi_I = -\frac{\sigma_0 r \ln r}{\varepsilon_0}, \quad E_I = 0,$$

$$\varphi_{II} = -\frac{\sigma_0 r \ln \rho}{\varepsilon_0} - E_a \rho \cos\theta + \frac{E_a r^2}{\rho} \cos\theta,$$

$$E_{II} = \left(\frac{\sigma_0 r}{\varepsilon_0 \rho} + E_a \cos\theta + \frac{E_a r^2}{\rho^2} \cos\theta\right) e_\rho$$

$$- E_a\left(1 - \frac{r^2}{\rho^2}\right) \sin\theta e_\theta,$$

$$\rho(\theta) = D_{II_\rho}|_{\rho=r} = \sigma_0 + 2\varepsilon_0 E_a \cos\theta.$$

(a) The difference between this case and the case of the cylindrical conductor lies in the fact that $\sigma(\theta)$ can be positive or negative for a conductor, while in this case $\sigma(\theta) \leq 0$. However, when $|E_a| < |\frac{\sigma_0}{2\varepsilon_0}|$, the two problems have the same solution.

(b) For the case of a conducting cylinder the electrostatic field must satisfy the following:

(1) Inside the conductor $E_I = 0$ and φ_I is a constant.
(2) Outside the conductor

$$\nabla^2 \varphi_{II} = 0,$$

$$\left(\frac{\partial \varphi_{II}}{\partial \theta}\right)_{\rho=r} = 0, \quad \varphi_{II}|_{\rho\to\infty} \to -E_a\rho\cos\theta.$$

The solution for φ_{II} is the same as before. For the solution of the conductor to fit the case of an insulating cylinder, the necessary condition is $|E_a| \leq |\frac{\sigma_0}{2\varepsilon_0}|$, which ensures that the surface charge density on the cylinder is negative everywhere.

1073

Two semi-infinite plane grounded aluminium sheets make an angle of $60°$. A single point charge $+q$ is placed as shown in Fig. 1.33. Make a large drawing indicating clearly the position, size of all image charges. In two or three sentences explain your reasoning.

(*Wisconsin*)

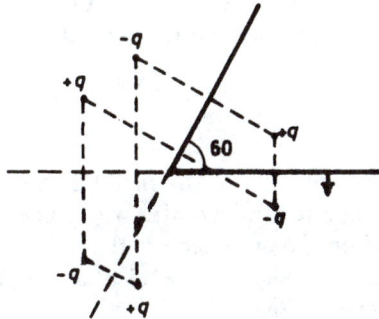

Fig. 1.33

Solution:

As in Fig. 1.33, since the planes are grounded, the image charges are distributed symmetrically on the two sides of each plane.

1074

A charge placed in front of a metallic plane:

(a) is repelled by the plane,

(b) does not know the plane is there,

(c) is attracted to the plane by a mirror image of equal and opposite charge.

(CCT)

Solution:

The answer is (c).

1075

The potential at a distance r from the axis of an infinite straight wire of radius a carrying a charge per unit length σ is given by

$$\frac{\sigma}{2\pi} \ln \frac{1}{r} + \text{const}.$$

This wire is placed at a distance $b \gg a$ from an infinite metal plane, whose potential is maintained at zero. Find the capacitance per unit length of the wire of this system.

(Wisconsin)

Solution:

In order that the potential of the metal plane is maintained at zero, we imagine that an infinite straight wire with linear charge density $-\sigma$ is symmetrically placed on the other side of the plane. Then the capacitance between the original wire and the metal plane is that between the two straight wires separated at $2b$.

The potential $\varphi(r)$ at a point between the two wires at distance r from the original wire (and at distance $2b - r$ from the image wire) is then

$$\varphi(r) = \frac{\sigma}{2\pi} \ln \frac{1}{r} - \frac{\sigma}{2\pi} \ln \frac{1}{2b - r}.$$

So the potential difference between the two wires is

$$V = \varphi(a) - \varphi(2b - a) = \frac{\sigma}{\pi} \ln \left(\frac{2b - a}{a} \right) \approx \frac{\sigma}{\pi} \ln \frac{2b}{a}.$$

Thus the capacitance of this system per unit length of the wire is

$$C = \frac{\sigma}{V} = \pi / \ln \frac{2b}{a}.$$

1076

A charge $q = 2\,\mu C$ is placed at $a = 10$ cm from an infinite grounded conducting plane sheet. Find

(a) the total charge induced on the sheet,

(b) the force on the charge q,

(c) the total work required to remove the charge slowly to an infinite distance from the plane.

(Wisconsin)

Solution:

(a) The method of images requires that an image charge $-q$ is placed symmetrically with respect to the plane sheet. This means that the total induced charge on the surface of the conductor is $-q$.

(b) The force acting on $+q$ is

$$F = \frac{1}{4\pi\varepsilon_0} \frac{q^2}{(2a)^2} = 9 \times 10^9 \times \frac{(2 \times 10^{-6})^2}{0.2^2} = 0.9 \text{ N},$$

where we have used $\varepsilon_0 = \frac{1}{4\pi \times 9 \times 10^9}$ C^2/(N·m^2).

(c) The total work required to remove the charge to infinity is

$$W = \int_a^\infty F dr = \int_a^\infty \frac{1}{4\pi\varepsilon_0} \frac{q^2}{(2r)^2} dr = \frac{q^2}{16\pi\varepsilon_0 a} = 0.09 \text{ J}.$$

1077

Charges $+q, -q$ lie at the points $(x, y, z) = (a, 0, a), (-a, 0, a)$ above a grounded conducting plane at $z = 0$. Find

(a) the total force on charge $+q$,

(b) the work done against the electrostatic forces in assembling this system of charges,

(c) the surface charge density at the point $(a, 0, 0)$.

(*Wisconsin*)

Solution:

(a) The method of images requires image charges $+q$ at $(-a, 0, -a)$ and $-q$ at $(a, 0, -a)$ (see Fig. 1.34). The resultant force exerted on $+q$ at $(a, 0, a)$ is thus

$$\mathbf{F} = \frac{q^2}{4\pi\varepsilon_0} \left[-\frac{1}{(2a)^2}\mathbf{e}_x - \frac{1}{(2a)^2}\mathbf{e}_z + \frac{1}{(2\sqrt{2}a)^2}\left(\frac{1}{\sqrt{2}}\mathbf{e}_x + \frac{1}{\sqrt{2}}\mathbf{e}_z\right) \right]$$

$$= \frac{q^2}{4\pi\varepsilon_0 a^2} \left[\left(-\frac{1}{4} + \frac{1}{8\sqrt{2}}\right)\mathbf{e}_x + \left(-\frac{1}{4} + \frac{1}{8\sqrt{2}}\right)\mathbf{e}_z \right].$$

This force has magnitude

$$F = \frac{(\sqrt{2} - 1)q^2}{32\pi\varepsilon_0 a^2}.$$

It is in the xz-plane and points to the origin along a direction at angle $45°$ to the x-axis as shown in Fig. 1.34.

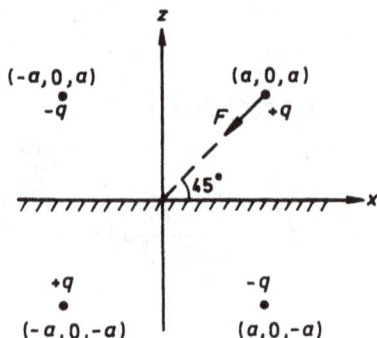

Fig. 1.34

(b) We can construct the system by slowly bringing the charges $+q$ and $-q$ from infinity by the paths

$$L_1 : z = x,\ y = 0,$$
$$L_2 : z = -x,\ y = 0,$$

symmetrically to the points $(a, 0, a)$ and $(-a, 0, a)$ respectively. When the charges are at $(l, 0, l)$ on path L_1 and $(-l, 0, l)$ on path L_2 respectively, each suffers a force $\frac{(\sqrt{2}-1)q^2}{32\pi\varepsilon_0 l^2}$ whose direction is parallel to the direction of the path so that the total work done by the external forces is

$$W = -2 \int_\infty^a F dl = 2 \int_a^\infty \frac{(\sqrt{2}-1)q^2}{32\pi\varepsilon_0 l^2} dl = \frac{(\sqrt{2}-1)q^2}{16\pi\varepsilon_0 a}.$$

(c) Consider the electric field at a point $(a, 0, 0^+)$ just above the conducting plane. The resultant field intensity $\mathbf{E}_1$ produced by $+q$ at $(a, 0, a)$ and $-q$ at $(a, 0, -a)$ is

$$\mathbf{E}_1 = -\frac{2q}{4\pi\varepsilon_0 a^2}\mathbf{e}_z.$$

The resultant field $\mathbf{E}_2$ produced by $-q$ at $(-a, 0, a)$ and $+q$ at $(-a, 0, -a)$ is

$$\mathbf{E}_2 = \frac{2q}{4\pi\varepsilon_0 a^2}\frac{1}{5\sqrt{5}}\mathbf{e}_z.$$

Hence the total field at $(a, 0, 0^+)$ is

$$\mathbf{E} = \mathbf{E}_1 + \mathbf{E}_2 = \frac{q}{2\pi\varepsilon_0 a^2}\left(\frac{1}{5\sqrt{5}} - 1\right)\mathbf{e}_z,$$

and the surface charge density at this point is

$$\sigma = \varepsilon_0 E = \frac{q}{2\pi\varepsilon_0 a^2} \left(\frac{1}{5\sqrt{5}} - 1 \right).$$

1078

Suppose that the region $z > 0$ in three-dimensional space is filled with a linear dielectric material characterized by a dielectric constant ε_1, while the region $z < 0$ has a dielectric material ε_2. Fix a charge $-q$ at $(x, y, z) = (0, 0, a)$ and a charge q at $(0, 0, -a)$. What is the force one must exert on the negative charge to keep it at rest?

(Columbia)

Solution:

Consider first the simple case where a point charge q_1 is placed at $(0, 0, a)$. The method of images requires image charges q_1' at $(0, 0, -a)$ and q_1'' at $(0, 0, a)$. Then the potential (in Gaussian units) at a point (x, y, z) is given by

$$\varphi_1 = \frac{q_1}{\varepsilon_1 r_1} + \frac{q_1'}{\varepsilon_1 r_2}, \quad (z \geq 0), \qquad \varphi_2 = \frac{q_1''}{\varepsilon_2 r_1}, \quad (z < 0),$$

where

$$r_1 = \sqrt{x^2 + y^2 + (z - a)^2}, \quad r_2 = \sqrt{x^2 + y^2 + (z + a)^2}.$$

Applying the boundary conditions at $(x, y, 0)$:

$$\varphi_1 = \varphi_2, \quad \varepsilon_1 \frac{\partial \varphi_1}{\partial z} = \varepsilon_2 \frac{\partial \varphi_2}{\partial z},$$

we obtain

$$q_1' = q_1'' = \frac{\varepsilon_1 - \varepsilon_0}{\varepsilon_1(\varepsilon_1 + \varepsilon_2)} q_1.$$

Similarly, if a point charge q_2 is placed at $(0, 0, -a)$ inside the dielectic ε_2, its image charges will be q_2' at $(0, 0, a)$ and q_2'' at $(0, 0, -a)$ with magnitudes

$$q_2' = q_2'' = \frac{\varepsilon_2 - \varepsilon_1}{\varepsilon_2(\varepsilon_1 + \varepsilon_2)} q_2.$$

When both q_1 and q_2 exist, the force on q_1 will be the resultant due to q_2, q_1' and q_1''. It follows that

$$F = \frac{q_1 q_1'}{4a^2 \varepsilon_1} + \frac{q_1 q_2}{4a^2 \varepsilon_2} + \frac{q_1 q_1''}{4a^2 \varepsilon_2}$$

$$= \frac{\varepsilon_1 - \varepsilon_2}{\varepsilon_1(\varepsilon_1 + \varepsilon_2)} \cdot \frac{q_1^2}{4a^2} + \frac{q_1 q_2}{2(\varepsilon_1 + \varepsilon_2)a^2} \; .$$

In the present problem $q_1 = -q, q_2 = +q$, and one has

$$F = \frac{\varepsilon_1 - \varepsilon_2}{\varepsilon_1(\varepsilon_1 + \varepsilon_2)} \frac{q^2}{4a^2} - \frac{q^2}{2(\varepsilon_1 + \varepsilon_2)a^2} = -\frac{q^2}{4\varepsilon_1 a^2} \; .$$

Hence, a force $-F$ is required to keep on $-q$ at rest.

1079

When a cloud passes over a certain spot on the surface of the earth a vertical electric field of $E = 100$ volts/meter is recorded here. The bottom of the cloud is a height $d = 300$ meters above the earth and the top of the cloud is a height $d = 300$ meters above the bottom. Assume that the cloud is electrically neutral but has charge $+q$ at its top and charge $-q$ at its bottom. Estimate the magnitude of the charge q and the external electrical force (direction and magnitude) on the cloud. You may assume that there are no charges in the atmosphere other than those on the cloud.

(*Wisconsin*)

Solution:

We use the method of images. The positions of the image charges are shown in Fig. 1.35. Then the electric field intensity at the point 0 on the surface of the earth is

$$E = 2 \cdot \frac{1}{4\pi\varepsilon_0} \frac{q}{d^2} - 2 \cdot \frac{1}{4\pi\varepsilon_0} \frac{q}{(2d)^2} \; ,$$

whence we get

$$q = \frac{8\pi\varepsilon_0 d^2 E}{3} = 6.7 \times 10^{-4} \text{ C} .$$

The external force acting on the cloud is the electrostatic force between the image charges and the charges in the cloud, i.e.,

$$F = \frac{q^2}{4\pi\varepsilon_0} \left[-\frac{1}{(2d)^2} + \frac{1}{(3d)^2} + \frac{1}{(3d)^2} - \frac{1}{(4d)^2} \right]$$

$$= \frac{q^2}{4\pi\varepsilon_0 d^2} \left[\frac{2}{9} - \frac{1}{4} - \frac{1}{16} \right] = -4.05 \times 10^{-3} \text{ N} .$$

This force is an attraction, as can be seen from Fig. 1.35.

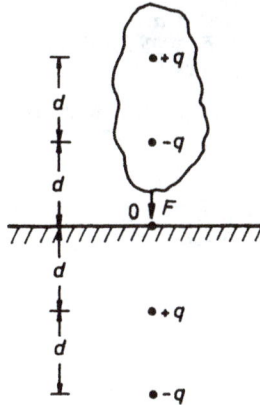

Fig. 1.35

1080

A point charge q is located at radius vector $\mathbf{s}$ from the center of a perfectly conducting grounded sphere of radius a.

(a) If (except for q) the region outside the sphere is a vacuum, calculate the electrostatic potential at an arbitrary point $\mathbf{r}$ outside the sphere. As usual, take the reference ground potential to be zero.

(b) Repeat (a) if the vacuum is replaced by a dielectric medium of dielectric constant ε.

(*CUSPEA*)

Solution:

We use the method of images.

(a) As shown in Fig. 1.36, the image charge q' is to be placed on the line oq at distance s' from the spherical center. Letting $\mathbf{n} = \frac{\mathbf{r}}{r}, \mathbf{n}' = \frac{\mathbf{s}}{s} = \frac{\mathbf{s}'}{s'}$, the potential at $\mathbf{r}$ is

$$\phi(\mathbf{r}) = \frac{1}{4\pi\varepsilon_0} \left[\frac{q}{|\mathbf{r} - \mathbf{s}|} + \frac{q'}{|\mathbf{r} - \mathbf{s}'|} \right]$$

$$= \frac{1}{4\pi\varepsilon_0} \left[\frac{q}{|r\mathbf{n} - s\mathbf{n}'|} + \frac{q'}{|r\mathbf{n} - s'\mathbf{n}'|} \right].$$

The boundary condition requires $\phi(r = a) = 0$. This can be satisfied if

$$q' = -\frac{a}{s}q, \quad s' = \frac{a^2}{s}.$$

The electrostatic uniqueness theorem then gives the potential at a point $\mathbf{r}$ outside the sphere as

$$\phi(\mathbf{r}) = \frac{q}{4\pi\varepsilon_0}\left[\frac{1}{|\mathbf{r}-\mathbf{s}|} - \frac{a/s}{|\mathbf{r}-\frac{a^2}{s^2}\mathbf{s}|}\right].$$

(b) When the outside of the sphere is filled with a dielectric medium of dielectric constant ε, we simply replace ε_0 in (a) with ε. Thus

$$\phi(\mathbf{r}) = \frac{q}{4\pi\varepsilon}\left[\frac{1}{|\mathbf{r}-\mathbf{s}|} - \frac{a/s}{|\mathbf{r}-\frac{a^2}{s^2}\mathbf{s}|}\right].$$

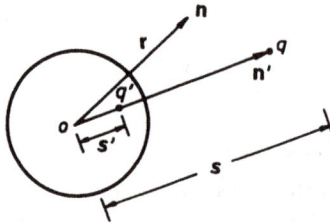

Fig. 1.36

1081

Two similar charges are placed at a distance $2b$ apart. Find, approximately, the minimum radius a of a grounded conducting sphere placed midway between them that would neutralize their mutual repulsion.

(*SUNY, Buffalo*)

Solution:

We use the method of images. The electric field outside the sphere corresponds to the resultant field of the two given charges $+q$ and two image charges $+q'$. The magnitudes of the image charges are both $q' = -q\frac{a}{b}$, and they are to be placed at two sides of the center of the sphere at the same distance $b' = \frac{a^2}{b}$ from it (see Fig. 1.37).

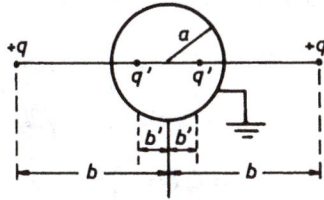

Fig. 1.37

For each given charge $+q$, apart from the electric repulsion acted on it by the other given charge $+q$, there is also the attraction exerted by the two image charges. For the resultant force to vanish we require

$$\frac{q^2}{4b^2} = \frac{q^2 \frac{a}{b}}{(b - \frac{a^2}{b})^2} + \frac{q^2 \frac{a}{b}}{(b + \frac{a^2}{b})^2}$$

$$= \frac{2q^2 a}{b^3}\left[1 + 3\left(\frac{a}{b}\right)^4 + 5\left(\frac{a}{b}\right)^8 + \ldots\right] \approx \frac{2q^2 a}{b^3},$$

The value of $a(a < b)$ that satisfies the above requirement is therefore approximately

$$a \approx \frac{b}{8}.$$

1082

(a) Two equal charges $+Q$ are separated by a distance $2d$. A grounded conducting sphere is placed midway between them. What must the radius of the sphere be if the two charges are to experience zero total force?

(b) What is the force on each of the two charges if the same sphere, with the radius determined in part (a), is now charged to a potential V?

(*Columbia*)

Solution:

(a) Referring to Problem **1081**, we have $r_0 \approx d/8$.

(b) When the sphere is now charged to a potential V, the potential outside the sphere is correspondingly increased by

$$\phi = \frac{V r_0}{r} \approx \frac{V d}{8r},$$

where r is the distance between the field point and the center of the sphere.

An additional electric field is established being

$$E = -\nabla\phi = \frac{Vd}{8r^2}e_r \, .$$

Therefore, the force exerted on each charge $+Q$ is

$$F = QE = \frac{QV}{8d} \, .$$

The direction of the force is outwards from the sphere along the line joining the charge and the center.

1083

A charge q is placed inside a spherical shell conductor of inner radius r_1 and outer radius r_2. Find the electric force on the charge. If the conductor is isolated and uncharged, what is the potential of its inner surface?

(Wisconsin)

Solution:

Apply the method of images and let the distance between q and the center of the shell be a. Then an image charge $q' = -\frac{r_1}{a}q$ is to be placed at $b = \frac{r_1^2}{a}$ (see Fig. 1.38). Since the conductor is isolated and uncharged, it is an equipotential body with potential $\varphi = \varphi_0$, say. Then the electric field inside the shell $(r < r_1)$ equals the field created by q and q'.

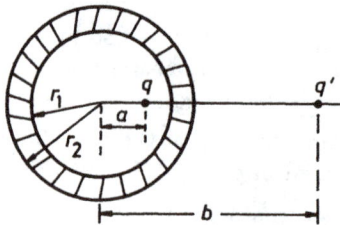

Fig. 1.38

The force on the charge q is that exerted by q':

$$
\begin{aligned}
F &= \frac{qq'}{4\pi\varepsilon_0(b-a)^2} = -\frac{\frac{r_1}{a}q^2}{4\pi\varepsilon_0(\frac{r_1^2}{a}-a)^2} \\
&= -\frac{ar_1q^2}{4\pi\varepsilon_0(r_1^2-a^2)^2} \, .
\end{aligned}
$$

In zone $r > r_2$ the potential is $\varphi_{\text{out}} = \frac{q}{4\pi\varepsilon_0 r}$. In particular, the potential of the conducting sphere at $r = r_2$ is

$$\varphi_{\text{sphere}} = \frac{q}{4\pi\varepsilon_0 r_2}.$$

Owing to the conductor being an equipotential body, the potential of the inner surface of the conducting shell is also $\frac{q}{4\pi\varepsilon_0 r_2}$.

1084

Consider an electric dipole $\mathbf{P}$. Suppose the dipole is fixed at a distance z_0 along the z-axis and at an orientation θ with respect to that axis (i.e., $\mathbf{P} \cdot \mathbf{e}_z = |\mathbf{P}| \cos \theta$). Suppose the xy plane is a conductor at zero potential. Give the charge density on the conductor induced by the dipole.

(*Columbia*)

Solution:

As shown in Fig. 1.39, the dipole is $\mathbf{P} = P(\sin\theta, 0, \cos\theta)$, and its image dipole is $\mathbf{P}' = P(-\sin\theta, 0, \cos\theta)$. In the region $z > 0$ the potential at a point $\mathbf{r} = (x, y, z)$ is

$$\varphi(\mathbf{r}) = \frac{1}{4\pi\varepsilon_0} \left\{ \frac{P[x\sin\theta + (z - z_0)\cos\theta]}{[x^2 + y^2 + (z - z_0)^2]^{3/2}} + \frac{P[-x\sin\theta + (z + z_0)\cos\theta]}{[x^2 + y^2 + (z + z_0)^2]^{3/2}} \right\}.$$

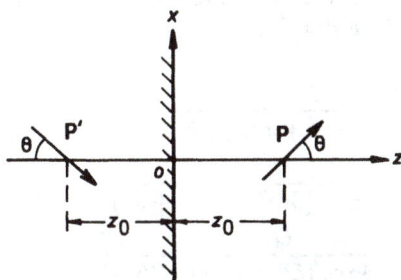

Fig. 1.39

The induced charge density on the surface of the conductor is then

$$\sigma = -\varepsilon_0 \frac{\partial\varphi}{\partial z}\bigg|_{z=0} = -\frac{P\cos\theta}{2\pi(x^2 + y^2 + z_0^2)^{3/2}}$$
$$+ \frac{3Pz_0(-x\sin\theta + z_0\cos\theta)}{2\pi(x^2 + y^2 + z_0^2)^{5/2}}.$$

1085

Two large flat conducting plates separated by a distance D are connected by a wire. A point charge Q is placed midway between the two plates, as in Fig. 1.40. Find an expression for the surface charge density induced on the lower plate as a function of D, Q and x (the distance from the center of the plate).

(Columbia)

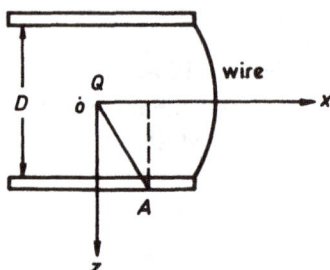

Fig. 1.40

Solution:

We use the method of images. The positions of the image charges are shown in Fig. 1.41. Consider an arbitrary point A on the lower plate. Choose the xz-plane to contain A. It can be seen that the electric field at A, which is at the surface of a conductor, has only the z-component and its magnitude is (letting $d = \frac{D}{2}$)

$$
\begin{aligned}
E_z &= \frac{Q}{4\pi\varepsilon_0(d^2 + x^2)} \cdot \frac{2d}{(d^2 + x^2)^{1/2}} \\
&\quad - \frac{Q}{4\pi\varepsilon_0[(3d)^2 + x^2]} \cdot \frac{2 \cdot 3d}{[(3d)^2 + x^2]^{1/2}} \\
&\quad + \frac{Q}{4\pi\varepsilon_0[(5d)^2 + x^2]} \cdot \frac{2 \cdot 5d}{[(5d)^2 + x^2]^{1/2}} - \cdots \\
&= \frac{QD}{4\pi\varepsilon_0} \sum_{n=0}^{\infty} \frac{(-1)^n(2n+1)}{[(n+\frac{1}{2})^2 D^2 + x^2]^{3/2}} .
\end{aligned}
$$

Accordingly,

$$
\sigma(x) = -\varepsilon_0 E_z = -\frac{QD}{4\pi} \sum_{n=0}^{\infty} \frac{(-1)^n(2n+1)}{[(n+\frac{1}{2})^2 D^2 + x^2]^{3/2}} .
$$

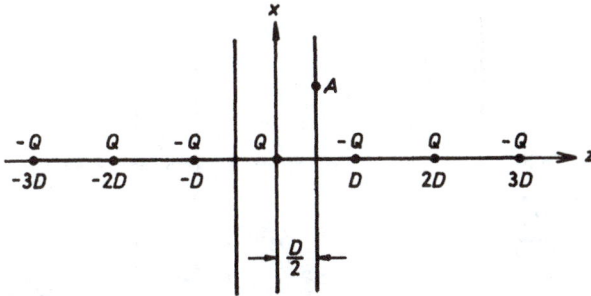

Fig. 1.41

1086

Two large parallel conducting plates are separated by a small distance $4x$. The two plates are gounded. Charges Q and $-Q$ are placed at distances x and $3x$ from one plate as shown in Fig. 1.42.

(a) How much energy is required to remove the two charges from between the plates and infinitely apart?

(b) What is the force (magnitude and direction) on each charge?

(c) What is the force (magnitude and direction) on each plate?

(d) What is the total induced charge on each plate? (Hint: What is the potential on the plane midway between the two charges?)

(e) What is the total induced charge on the inside surface of each plate after the $-Q$ charge has been removed, the $+Q$ remaining at rest?

(*MIT*)

Solution:

(a) The potential is found by the method of images, which requires image charges $+Q$ at $\cdots - 9x, -5x, 3x, 7x, 11x \cdots$ and image charges $-Q$ at $\cdots -7x, -3x, 5x, 9x, 13x, \cdots$ along the x-axis as shown in Fig. 1.43. Then the charge density of the system of real and image charges can be expressed as

$$\rho = \sum_{k=-\infty}^{\infty} (-1)^{k+1} Q\delta[x - (2k+1)x]$$

where δ is the one-dimensional Dirac delta function.

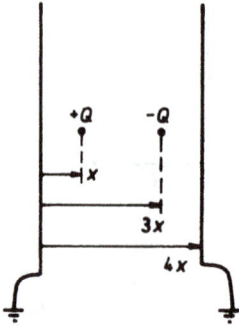

Fig. 1.42

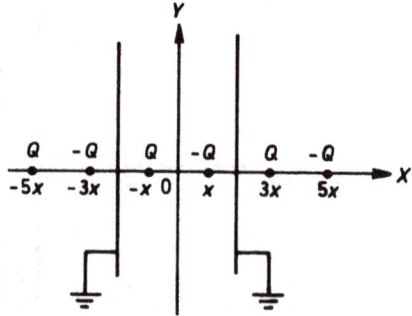

Fig. 1.43

The electrostatic field energy of the system is

$$W = \frac{1}{2}\sum QU = \frac{1}{2}QU_+ - \frac{1}{2}QU_- \,,$$

where U_+ is the potential at the $+Q$ charge produced by the other real and image charges not including the $+Q$ itself, while U_- is the potential at the $-Q$ charge produced by the other real and image charges not including the $-Q$ itself. As

$$U_+ = \frac{1}{4\pi\varepsilon_0}\left[-\frac{2Q}{(2x)} + \frac{2Q}{(4x)} - \frac{2Q}{(6x)} + \cdots \right]$$

$$= \frac{-Q}{4\pi\varepsilon_0 x}\cdot \sum_{k=1}^{+\infty}(-1)^{k-1}\cdot\frac{1}{k} = \frac{-Q\ln 2}{4\pi\varepsilon_0 x}\,,$$

$$U_- = -U_+ = \frac{Q\ln 2}{4\pi\varepsilon_0 x}\,,$$

we have

$$W = -\frac{Q^2}{4\pi\varepsilon_0 x}\ln 2\,.$$

Hence the energy required to remove the two charges to infinite distances from the plates and from each other is $-W$.

(b) The force acting on $+Q$ is just that exerted by the fields of all the other real and image charges produced by Q. Because of symmetry this force is equal to zero. Similarly the force on $-Q$ is also zero.

(c) Consider the force exerted on the left conducting plate. This is the resultant of all the forces acting on the image charges of the left plate (i.e., image charges to the left of the left plate) by the real charges $+Q, -Q$ and all the image charges of the right plate (i.e., image charges to the right side of right plate).

Let us consider first the force F_1 acting on the image charges of the left plate by the real charge $+Q$:

$$F_1 = \frac{Q^2}{4\pi\varepsilon_0(2x)^2} - \frac{Q^2}{4\pi\varepsilon_0(4x)^2} + \cdots = \frac{Q^2}{16\pi\varepsilon_0 x^2} \sum_{n=1}^{\infty} \frac{(-1)^{n-1}}{n^2},$$

taking the direction along $+x$ as positive.

We next find the force F_2 between the real charge $-Q$ and the image charges of the left plate:

$$F_2 = -\frac{Q^2}{4\pi\varepsilon_0(4x)^2} + \frac{Q^2}{4\pi\varepsilon_0(6x)^2} - \cdots = \frac{Q^2}{16\pi\varepsilon_0 x^2} \sum_{n=2}^{\infty} \frac{(-1)^{n-1}}{n^2}.$$

Finally consider the force F_3 acting on the image charges of the left plate by the image charges of the right plate:

$$F_3 = \frac{Q^2}{16\pi\varepsilon_0 x^2} \sum_{m=3}^{\infty} \sum_{n=m}^{\infty} \frac{(-1)^{n-1}}{n^2}.$$

Thus the total force exerted on the left plate is

$$F = F_1 + F_2 + F_3 = \frac{Q^2}{16\pi\varepsilon_0 x^2} \sum_{m=1}^{\infty} \sum_{n=m}^{\infty} \frac{(-1)^n}{n^2}$$

$$= \frac{Q^2}{16\pi\varepsilon_0 x^2} \left(1 - \frac{2}{2^2} + \frac{3}{3^2} - \frac{4}{4^2} + \cdots \right) = \frac{Q^2}{16\pi\varepsilon_0 x^2} \sum_{n=1}^{\infty} \frac{(-1)^{n-1}}{n^2}.$$

Using the identity

$$\sum_{n=1}^{\infty} \frac{(-1)^{n-1}}{n} = \ln 2,$$

we obtain

$$F = \frac{Q^2}{16\pi\varepsilon_0 x^2} \ln 2.$$

This force directs to the right. In a similar manner, we can show that the magnitude of the force exerted on the right plate is also equal to $\frac{Q^2}{16\pi\varepsilon_0 x^2} \ln 2$, its direction being towards the left.

(d) The potential on the plane $x = 0$ is zero, so only half of the lines of force emerging from the $+Q$ charge reach the left plate, while those emerging from the $-Q$ charge cannot reach the left plate at all. Therefore, the total induced charge on the left plate is $-\frac{Q}{2}$, and similarly that of the right plate is $\frac{Q}{2}$.

(e) When the $+Q$ charge alone exists, the sum of the total induced charges on the two plates is $-Q$. If the total induced charge is $-Q_x$ on the left plate, then the total reduced charge is $-Q + Q_x$ on the right plate. Similarly if $-Q$ alone exists, the total induced charge on the left plate is $Q - Q_x$ and that on the right plate is $+Q_x$, by reason of symmetry. If the two charges exist at the same time, the induced charge on the left plate is the superposition of the induced charges produced by both $+Q$ and $-Q$. Hence we have, using the result of (d),

$$Q - 2Q_x = -\frac{Q}{2} \, ,$$

or

$$Q_x = \frac{3}{4}Q \, .$$

Thus after $-Q$ has been removed, the total induced charge on the inside surface of the left plate is $-3Q/4$ and that of the right plate is $-Q/4$.

1087

What is the least positive charge that must be given to a spherical conducter of radius a, insulated and influenced by an external point charge $+q$ at a distance $r > a$ from its center, in order that the surface charge density on the sphere may be everywhere positive? What if the external point charge is $-q$?

(SUNY, Buffalo)

Solution:

Use Cartesian coordinates with the origin at the center of the sphere and the z-axis along the line joining the spherical center and the charge q. It is obvious that the greatest induced surface charge density, which is negative, on the sphere will occur at $(0, 0, a)$.

The action of the conducting spherical surface may be replaced by that of a point charge $(-\frac{a}{r}q)$ at $(0, 0, \frac{a^2}{r})$ and a point charge $(\frac{a}{r}q)$ at the spherical center $(0,0,0)$. Then, the field $\mathbf{E}$ at $(0,0,a_+)$ is

$$\mathbf{E} = \frac{1}{4\pi\varepsilon_0}\left[\frac{\frac{a}{r}q}{a^2} - \frac{q}{(r-a)^2} - \frac{\frac{a}{r}q}{(a-\frac{a^2}{r})^2}\right]\mathbf{e}_z$$

$$= \frac{1}{4\pi\varepsilon_0}\left[\frac{q}{ar} - \frac{q}{(r-a)^2} - \frac{\frac{r}{a}q}{(r-a)^2}\right]\mathbf{e}_z .$$

Hence, the maximum negative induced surface charge density is

$$\sigma_e = \varepsilon_0 E_n = \frac{q}{4\pi}\left[\frac{1}{ar} - \frac{1}{(r-a)^2} - \frac{r/a}{(r-a)^2}\right] .$$

If a positive charge Q is given to the sphere, it will distribute uniformly on the spherical surface with a surface density $Q/4\pi a^2$. In order that the total surface charge density is everywhere positive, we require that

$$Q \geq -\sigma \cdot 4\pi a^2 = a^2 q\left[\left(1+\frac{r}{a}\right)\frac{1}{(r-a)^2} - \frac{1}{ar}\right]$$

$$= \frac{a^2(3r-a)}{r(r-a)^2}q .$$

On the other hand, the field at point $(0,0,-a_-)$ is

$$\mathbf{E} = -\frac{q}{4\pi\varepsilon_0}\left[\frac{1}{ra} + \frac{1}{(r+a)^2} - \frac{r/a}{(r+a)^2}\right]\mathbf{e}_z .$$

If we replace q by $-q$, the maximum negative induced surface charge density will occur at $(0,0,-a)$. Then as above the required positive charge is

$$Q \geq -\sigma \cdot 4\pi a^2 = -\varepsilon_0 \cdot 4\pi a^2\left(-\frac{q}{4\pi\varepsilon_0}\right) \cdot \left[\frac{1}{ra} + \frac{1}{(r+a)^2} - \frac{r/a}{(r+a)^2}\right]$$

$$= q\left[\frac{a}{r} + \frac{a^2}{(r+a)^2} - \frac{ar}{(r+a)^2}\right] = \frac{qa^2(3r+a)}{r(r+a)^2} .$$

1088

(a) Find the electrostatic potential arising from an electric dipole of magnitude d situated a distance L from the center of a grounded conducting sphere of radius a; assume the axis of the dipole passes through the center of the sphere.

(b) Consider two isolated conducting spheres (radii a) in a uniform electric field oriented so that the line joining their centers, of length R, is parallel to the electric field. When R is large, qualitatively describe the fields which result, through order R^{-4}.

(*Wisconsin*)

Solution:

(a) Taking the spherical center as the origin and the axis of symmetry of the system as the z-axis, then we can write $\mathbf{P} = d\mathbf{e}_z$. Regarding $\mathbf{P}$ as a positive charge q and a negative charge $-q$ separated by a distance $2l$ such that $d = \lim\limits_{l \to 0} 2ql$, we use the method of images. As shown in Fig. 1.44, the coordinates of q and $-q$ are respectively given by

$$q: \quad z = -L + l, \quad -q: \quad z = -L - l.$$

Let q_1 and q_2 be the image charges of q and $-q$ respectively. For the spherical surface to be of equipotential, the magnitudes and positions of q_1 and q_2 are as follows (Fig. 1.44):

$$q_1 = -\frac{a}{L-l}q \quad \text{at} \quad \left(0,0,-\frac{a^2}{L-l}\right),$$

$$q_2 = \frac{a}{L+l}q \quad \text{at} \quad \left(0,0,-\frac{a^2}{L+l}\right).$$

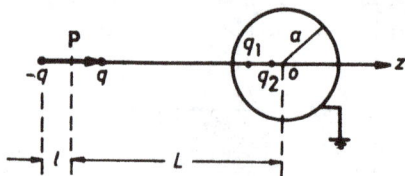

Fig. 1.44

As $L \gg 1$, by the approximation

$$\frac{1}{L \pm l} \approx \frac{1}{L} \mp \frac{l}{L^2}$$

the magnitudes and positions may be expressed as

$$q_1 = -\frac{a}{L}q - \frac{ad}{2L^2} \quad \text{at} \quad \left(0, 0, \frac{-a^2}{L} - \frac{a^2 l}{L^2}\right),$$

$$q_2 = \frac{a}{L}q - \frac{ad}{2L^2} \quad \text{at} \quad \left(0, 0, -\frac{a^2}{L} + \frac{a^2 l}{L^2}\right),$$

where we have used $d = 2ql$. Hence, an image dipole with dipole moment $\mathbf{P}' = \frac{a}{L}q \cdot \frac{2a^2 l}{L^2}\mathbf{e}_z = \frac{a^3}{L^3}\mathbf{P}$ and an image charge $q' = -\frac{ad}{L^2}$ may be used in place of the action of q_1 and q_2. Both $\mathbf{P}'$ and q' are located at $\mathbf{r}' = (0, 0, \frac{-a^2}{L})$ (see Fig. 1.45). Therefore, the potential at $\mathbf{r}$ outside the sphere is the superposition of the potentials produced by $\mathbf{P}, \mathbf{P}'$, and q', i.e.,

$$\varphi(\mathbf{r}) = \frac{1}{4\pi\varepsilon_0}\left[\frac{q'}{|\mathbf{r}-\mathbf{r}'|} + \frac{\mathbf{P}'\cdot(\mathbf{r}-\mathbf{r}')}{|\mathbf{r}-\mathbf{r}'|^3} + \frac{\mathbf{P}\cdot(\mathbf{r}+L\mathbf{e}_z)}{|\mathbf{r}+L\mathbf{e}_z|^3}\right]$$

$$= \frac{1}{4\pi\varepsilon_0}\left[-\frac{ad}{L^2(r^2 + \frac{a^2 r}{L}\cos\theta + \frac{a^4}{L^2})^{1/2}}\right.$$

$$+ \frac{a^3 d(r\cos\theta + \frac{a^2}{L})}{L^3(r^2 + \frac{a^2 r}{L}\cos\theta + \frac{a^4}{L^2})^{3/2}} + \left.\frac{d(r\cos\theta + L)}{(r^2 + 2rL\cos\theta + L^2)^{3/2}}\right].$$

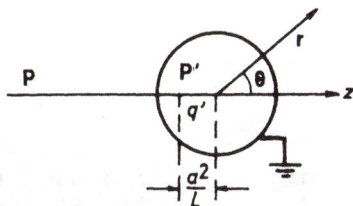

Fig. 1.45

(b) A conducting sphere of radius a in an external field $\mathbf{E}$ corresponds to an electric dipole with moment $\mathbf{P} = 4\pi\varepsilon_0 a^2 \mathbf{E}$ in respect of the field outside the sphere. The two isolated conducting spheres in this problem may be regarded as one dipole if we use the approximation of zero order. But when we apply the approximation of a higher order, the interaction between the two conducting spheres has to be considered. Now the action by the first sphere on the second is like the case (a) in this problem (as the two spheres are separated by a large distance). In other words, this action can be considered as that of the image dipole $\mathbf{P}' = \frac{a^3}{R^3}\mathbf{P}$ and image charge

$q' = -\frac{ap}{R^2}$. As $a^2 \ll L$, the image dipole and charge can be taken to be approximately located at the spherical center. Thus the electric field at a point outside the spheres is the resultant of the fields due to a point charge q' and a dipole of moment $\mathbf{P} + \mathbf{P}' = (1 + \frac{a^3}{R^3})\mathbf{P}$ at each center of the two spheres. The potential can then be expressed in terms through order $1/R^4$.

1089

An electric dipole of moment $\mathbf{P}$ is placed at a distance r from a perfectly conducting, isolated sphere of radius a. The dipole is oriented in the direction radially away from the sphere. Assume that $r \gg a$ and that the sphere carries no net charge.

(a) What are the boundary conditions on the $\mathbf{E}$ field at the surface of the sphere?

(b) Find the approximate force on the dipole.

(*MIT*)

Solution:

(a) Use spherical coordinates with the z-axis along the direction of the dipole $\mathbf{P}$ and the origin at the spherical center. The boundary conditions for $\mathbf{E}$ on the surface of the sphere are

$$E_n = \sigma/\varepsilon_0, \quad E_t = 0,$$

where σ is the surface charge density.

(b) The system of images is similar to that of the previous Problem 1088 and consists of an image dipole $\mathbf{P}' = (\frac{a}{r})^3\mathbf{P}$ and an image charge $q' = -\frac{aP}{r^2}$ at $r' = \frac{a^2}{r}\mathbf{e}_z$. In addition, an image charge $q'' = -q'$ is to be added at the center of the sphere as the conducting sphere is isolated and uncharged. However, since $r \gg a$, we can consider q' and q'' as composing an image dipole of moment $\mathbf{P}'' = \frac{aP}{r^2} \cdot \frac{a^2}{r} = \frac{a^3}{r^3}\mathbf{P}$.

As $r \gg a$, the image dipoles $\mathbf{P}'$ and $\mathbf{P}''$ may be considered as approximately located at the center of the sphere. That is, the total image dipole moment is

$$\mathbf{P}_{\text{image}} = \mathbf{P}' + \mathbf{P}'' = 2\left(\frac{a}{r}\right)^3\mathbf{P}.$$

The problem is then to find the force exerted on $\mathbf{P}$ by $\mathbf{P}_{\text{image}}$.

The potential at a point $\mathbf{r}$ produced by $\mathbf{P}_{\text{image}}$ is

$$\varphi(\mathbf{r}) = \frac{1}{4\pi\varepsilon_0} \frac{\mathbf{P}_{\text{image}} \cdot \mathbf{r}}{r^3},$$

and the corresponding electric field is

$$\mathbf{E}(\mathbf{r}) = -\nabla\varphi(\mathbf{r}) = \frac{3(\mathbf{r} \cdot \mathbf{P}_{\text{image}})\mathbf{r} - r^2\mathbf{P}_{\text{image}}}{4\pi\varepsilon_0 r^5}.$$

At the location of $\mathbf{P}$, $\mathbf{r} = r\mathbf{e}_z$, the field produced by $\mathbf{P}_{\text{image}}$ is then

$$\mathbf{E} = \frac{\mathbf{P}_{\text{image}}}{2\pi\varepsilon_0 r^3} = \frac{a^3 P}{\pi\varepsilon_0 r^6}\mathbf{e}_z = \frac{a^3}{\pi\varepsilon_0 r^6}\mathbf{P}.$$

The energy of $\mathbf{P}$ in this field is

$$W = -\mathbf{P} \cdot \mathbf{E} = -\frac{a^3 P^2}{\pi\varepsilon_0 r^6},$$

which gives the force on $\mathbf{P}$ as

$$\mathbf{F} = -\nabla W = -\frac{6a^3 P^2}{\pi\varepsilon_0 r^7}\mathbf{e}_z = -\frac{6a^3 PP}{\pi\varepsilon_0 r^7}.$$

1090

Suppose that the potential between two point charges q_1 and q_2 separated by r were actually $q_1 q_2 e^{-Kr}/r$, instead of $q_1 q_2/r$, with K very small. What would replace Poisson's equation for the electric potential? Give the conceptual design of a null experiment to test for a nonvanishing K. Give the theoretical basis for your design.

(Chicago)

Solution:

With the assumption given, Poisson's equation is to be replaced by

$$\nabla^2\phi + K^2\phi = -4\pi\rho$$

in Gaussian units, where ρ is the charge density.

To test for a nonvanishing K, consider a Faraday cage in the form of a conducting spherical shell S, of volume V, enclosing and with the center at q_1, as shown in Fig. 1.46. Let the radius vector of q_2 be $\mathbf{r}_0'$. Denoting a source point by $\mathbf{r}'$ and a field point by $\mathbf{r}$, use Green's theorem

$$\int_V (\psi\nabla'^2\phi - \phi\nabla'^2\psi)dV' = \oint_S (\psi\nabla'\phi - \phi\nabla'\psi)\cdot d\mathbf{S}'\,.$$

Choose for ϕ the potential interior to S due to q_2, which is external to S, given by

$$\nabla^2\phi + K^2\phi = 0\,,$$

as $\rho = 0$ (being the charge density corresponding to the distribution of q_2) inside S, and for ψ a Green's function $G(\mathbf{r},\mathbf{r}')$ satisfying

$$\nabla^2 G = -4\pi\delta(\mathbf{r}-\mathbf{r}')\,,$$
$$G = 0 \quad \text{on} \quad S\,,$$

where $\delta(\mathbf{r}-\mathbf{r}')$ is Dirac's delta function.

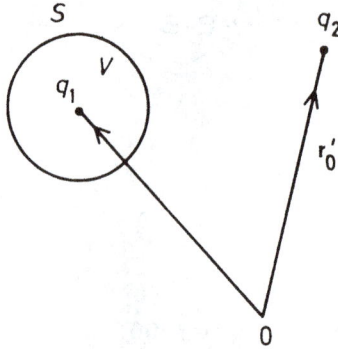

Fig. 1.46

The integrals in the integral equation are as follows:

$$\int_V \psi\nabla'^2\phi dV' = \int_V G\nabla'^2\phi dV' = -K^2\int_V G\phi dV'\,,$$

$$\int_V \phi\nabla'^2\psi dV' = \int_V \phi\nabla'^2 G dV' = -4\pi\int_V \phi(\mathbf{r},\mathbf{r}')\delta(\mathbf{r}-\mathbf{r}')dV' = -4\pi\phi(\mathbf{r})\,,$$

$$\oint_S \psi\nabla'\phi\cdot d\mathbf{S}' = \oint_S G\nabla'\phi\cdot d\mathbf{S}' = 0 \quad \text{as} \quad G = 0 \quad \text{on} \quad S\,,$$

$$\oint_S \phi\nabla'\psi\cdot d\mathbf{S}' = \phi_S\oint_S \nabla'G\cdot d\mathbf{S}' = \phi_S\int_V \nabla'^2 G dV'$$

$$= -4\pi\phi_S\int_V \delta(\mathbf{r}-\mathbf{r}')dV' = -4\pi\phi_S$$

as ϕ =const.= ϕ_S, say, for S a conductor. Note that the divergence theorem has been used in the last equation.

The integral equation then gives

$$\phi(\mathbf{r}) = \frac{1}{4\pi}\left(4\pi\phi_S + K^2\int_V G\phi dV'\right)$$
$$= \phi_S + \frac{K^2}{4\pi}\int_V G(\mathbf{r},\mathbf{r}')\phi(\mathbf{r},\mathbf{r}')dV'.$$

If $K = 0$, then $\phi(\mathbf{r}) = \phi_S$, i.e., the sphere V is an equipotential volume so that no force will be experienced by q_1. If $K \neq 0$, then $\phi(\mathbf{r})$ will depend on $\mathbf{r}$ so that q_1 will experience a force $-q_1\nabla\phi$.

Hence measuring the force on q_1 will determine whether or not $K = 0$.

1091

A very long conducting pipe has a square cross section of its inside surface, with side D as in Fig. 1.47. Far from either end of the pipe is suspended a point charge located at the center of the square cross section.

(a) Determine the eletric potential at all points inside the pipe, perhaps in the form of an infinite series.

(b) Give the asymptotic expression for this potential for points far from the point charge.

(c) Sketch some electric field lines in a region far from the point charge. (Hint: avoid using images.)

(*UC, Berkeley*)

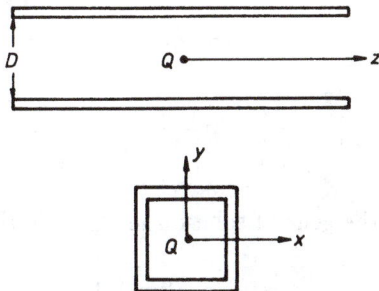

Fig. 1.47

Solution:

(a) Poisson's equation for the potential and the boundary conditions can be written as follows:

$$\left.\begin{aligned}
\nabla^2\varphi &= -\frac{Q}{\varepsilon_0}\delta(x)\delta(y)\delta(z)\,, \\
\varphi\big|_{x=\pm D/2} &= 0\,, \\
\varphi\big|_{y=\pm D/2} &= 0\,.
\end{aligned}\right\}$$

By Fourior transform

$$\bar{\varphi}(x,y,k) = \int_{-\infty}^{\infty} \varphi(x,y,z)e^{ikz}\,dz\,,$$

the above become

$$\left.\begin{aligned}
\left(\frac{\partial^2}{\partial x^2} + \frac{\partial^2}{\partial y^2} - k^2\right)\bar{\varphi}(x,y,k) &= -\frac{Q}{\varepsilon_0}\delta(x)\delta(y)\,, \\
\bar{\varphi}(x,y,k)\big|_{x=\pm D/2} &= 0\,, \\
\bar{\varphi}(x,y,k)\big|_{y=\pm D/2} &= 0\,.
\end{aligned}\right\} \qquad (1)$$

Use $F(\Omega)$ to denote the functional space of the functions which are equal to zero at $x = \pm\frac{D}{2}$ or $y = \pm\frac{D}{2}$. A set of unitary and complete basis in this functional space is

$$\left.\begin{aligned}
\frac{2}{D}\cos\frac{(2m+1)\pi x}{D}\cos\frac{(2m'+1)\pi y}{D}\,, \quad \frac{2}{D}\cos\frac{(2m+1)\pi x}{D}\sin\frac{2n'\pi y}{D}\,, \\
\frac{2}{D}\sin\frac{2n\pi x}{D}\cos\frac{(2m'+1)\pi y}{D}\,, \quad \frac{2}{D}\sin\frac{2n\pi x}{D}\sin\frac{2n'\pi x}{D}\,. \\
m, m' \geq 0\,, \; n, n' \geq 1\,.
\end{aligned}\right\}$$

In this functional space $\delta(x)\delta(y)$ may be expanded as

$$\delta(x)\delta(y) = \left(\frac{2}{D}\right)^2 \sum_{m,m'=0}^{\infty} \cos\frac{(2m+1)\pi x}{D}\cos\frac{(2m'+1)\pi y}{D}\,. \qquad (2)$$

Letting $\bar{\varphi}(x,y,k)$ be the general solution in the following form,

$$\bar{\varphi}(x,y,k) = \sum_{m,m'=0}^{\infty} \bar{\varphi}_{mm'}(k)\cos\frac{(2m+1)\pi x}{D}\cos\frac{(2m'+1)\pi y}{D}\,, \qquad (3)$$

and substituting Eq. (3) into Eq. (1), we find from Eq. (2) that

$$\bar{\varphi}(x,y,k) = \frac{4Q}{\varepsilon D^2} \sum_{m,m'=0}^{\infty} \frac{\cos\frac{(2m+1)\pi x}{D}\cos\frac{(2m'+1)\pi y}{D}}{k^2 + (\frac{(2m+1)\pi}{D})^2 + (\frac{(2m'+1)\pi}{D})^2}.$$

Applying the integral formula

$$\int_{-\infty}^{\infty} \frac{e^{ikz}}{k^2 + \lambda^2} dk = \frac{\pi}{\lambda} e^{-\lambda|z|}, \quad (\lambda > 0),$$

we finally obtain

$$\varphi(x,y,z) = \frac{2Q}{\pi\varepsilon_0 D} \sum_{m,m'=0}^{\infty} \frac{\cos\frac{(2m+1)\pi x}{D}\cos\frac{(2m'+1)\pi y}{D}}{\sqrt{(2m+1)^2 + (2m'+1)^2}}$$
$$\cdot e^{-\frac{\pi|z|}{D}\sqrt{(2m+1)^2+(2m'+1)^2}}.$$

(b) For points far from the point charge we need only choose the terms with $m = m' = 0$ for the potential, i.e.,

$$\varphi = \frac{\sqrt{2}Q}{\varepsilon_0 \pi D} \cos\frac{\pi x}{D} \cdot \cos\frac{\pi y}{D} e^{-\frac{\sqrt{2}\pi}{D}|z|}.$$

(c) For the region $z > 0$, the asymptotic expression of the electric field for $z \gg D$ is

$$\left.\begin{array}{l} E_x = -\frac{\partial\varphi}{\partial x} = \frac{\sqrt{2}Q}{\varepsilon_0 D^2} \sin\frac{\pi x}{D}\cos\frac{\pi y}{D}e^{-\frac{\sqrt{2}\pi}{D}z}, \\[2mm] E_y = -\frac{\partial\varphi}{\partial y} = \frac{\sqrt{2}Q}{\varepsilon_0 D^2}\cos\frac{\pi x}{D}\sin\frac{\pi y}{D}e^{-\frac{\sqrt{2}\pi}{D}z}, \\[2mm] E_z = \frac{2Q}{\varepsilon_0 D^2}\cos\frac{\pi x}{D}\cos\frac{\pi y}{D}e^{-\frac{\sqrt{2}\pi}{D}z}. \end{array}\right\}$$

The electric lines of force far from the point charge are shown in Fig. 1.48.

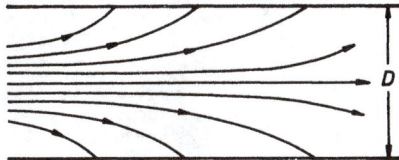

Fig. 1.48

1092

Consider two dipoles $\mathbf{P}_1$ and $\mathbf{P}_2$ separated by a distance d. Find the force between them due to the electrostatic interaction between the two dipole moments, for arbitrary orientation of the directions of $\mathbf{P}_1$ and $\mathbf{P}_2$. For the special case in which $\mathbf{P}_1$ is parallel to the direction between the two dipoles, determine the orientation of $\mathbf{P}_2$ which gives the maximum attraction force.

(*Columbia*)

Solution:

In Fig. 1.49 the radius vector $\mathbf{r}$ is directed from $\mathbf{P}_1$ to $\mathbf{P}_2$. Taking the electric field produced by $\mathbf{P}_1$ as the external field, its intensity at the position of $\mathbf{P}_2$ is given by

$$\mathbf{E}_e = \frac{3(\mathbf{P}_1 \cdot \mathbf{r})\mathbf{r} - r^2\mathbf{P}_1}{4\pi\varepsilon_0 r^5}.$$

Hence the force on $\mathbf{P}_2$ is

$$\mathbf{F} = (\mathbf{P}_2 \cdot \nabla)\mathbf{E}_e$$
$$= \frac{3}{4\pi\varepsilon_0 r^7}\{r^2[(\mathbf{P}_1 \cdot \mathbf{P}_2)\mathbf{r} + (\mathbf{P}_1 \cdot \mathbf{r})\mathbf{P}_2 + (\mathbf{P}_2 \cdot \mathbf{r})\mathbf{P}_1] - 5(\mathbf{P}_1 \cdot \mathbf{r})(\mathbf{P}_2 \cdot \mathbf{r})\mathbf{r}\}.$$

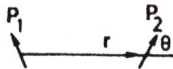

Fig. 1.49

If $\mathbf{P}_1 \| \mathbf{r}$, let $\mathbf{P}_2 \cdot \mathbf{r} = P_2 r \cos\theta$. Then $\mathbf{P}_1 \cdot \mathbf{P}_2 = P_1 P_2 \cos\theta$ and the force between $\mathbf{P}_1$ and $\mathbf{P}_2$ becomes

$$\mathbf{F} = \frac{3}{4\pi\varepsilon_0 r^5}\{-3P_1 P_2 \cos\theta\, \mathbf{r} + P_1 r \mathbf{P}_2\}.$$

The maximum attraction is obviously given by $\theta = 0°$, when $\mathbf{P}_2$ is also parallel to $\mathbf{r}$. This maximum is

$$\mathbf{F}_{max} = -\frac{3P_1 P_2 \mathbf{r}}{2\pi\varepsilon_0 r^5}.$$

Note that the negative sign signifies attraction.

1093

An electric dipole with dipole moment $\mathbf{P}_1 = P_1 \mathbf{e}_z$ is located at the origin of the coordinate system. A second dipole of dipole moment $\mathbf{P}_2 = P_2 \mathbf{e}_z$ is located at (a) on the $+z$ axis a distance r from the origin, or (b) on the $+y$ axis a distance r from the origin. Show that the force between the two dipoles is attractive in Fig. 1.50(a) and repulsive in Fig. 1.50(b). Calculate the magnitude of the force in the two cases.

(*Columbia*)

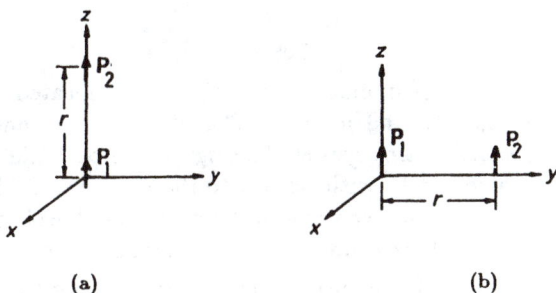

(a)   (b)

Fig. 1.50

Solution:

The electric field produced by $\mathbf{P}_1$ is

$$\mathbf{E}_1 = -\nabla\left(\frac{\mathbf{P}_1 \cdot \mathbf{r}}{4\pi\varepsilon_0 r^3}\right) = -\frac{\mathbf{P}_1}{4\pi\varepsilon_0 r^3} + \frac{3P_1\cos\theta}{4\pi\varepsilon_0 r^3}\mathbf{e}_r,$$

where θ is the angle between $\mathbf{P}_1$ and $\mathbf{r}$. The interaction energy between $\mathbf{P}_2$ and $\mathbf{P}_1$ is

$$W_e = -\mathbf{P}_2 \cdot \mathbf{E}_1 = \frac{P_1 P_2}{4\pi\varepsilon_0 r^3} - \frac{3P_1 P_2}{4\pi\varepsilon_0 r^3}\cos^2\theta.$$

Hence the components of the force acting on $\mathbf{P}_2$ are

$$F_r = -\frac{\partial W_e}{\partial r} = \frac{3P_1 P_2}{4\pi\varepsilon_0 r^4} - \frac{9P_1 P_2}{4\pi\varepsilon_0 r^4}\cos^2\theta,$$

$$F_\theta = -\frac{1}{r}\frac{\partial W_e}{\partial\theta} = -\frac{3P_1 P_2}{2\pi\varepsilon_0 r^4}\sin\theta\cos\theta.$$

(a) In this case $\theta = 0$ and we have

$$F_\theta = 0, \quad F_r = -\frac{3P_1 P_2}{2\pi\varepsilon_0 r^4}.$$

The negative sign denotes an attractive force.

(b) In this case $\theta = \frac{\pi}{2}$ and we have

$$F_\theta = 0, \quad F_r = \frac{3P_1 P_2}{4\pi\varepsilon_0 r^4}.$$

The positive sign denotes a repulsive force.

1094

An electric dipole of moment $\mathbf{P} = (P_x, 0, 0)$ is located at the point $(x_0, y_0, 0)$, where $x_0 > 0$ and $y_0 > 0$. The planes $x = 0$ and $y = 0$ are conducting plates with a tiny gap at the origin. The potential of the plate at $x = 0$ is maintained at V_0 with respect to the plate $y = 0$. The dipole is sufficiently weak so that you can ignore the charges induced on the plates. Figure 1.51 is a sketch of the conductors of constant electrostatic potentials.

(a) Based on Fig. 1.51, deduce a simple expression for the electrostatic potential $\phi(x, y)$.

(b) Calculate the force on the dipole.

(MIT)

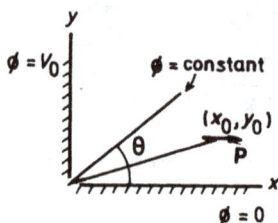

Fig. 1.51

Solution:

(a) Any plane passing through the z-axis is an equipotential surface whose potential only depends on the angle θ it makes with the $y = 0$ plane:

$$\phi(x, y) = \phi(\theta).$$

Accordingly, Laplace's equation is reduced to one dimension only:

$$\frac{d^2\phi}{d\theta^2} = 0,$$

with the solution

$$\phi(\theta) = \frac{2V_0}{\pi}\theta,$$

taking into account the boundary conditions $\phi = 0$ for $\theta = 0$ and $\phi = V_0$ for $\theta = \frac{\pi}{2}$. This can also be written in Cartesian coordinates as

$$\phi(x, y) = \frac{2V_0}{\pi} \arctan\left(\frac{y}{x}\right).$$

(b) The field is then

$$\mathbf{E} = -\nabla\phi = \frac{2V_0}{\pi}\left(\frac{y}{x^2 + y^2}\mathbf{e}_x - \frac{x}{x^2 + y^2}\mathbf{e}_y\right).$$

Hence, the force acting on the dipole $(\mathbf{P}_x, 0, 0)$ is

$$\mathbf{F} = (\mathbf{P} \cdot \nabla)\mathbf{E} = P_x\frac{\partial\mathbf{E}}{\partial x}\bigg|_{x=x_0, y=y_0}$$

$$= \frac{2V_0 P_x}{\pi}\left(-\frac{2x_0 y_0}{(x_0^2 + y_0^2)^2}\mathbf{e}_x + \frac{x_0^2 - y_0^2}{(x_0^2 + y_0^2)^2}\mathbf{e}_y\right).$$

1095

Inside a smoke precipitator a long wire of radius R has a static charge λ Coulombs per unit length. Find the force of attraction between this wire and an uncharged spherical dielectric smoke particle of dielectric constant ε and radius a just before the particle touches the wire (assume $a < R$). Show all work and discuss in physical terms the origin of the force.

(*SUNY, Buffalo*)

Solution:

As $a \ll R$, we can consider the smoke particle to lie in a uniform field. In Gaussian units the field inside a spherical dielectric in a uniform external field is (see Problem 1062)

$$\mathbf{E}_{\text{in}} = \frac{3}{2 + \varepsilon}\mathbf{E}_{\text{ex}}.$$

The small sphere can be considered an electric dipole of moment

$$\mathbf{P} = \frac{4}{3}\pi a^3 \cdot \frac{3(\varepsilon - 1)}{\varepsilon + 2}\mathbf{E}_{\text{ex}} = \frac{4\pi a^3(\varepsilon - 1)}{\varepsilon + 2}\mathbf{E}_{\text{ex}}.$$

The energy of the polarized smoke particle in the external field is

$$W = -\frac{1}{2}\mathbf{P} \cdot \mathbf{E}_{ex} = -\frac{2\pi a^3(\varepsilon - 1)}{\varepsilon + 2}E_{ex}^2 .$$

$\mathbf{E}_{ex}$ is radial from the axis of the wire and is given by Gauss' flux theorem as

$$\mathbf{E}_{ex} = \frac{\lambda}{2\pi r}\mathbf{e}_r .$$

Hence

$$W = -\frac{(\varepsilon - 1)a^3\lambda^2}{2\pi(\varepsilon + 2)r^2} ,$$

and the force exerting on the smoke particle is

$$\mathbf{F} = -\nabla W = -\frac{(\varepsilon - 1)a^3\lambda^2}{\pi(\varepsilon + 2)r^3}\mathbf{e}_r .$$

Just before the particle touches the wire, $r = R$ and the force is

$$\mathbf{F} = -\frac{(\varepsilon - 1)a^3\lambda^2}{\pi(\varepsilon + 2)R^3}\mathbf{e}_r .$$

The negative sign shows that this force is an attraction. This force is caused by the nonuniformity of the radial field since it is given by $-\nabla W$. The polarization of the smoke particle in the external field makes it act like an electric dipole, which in a nonuniform field will suffer an electric force.

5. MISCELLANEOUS APPLICATIONS (1096–1108)

1096

A sphere of radius a has a bound charge Q distributed uniformly over its surface. The sphere is surrounded by a uniform fluid dielectric medium with fixed dielectric constant ε as in Fig. 1.52. The fluid also contains a free charge density given by

$$\rho(\mathbf{r}) = -kV(\mathbf{r}) ,$$

where k is a constant and $V(\mathbf{r})$ is the electric potential at $\mathbf{r}$ relative to infinity.

(a) Compute the potential everywhere, letting $V = 0$ at $\mathbf{r} \to \infty$.

(b) Compute the pressure as function of $\mathbf{r}$ in the dielectric.

(*Princeton*)

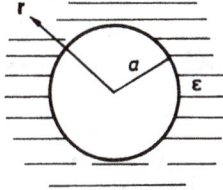

Fig. 1.52

Solution:

The electric potential satisfies Poisson's equation

$$\nabla^2 V(\mathbf{r}) = -\frac{\rho}{\varepsilon} = \frac{kV(\mathbf{r})}{\varepsilon}, \quad r > a. \tag{1}$$

Considering the spherical symmetry of this problem, we have $V(\mathbf{r}) = V(r)$. Equation (1) then becomes

$$\frac{1}{r^2}\frac{d}{dr}\left(r^2\frac{dV}{dr}\right) = \frac{k}{\varepsilon}V(r), \quad r > a.$$

Writing $V = u/r$, one has

$$\frac{d^2 u}{dr^2} = \frac{k}{\varepsilon}u. \tag{2}$$

The solutions of Eq. (2) can be classified according to the values of k:

(1) If $k > 0$, the solution is

$$u = A\exp\left(\pm\sqrt{\frac{k}{\varepsilon}}r\right).$$

Accordingly,

$$V = \frac{A}{r}\exp\left(\pm\sqrt{\frac{k}{\varepsilon}}r\right).$$

The condition $V = 0$ for $r \to \infty$ indicates that only the negative exponent is allowed. Gauss' theorem for the spherical surfaces,

$$-\oint_S \frac{\partial V}{\partial r}dS = Q,$$

then determines the coefficient A as

$$A = \frac{Qe^{\alpha a}}{4\pi(\alpha a + 1)},$$

where $\alpha = \sqrt{k/\varepsilon}$. On the other hand, as there is no electric field inside the sphere, the potential inside is a constant equal to the potential on the surface. Therefore

$$V(r) = \begin{cases} \frac{Qe^{\alpha(a-r)}}{4\pi(\alpha a + 1)r}, & r > a, \\ \frac{Q}{4\pi a(\alpha a + 1)}, & r \leq a. \end{cases}$$

Stability of the fluid means that

$$p\mathbf{n} + \mathbf{n} \cdot \mathbf{T} = \text{const},$$

where $\mathbf{n} = \mathbf{e}_r$, $\mathbf{T}$ is Maxwell's stress tensor. If the fluid is still, the constant is equal to zero and one has

$$p\mathbf{e}_r = -\mathbf{e}_r \cdot \mathbf{T}.$$

As ε is fixed, we further have

$$\mathbf{T} = \mathbf{DE} - \frac{1}{2}(\mathbf{D} \cdot \mathbf{E})\mathbf{I} = \frac{\varepsilon}{2} \begin{pmatrix} (\Delta V)^2 & 0 & 0 \\ 0 & 0 & 0 \\ 0 & 0 & 0 \end{pmatrix}.$$

Hence, the pressure is

$$p = -\frac{\varepsilon}{2}(\nabla V)^2 = -\frac{\varepsilon}{2}\frac{(1 + \alpha r)^2}{r^2}V^2(r).$$

(2) If $k < 0$, with $\beta^2 = -k/\varepsilon$, the solution of Eq. (2) becomes

$$V(r) = \frac{B}{r}e^{i\beta r},$$

with real part

$$V(r) = \frac{B}{r}\cos(\beta r).$$

Substitution in Gauss' theorem

$$-\oint_S \frac{\partial V}{\partial r} dS = Q$$

gives

$$B = \frac{Q}{4\pi(\beta a \sin\beta a + \cos\beta a)} \ .$$

Hence the electric potential is

$$V = \begin{cases} \frac{Q\cos\beta a}{4\pi(\beta a \sin\beta a + \cos\beta a)r} \ , & r > a, \\ \frac{Q}{4\pi a(\beta a \tan\beta a + 1)} \ , & r \le a, \end{cases}$$

and the pressure is

$$p = -\frac{\varepsilon}{2}(\nabla V)^2 = -\frac{\varepsilon}{2}\frac{(\beta r + \tan\beta r + 1)^2}{r^2}V^2(r) \ .$$

1097

Flat metallic plates P, P', and P'' (see Fig. 1.53) are vertical and the plate P, of mass M, is free to move vertically between P' and P''. The three plates form a double parallel-plate capacitor. Let the charge on this capacitor be q. Ignore all fringing-field effects. Assume that this capacitor is discharging through an external load resistor R, and neglect the small internal resistance. Assume that the discharge is slow enough so that the system is in static equilibrium at all times.

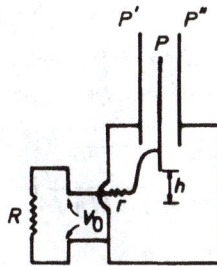

Fig. 1.53

(a) How does the gravitational energy of the system depend on the height h of P?

(b) How does the electrostatic energy of the system depend on h and on the charge q?

(c) Determine h as a function of q.

(d) Does the output voltage increase, decrease, or stay the same as this capacitor discharges?

(Wisconsin)

Solution:

(a) The gravitational energy of the system is

$$W_g = Mgh.$$

(b) We suppose that the distance between P and P' and that between P and P'' are both d. Also suppose that each of the three plates has width a and length l, and when $h = h_0$, the top of plate P coincides with those of plates P' and P''. The system may be considered as two capacitors in parallel, each with charge $q/2$ and capacitance

$$C = \frac{\varepsilon_0 a(l + h - h_0)}{d}$$

when the height of P is h.

The electrostatic energy stored in the system is then

$$W_e = 2 \cdot \frac{1}{2}\left(\frac{q}{2}\right)^2 \frac{1}{C} = \frac{q^2 d}{4\varepsilon_0 a(l + h - h_0)}.$$

(c) The total energy of the system is

$$W = W_g + W_e = \frac{q^2 d}{4\varepsilon_0 a(l + h - h_0)} + Mgh.$$

Since the discharge process is slow, P for each q will adjust to an equilibrium position h where the energy of the system is minimum. Thus for each q, $\frac{\partial W}{\partial h}\big|_q = 0$, giving

$$h = \frac{q}{2}\sqrt{\frac{d}{\varepsilon_0 a M g}} + h_0 - l.$$

Therefore, h varies linearly with q.

(d) As the system is discharging through R, q decreases and h decreases also. Hence the output voltage

$$V_0 = \frac{q}{2}\bigg/\frac{\varepsilon_0 a(l + h - h_0)}{d} = \sqrt{\frac{Mgd}{\varepsilon_0 a}}$$

does not vary with q, i.e., V_0 remains constant as the capacitor discharges.

1098

A capacitor consisting of two plane parallel plates separated by a distance d is immersed vertically in a dielectric fluid of dielectric constant K and density ρ. Calculate the height to which the fluid rises between the plates

(a) when the capacitor is connected to a battery that maintains a constant voltage V across the plates, and

(b) when the capacitor carries a charge Q, but is not connected to a battery.

Explain physically the mechanism of the effect and indicate explicitly how it is incorporated in your calculation. (You may neglect effects of surface tension and the finite size of the capacitor plates.)

(*UC, Berkeley*)

Solution:

When the capacitor is charged, it has a tendency to attract the dielectric fluid. When the electrostatic attraction is balanced by the weight of the excess dielectric fluid, the fluid level will rise no further. As shown in Fig. 1.54, let b be the width and a the length of the plates, x be the height of the capacitor in contact with the fluid, and h be the height to which the fluid between the plates rises from the fluid surface. Then the capacitance of the capacitor (in Gaussian units) is

$$C = \frac{b}{4\pi d}\left[Kx + (a - x)\right] = \frac{b}{4\pi d}\left[(K - 1)x + a\right],$$

where K is the dielectric constant of the fluid.

(a) If the voltage V does not change, as shown in Problem **1051** (a), the dielectric will be acted upon by an upward electrostatic force

$$F_e = \frac{(K - 1)bV^2}{8\pi a}.$$

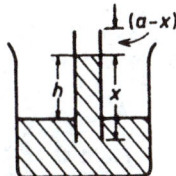

Fig. 1.54

This force is balanced by the weight $mg = \rho g h b d$ in equilibrium. Hence the rise is

$$h = \frac{V^2(K-1)}{8\pi \rho g d^2} .$$

(b) If the charge Q is kept constant instead, then according to Problem **1051** (b) the electrostatic force is

$$F = \frac{Q^2}{2C^2} \frac{dC}{dx} = \frac{2\pi d Q^2 (K-1)}{b[(K-1)x+a]^2} .$$

At equilibrium the fluid level will rise to a height

$$h = \frac{2\pi Q^2 (K-1)}{\rho g b^2 [(K-1)x+a]^2} .$$

1099

A cylindrical capacitor is composed of a long conducting rod of radius a and a long conducting shell of inner radius b. One end of the system is immersed in a liquid of dielectric constant ε and density ρ as shown in Fig. 1.55. A voltage difference V_0 is switched on across the capacitor. Assume that the capacitor is fixed in space and that no conduction current flows in the liquid. Calculate the equilibrium height of the liquid column in the tube.

(*MIT*)

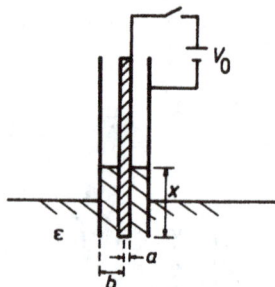

Fig. 1.55

Solution:

Let l be the length of the cylinder, and x the length of the dielectric contained in the cylinder. Neglecting edge effects, the capacitance is

$$C = \frac{2\pi[(\varepsilon - \varepsilon_0)x + \varepsilon_0 l]}{\ln(b/a)}.$$

As the voltage difference V_0 across the capacitor is kept constant, according to Problem **1051** the upward force acting on the dielectric is

$$F = \frac{V_0^2}{2}\frac{dC}{dx} = \frac{\pi(\varepsilon - \varepsilon_0)V_0^2}{\ln(b/a)}.$$

This force is in equilibrium with the gravity force:

$$\frac{\pi(\varepsilon - \varepsilon_0)V_0^2}{\ln(b/a)} = \rho g \cdot \pi(b^2 - a^2)h,$$

giving

$$h = \frac{(\varepsilon - \varepsilon_0)V_0^2}{\rho g(b^2 - a^2)\ln(b/a)}.$$

1100

As in Fig. 1.56, the central plate, bearing total charge Q, can move as indicated but makes a gastight seal where it slides on the walls. The air on both sides of the movable plate is initially at the same pressure p_0. Find value(s) of x where the plate can be in stable equilibrium.

(UC, Berkeley)

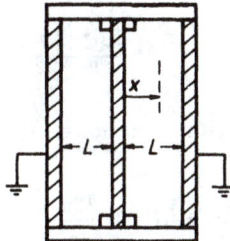

Fig. 1.56

Solution:

Initially, as the voltages on the two sides of the central plate are the same, we can consider the three plates as forming two parallel capacitors with capacitances C_1 and C_2. When the central plate is located at position x, the total capacitance of the parallel capacitors is

$$C = C_1 + C_2 = \frac{A}{\varepsilon_0(L+x)} + \frac{A}{\varepsilon_0(L-x)} = \frac{2AL}{\varepsilon_0(L^2 - x^2)} .$$

Hence the electrostatic energy of the system is

$$W_e = \frac{1}{2}\frac{Q^2}{C} = \frac{\varepsilon_0 Q^2 (L^2 - x^2)}{4AL} .$$

As the charge Q is distributed over the central plate, when the plate moves work is done against the electrostatic force. Hence the latter is given by

$$F_e = -\frac{\partial W_e}{\partial x} = \frac{Q^2 \varepsilon_0 x}{2AL} .$$

As $F > 0$, the force is in the direction of increasing x. As an electric conductor is also a good thermal conductor, the motion of the central plate can be considered isothermal. Let the pressures exerted by air on the left and right sides on the central plate be p_1 and p_2 respectively. We have by Boyle's law

$$p_1 = \frac{p_0 L}{L+x}, \quad p_2 = \frac{p_0 L}{L-x} .$$

When the central plate is in the equilibrium position, the electrostatic force is balanced by the force produced by the pressure difference, i.e.,

$$F_e = (p_2 - p_1)A ,$$

or

$$\frac{Q^2 \varepsilon_0 x}{2AL} = \frac{2Ap_0 Lx}{L^2 - x^2} .$$

This determines the equilibrium positions of the central plate as

$$x = \pm L \left(1 - \frac{4p_0 A^2}{\varepsilon_0 Q^2} \right)^{\frac{1}{2}} .$$

1101

Look at the person nearest to you. If he (or she) is not already spherical, assume that he (or she) is. Assign him (or her) an effective radius R, and recall that he (or she) is a pretty good electrical conductor. The room is in equilibrium at temperature T and is electromagnetically shielded. Make a rough estimate of the rms electrical charge on that person.

(*Princeton*)

Solution:

The capacitance of a conducting sphere of radius R is $C = 4\pi\varepsilon_0 R$. If the sphere carries charge Q, then its electric energy is $Q^2/2C$. According to the classical principle of equipartition of energy,

$$\frac{\overline{Q^2}}{2C} = \frac{1}{2}kT,$$

or

$$\sqrt{\overline{Q^2}} = \sqrt{CkT},$$

where k is Boltzmann's constant.
Taking $R = 0.5$ m, $T = 300$ K, we get

$$\sqrt{\overline{Q^2}} = \sqrt{4\pi\varepsilon_0 RkT}$$
$$= \sqrt{4\pi \times 8.85 \times 10^{-12} \times 0.5 \times 1.38 \times 10^{-23} \times 300}$$
$$= 4.8 \times 10^{-16} \text{ C}.$$

1102

An isolated conducting sphere of radius a is located with its center at a distance z from a (grounded) infinite conductor plate. Assume $z \gg a$ find

(a) the leading contribution to the capacitance between sphere and plane;

(b) the first (non-vanishing) correction to this value, when the capacitance is expressed in terms of a power-series expansion in a/z;

(c) to leading order the force between sphere and plane, when the sphere carries a charge Q. What is the energy of complete separation of the sphere from the plane? How does this energy compare with the energy of complete separation of two such spheres, with charges $+Q$ and $-Q$, initially spaced apart by a distance $2z$? Explain any difference between these two values.

(Columbia)

Solution:

(a) To leading order, we can regard the distance between the conducting sphere and the conductor plane as infinite. Then the capacitance of the whole system corresponds to that of an isolated conducting sphere of radius a, its value being

$$C = 4\pi\varepsilon_0 a .$$

(b) To find the first correction, we consider the field established by a point charge Q at the spherical center and its image charge $-Q$ at distance z from and on the other side of the plane. At a point on the line passing through the spherical center and normal to the plane the magnitude of this field is

$$E = \frac{Q}{4\pi\varepsilon_0(z-h)^2} - \frac{Q}{4\pi\varepsilon_0(z+h)^2} ,$$

where h is the distance from this point to the plane. The potential of the sphere is then

$$V = -\int_0^{z-a} E\,dh = \frac{Q}{4\pi\varepsilon_0(z-h)}\Big|_0^{z-a} - \frac{Q}{4\pi\varepsilon_0(z+h)}\Big|_0^{z-a}$$

$$= \frac{Q}{4\pi\varepsilon_0 a}\left[1 - \frac{a}{2z-a}\right] \approx \frac{Q}{4\pi\varepsilon_0 a}\left(1 - \frac{a}{2z}\right) .$$

Hence the capacitance is

$$C = \frac{Q}{V} \approx 4\pi\varepsilon_0 a\left(1 + \frac{a}{2z}\right)$$

and the first correction is $2\pi\varepsilon_0 a^2/z$.

(c) When the sphere carries charge Q, the leading term of the force between it and the conducting plane is just the attraction between two point charges Q and $-Q$ separated by $2z$. It follows that

$$F = -\frac{Q^2}{16\pi\varepsilon_0 z^2} .$$

The energy required to completely separate the sphere from the plane is

$$W_1 = -\int_z^\infty F\,dr = \int_z^\infty \frac{Q^2}{16\pi\varepsilon_0 r^2}\,dr = \frac{Q^2}{16\pi\varepsilon_0 z} .$$

On the other hand, the energy of complete separation of the two charges Q and $-Q$, initially spaced by a distance $2z$, is

$$W_2 = -\int_{2z}^{\infty} F\, dr = \int_{2z}^{\infty} \frac{Q^2}{4\pi\varepsilon_0 r^2}\, dr = \frac{Q^2}{8\pi\varepsilon_0 z} = 2W_1 .$$

The difference in the required energy is due to the fact that in the first case one has to move Q from z to ∞ while in the second case one has to move Q from z to ∞ and $-Q$ from $-z$ to $-\infty$, with the same force $-Q^2/4\pi\varepsilon_0 r^2$ applying to all the three charges.

1103

A dipole of fixed length $2R$ has mass m on each end, charge $+Q_2$ on one end and $-Q_2$ on the other. It is in orbit around a fixed point charge $+Q_1$. (The ends of the dipole are constrained to remain in the orbital plane.) Figure 1.57 shows the definitions of the coordinates r, θ, α. Figure 1.58 gives the radial distances of the dipole ends from $+Q_1$.

(a) Using the Lagrangian formulation, determine the equations of motion in the (r, θ, α) coordinate system, making the approximation $r \gg R$ when evaluating the potential.

(b) The dipole is in a circular orbit about Q_1 with $\dot{r} \approx \ddot{r} \approx \ddot{\theta} \approx 0$ and $\alpha \ll 1$. Find the period of small oscillations in the α coordinate.

(*Wisconsin*)

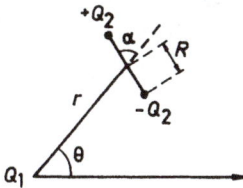

Fig. 1.57

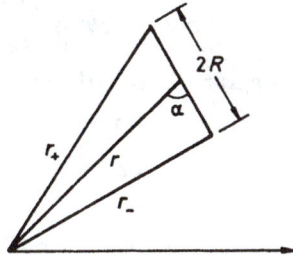

Fig. 1.58

Solution:

(a) The angle between the dipole and the polar axis is $(\theta + \alpha)$, so the angular velocity of the dipole about its center of mass is $(\dot{\theta} + \dot{\alpha})$. The kinetic energy of the dipole is then

$$T = \frac{1}{2} \cdot 2m \cdot (\dot{r}^2 + r^2\dot{\theta}^2) + \frac{1}{2} \cdot 2mR^2 \cdot (\dot{\theta} + \dot{\alpha})^2$$
$$= m\dot{r}^2 + m(r^2 + R^2)\dot{\theta}^2 + mR^2\dot{\alpha}^2 + 2mR^2\dot{\theta}\dot{\alpha}.$$

Moreover, the potential energy of the dipole is

$$V = \frac{1}{4\pi\varepsilon_0}\frac{Q_1Q_2}{r_+} - \frac{1}{4\pi\varepsilon_0}\frac{Q_1Q_2}{r_-}.$$

As

$$r_\pm = \sqrt{r^2 + R^2 \pm 2rR\cos\alpha} = r\sqrt{1 \pm 2\frac{R}{r}\cos\alpha + (\frac{R}{r})^2}$$
$$\approx r\sqrt{1 \pm 2\frac{R}{r}\cos\alpha} \approx r\left(1 \pm \frac{1}{2} \cdot 2\frac{R}{r}\cos\alpha\right) = r \pm R\cos\alpha,$$
$$\frac{1}{r_+} - \frac{1}{r_-} = -\frac{2R\cos\alpha}{r^2 - R^2\cos^2\alpha} \approx -\frac{2R\cos\alpha}{r^2},$$

and the potential energy is

$$V = -\frac{Q_1Q_2}{4\pi\varepsilon_0} \cdot \frac{2R\cos\alpha}{r^2}.$$

The above give the Lagrangian $L = T - V$. Lagrange's equation

$$\frac{d}{dt}\left(\frac{\partial L}{\partial \dot{r}}\right) - \frac{\partial L}{\partial r} = 0$$

gives

$$m\ddot{r} - mr\dot{\theta}^2 + \frac{Q_1Q_2}{4\pi\varepsilon_0} \cdot \frac{2R\cos\alpha}{r^2} = 0; \qquad (1)$$

of the other Lagrange's equations:

$$\frac{d}{dt}\left(\frac{\partial L}{\partial \dot{\theta}}\right) - \frac{\partial L}{\partial \theta} = 0$$

gives

$$(r^2 + R^2)\ddot{\theta} + R^2\ddot{\alpha} + 2mr\dot{r}\dot{\theta} = 0, \qquad (2)$$

and $\frac{d}{dt}\left(\frac{\partial L}{\partial \dot{\alpha}}\right) - \frac{\partial L}{\partial \alpha} = 0$ gives

$$mR(\ddot{\alpha} + \ddot{\theta}) + \frac{Q_1Q_2}{4\pi\varepsilon_0} \cdot \frac{\sin\alpha}{r^2} = 0. \qquad (3)$$

Equations (1)–(3) are the equations of the motion of the dipole.

(b) As $\dot{r} \approx \ddot{r} \approx 0$, r is a constant. Also with $\ddot{\theta} = 0, \alpha \ll 1$ (i.e., $\sin \alpha \approx \alpha$), Eq. (3) becomes

$$m R \ddot{\alpha} + \frac{Q_1 Q_2}{4\pi\varepsilon_0} \cdot \frac{\alpha}{r^2} = 0.$$

This shows that the motion in α is simple harmonic with angular frequency

$$\omega = \sqrt{\frac{Q_1 Q_2}{4\pi\varepsilon_0} \cdot \frac{1}{m R r^2}}.$$

The period of such small oscillations is

$$T = \frac{2\pi}{\omega} = 2\pi \sqrt{\frac{4\pi\varepsilon_0}{Q_1 Q_2} \cdot m R r^2}.$$

1104

The Earth's atmosphere is an electrical conductor because it contains free charge carriers that are produced by cosmic ray ionization. Given that this free charge density is constant in space and time and independent of the horizontal position.

(a) Set up the equations and boundary conditions for computing the atmospheric electric field as a function of altitude if the near-surface field is constant in time and vertical, has no horizontal variation, and has a magnitude of 100 volts/meter. You may assume that the surface of the Earth is perfectly flat if you wish.

(b) Estimate the altitude dependence of the conductivity.

(c) Solve the equations of part (a) above.

(*UC, Berkeley*)

Solution:

(a) This problem is that of a steady field in a conductor. The continuity equation $\nabla \cdot \mathbf{E} + \frac{\partial \rho}{\partial t} = 0$ and Ohm's law $j = \sigma E$ give the basic equation

$$\frac{dj}{dz} = 0, \tag{1}$$

taking the z-axis along the vertical.

The given boundary condition is

$$E|_{z=0} = 100 \text{ V/m}.$$

(b) Since the frequency of collision between a free charge and the atmospheric molecules is proportional to the density of the latter, while the conductivity is inversely proportional to the collision frequency, the conductivity will be inversely proportional to the density of atmospheric molecules. For an isothermal atmosphere the number density of the atmospheric molecules is

$$n = n_0 e^{-\frac{mgz}{kT}},$$

where m is the average mass of a molecule, g is the acceleration of gravity, k is Boltzmann's constant, and T is the absolute temperature. Besides, the conductivity is also proportional to the number density of the free charges. As this density is assumed to be independent of altitude, the altitude dependence of the conductivity can be given as

$$\sigma = \sigma_0 e^{\frac{mgz}{kT}}. \tag{2}$$

(c) Equation (1) gives

$$\frac{dE}{dz} = -\frac{E}{\sigma}\frac{d\sigma}{dz}.$$

Using Eq. (2) and integrating we have

$$E = E_0 e^{-\frac{mgz}{kT}},$$

where $E_0 = 100 \text{ V/m}$.

1105

Two flat plates, each 5 cm in diameter, one copper, one zinc (and both fitted with insulating handles), are placed in contact (see Fig. 1.59) and then briskly separated.

(a) Estimate the maximum charge one might expect to find on each plate after complete ($\gg$ 5 cm) separation.

(b) Volta in experiments of this sort (c.1795) observed charges of the order (in our units) of 10^{-9} Coulomb. Compare this result with your estimate in (a), reconciling any difference.

(c) What charge would be expected if the plates, before separation, were arranged as in Fig. 1.60?

(Columbia)

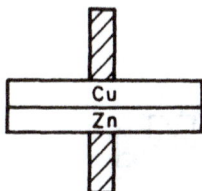

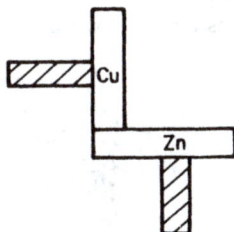

Fig. 1.59   Fig. 1.60

Solution:

(a) When the two plates are in contact, they can be taken as a parallel-plate capacitor. Letting δ be their separation and V be the potential difference, the magnitude of charge on each plate is

$$Q = CV = \frac{\pi \varepsilon_0 (\frac{d}{2})^2 V}{\delta}.$$

As $d = 0.05$ m, taking the contact potential as $V \sim 10^{-3}$ V and $\delta \sim 10^{-10}$ m we obtain

$$Q \approx 1.7 \times 10^{-7} \text{ C}.$$

(b) The above estimated value is greater than the experimental results of Volta ($\approx 10^{-9}$ C). This is probably caused by the following. First of all, due to the roughness of the plates' surfaces, their average separation might be larger than 10^{-10} m. Secondly, in the separating process the charges might accumulate on some ridges of the plates (also because of roughness), so that some of the charges might cancel between the two plates.

(c) According to Fig. 1.60, the contact area is less than that of case (a), hence the corresponding charges after separation will be much diminished.

1106

An ionization chamber is made of a metal cylinder of radius a and length L with a wire of radius b along the cylinder axis. The cylinder is connected to negative high voltage $-V_0$ and the wire is connected to ground by a resistor R, as shown in Fig. 1.61. The ionization chamber is filled with argon at atmospheric pressure. Describe (quantitatively) as a function of time the voltage ΔV across the resistor R for the case where an ionizing particle traverses the tube parallel to the axis at a distance $r = a/2$ from the central axis and creates a total of $N = 10^5$ ion-electron pairs.

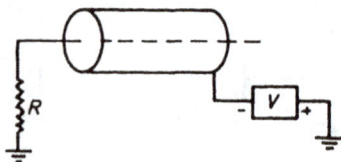

Fig. 1.61

Given: $a= 1$ cm, $b = 0.1$ mm, $L = 50$ cm, $V_0 = 1000$ V, $R = 10^5\Omega$,

mobility of argon ions $\quad \mu_+ = 1.3 \; \dfrac{\text{cm}}{\text{s}} \cdot \dfrac{\text{cm}}{\text{V}}$.

mobility of electrons $\quad \mu_- = 6 \times 10^3 \; \dfrac{\text{cm}}{\text{s}} \cdot \dfrac{\text{cm}}{\text{V}}$.

(Hint: In order to make reasonable approximation, you might have to calculate the RC time constant of this system.)

The voltage (1000 volts) is insufficient to produce ion multiplication near the wire (i.e., this is not a proportional counter).

Note that the detailed shape of the pulse rise is important.

(*Princeton*)

Solution:

Use cylindrical coordinates (r, φ, z) with the z-axis along the cylindrical axis. The electric field at a point (r, φ, z) satisfies $\mathbf{E} \propto \frac{1}{r}\mathbf{e}_r$ according to Gauss' flux theorem. From

$$-\int_b^a E(r)dr = -V_0$$

we get

$$E(r) = \frac{V_0}{r\ln(\frac{a}{b})}\mathbf{e}_r \ .$$

If Q_0 is the charge on the wire, Gauss' theorem gives

$$E(r) = \frac{Q_0}{2\pi\varepsilon_0 Lr}\mathbf{e}_r \ .$$

The capacitance of the chamber is accordingly

$$C = Q_0/V_0 = 2\pi\varepsilon_0 L/\ln(a/b)$$

$$= 2\pi \times 8.85 \times 10^{-12} \times 0.5 \Big/ \ln\left(\frac{1}{0.01}\right)$$

$$= 6 \times 10^{-12} \text{ F}.$$

Hence the time constant of the circuit is

$$RC = 10^5 \times 6 \times 10^{-12} = 6 \times 10^{-7} \text{ s}.$$

The mobility of a charged particle is defined as $\mu = \frac{1}{E}\frac{dr}{dt}$, or $dt = \frac{dr}{\mu E}$. Hence the time taken for the particle to drift from r_1 to r_2 is

$$\Delta t = \int_{r_1}^{r_2} \frac{dr}{\mu_0 \frac{V_0}{r\ln(a/b)}} = \frac{\ln(a/b)}{2\mu V_0}(r_2^2 - r_1^2).$$

For an electron to drift from $r = a/2$ to the wire, we have

$$\Delta t_- \approx \frac{\ln(\frac{a}{b})}{2\mu_- V_0}\left[\left(\frac{a}{2}\right)^2 - b^2\right] \approx \frac{\ln(\frac{a}{b})}{2\mu_- V_0} \cdot \frac{a^2}{4}$$

$$= \frac{\ln 100}{2 \times 6 \times 10^3 \times 10^{-4} \times 1000} \times \frac{10^{-4}}{4} = 9.6 \times 10^{-8} \text{ s}$$

and for a positive ion to reach the cylinder wall, we have

$$\Delta t_+ = \frac{\ln(\frac{a}{b})}{2\mu_+ V_0}\left[a^2 - \left(\frac{a}{2}\right)^2\right] = \frac{\ln\frac{a}{b}}{2\mu_+ V_0} \cdot \frac{3a^2}{4}$$

$$= \frac{\ln 100}{2 \times 1.3 \times 10^{-4} \times 1000} \times \frac{3 \times 10^{-4}}{4} = 1.3 \times 10^{-3} \text{ s}.$$

It follows that $\Delta t_- \ll RC \ll \Delta t_+$. When the electrons are drifting from $r = a/2$ to the anode wire at $r = b$, the positive ions remain essentially stationary at $r = a/2$, and the discharge through resistor R is also negligible. The output voltage ΔV of the anode wire at $t \leq \Delta t_-$ (taking $t = 0$ at the instant when the ionizing particle enters the chamber) can be derived from energy conservation. When a charge q in the chamber displaces by dr, the work done by the field is $q\mathbf{E} \cdot d\mathbf{r}$ corresponding to a decrease of the energy stored in the capacitance of $d(CV^2/2)$. Thus $CVdV = -q\mathbf{E} \cdot d\mathbf{r} = -qEdr$. Since $\Delta V \ll V_0, V \approx V_0$, and we can write

$$CV_0 dV = -qEdr.$$

Integrating, we have

$$CV_0\Delta V = -q \int_{a/2}^{r} Edr = -q \int_{a/2}^{r} \frac{V_0}{r\ln(\frac{a}{b})}dr$$

$$= -q\frac{V_0}{\ln(\frac{a}{b})}\ln\left(\frac{2r}{a}\right).$$

Noting that

$$\Delta t_- \approx \frac{\ln(\frac{a}{b})}{2\mu_- V_0}\left[\left(\frac{a}{2}\right)^2 - b^2\right],$$

we have

$$t = \frac{\ln(\frac{a}{b})}{2\mu_- V_0}\left[\left(\frac{a}{2}\right)^2 - r^2\right],$$

or

$$r = \frac{a}{2}\left\{1 - \frac{t}{\Delta t_-}\left(1 - \frac{2b}{a}\right)^2\right\}^{1/2},$$

and, as $q = -Ne$,

$$\Delta V = \frac{Ne}{C}\ln\left\{1 - \frac{[1 - (\frac{2b}{a})^2]t}{\Delta t_-}\right\}^{1/2} \bigg/ \ln\left(\frac{a}{b}\right), \quad 0 \le t \le \Delta t_-.$$

At $t = \Delta t_-$,

$$\Delta V = \frac{Ne}{C}\ln\left(\frac{2b}{a}\right)\bigg/\ln\left(\frac{a}{b}\right)$$

$$= -\frac{10^5 \times 1.6 \times 10^{-19}}{6 \times 10^{-12}} \times \frac{\ln 50}{\ln 100}$$

$$= -2.3 \times 10^{-3} \text{ V}.$$

This voltage is then discharged through the RC circuit. Therefore, the variation of ΔV with time is as follows:

$$\Delta V = 5.86 \times 10^{-3}\ln\left[1 - \frac{(1 - \frac{1}{50^2})^{1/2}t}{9.6 \times 10^{-8}}\right] \text{ V}, \quad \text{for } 0 \le t \le 9.6 \times 10^{-8} \text{ s};$$

$$\Delta V = -2.3 \times 10^{-3}\exp\left(-\frac{t}{6 \times 10^{-7}}\right) \text{ V}, \quad \text{for } t > 9.6 \times 10^{-8} \text{ s}.$$

This means that the voltage across the two ends of R decreases to -2.3 mV in the time Δt_-, and then increases to zero with the time constant RC. A final remark is that, as the ions drift only slowly and the induced charges on the two electrodes of the chamber are discharged quickly through the *RC* circuit, their influence on the wave form of ΔV can be completely ignored.

1107

An intense energetic electron beam can pass normally through a grounded metal foil. The beam is switched on at $t = 0$ at a current $I = 3 \times 10^6$ amp and a cross sectional area $A = 1000$ cm^2. After the beam has run for 10^{-8} sec, calculate the electric field at the point P on the output face on the foil and near the beam axis due to the space charge of the beam.

(*Wisconsin*)

Solution:

At $t = 10^{-8}$s, the beam forms a charge cylinder on the right side of the foil with a cross sectional area $A = 1000$ cm^2 as shown in Fig. 1.62. The length h of the cylinder is $ct = 3 \times 10^8 \times 10^{-8} = 3$ m, assuming the electrons to have sufficiently high energy so that their speed is close to the velocity of light. We may consider a total charge of

$$-Q = -It = -3 \times 10^6 \times 10^{-8} = -3 \times 10^{-2} \text{ C}$$

being uniformly distributed in this cylinder. As the charge on the left side of the foil does not contribute to the electric field at point P (Shielding effect), the action of the grounded metal foil can be replaced by an image charge cylinder. This image cylinder and the real cylinder are symmetrical with respect to the metal foil and their charges are opposite in sign (see Fig. 1.62).

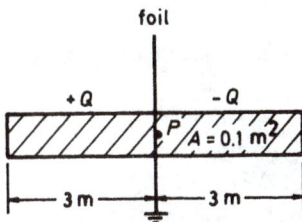

Fig. 1.62

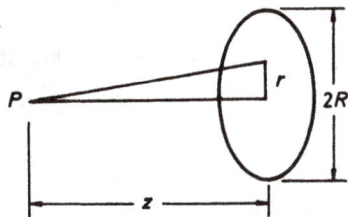

Fig. 1.63

We first calculate the electric field at point P on the axis of a uniformly charge disc of surface charge density σ as shown in Fig. 1.63. The potential is

$$\varphi_P = \frac{1}{4\pi\varepsilon_0} \int_0^R \frac{\sigma \cdot 2\pi r dr}{\sqrt{z^2 + r^2}} = \frac{1}{4\pi\varepsilon_0} \int_0^R \frac{\sigma \pi dr^2}{\sqrt{z^2 + r^2}}$$

$$= \frac{\pi\sigma}{4\pi\varepsilon_0} 2\sqrt{z^2 + r^2}\Big|_0^R = \frac{\sigma}{2\varepsilon_0}\left[\sqrt{z^2 + R^2} - z\right],$$

and the field intensity is

$$E_P = -\frac{\partial\varphi_P}{\partial z} = -\frac{\sigma}{2\varepsilon_0}\left(\frac{z}{\sqrt{z^2 + R^2}} - 1\right).$$

Refer now to Fig. 1.62. The field at the point P produced by the right charge cylinder is

$$E_P = \frac{1}{2\varepsilon_0} \int_0^h \left(\frac{Q}{h\pi R^2} dz\right)\left(\frac{z}{\sqrt{z^2 + R^2}} - 1\right)$$

$$= \frac{Q}{2\pi\varepsilon_0 h R^2} \int_0^h \left(\frac{\frac{1}{2}dz^2}{\sqrt{z^2 + R^2}} - dz\right)$$

$$= \frac{Q}{2\pi\varepsilon_0 h R^2}\left[\sqrt{z^2 + R^2}\Big|_0^h - h\right]$$

$$= \frac{Q}{2\pi\varepsilon_0 h R^2}\left[\sqrt{h^2 + R^2} - R - h\right].$$

Hence the total electric field at P is

$$E_P = \frac{-Q}{\pi\varepsilon_0 h R^2}\left[R + h - \sqrt{R^2 + h^2}\right]$$

$$= \frac{-3 \times 10^{-2}}{\pi \times 8.85 \times 10^{-12} \times 3 \times \frac{0.1}{\pi}} \times \left[3 + \sqrt{\frac{0.1}{\pi}} - \sqrt{3^2 + \frac{0.1}{\pi}}\right]$$

$$= -1.42 \times 10^9 \text{ V/m}.$$

The minus sign indicates that the field intensity points to the right.

1108

The Fig. 1.64 could represent part of a periodic structure of alternating metallic electrodes and gaps found in an electrostatic accelerator. The voltage on any electrode is higher by V_0 than that on the previous electrode. The structure is two-dimensional in that the electrodes have infinite extent in the z direction. The object of this problem is to find the electric field in the region $|y| < W$.

(a) For the purpose of mathematical simplicity we will assume that **E** along lines such as that between points a and b is constant and has no y component. What does this imply about the electrostatic potential between a and b? How might one try to achieve such a boundary condition in practice?

(b) As a guide to subsequent calculations, use physical reasoning to make a sketch of the electric field lines (with directions) in the structure.

(c) Find an expression for the electrostatic potential $\phi(x,y)$ in the region $|y| < W$ as an infinite sum over individual solutions of Laplace's equation.

(d) Find the electric field $\mathbf{E}(x,y)$.

(*MIT*)

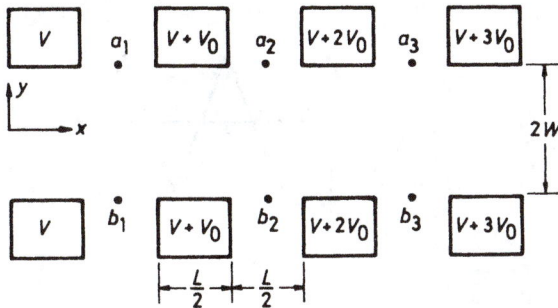

Fig. 1.64

Solution:

(a) Use the coordinates shown in Fig. 1.64. The electric field between the points a_i and b_i is constant, and has no y component. This shows that the electric field lines between a_i and b_i are parallel to the x-axis. Mathematically the potential can be expressed as

$$\varphi(x, \pm W) = \begin{cases} V + nV_0, & x \in \left[\frac{2n}{2}L, \frac{2n+1}{2}L\right], \\ V - nV_0, & x \in \left[\frac{2n-1}{2}L, \frac{2n}{2}L\right], \end{cases}$$

$$= \frac{V_0 x}{L} - \frac{V_0}{4} + V$$

$$+ \begin{cases} nV_0 + \frac{V_0}{4} - \frac{V_0 x}{L}, & x \in \left[\frac{2nL}{2}, \frac{2n+1}{2}L\right] \\ -nV_0 + \frac{V_0}{4} + \frac{V_0 x}{L}, & x \in \left[-\frac{2n-1}{2}L, \frac{2nL}{2}\right] \end{cases} \quad (n = 0, \pm 1, \ldots)$$

$$= \varphi_1(x) + \varphi_2(x).$$

Here $\varphi_2(x)$ represents the sawtooth wave shown in Fig. 1.65. Its Fourier cosine series, of period L, is

$$\varphi_2(x, \pm W) = \frac{2}{L} \sum_{m=1}^{\infty} \cos \frac{2m\pi x}{L} \int_0^L \varphi_2(x, \pm W) \cos \frac{2m\pi x}{L} dx$$

$$= \frac{2V_0}{\pi^2} \sum_{m=0}^{\infty} \frac{1}{(2m+1)^2} \cos \frac{2(2m+1)\pi x}{L}.$$

To achieve these conditions in an electrostatic accelerator, the separation of the electrodes in the y direction must be much greater than L.

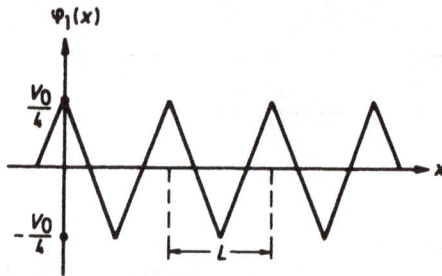

Fig. 1.65

(b) The electric field lines in the accelerator is shown in Fig. 1.66. (Here we only show the electrostatic lines of force between two neighboring electrodes; the pattern repeats itself.)

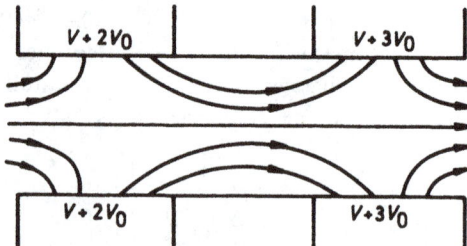

Fig. 1.66

(c) The electric potential in the region $|y| < W$ satisfies the following equation and boundary condition:

$$\begin{cases} \left(\frac{\partial^2}{\partial x^2} + \frac{\partial^2}{\partial y^2}\right)\phi(x,y) = 0, \\ \phi(x, \pm W) = \varphi_1(x) + \varphi_2(x). \end{cases}$$

Defining $\psi(x,y) = \phi(x,y) + V - \frac{V_0}{4} + \frac{V_0 x}{L}$, it satisfies the following equation and boundary condition

$$\begin{cases} \left(\frac{\partial^2}{\partial x^2} + \frac{\partial^2}{\partial y^2}\right)\psi(x,y) = 0, \\ \psi(x, \pm W) = \varphi_2(x). \end{cases}$$

Since $\varphi_2(x)$ in the boundary condition is an even periodic function with period L (see Fig. 1.65), the solution must also be an even periodic function of x. Thus we can write

$$\psi(x,y) = \psi_0(y) + \sum_{m=1}^{+\infty} \cos\left(\frac{2m\pi x}{L}\right)\psi_m(y)$$

and substitute it in the equation for $\psi(x,y)$. We immediately obtain

$$\psi_m(y) = a_m \cosh\frac{2m\pi y}{L} + b_m \sinh\frac{2m\pi y}{L}.$$

Also, substituting $\psi(x,y)$ in the boundary condition and comparing the coefficients, we get

$$\psi(x,y) = \frac{2V_0}{\pi^2}\sum_{n=0}^{+\infty}\frac{1}{(2n+1)^2}\frac{\cosh\left[\frac{2(2n+1)}{L}\pi y\right]}{\cosh\left[\frac{2(2n+1)}{L}\pi W\right]}\cosh\left[\frac{2(2n+1)}{L}\pi x\right].$$

Hence

$$\phi(x,y) = V - \frac{V_0}{L} + \frac{V_0 x}{L} + \frac{2V_0}{\pi^2}\sum_{n=0}^{+\infty}\frac{1}{(2n+1)^2 \cosh[\frac{2(2n+1)}{L}\pi W]}$$
$$\cdot \cosh\left[\frac{2(2n+1)}{L}\pi y\right]\cos\left[\frac{2(2n+1)}{L}\pi x\right].$$

(d) Using $\mathbf{E} = -\nabla\phi$, we obtain the field components

$$\begin{cases} E_x = -\frac{V_0}{L} + \frac{4V_0}{\pi L}\sum_{n=0}^{+\infty}\frac{\cosh[\frac{2(2n+1)}{L}\pi y]\sin[\frac{2(2n+1)}{L}\pi x]}{(2n+1)\cosh[\frac{2(2n+1)}{L}\pi W]}, \\ E_y = -\frac{4V_0}{\pi L}\cdot\sum_{n=0}^{+\infty}\frac{\sinh[\frac{2(2n+1)}{L}\pi y]\cos[\frac{2(2n+1)}{L}\pi x]}{(2n+1)\cosh[\frac{2(2n+1)}{L}\pi W]}. \end{cases}$$

PART 2

MAGNETOSTATIC FIELD AND QUASI-STATIONARY ELECTROMAGNETIC FIELD

1. MAGNETIC FIELD OF CURRENTS (2001–2038)

A cylindrical wire of permeability μ carries a steady current I. If the radius of the wire is R, find **B** and **H** inside and outside the wire.

(Wisconsin)

Solution:

Use cylindrical coordinates with the z-axis along the axis of the wire and the positive direction along the current flow, as shown in Fig. 2.1. On account of the uniformity of the current the current density is

$$\mathbf{j} = \frac{I}{\pi R^2}\mathbf{e}_z \, .$$

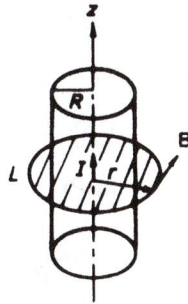

Fig. 2.1

Consider a point at distance r from the axis of the wire. Ampère's circuital law

$$\oint_L \mathbf{H} \cdot d\mathbf{l} = I \, ,$$

where L is a circle of radius r with centre on the z-axis, gives for $r > R$,

$$\mathbf{H}(r) = \frac{I}{2\pi r}\mathbf{e}_\theta \, ,$$

or

$$\mathbf{B}(r) = \frac{\mu_0 I}{2\pi r}\mathbf{e}_\theta$$

since by symmetry $\mathbf{H}(r)$ and $\mathbf{B}(r)$ are independent of θ. For $r < R$,

$$I(r) = \pi r^2 j = \frac{Ir^2}{R^2}$$

147

and the circuital law gives

$$H(r) = \frac{Ir}{2\pi R^2}\mathbf{e}_\theta, \quad B(r) = \frac{\mu Ir}{2\pi R^2}\mathbf{e}_\theta.$$

2002

A long non-magnetic cylindrical conductor with inner radius a and outer radius b carries a current I. The current density in the conductor is uniform. Find the magnetic field set up by this current as a function of radius

(a) inside the hollow space ($r < a$);

(b) within the conductor ($a < r < b$);

(c) outside the conductor ($r > b$).

(*Wisconsin*)

Solution:

Use cylindrical coordinates as in Problem **2001**. The current density in the conductor is

$$j = \frac{I}{\pi(b^2 - a^2)}.$$

The current passing through a cross-section enclosed by a circle of radius r, where $a < r < b$, is

$$I(r) = \pi(r^2 - a^2)j = \frac{I(r^2 - a^2)}{b^2 - a^2}.$$

By symmetry, Ampère's circuital law gives

(a) $\mathbf{B} = 0$, ($r < a$).

(b) $\mathbf{B}(r) = \frac{\mu_0 I}{2\pi r} \cdot \frac{r^2 - a^2}{b^2 - a^2}\mathbf{e}_\theta$, ($a < r < b$).

(c) $\mathbf{B}(r) = \frac{\mu_0 I}{2\pi r}\mathbf{e}_\theta$, ($r > b$).

2003

The direction of the magnetic field of a long straight wire carrying a current is:

(a) in the direction of the current

(b) radially outward

(c) along lines circling the current

(CCT)

Solution:

The answer is (c).

2004

What is the magnetic field due to a long cable carrying 30,000 amperes at a distance of 1 meter?

(a) 3×10^{-3} Tesla, (b) 6×10^{-3} Tesla, (c) 0.6 Tesla.

(CCT)

Solution:

The answer is (b).

2005

A current element idl is located at the origin; the current is in the direction of the z-axis. What is the x component of the field at a point $P(x, y, z)$?

(a) 0, (b) $-iydl/(x^2 + y^2 + z^2)^{3/2}$, (c) $ixdl/(x^2 + y^2 + z^2)^{3/2}$.

(CCT)

Solution:

The answer is (b).

2006

Consider 3 straight, infinitely long, equally spaced wires (with zero radius), each carrying a current I in the same direction.

(a) Calculate the location of the two zeros in the magnetic field.

(b) Sketch the magnetic field line pattern.

(c) If the middle wire is rigidly displaced a very small distance x ($x \ll d$) upward while the other 2 wires are held fixed, describe qualitatively the subsequent motion of the middle wire.

(Wisconsin)

Solution:

(a) Assume the three wires are coplanar, then the points of zero magnetic field must also be located in the same plane. Let the distance of such a point from the middle wire be x. Then the distance of this point from the other two wires are $d \pm x$. Applying Ampère's circuital law we obtain for a point of zero magnetic field

$$\frac{\mu_0 I}{2\pi(d-x)} = \frac{\mu_0 I}{2\pi x} + \frac{\mu_0 I}{2\pi(d+x)} .$$

Two solutions are possible, namely

$$x = \pm\frac{1}{\sqrt{3}}d ,$$

corresponding to two points located between the middle wire and each of the other 2 wires, both having distance $\frac{1}{\sqrt{3}}d$ from the middle wire.

(b) The magnetic field lines are as shown in Fig. 2.2(a).

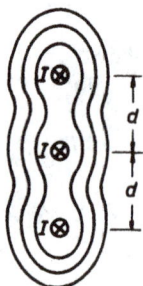

Fig. 2.2(a)

(c) When the middle wire is displaced a small distance x in the same plane, the resultant force per unit length on the wire is

$$f = \frac{\mu_0 I^2}{2\pi(d+x)} - \frac{\mu_0 I^2}{2\pi(d-x)} .$$

As $x \ll d$, this force is approximately

$$f \approx -\frac{\mu_0 I^2}{\pi d^2}x .$$

That is, the force is proportional but opposite to the displacement. Hence, the motion is simple harmonic about the equilibrium position with a period $T = 2\pi \sqrt{\frac{\pi m}{\mu_0} \frac{d}{I}}$, where m is the mass per unit length of the middle wire.

This however is only one of the normal modes of oscillation of the middle wire. The other normal mode is obtained when the wire is displaced a small distance x out from and normal to the plane as shown in Fig. 2.2(b). The resultant force on the wire is in the negative x direction, being

$$f = -2 \cdot \frac{\mu_0 I^2}{2\pi\sqrt{d^2 + x^2}} \frac{x}{\sqrt{d^2 + x^2}} \approx -\frac{\mu_0 I^2}{\pi d^2} x \ .$$

This motion is also simple harmonic with the same period.

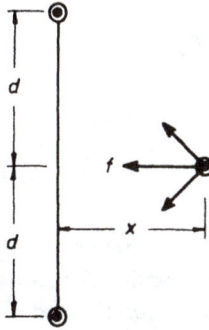

Fig. 2.2(b)

2007

As in Fig. 2.3, an infinitely long wire carries a current $I = 1$ A. It is bent so as to have a semi-circular detour around the origin, with radius 1 cm. Calculate the magnetic field at the origin.

(UC, Berkeley)

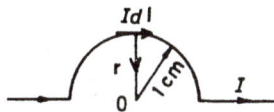

Fig. 2.3

Solution:

The straight parts of the wire do not contribute to the magnetic field at O since for them $I d\mathbf{l} \times \mathbf{r} = 0$. We need only to consider the contribution of the semi-circular part. The magnetic field at O produced by a current element $I d\mathbf{l}$ is

$$d\mathbf{B} = \frac{\mu_0}{4\pi} \frac{I d\mathbf{l} \times \mathbf{r}}{r^3} .$$

As $I d\mathbf{l}$ and $\mathbf{r}$ are mutually perpendicular for the semi-circular wire, $d\mathbf{B}$ is always pointing into the page. The total magnetic field of the semi-circular wire is then

$$B = \int dB = \frac{\mu_0 I}{4\pi r} \int_0^\pi d\theta = \frac{\mu_0 I}{4r} .$$

With $I = 1$ A, $r = 10^{-2}$ m, the magnetic induction at O is

$$B = 3.14 \times 10^{-5} \text{ T} ,$$

pointing perpendicularly into the page.

2008

A semi-infinite solenoid of radius R and n turns per unit length carries a current I. Find an expression for the radial component of the magnetic field $B_r(z_0)$ near the axis at the end of the solenoid where $r \ll R$ and $z = 0$.

(*MIT*)

Solution:

We first find an expression for the magnetic induction at a point on the axis of the solenoid. As shown in Fig. 2.4, the field at point z_0 on the axis is given by

$$B(z_0) = \frac{\mu_0}{4\pi} \int_0^\infty \frac{2\pi R^2 n I dz}{[R^2 + (z - z_0)^2]^{3/2}} .$$

Let $z - z_0 = R \tan \theta$. Then $dz = R \sec^2 \theta d\theta$ and we get

$$B(z_0) = \frac{\mu_0}{4\pi} \int_{-\theta_0}^{\frac{\pi}{2}} \frac{2\pi R^2 n I \cdot R \sec^2 \theta d\theta}{R^3 \sec^3 \theta}$$

$$= \frac{\mu_0}{4\pi} \int_{-\theta_0}^{\frac{\pi}{2}} 2\pi n I \cos \theta d\theta = \frac{\mu_0}{4\pi} \cdot 2\pi n I \sin \theta \Big|_{-\theta_0}^{\frac{\pi}{2}} .$$

As

$$R\tan\theta_0 = z_0 \, ,$$

we have

$$\sin^2\theta_0 = \frac{1}{\cot^2\theta_0 + 1} = \frac{1}{(\frac{R}{z_0})^2 + 1} = \frac{z_0^2}{R^2 + z_0^2} \, ,$$

or

$$\sin\theta_0 = \frac{z_0}{\sqrt{R^2 + z_0^2}} \, .$$

Hence

$$B(z_0) = \frac{1}{2}\mu_0 nI\left(1 + \frac{z_0}{\sqrt{R^2 + z_0^2}}\right) \, .$$

Next, we imagine a short cylinder of thickness dz_0 and radius r along the z-axis as shown in Fig. 2.5. Applying Maxwell's equation

$$\oint \mathbf{B} \cdot d\mathbf{S} = 0$$

to its surface S we obtain

$$[B_z(z_0 + dz) - B_z(z_0)] \cdot \pi r^2 = B_r(z_0)2\pi r dz_0 \, .$$

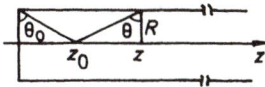

Fig. 2.4

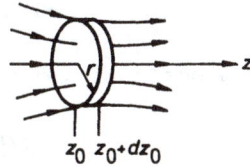

Fig. 2.5

For $r \ll R$, we can take $B_z(z_0) = B(z_0)$. The above equation then gives

$$\frac{dB(z_0)}{dz_0}\pi r^2 = B_r(z_0) \cdot 2\pi r \, ,$$

or

$$B_r(z_0) = \frac{r}{2}\frac{dB(z_0)}{dz_0} = \frac{\mu_0 nIrR^2}{4(R^2 + z_0^2)^{3/2}} \, .$$

At the end of the solenoid, where $z_0 = 0$, the radial component of the magnetic field is

$$B_r(0) = \frac{\mu_0 nIr}{4R} \, .$$

2009

A very long air-core solenoid of radius b has n turns per meter and carries a current $i = i_0 \sin \omega t$.

(a) Write an expression for the magnetic field $\mathbf{B}$ inside the solenoid as a function of time.

(b) Write expressions for the electric field $\mathbf{E}$ inside and outside the solenoid as functions of time. (Assume that $\mathbf{B}$ is zero outside the solenoid.) Make a sketch showing the shape of the electric field lines and also make a graph showing how the magnitude of $\mathbf{E}$ depends on the distance from the axis of the solenoid at time $t = \frac{2\pi}{\omega}$.

(*Wisconsin*)

Solution:

(a) Inside the solenoid the field $\mathbf{B}$ is uniform everywhere and is in the axial direction, i.e.,

$$\mathbf{B}(t) = \mu_0 n i(t)\mathbf{e}_z = \mu_0 n i_0 \sin(\omega t)\mathbf{e}_z .$$

(b) Using $\oint \mathbf{E} \cdot d\mathbf{l} = -\int \frac{\partial \mathbf{B}}{\partial t} \cdot d\mathbf{S}$ and the axial symmetry we can find the electric field at points distance r from the axis inside and outside the solenoid. For $r < b$, one has $E \cdot 2\pi r = -\pi r^2 \frac{dB}{dt} = -\pi r^2 \cdot \mu_0 n i_0 \omega \cos \omega t$, giving

$$E(t) = -\frac{\mu_0}{2} n i_0 \omega r \cos \omega t .$$

For $r > b$, one has $E \cdot 2\pi r = -\pi b^2 \cdot \mu_0 n i_0 \omega \cos \omega t$, giving

$$E(t) = -\frac{b^2}{2r} n i_0 \omega \cos \omega t .$$

In the vector form, we have

$$\mathbf{E}(t) = \begin{cases} -\frac{\mu_0}{2} n i_0 \omega r \cos(\omega t)\mathbf{e}_\theta , & (r < b) \\ -\frac{b^2}{2r} n i_0 \omega \cos(\omega t)\mathbf{e}_\theta , & (r > b) \end{cases}$$

At $t = \frac{2\pi}{\omega}$, $\cos \omega t = 1$ and we have

$$\mathbf{E}\left(\frac{2\pi}{\omega}\right) = \begin{cases} -\frac{\mu_0}{2} n i_0 \omega r \mathbf{e}_\theta & (r < b), \\ -\frac{b^2}{2r} n i_0 \omega \mathbf{e}_\theta & (r > b). \end{cases}$$

The relation between $|E|$ and r is shown in Fig. 2.6. Up to $r = b$ the electric field lines are a series of concentric circles as shown in Fig. 2.7.

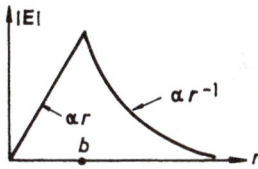

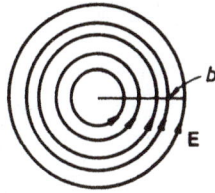

Fig. 2.6   Fig. 2.7

2010

Assume that the earth's magnetic field is caused by a small current loop located at the center of the earth. Given that the field near the pole is 0.8 gauss, that the radius of the earth is $R = 6 \times 10^6$ m, and that $\mu_0 = 4\pi \times 10^{-7}$ H/m, use the Biot-Savart law to calculate the strength of the mangnetic moment of the small current loop.

(*Wisconsin*)

Solution:

Assume that the axis of the current loop, of small radius a, coincides with the axis of rotation of the earth, which is taken to be the z-axis as shown in Fig. 2.8. The contribution of a current element $I d\mathbf{l}$ to the magnetic induction $\mathbf{B}$ at an axial point z is, according to the Biot-Savart law,

$$d\mathbf{B} = \frac{\mu_0}{4\pi} \frac{I d\mathbf{l} \times \mathbf{r}}{r^3} .$$

$d\mathbf{B}$ is in the plane containing the z-axis and $\mathbf{r}$ and is perpendicular to $\mathbf{r}$. Summing up the contributions of all current elements of the loop, by symmetry the resultant $\mathbf{B}$ will be along the z-axis, i.e.,

$$\mathbf{B} = B_z \mathbf{e}_z , \quad \text{or}$$

$$dB_z = dB \cdot \frac{a}{r} .$$

At the pole, $z = R$. As $R \gg a, r = \sqrt{R^2 + a^2} \approx R$ and

$$B_z = \frac{\mu_0}{4\pi} \frac{I a}{R^3} \oint dl = \frac{\mu_0}{4\pi} \frac{I a}{R^3} \cdot 2\pi a$$

$$= \frac{\mu_0}{2\pi} \frac{I S}{R^3} ,$$

where $S = \pi a^2$ is the area of the current loop.

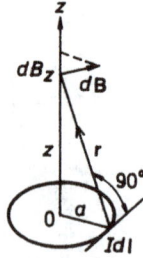

Fig. 2.8

The magnetic moment of the loop is $\mathbf{m} = IS\mathbf{e}_z$, thus

$$m = \frac{2\pi R^3}{\mu_0} B_z .$$

Using the given data $R = 6 \times 10^6$ m, $B_z = 0.8$ Gs, we obtain

$$m \approx 8.64 \times 10^{-26} \text{ Am}^2 .$$

2011

A capacitor (in vacuum) consists of two parallel circular metal plates each of radius r separated by a small distance d. A current i charges the capacitor. Use the Poynting vector to show that the rate at which the electromagnetic field feeds energy into the capacitor is just the time rate of change of the electrostatic field energy stored in the capacitor. Show that the energy input is also given by iV, where V is the potential difference between the plates. Assume that the electric field is uniform out to the edges of the plates.

(Wisconsin)

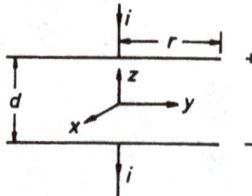

Fig. 2.9

Solution:

Use coordinates as shown in Fig. 2.9. When the positive plate carries charge Q, the electric field between the plates is

$$E = \frac{Q}{\pi r^2 \varepsilon_0}(-e_z).$$

As Q is changing, so is E, producing a magnetic field between the plates. Applying

$$\oint_C H \cdot dl = \int_S \frac{\partial D}{\partial t} \cdot dS$$

to a loop C between the plates, of area S parallel and equal to that of the plates, by symmetry we obtain

$$H \cdot 2\pi r = \frac{dQ}{dt} = i,$$

or

$$H = \frac{i}{2\pi r}(-e_\theta).$$

Hence the Poynting vector of the electromagnetic field is

$$N = E \times H = \frac{Q}{\pi r^2 \varepsilon_0}(-e_z) \times \frac{i}{2\pi r}(-e_\theta) = -\frac{iQ}{2\pi^2 r^3 \varepsilon_0}e_r.$$

Th energy flux enters the capacitor through the curved sides of the capacitor. The flow per unit time is then

$$P = N \cdot 2\pi r d = \frac{iQ}{\pi r^2 \varepsilon_0}d.$$

The electrostatic energy stored in the capacitor is

$$W_e = \frac{1}{2}\varepsilon_0 E^2 \cdot \pi r^2 d = \frac{1}{2}\varepsilon_0 \left(\frac{Q}{\pi r^2 \varepsilon_0}\right)^2 \pi r^2 d = \frac{Q^2 d}{2\pi r^2 \varepsilon_0},$$

and the rate of increase is

$$\frac{dW_e}{dt} = \frac{d}{2\pi r^2 \varepsilon_0} \cdot 2Q\frac{dQ}{dt} = \frac{iQ}{\pi r^2 \varepsilon_0}d.$$

Thus we have

$$P = \frac{dW_e}{dt}.$$

On the other hand, $Q = CV = \frac{\varepsilon_0 \pi r^2}{d}V$, or $\frac{Qd}{\varepsilon_0 \pi r^2} = V$. Hence $P = iV$ also.

2012

A parallel-plate capacitor is made of circular plates as shown in Fig. 2.10. The voltage across the plates (supplied by long resistanceless lead wires) has the time dependence $V = V_0 \cos \omega t$. Assume $d \ll a \ll c/\omega$, so that fringing of the electric field and retardation may be ignored.

(a) Use Maxwell's equations and symmetry arguments to determine the electric and magnetic fields in region I as functions of time.

(b) What current flows in the lead wires and what is the current density in the plates as a function of time?

(c) What is the magnetic field in region II? Relate the discontinuity of **B** accross a plate to the surface current in the plate.

(*CUSPEA*)

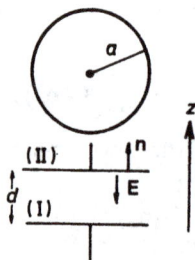

Fig. 2.10

Solution:

(a) Because $d \ll a$, the electric field in region I is approximately $\mathbf{E}^{(I)} = E_z^{(I)} \mathbf{e}_z$, where

$$E_z^{(I)} = -\frac{V_0}{d} \cos \omega t$$

at time t.

Apply Maxwell's equation

$$\oint_L \mathbf{B} \cdot d\mathbf{l} = \mu_0 \varepsilon_0 \int_S \frac{\partial \mathbf{E}}{\partial t} \cdot d\mathbf{S}$$

to a circle L of radius r centered at the line joining the centers of the two plates. By symmetry, $\mathbf{B}^{(I)} = B_\phi^{(I)} \mathbf{e}_\phi$. Thus one has

$$2\pi r B_\phi^{(I)} = \mu_0 \varepsilon_0 \pi r^2 \left(\frac{V_0 \omega}{d} \sin \omega t \right),$$

or

$$B_\phi^{(I)} = \frac{\mu_0 \varepsilon_0 V_0 \omega}{2d} r \sin \omega t .$$

(b) Let σ be the surface charge density of the upper plate which is the interface between regions I and II. We have

$$\sigma = -\varepsilon_0 E_z^{(I)} = \frac{\varepsilon_0 V_0}{d} \cos \omega t .$$

Then the total charge on the plate is

$$Q = \pi a^2 \sigma = \frac{\pi a^2 \varepsilon_0 V_0}{d} \cos \omega t .$$

Note that σ is uniform because $E_z^{(I)}$ is uniform for any instant t. The time variation of Q shows that an alternating current I passes through the lead wires:

$$I = \frac{dQ}{dt} = -\frac{\pi a^2 \varepsilon_0 V_0 \omega}{d} \sin \omega t .$$

As the charge Q on the plate changes continuously with time, there will be surface current flowing in the plate. As shown in Fig. 2.11, this current flows towards the center of the plate along radial directions. The total current flowing through the shaded loop is

$$i = -\frac{d}{dt}\left[\pi(a^2 - r^2)\sigma\right] = \frac{\pi(a^2 - r^2)\varepsilon_0 V_0 \omega}{d} \sin \omega t .$$

Hence, the linear current density (current per unit width) in the plate is

$$j_l(r) = \frac{i}{2\pi r}e_r = \frac{(a^2 - r^2)\varepsilon_0 V_0 \omega}{2dr} \sin(\omega t)e_r .$$

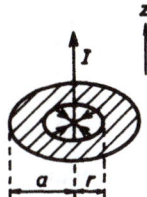

Fig. 2.11

(c) In Ampere's circuital law

$$\oint_L \mathbf{B}^{(II)} \cdot d\mathbf{l} = \mu_0 I \,,$$

the direction of flow of I and the sense of traversing L follow the right-handed screw rule. At time t, I flows along the $-z$-direction and by axial symmetry

$$\mathbf{B}^{(II)} = B_\phi^{(II)} \mathbf{e}_\phi \,.$$

Hence

$$B_\phi^{(II)} = -\frac{\mu_0 I}{2\pi r} = \frac{\mu_0 \varepsilon_0 a^2 V_0 \omega}{2dr} \sin \omega t \,.$$

Thus

$$\mathbf{B}_\phi^{(II)} - \mathbf{B}_\phi^{(I)} = \frac{\mu_0 \varepsilon_0 (a^2 - r^2) V_0 \omega}{2dr} \sin(\omega t) \mathbf{e}_r = \mu_0 j_l \,,$$

or

$$\mathbf{n} \times (\mathbf{B}^{(II)} - \mathbf{B}^{(I)}) = \mu_0 \mathbf{j} \,.$$

This is just the boundary condition for the magnetic field intensity

$$H_t^{(II)} - H_t^{(I)} = j_l \,.$$

2013

A capacitor having circular disc plates of radius R and separation $d \ll R$ is filled with a material having a dielectric constant K_e. A time varying potential $V = V_0 \cos \omega t$ is applied to the capacitor.

(a) As a function of time find the electric field (magnitude and direction) and free surface charge density on the capacitor plates. (Ignore magnetic and fringe effects.)

(b) Find the magnitude and direction of the magnetic field between the plates as a function of distance from the axis of the disc.

(c) Calculate the flux of the Poynting vector from the open edges of the capacitor.

(Wisconsin)

Solution:

This problem is similar to Problem **2012**. The answers for (a) and (b) are

(a) $E = \dfrac{V_0}{d}\cos(\omega t)e_z$, $\sigma = \pm k_e \varepsilon_0 \dfrac{V_0}{d}\cos(\omega t)$,

(b) $B = \dfrac{k_e \varepsilon_0 \mu_0 \omega V_0}{2d} r \sin(\omega t)e_\theta$.

(c) The Poynting vector at $r = R$ is

$$N = \frac{1}{\mu_0}(E \times B)|_{r=R} = -\frac{k_e \varepsilon_0 \omega V_0^2 R}{2d^2}\sin(\omega t)\cos(\omega t)e_r .$$

Thus N is radial in the cylindrical coordinates. Hence the flux of the Poynting vector from the open edges of the capacitor (i.e., the curved surface of a cylinder of height d and radius R) is

$$\phi = 2\pi R d N = \frac{\pi k_e \varepsilon_0 \omega V_0^2 R^2 \sin 2\omega t}{2d} .$$

2014

A parallel plate capacitor has circular plates of radius R and separation $d \ll R$. The potential difference V across the plates varies as $V = V_0 \sin \omega t$. Assume that the electric field between the plates is uniform and neglect edge effects and radiation.

(a) Find the direction and magnitude of the magnetic induction **B** at point P which is at a distance r ($r < R$) from the axis of the capacitor.

(b) Suppose you wish to measure the magnetic field **B** at the point P using a piece of wire and a sensitive high-impedance oscilloscope. Make a sketch of your experimental arrangement and estimate the signal detected by the oscilloscope.

(*Wisconsin*)

Solution:

(a) Referring to Problem **2012**, the magnetic induction at point P is

$$B(r,t) = \frac{\varepsilon_0 \mu_0 V_0 \omega r}{2d}\cos(\omega t)e_\theta .$$

(b) Figure 2.12 shows the experimental arrangement. A small square loop of area ΔS made of a wire, whose two ends are connected to the oscilloscope, is placed at P such that the plane of the loop contains the axis of the capacitor. A sinusoidal wave will appear on the oscilloscope, whose amplitude and frequency are measured. These correspond to the amplitude and frequency of the electromotive force ε. Then from

$$|\varepsilon| = \left|\frac{\partial \phi}{\partial t}\right| = \omega |\mathbf{B}| \Delta S$$

we can find the amplitude of the magnetic induction $\mathbf{B}$.

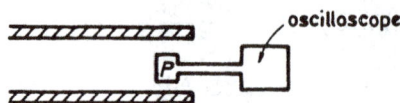

Fig. 2.12

2015

What is the drift velocity of electrons in a 1 mm Cu wire carrying 10 A? $10^{-5}, 10^{-2}, 10^{1}, 10^{5}$ cm/sec.

(*Columbia*)

Solution:

It is 10^{-2}cm/sec.

2016

What is the average random speed of electrons in a conductor? $10^{2}, 10^{4}, 10^{6}, 10^{8}$ cm/sec.

(*Columbia*)

Solution:

It is 10^{6} cm/sec.

2017

Which is the correct boundary condition in magnetostatics at a boundary between two different media?

(a) The component of **B** normal to the surface has the same value.

(b) The component of **H** normal to the surface has the same value.

(c) The component of **B** parallel to the surface has the same value.

$$(CCT)$$

Solution:

The answer is (a).

2018

A system of conductors has a cross section given by the intersection of two circles of radius b with centers separated by $2a$ as shown in Fig. 2.13. The conducting portion is shown shaded, the unshaded lens-shaped region being a vacuum. The conductor on the left carries a uniform current density J going into the page, and the conductor on the right carries a uniform current density J coming out of the page. Assume that the magnetic permeability of the conductor is the same as that of the vacuum. Find the magnetic field at all points x, y in the vacuum enclosed betwen the two conductors.

$$(MIT)$$

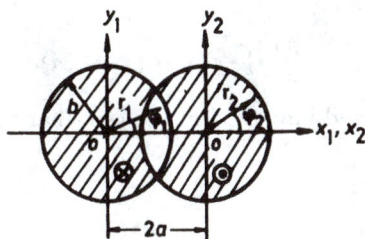

Fig. 2.13

Solution:

As the magnetic permeability of the conductor is the same as that of the vacuum, we can think of the lens-shaped region as being filled with the same conductor without affecting either the magnetic property of the conducting system or the distribution of the magnetic field. We can then consider this region as being traversed by two currents of densities $\pm J$, i.e., having the same magnitude but opposite directions. Thus we have two cylindrical conductors, each having a uniform current distribution, and the magnetic induction in the region is the sum of their contributions. In their own cylindrical coordinates Ampère's circuital law yields

$$\mathbf{B}_1 = -\frac{\mu_0}{2} J r_1 \mathbf{e}_{\varphi_1}, \quad (r_1 \leq b)$$

$$\mathbf{B}_2 = \frac{\mu_0}{2} J r_2 \mathbf{e}_{\varphi_2}. \quad (r_2 \leq b)$$

As

$$\mathbf{e}_{\varphi_1} = (-\sin\varphi_1, \cos\varphi_1) \begin{pmatrix} \mathbf{e}_x \\ \mathbf{e}_y \end{pmatrix}, \quad \mathbf{e}_{\varphi_2} = (-\sin\varphi_2, \cos\varphi_2) \begin{pmatrix} \mathbf{e}_x \\ \mathbf{e}_y \end{pmatrix},$$

we have

$$\mathbf{B}_1 = \frac{\mu_0}{2} J(y_1 \mathbf{e}_x - x_1 \mathbf{e}_y),$$
$$\mathbf{B}_2 = \frac{\mu_0}{2} J(-y_2 \mathbf{e}_x + x_2 \mathbf{e}_y).$$

Using the transformation

$$\begin{cases} x_2 = x_1 - 2a \\ y_2 = y_1 \end{cases}$$

we have

$$\mathbf{B}_2 = \frac{\mu_0}{2} J[-y_1 \mathbf{e}_x + (x_1 - 2a)\mathbf{e}_y].$$

Hence the magnetic field induction in the lens-shaped region is

$$\mathbf{B} = \mathbf{B}_1 + \mathbf{B}_2 = \frac{\mu_0}{2} J[(y_1 - y_2)\mathbf{e}_x + (x_1 - x_2 - 2a)\mathbf{e}_y] = -\mu_0 a J \mathbf{e}_y.$$

This means that the field is uniform and is in the $-\mathbf{e}_y$ direction.

2019

A cylindrical thin shell of electric charge has length l and radius a, where $l \gg a$. The surface charge density on the shell is σ. The shell rotates about its axis with an angular velocity ω which increases slowly with time as $\omega = kt$, where k is a constant and $t \geq 0$, as in Fig. 2.14. Neglecting fringing effects, determine:

(a) The magnetic field inside the cylinder.

(b) The electric field inside the cylinder.

(c) The total electric field energy and the total magnetic field energy inside the cylinder.

(Wisconsin)

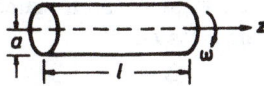

Fig. 2.14

Solution:

(a) Use cylindrical coordinates (ρ, φ, z) with the z-axis along the axis of the cylinder. The surface current density (surface current per unit width) on the cylindrical shell is $\alpha = \sigma\omega a e_\varphi$. It can be expressed as a volume current density (current per unit cross-sectional area) $\mathbf{J} = \sigma\omega a\delta(\rho - a)e_\varphi$. By symmetry we have $\mathbf{B} = B_z(\rho)e_z$. Then the equation $\nabla \times \mathbf{B} = \mu_0\mathbf{J}$ reduces to

$$-\frac{\partial B_z}{\partial \rho} = \mu_0\sigma\omega a\delta(\rho - a),$$

which gives

$$\mathbf{B}(\rho) = \mu_0\sigma\omega a e_z, \quad (\rho < a).$$

(b) Apply Maxwell's equation

$$\oint_L \mathbf{E} \cdot d\mathbf{l} = -\frac{d}{dt}\int_S \mathbf{B} \cdot d\mathbf{S}$$

to a circle of radius ρ in a plane perpendicular to the z-axis and with the center at the axis. On this circle, $\mathbf{E}$ is tangential and has the same magnitude, i.e., $\mathbf{E} = E(\rho)e_\varphi$. Hence, noting that $\omega = kt$ we have

$$\mathbf{E}(\rho) = -\frac{\mu_0\sigma a k\rho}{2}e_\varphi. \quad (\rho < a)$$

(c) $U_E = \int \frac{1}{2}\varepsilon_0 E^2 dV = \frac{1}{2}\varepsilon_0 l \int_0^a \left(\frac{\mu_0\sigma a k\rho}{2}\right)^2 2\pi\rho d\rho = \frac{\pi\varepsilon_0\mu_0^2\sigma^2 k^2 a^6 l}{16}.$

$$U_B = \int \frac{B^2}{2\mu_0}dV = \frac{1}{2\mu_0}(\mu_0\sigma\omega a)^2 \cdot \pi a^2 l = \frac{\pi\mu_0\sigma^2 a^4 k^2 l t^2}{2}.$$

2020

A long, solid dielectric cylinder of radius a is permanently polarized so that the polarization is everywhere radially outward, with a magnitude proportional to the distance from the axis of the cylinder, i.e., $\mathbf{P} = \frac{1}{2}P_0 r \mathbf{e}_r$.

(a) Find the charge density in the cylinder.

(b) If the cylinder is rotated with a constant angular velocity ω about its axis without change in $\mathbf{P}$, what is the magnetic field on the axis of the cylinder at points not too near its ends?

(SUNY, Buffalo)

Solution:

(a) Using cylindrical coordinates (r, θ, z), we have $P = P_r = P_0 r/2$. The bound charge density is

$$\rho = -\nabla \cdot \mathbf{P} = -\frac{1}{r}\frac{\partial}{\partial r}\left(r \cdot \frac{P_0 r}{2}\right) = -P_0 .$$

(b) As $\boldsymbol{\omega} = \omega \mathbf{e}_z$, the volume current density at a point $\mathbf{r} = r\mathbf{e}_r + z\mathbf{e}_z$ in the cylinder is

$$\mathbf{j}(\mathbf{r}) = \rho \mathbf{v} = \rho \boldsymbol{\omega} \times \mathbf{r} = -P_0 \omega \mathbf{e}_z \times (r\mathbf{e}_r + z\mathbf{e}_z) = -P_0 \omega r \mathbf{e}_\theta .$$

On the surface of the cylinder there is also a surface charge distribution, of density

$$\sigma = \mathbf{n} \cdot \mathbf{P} = \mathbf{e}_r \cdot \frac{P_0 r}{2}\Big|_{r=a} = \frac{P_0 a}{2} .$$

This produces a surface current density of

$$\boldsymbol{\alpha} = \sigma \mathbf{v} = \frac{P_0}{2}\omega a^2 \mathbf{e}_\theta .$$

To find the magnetic field at a point on the axis of the cylinder not too near its ends, as the cylinder is very long we can take this point as the origin and regard the cylinder as infinitely long. Then the magnetic induction at the origin is given by

$$\mathbf{B} = -\frac{\mu_0}{4\pi}\left(\int_V \frac{\mathbf{j}(\mathbf{r}') \times \mathbf{r}'}{r'^3}dV' + \int_S \frac{\boldsymbol{\alpha}(\mathbf{r}') \times \mathbf{r}'}{r'^3}dS'\right),$$

where V and S are respectively the volume and curved surface area of the cylinder and $\mathbf{r}' = (r, \theta, z)$ is a source point. Note the minus sign arises because $\mathbf{r}'$ directs from the field point to a source point, rather than the other way around. Consider the volume integral

$$\int_V \frac{\mathbf{j}(\mathbf{r}') \times \mathbf{r}'}{r'^3} dV' = \int_V \frac{-P_0\omega r\mathbf{e}_\theta \times (r\mathbf{e}_r + z\mathbf{e}_z)}{(r^2 + z^2)^{3/2}} rdrd\theta dz$$

$$= P_0\omega\left[\int_V \frac{r^3 drd\theta dz}{(r^2 + z^2)^{3/2}}\mathbf{e}_z - \int_V \frac{r^2 drd\theta dz}{(r^2 + z^2)^{3/2}}\mathbf{e}_r\right].$$

As the cylinder can be considered infinitely long, by symmetry the second integral vanishes. For the first integral we put $z = r\tan\beta$. We then have

$$\int_V \frac{\mathbf{j} \times \mathbf{r}'}{r'^3}dV' = P_0\omega\int_0^{2\pi} d\theta \int_0^a rdr \int_{-\frac{\pi}{2}}^{\frac{\pi}{2}} \cos\beta\, d\beta\, \mathbf{e}_z = 2\pi P_0\omega a^2\mathbf{e}_z.$$

Similarly, the surface integral gives

$$\int_S \frac{\alpha(\mathbf{r}') \times \mathbf{r}'}{r'^3}dS' = \int_S \frac{\frac{P_0}{2}\omega a^2\mathbf{e}_\theta \times (a\mathbf{e}_r + z\mathbf{e}_z)}{(a^2 + z^2)^{3/2}}dS'$$

$$= -\frac{P_0}{2}\omega a^4\int_S \frac{d\theta dz}{(a^2 + z^2)^{3/2}}\mathbf{e}_z = -2\pi P_0\omega a^2\mathbf{e}_z.$$

Hence, the magnetic induction **B** vanishes at points of the cylindrical axis not too near the ends.

2021

A cylinder of radius R and infinite length is made of permanently polarized dielectric. The polarization vector **P** is everywhere proportional to the radial vector $\mathbf{r}$, $\mathbf{P} = a\mathbf{r}$, where a is a positve constant. The cylinder rotates around its axis with an angular velocity ω. This is a non-relativistic problem $-\omega R \ll c$.

(a) Find the electric field **E** at a radius r both inside and outside the cylinder.

(b) Find the magnetic field **B** at a radius r both inside and outside the cylinder.

(c) What is the total electromagnetic energy stored per unit length of the cylinder,

(i) before the cyliner started spinning?

(ii) while it is spinning?

Where did the extra energy come from?

(*UC, Berkeley*)

Solution:

(a) Use cylindrical coordinats (r, θ, z) with the axis of the cylinder along the z direction, the rotational angular velocity of the cylinder is $\boldsymbol{\omega} = \omega \mathbf{e}_z$. The volume charge density inside the cylinder is

$$\rho = -\nabla \cdot \mathbf{P} = -\nabla \cdot (ar) = -2a.$$

The surface charge density on the cylinder is then

$$\sigma = \mathbf{n} \cdot \mathbf{P} = \mathbf{e}_r \cdot (ar)|_{r=R} = aR.$$

The total charge per unit length is therefore $-2a \cdot \pi R^2 + 2\pi R \cdot aR = 0$. Thus the net total charge of the cylinder is zero. From Gauss' flux theorem $\oint \mathbf{E} \cdot d\mathbf{S} = Q/\varepsilon_0$ and the axial symmetry we find that

$$\mathbf{E} = \begin{cases} -\frac{ar}{\varepsilon_0}\mathbf{e}_r, & r < R, \\ 0 & r > R. \end{cases}$$

(b) The volume current density is $\mathbf{j} = \rho\mathbf{v} = -2a\omega r\mathbf{e}_\theta$, and the surface current density is $\boldsymbol{\alpha} = \sigma\mathbf{v} = a\omega R^2\mathbf{e}_\theta$. If the cylinder is infinitely long, by symmetry $\mathbf{B} = B(r)\mathbf{e}_z$. The equation and boundary condition to be satisfied by $\mathbf{B}$ are

$$\nabla \times \mathbf{B} = \mu_0\mathbf{j}, \quad \mathbf{n} \times (\mathbf{B}_2 - \mathbf{B}_1) = \mu_0\boldsymbol{\alpha}.$$

Here $\mathbf{B}_1$ and $\mathbf{B}_2$ are the magnetic inductions inside and outside the cylinder, respectively. The equation gives

$$-\frac{\partial B_1}{\partial r} = -2\mu_0 a\omega r, \quad -\frac{\partial B_2}{\partial r} = 0.$$

Thus B_2 is a constant. As $B_2 \to 0$ for $r \to \infty$, the constant is zero. The boundary conditon at $r = R$,

$$-[B_2(R) - B_1(R)] = \mu_0 a\omega R^2,$$

yields

$$B_1(R) = \mu_0 a\omega R^2.$$

Integrating the differential equation for B_1 from r to R, we obtain

$$B_1(r) = B_1(R) - \mu_0 a\omega(R^2 - r^2) = \mu_0 a\omega r^2 .$$

Hence the magnetic fields inside and outside the cylinder are

$$\mathbf{B} = \begin{cases} \mu_0 a\omega r^2 \mathbf{e}_z & , \quad r < R, \\ 0 & , \quad r > R. \end{cases}$$

(c) (i) Before the cylinder starts spinning, only the electric energy exists, the total being

$$W_e = \int_\infty \frac{\varepsilon_0}{2} E^2 dV .$$

So the energy stored per unit length of the cylinder is

$$\frac{dW_e}{dz} = \int_0^{2\pi} d\theta \int_0^R \frac{\varepsilon_0}{2} \left(\frac{ar}{\varepsilon_0}\right)^2 r\,dr = \frac{\pi a^2 R^2}{4\varepsilon_0} .$$

(ii) When the cylinder is spinning, both electric and magnetic energies exist. The electric energy is the same as for case (i), and the magnetic energy stored per unit length of the cylinder is

$$\frac{dW_m}{dz} = \int_\infty \frac{B^2}{2\mu_0} dV = \int_0^{2\pi} d\theta \int_0^R \frac{1}{2\mu_0} (\mu_0 a\omega r^2)^2 r\,dr = \frac{\pi \mu_0 a^2 \omega^2 R^6}{6} .$$

Therefore the total energy stored per unit length is

$$\frac{dW}{dz} = \frac{dW_e}{dz} + \frac{dW_m}{dz} = \frac{\pi a^2 R^4}{4\varepsilon_0} + \frac{\mu_0 \pi a^2 \omega^2 R^6}{6}$$

$$= \frac{\pi a^2 R^4}{4\varepsilon_0} \left[1 + \frac{2\omega^2 R^2}{3c^2}\right] .$$

The extra energy, the magnetic energy, comes from the work done by external agency to initiate the rotation of the cylinder from rest.

2022

A long coaxial cable consists of a solid inner cylindrical conductor of radius R_1 and a thin outer cylindrical conducting shell of radius R_2. At one end the two conductors are connected together by a resistor and at the other end they are connected to a battery. Hence, there is a current i in the conductors and a potential difference V between them. Neglect the resistance of the cable itself.

(a) Find the magnetic field **B** and the electric field **E** in the region $R_2 > r > R_1$, i.e., between the conductors.

(b) Find the magnetic energy and electric energy per unit length in the region between the conductors.

(c) Assuming that the magnetic energy in the inner conductor is negligible, find the inductance per unit length L and the capacitance per unit length C.

(*Wisconsin*)

Solution:

(a) Use cylindrical coordinates (r, θ, z) where the z axis is along the axis of the cable and its positive direction is the same as that of the current in the inner conductor. From $\oint_C \mathbf{B} \cdot d\mathbf{l} = \mu_0 i$ and the axial symmetry we have

$$\mathbf{B} = \frac{\mu_0 i}{2\pi r} \mathbf{e}_\theta .$$

From $\oint_S \mathbf{E} \cdot d\mathbf{S} = \frac{Q}{\varepsilon_0}$ and the axial symmetry we have

$$\mathbf{E} = \frac{\lambda}{2\pi\varepsilon_0 r} \mathbf{e}_r ,$$

where λ is the charge per unit length of the inner conductor. The voltage between the conductors is $V = -\int_{R_2}^{R_1} \mathbf{E} \cdot d\mathbf{r}$, giving

$$\lambda = 2\pi\varepsilon_0 V / \ln \frac{R_2}{R_1} .$$

Accordingly,

$$\mathbf{E} = \frac{V}{r \ln \frac{R_2}{R_1}} \mathbf{e}_r .$$

(b) The magnetic energy density is $w_m = \frac{B^2}{2\mu_0} = \frac{\mu_0}{2}\left(\frac{i}{2\pi r}\right)^2$. Hence the magnetic energy per unit length is

$$\frac{dW_m}{dz} = \int w_m \, dS = \int_{R_1}^{R_2} \frac{\mu_0}{2}\left(\frac{i}{2\pi r}\right)^2 \cdot 2\pi r \, dr = \frac{\mu_0 i^2}{4\pi} \ln \frac{R_2}{R_1} .$$

The electric energy density is $w_e = \frac{\varepsilon_0 E^2}{2} = \frac{\varepsilon_0}{2}(\frac{V}{2\ln\frac{R_2}{R_1}})^2$. Hence the electric energy per unit length is

$$\frac{dW_e}{dz} = \int w_e dS = \frac{\pi\varepsilon_0 V^2}{\ln\frac{R_2}{R_1}}.$$

(c) From $\frac{dW_m}{dz} = \frac{1}{2}(\frac{dL}{dz})i^2$, the inductance per unit length is

$$\frac{dL}{dz} = \frac{\mu_0}{2\pi}\ln\frac{R_2}{R_1}.$$

From $\frac{dW_e}{dz} = \frac{1}{2}(\frac{dC}{dz})V^2$, the capacitance per unit length is

$$\frac{dC}{dz} = \frac{2\pi\varepsilon_0}{\ln\frac{R_2}{R_1}}.$$

2023

The conductors of a coaxial cable are connected to a battery and resistor as shown in Fig. 2.15. Starting from first principles find, in the region between r_1 and r_2,

(a) the electric field in terms of V, r_1 and r_2,

(b) the magnetic field in terms of V, R, r_1 and r_2,

(c) the Poynting vector.

(d) Show by integrating the Poynting vector that the power flow between r_1 and r_2 is V^2/R.

(*Wisconsin*)

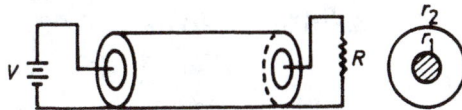

Fig. 2.15

Solution:

(a), (b) Referring to Problem **2022**, we have

$$E = \frac{V}{r\ln\frac{r_2}{r_1}}e_r, \quad B = \frac{\mu_0 I}{2\pi r}e_\theta.$$

As $I = V/R$

$$\mathbf{B} = \frac{\mu_0 V}{2\pi r R} \mathbf{e}_\theta .$$

(c) $\mathbf{N} = \mathbf{E} \times \mathbf{H} = \mathbf{E} \times \dfrac{\mathbf{B}}{\mu_0} = \dfrac{V}{r \ln \frac{r_2}{r_1}} \mathbf{e}_r \times \dfrac{V}{2\pi r R} \mathbf{e}_\theta = \dfrac{V^2}{2\pi r^2 R \ln \frac{r_2}{r_1}} \mathbf{e}_z .$

(d) $P = \int_{r_1 < r < r_2} \mathbf{N} \cdot d\mathbf{S} = \int_{r_1}^{r_2} \dfrac{V^2}{2\pi R \ln \frac{r_2}{r_1}} \dfrac{1}{r^2} \cdot 2\pi r dr = \dfrac{V^2}{R} .$

2024

Suppose the magnetic field on the axis of a right circular cylinder is given by

$$\mathbf{B} = B_0(1 + \nu z^2)\mathbf{e}_z .$$

Suppose the θ-component of $\mathbf{B}$ is zero inside the cylinder.

(a) Calculate the radial component of the field $B_r(r, z)$ for points near the axis.

(b) What current density $j(r, z)$ is required inside the cylinder if the field described above is valid for all radii r?

(*Wisconsin*)

Solution:

(a) As in Fig. 2.16, consider a small cylinder of thickness dz and radius r at and perpendicular to the z-axis and apply Maxwell's equation $\oint_S \mathbf{B} \cdot d\mathbf{S} = 0$. As r is very small, we have

$$B_z(r, z) \approx B_z(0, z) .$$

Hence

$$[B_z(0, z + dz) - B_z(0, z)]\pi r^2 + B_r(r, z)2\pi r dz = 0 ,$$

or

$$-\frac{\partial B(0, z)}{\partial z} dz \cdot \pi r^2 = B_r(r, z) \cdot 2\pi r dz ,$$

giving

$$B_r(r, z) = -\frac{r}{2} \frac{\partial B(0, z)}{\partial z} = -\frac{r}{2} \frac{d}{dz}[B_0(1 + \nu z^2)] = -\nu B_0 rz .$$

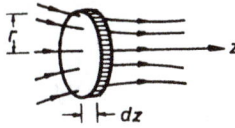

Fig. 2.16

(b) Suppose the following are valid everywhere:

$$B_r(r, z) = -\nu B_0 r z \,,$$
$$B_z(r, z) = B_0(1 + \nu z^2) \,.$$

For a conductor $\dot{\mathbf{D}}$ can be neglected and Maxwell's equation reduces to $\mathbf{j} = \frac{1}{\mu_0} \nabla \times \mathbf{B}$. Noting that $B_\theta = 0, \frac{\partial B_r}{\partial \theta} = \frac{\partial B_z}{\partial \theta} = 0$, we have

$$\mathbf{j}(r, z) = \frac{1}{\mu_0}\left[\frac{\partial B_r}{\partial z} - \frac{\partial B_z}{\partial r}\right]\mathbf{e}_\theta = \frac{1}{\mu_0}\frac{\partial B_r}{\partial z}\mathbf{e}_\theta = -\frac{1}{\mu_0}\nu B_0 r \mathbf{e}_\theta \,.$$

This is the current density required.

2025

A toroid having an iron core of square cross section (Fig. 2.17) and permeability μ is wound with N closely spaced turns of wire carrying a current I. Find the magnitude of the magnetization M everywhere inside the iron.

(*Wisconsin*)

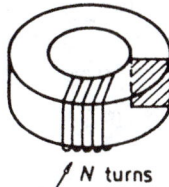

N turns

Fig. 2.17

Solution:

According to Ampère's circuital law

$$\oint \mathbf{H} \cdot d\mathbf{l} = NI ,$$

$$H = \frac{NI}{2\pi r} ,$$

where r is the distance from the axis of the toroid.

The magnetization M inside the iron is

$$M = \frac{B}{\mu_0} - H = \frac{\mu H}{\mu_0} - H = \frac{\mu - 1}{\mu_0} \cdot \frac{NI}{2\pi r} .$$

2026

A C-magnet is shown in Fig. 2.18. All dimensions are in cm. The relative permeability of the soft Fe yoke is 3000. If a current $I = 1$ amp is to produce a field of about 100 gauss in the gap, how many turns of wire are required?

(Wisconsin)

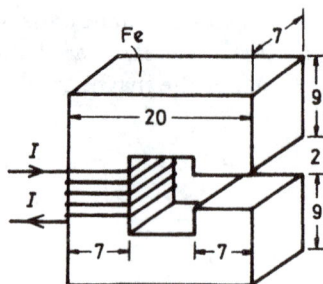

Fig. 2.18

Solution:

Consider a cross section of the magnet parallel to the plane of the paper and denote its periphery, which is (including the gap) a square of sides $l = 20$ cm, by L. As the normal component of $\mathbf{B}$ is continuous, the magnetic intensity in the gap is B/μ_0, while that inside the magnet is $B/\mu_0\mu_r$, where μ_r is the relative permeability of the iron. Ampere's circuital law

$$\oint_L \mathbf{H} \cdot d\mathbf{l} = NI$$

applied to L becomes

$$\frac{B}{\mu_0}d + \frac{B}{\mu_0\mu_r}(4l - d) = NI,$$

where $d = 2$ cm is the width of the gap. Hence

$$
\begin{aligned}
N &= \frac{B}{\mu_0 I}\left[d + \frac{1}{\mu_r}(4l - d)\right] \\
&= \frac{100 \times 10^{-4}}{4\pi \times 10^{-7} \times 1}\left(0.02 + \frac{0.2 \times 4 - 0.02}{3000}\right) \\
&= 161 \text{ turns}
\end{aligned}
$$

are required.

2027

An electromagnet is made by wrapping a current carrying coil N times around a C-shaped piece of iron ($\mu \gg \mu_0$) as shown in Fig. 2.19. If the cross sectional area of the iron is A, the current is i, the width of the gap is d, and the length of each side of the "C" is l, find the B-field in the gap.

(*Columbia*)

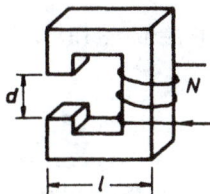

Fig. 2.19

Solution:

Putting $\mu_r = \mu/\mu_0$ in the result of Problem **2026**, we find

$$B = \frac{NI\mu_0\mu}{d(\mu - \mu_0) + 4l\mu_0}.$$

2028

Design a magnet (using a minimum mass of copper) to produce a field of 10,000 gauss in a 0.1 meter gap having an area of 1m×2m. Assume very high permeability iron. Calculate the power required and the weight of the necessary copper. (The resistivity of copper is 2×10^{-6} ohm-cm; its density is 8 g/cm^3 and its maximum current density is 1000 amp/cm^2.) What is the force of attraction between the poles of the magnet?

(*Princeton*)

Solution:

L is the periphery of a cross section of the magnet parallel to the plane of the diagram as shown in Fig. 2.20. Ampère's circuital law becomes

$$\oint_L \mathbf{H} \cdot d\mathbf{l} = \frac{B}{\mu_0}x + \frac{B}{\mu}(L-x) = NI,$$

where x is the width of the gap. As $\mu \gg \mu_0$, the second term in the middle may be neglected. Denoting the cross section of the copper wire by S, the current crossing S is $I = jS$. Together we have

$$N = \frac{Bx}{\mu_0 jS}.$$

The power dissipated in the wire, which is the power required, is

$$P = I^2 R = I^2 \rho \frac{2N(a+b)}{S} = 2j\rho(a+b)\frac{B}{\mu_0}x,$$

where ρ is the resistivity of copper. Using the given data, we get

$$P = 9.5 \times 10^4 \text{ W}.$$

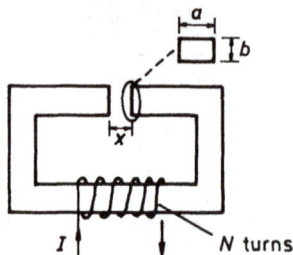

Fig. 2.20

Let δ be the density of copper, then the necessary weight of the copper is

$$2N(a+b)S\delta = 2(a+b)\frac{B}{j\mu_0}x\delta = 3.8 \text{ kg}.$$

The cross section of the gap is $A = a \cdot b$. Hence the force of attraction between the plates is

$$F = \frac{AB^2}{2\mu_0} = 8 \times 10^5 \text{ N}.$$

2029

A cylindrical soft iron rod of length L and diameter d is bent into a circular shape of radius R leaving a gap where the two ends of the rod almost meet. The gap spacing s is constant over the face of the ends of the rod. Assume $s \ll d, d \ll R$. N turns of wire are wrapped tightly around the iron rod and a current I is passed through the wire. The relative permeability of the iron is μ_r. Neglecting fringing, what is the magnetic field B in the gap?

(*MIT*)

Solution:

As $s \ll d \ll R$, magnetic leakage in the gap may be neglected. The magnetic field in the gap is then the same as that in the rod. From Ampère's circuital law

$$\oint \mathbf{H} \cdot d\mathbf{l} = NI$$

we obtain

$$B = \frac{\mu_r \mu_0 NI}{2\pi R + (\mu_r - 1)s}.$$

2030

The figure 2.21 shows the cross section of an infinitely long circular cylinder of radius $3a$ with an infinitely long cylindrical hole of radius a displaced so that its center is at a distance a from the center of the big cylinder. The solid part of the cylinder carries a current I, distributed uniformly over the cross section, and out from the plane of the paper.

(a) Find the magnetic field at all points on the plane P containing the axes of the cylinders.

(b) Determine the magnetic field throughout the hole; it is of a particularly simple character.

(UC, Berkeley)

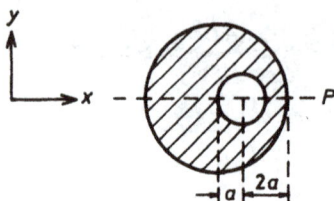

Fig. 2.21

Solution:

(a) According to the principle of superposition the field can be regarded as the difference of two fields $\mathbf{H_2}$ and $\mathbf{H_1}$, where $\mathbf{H_2}$ is the field produced by a solid (without the hole) cylinder of radius $3a$ and $\mathbf{H_1}$ is that produced by a cylinder of radius a at the position of the hole. The current in each of these two cylinders is uniformly distributed over the cross section. The currents I_1 and I_2 in the small and large cylinders have current densities $-j$ and $+j$ respectively. Then as $I = I_2 - I_1 = 9\pi a^2 j - \pi a^2 j = 8\pi a^2 j$, we have $j = \frac{I}{8\pi a^2}$ and

$$I_1 = \pi a^2 j = \frac{I}{8}, \quad I_2 = 9\pi a^2 j = \frac{9}{8}I.$$

Take the z-axis along the axis of the large cylinder with its positive direction in the direction of I_2, which we assume to be upwards from the plane of the paper. Take the x-axis crossing the axis of the small cylinder as shown in Fig. 2.21. Then the plane P is the xz plane, i.e., the plane $y = 0$. Ampere's law gives $\mathbf{H_1}$ and $\mathbf{H_2}$ as follows (noting $r = \sqrt{x^2 + y^2}, r_1 = \sqrt{(x-a)^2 + y^2}$, being the distances of the field point from the cylinder and hole respectively):

$$H_{2x} = -\frac{Iy}{16\pi a^2}, \qquad H_{2y} = \frac{Ix}{16\pi a^2}, \qquad (r \le 3a)$$

$$H_{2x} = -\frac{9Iy}{16\pi(x^2 + y^2)}, \qquad H_{2y} = \frac{9Ix}{16\pi(x^2 + y^2)}, \qquad (r > 3a)$$

$$H_{1x} = -\frac{Iy}{16\pi a^2}, \qquad H_{1y} = \frac{I(x-a)}{16\pi a^2}, \qquad (r_1 \le a)$$

$$H_{1x} = -\frac{Iy}{16\pi[(x-a)^2 + y^2]}, \quad H_{1y} = \frac{I(x-a)}{16\pi[(x-a)^2 + y^2]}, \quad (r_1 > a).$$

On the plane P, $H_{2x} = H_{1x} = 0$. Hence $H_x = 0$, $H_y = H_{2y} - H_{1y}$. We therefore have the following:

(1) Inside the hole ($0 < x < 2a$),

$$H_y = \frac{Ix}{16\pi a^2} - \frac{I(x-a)}{16\pi a^2} = \frac{Ia}{16\pi a^2} = \frac{I}{16\pi a}.$$

(2) Inside the solid part ($2a \le x \le 3a$ or $-3a \le x \le 0$),

$$H_y = \frac{Ix}{16\pi a^2} - \frac{I(x-a)}{16\pi[(x-a)^2 + y^2]} = \frac{I(x^2 - ax - a^2)}{16\pi a^2(x-a)}.$$

(3) Outside the cylinder ($|x| > 3a$),

$$H_y = \frac{9Ix}{16\pi(x^2 + y^2)} - \frac{I(x-a)}{16\pi[(x-a)^2 + y^2]} = \frac{(8x - 9a)I}{16\pi x(x-a)}.$$

(b) The magnetic field at all points inside the hole ($r_1 \le a$) is

$$H_x = -\frac{Iy}{16\pi a^2} + \frac{Iy}{16\pi a^2} = 0,$$

$$H_y = \frac{Ix}{16\pi a^2} - \frac{I(x-a)}{16\pi a^2} = \frac{I}{16\pi a}.$$

This field is uniform inside the hole and is along the positive y-direction.

2031

(a) A sphere of radius r is at a potential V and is immersed in a conducting medium of conductivity σ. Calculate the current flowing from the sphere to infinity.

(b) Two spheres, with potentials $+V$ and 0, have their centers at positions $x = \pm d$, where $d \gg r$. For points equidistant from the two spheres (i.e., on the yz plane) and far away ($\gg d$) calculate the current density **J**.

(c) For the same geometry as in (b), calculate the magnetic field on the yz plane for distant points.

(UC, Berkeley)

Solution:

(a) If the sphere carries charge Q, the potential on its surface is

$$V = \frac{Q}{4\pi\varepsilon_0 r},$$

i.e., $Q = 4\pi\varepsilon_0 rV$. When the sphere is immersed in a conducting medium of conductivity σ, the current that starts to flow out from the sphere is

$$I = \oint_S \mathbf{J} \cdot d\mathbf{S} = \sigma \oint_S \mathbf{E} \cdot d\mathbf{S} = \sigma \frac{Q}{\varepsilon_0} = 4\pi\sigma rV,$$

where S is the spherical surface and we have assumed the medium to be Ohmic. If the potential V is maintained, the current I is steady.

(b) As $d \gg r$, we can regard the spheres as point charges. Suppose the sphere with potential V carries a net charge $+Q$ and that with potential 0, $-Q$. Take the line joining the two spherical centers as the x-axis and the mid-point of this line as the origin. Then the potential of an arbitrary point x on the line is

$$V(x) = \frac{Q}{4\pi\varepsilon_0}\left(\frac{1}{d-x} - \frac{1}{d+x}\right).$$

The potential difference between the two spherical surfaces is then

$$
\begin{aligned}
V &= \frac{Q}{4\pi\varepsilon_0}\left(\frac{1}{d-x} - \frac{1}{d+x}\right)\Big|_{-d+r}^{d-r} \\
&= \frac{Q}{4\pi\varepsilon_0} \cdot \frac{4(d-r)}{r(2d-r)} \approx \frac{Q}{2\pi\varepsilon_0 r}
\end{aligned}
$$

as $d \gg r$. Hence

$$Q = 2\pi\varepsilon_0 rV.$$

On the yz plane the points which are equidistant from the two spheres will constitute a circle with center at the origin. By symmetry the magnitudes of the electric and magnetic fields at these points are the same, so we need only calculate them for a point, say the intersection of the circle and the z-axis (see Fig. 2.22). Let R be the radius of the circle, $\mathbf{E}_1$ and $\mathbf{E}_2$ be the electric fields produced by $+Q$ and $-Q$ respectively. The resultant of these fields is along the $-x$ direction:

$$\mathbf{E} = -\frac{2Q}{4\pi\varepsilon_0(R^2 + d^2)}\cos\theta\, \mathbf{e}_x = -\frac{Qd}{2\pi\varepsilon_0(R^2 + d^2)^{3/2}}\mathbf{e}_x.$$

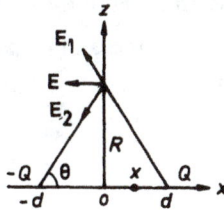

Fig. 2.22

The current density at this point is then

$$\mathbf{J} = \sigma \mathbf{E} = -\frac{\sigma Q d}{2\pi\varepsilon_0 (R^2 + d^2)^{3/2}} \mathbf{e}_x = -\frac{V r d}{(R^2 + d^2)^{3/2}} \mathbf{e}_x \,.$$

As the choice of z-axis is arbitrary, the above results apply to all points of the circle.

(c) Using a circle of radius R as the loop L, in Ampère's circuital law

$$\oint_L \mathbf{B} \cdot d\mathbf{l}' = \mu_0 \int_S \mathbf{J} \cdot d\mathbf{S}' \,, \quad d\mathbf{S}' = r' dr' d\theta' \mathbf{e}_x \,,$$

we have

$$2\pi R B = -\mu_0 \int_0^{2\pi} d\theta' \int_0^R \frac{V r d}{(r'^2 + d^2)^{3/2}} dr'$$

$$= 2\pi\mu_0 V r d \left(\frac{1}{\sqrt{R^2 + d^2}} - \frac{1}{d} \right) .$$

For distant points, $R \gg d$ and we obtain to good approximation

$$B = \frac{\mu_0 V r}{R} \,.$$

Note $\mathbf{B}$ is tangential to the circle R and is clockwise when viewed from the side of positive x.

2032

Consider a thin spherical shell of dielectric which has a radius R and rotates with an angular velocity ω. A constant surface charge of density σ

is placed on the sphere, and this produces a uniform magnetic field which is proportional to ω. Suppose that the mass of the shell is negligible.

(a) Find the magnetic field inside and outside the rotating shell.

(b) A constant torque **N** is applied parallel to ω. How long does it take for the shell to stop?

(UC, Berkeley)

Solution:

Use coordinates with the z-axis along the rotating axis and the origin at the center of the sphere (Fig. 2.23). The surface current density on the spherical shell in spherical coordinates is

$$\alpha = R\sigma\omega \sin\theta \mathbf{e}_\varphi$$

or, expressed as a volume current density,

$$\mathbf{J} = \alpha\delta(r - R).$$

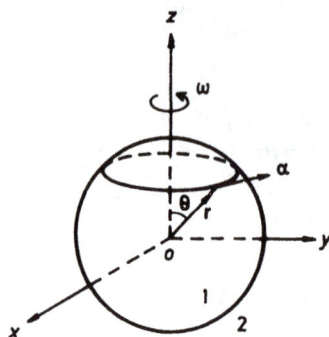

Fig. 2.23

The magnetic dipole moment of the sphere is then

$$\mathbf{m} = \frac{1}{2}\int \mathbf{r}' \times \mathbf{J}dV' = \frac{1}{2}\mathbf{e}_z \int\int \int r \cdot R\sigma\omega \sin\theta\delta(r - R)$$
$$\cdot 2\pi r \sin\theta \cdot rd\theta \cdot dr \cdot \sin\theta$$
$$= \mathbf{e}_z \pi R^4\sigma\omega \int_0^\pi \sin^3\theta d\theta = \frac{4\pi}{3}R^4\sigma\omega\mathbf{e}_z.$$

Note that for any pair of symmetrical points on a ring as shown in Fig. 2.23 the total contribution to **m** is the z direction. Hence the extra $\sin\theta$ in the integral above.

The magnetization of the sphere is

$$\mathbf{M} = \frac{\mathbf{m}}{\frac{4}{3}\pi R^3} = \sigma\omega R\mathbf{e}_z\,.$$

Since there is no free current inside and outside the sphere, we can apply the method of the magnetic scalar potential. The inside and outside potentials satisfy Laplace's equation:

$$\nabla^2\varphi_1 = \nabla^2\varphi_2 = 0\,.$$

We require that $\varphi_1|_0$ is finite and $\varphi_2|_\infty \to 0$. By separating the variables we obtain the solutions

$$\varphi_1 = \sum_n a_n r^n P_n(\cos\theta)\,, \quad \varphi_2 = \sum_n \frac{b_n}{r^{n+1}} P_n(\cos\theta)\,.$$

On the spherical surface the following conditions apply:

$$\varphi_1 = \varphi_2|_{r=R}\,, \quad \frac{\partial\varphi_2}{\partial r} - \frac{\partial\varphi_1}{\partial r}\bigg|_{r=R} = -\sigma R\omega\cos\theta\,.$$

These give

$$a_1 = \frac{\sigma R\omega}{3}\,, \quad b_1 = \frac{\sigma R^3\omega}{3}\,,$$

all other coefficients being zero. The magnetic scalar potentials are then

$$\varphi_1 = \frac{1}{3}\sigma R\omega\cdot\mathbf{r}\,, \quad \varphi_2 = \frac{1}{3}\sigma R^3\omega\cdot\frac{\mathbf{r}}{r^3}\,.$$

Hence the magnetic fields inside and outside the sphere are

$$\mathbf{H}_1 = -\nabla\varphi_1 = -\frac{1}{3}\sigma R\omega\mathbf{e}_z\,,$$

$$\mathbf{B}_1 = \mu_0(\mathbf{H}_1 + \mathbf{M}) = \frac{2}{3}\mu_0\sigma R\omega\mathbf{e}_z\,, \quad (r \le R)$$

$$\mathbf{B}_2 = \mu_0\mathbf{H}_2 = -\nabla\varphi_2 = \frac{\mu_0}{3}\sigma R^3\left(\frac{3(\boldsymbol{\omega}\cdot\mathbf{r})\mathbf{r}}{r^5} - \frac{\boldsymbol{\omega}}{r^3}\right)$$

$$= \frac{\mu_0}{3}\frac{R^4}{r^3}\sigma\omega(3\cos\theta\mathbf{e}_r - \mathbf{e}_z)\,, \quad (r > R)\,.$$

Before the application of the constant torque **N**, the total magnetic energy of the system is

$$W_m = \int_\infty \frac{B^2}{2\mu_0} dV = \int_{V_{in}} \frac{B^2}{2\mu_0} dV + \int_{V_{out}} \frac{B^2}{2\mu_0} dV$$

$$= \frac{1}{2\mu_0} \cdot \left(\frac{2}{3}\mu_0 \sigma R\omega\right)^2 \cdot \frac{4}{3}\pi R^3$$

$$+ \frac{1}{2\mu_0} \left(\frac{\mu_0}{3} R^4 \sigma\omega\right)^2 \int_{V_{out}} \frac{(1+3\cos^2\theta)}{r^6} dV,$$

where V_{in} and V_{out} refer respectively to the space inside and outside the sphere. Noting

$$\int_{V_{out}} \left(\frac{1+3\cos^2\theta}{r^6}\right) dV = \int_R^\infty dr \int_0^\pi d\theta \int_0^{2\pi} d\varphi \frac{(1+3\cos^2\theta)}{r^4} \sin\theta = \frac{4\pi}{3R^4},$$

we have

$$W_m = \frac{4\pi}{9}\mu_0 \sigma^2 \omega^2 R^5.$$

Suppose the rotation stops after time t due to the action of the constant torque **N**. Conservation of energy requires

$$\frac{dW_m}{dt} = N\omega.$$

With N constant we get

$$t = \frac{W_m}{N\omega} = \frac{4\pi\mu_0 \sigma^2 \omega R^5}{N}.$$

2033

A thin spherical shell of radius R carries a uniform surface charge density σ. The shell is rotated at constant angular velocity ω about a diameter.

(a) Write down the boundary conditions which relate the magnetic field just inside the shell to that just outside the shell.

(b) The magnetic field which satisfies these conditions is uniform inside the shell and of dipole form outside the shell. Determine the magnitude of the inside magnetic field.

(*CUSPEA*)

Solution:

(a) Call the inside of the shell region 1, the outside region 2. Take the rotating axis as the z-axis. The current density on the spherical shell is

$$\alpha = \sigma R\omega \sin\theta \mathbf{e}_\varphi .$$

The boundary relations on the spherical surface are as follows:

In the tangential direction: $B_{1r} = B_{2r}\big|_{r=R}$,

In the normal direction: $\mathbf{e}_r \times \left(\dfrac{\mathbf{B}_2}{\mu_0} - \dfrac{\mathbf{B}_1}{\mu_0}\right)\bigg|_{r=R} = \alpha$,

or

$$B_{2\theta} - B_{1\theta}\big|_{r=R} = \mu_0 \sigma\omega R\sin\theta .$$

(b) Referring to Problem **2032**, we see that the magnetic fields inside and outside the shell are

$$\mathbf{B}_1 = \frac{2}{3}\mu_0\sigma\omega R\mathbf{e}_z , \quad (r \leq R) \tag{1}$$

$$\mathbf{B}_2 = \frac{\mu_0}{3}\sigma R^4\left(\frac{3(\boldsymbol{\omega}\cdot\mathbf{r})\mathbf{r}}{r^5} - \frac{\boldsymbol{\omega}}{r^3}\right), \quad (r > R) \tag{2}$$

where $\boldsymbol{\omega} = \omega\mathbf{e}_z$. Note that the magnetic field inside the sphere is a uniform field. Also as the magnetic field produced by a magnetic dipole of moment $\mathbf{m}$ can be expressed as $\frac{\mu_0}{4\pi}[\frac{3(\mathbf{m}\cdot\mathbf{r})\mathbf{r}}{r^5} - \frac{\mathbf{m}}{r^3}]$, Eq. (2) shows that the magnetic field outside the shell is that of a dipole of moment

$$\mathbf{m} = \frac{4\pi}{3}\sigma R^4\omega\mathbf{e}_z .$$

From $\mathbf{e}_z = \cos\theta\mathbf{e}_r - \sin\theta\mathbf{e}_\theta$, we can rewrite Eqs. (1) and (2) as

$$\mathbf{B}_1 = \frac{2}{3}\mu_0\sigma\omega R(\cos\theta\mathbf{e}_r - \sin\theta\mathbf{e}_\theta),$$

$$\mathbf{B}_2 = \frac{\mu_0\sigma\omega R^4}{3r^3}(2\cos\theta\mathbf{e}_r + \sin\theta\mathbf{e}_\theta).$$

Clearly, these expressions satisfy the boundary conditions stated in (a).

2034

Consider a spherical volume of radius R within which it is desired to have a uniform magnetic field **B**. What current distribution on the surface of the sphere is required to generate this field?

(UC, Berkeley)

Solution:

By analogy with a uniform polarized sphere, we deduce that the magnetic field inside a uniform magnetized sphere is uniform. Let **M** be the magnetization, then the surface current density is $\alpha_S = -\mathbf{n} \times \mathbf{M}$. Take the z-axis along **M** so that $\mathbf{M} = M\mathbf{e}_z$. In spherical coordinates

$$\mathbf{e}_z = \cos\theta \mathbf{e}_r - \sin\theta \mathbf{e}_\theta \quad (\mathbf{n} = \mathbf{e}_r),$$

so that

$$\alpha_S = -\mathbf{e}_r \times M(\cos\theta \mathbf{e}_r - \sin\theta \mathbf{e}_\theta) = M\sin\theta \mathbf{e}_\varphi.$$

Then making use of the magnetic scalar potential, we find the magnetic field inside the sphere: $\mathbf{B} = \frac{2\mu_0}{3}\mathbf{M}$ (refer to Problem **2033**.) Hence

$$\alpha_S = \frac{3B}{2\mu_0}\sin\theta \mathbf{e}_\varphi.$$

2035

As in Fig. 2.24, a thin spherical shell of radius R has a fixed charge $+q$ distributed uniformly over its surface.

(a) A small circular section (radius $r \ll R$) of charge is removed from the surface. Find the electric fields just inside and just outside the sphere at the hole.

The cut section is replaced and the sphere is set in motion rotating with constant angular velocity $\omega = \omega_0$ about the z-axis.

(b) Calculate the line integral of the electric field along the z-axis from $-\infty$ to $+\infty$.

(c) Calculate the line integral of the magnetic field along the same path.

Now the sphere's angular velocity increases linearly with time:

$$\omega = \omega_0 + kt.$$

(d) Calculate the line integral of the electric field around the circular path P (shown in Fig. 2.25) located at the center of the sphere. Assume that the normal to the plane containing the path is along the $+z$ axis and that its radius is $r_P \ll R$.

(Chicago)

Fig. 2.24

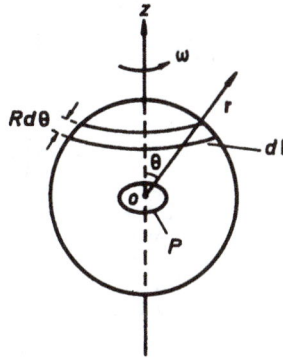

Fig. 2.25

Solution:

(a) Before the small circular section of charge is removed, the electric field inside the sphere is zero, while the field outside the sphere is $E = \frac{q}{4\pi\epsilon_0 R^2}$. Referring to Problem **1021**, the electric field produced by this small circular section is $\frac{\sigma}{2\epsilon_0} = \frac{q}{8\pi\epsilon_0 R^2}$. Therefore, after the small section is removed the electric fields just inside and just outside the sphere at the hole are both $\frac{q}{8\pi\epsilon_0 R^2}$.

(b) By symmetry, $\int_{-\infty}^{\infty} E dz = 0$.

(c) Ampère's law gives

$$\oint \mathbf{B} \cdot d\mathbf{l} = \int_{-\infty}^{\infty} B dz = \mu_0 I .$$

As the electric current is $I = \frac{q\omega_0}{2\pi}$, we have

$$\int_{-\infty}^{\infty} B dz = \frac{q\mu_0\omega_0}{2\pi} .$$

(d) Consider a ring of width $Rd\theta$ as shown in Fig. 2.25. The surface current density on the ring is $\frac{q}{4\pi R^2} \cdot \omega R \sin\theta$. The contributions of a pair of symmetrical points on the ring to the magnetic field at the center of the sphere will sum up to a resultant in the z-direction. Thus the total contribution of the ring to the magnetic field is given by the Biot-Savart law as

$$dB_z = \frac{\mu_0}{4\pi} \int_{\varphi=0}^{2\pi} \frac{q}{4\pi R^2} \omega R \sin\theta \cdot \frac{R}{R^3} \sin\theta \, R d\theta \cdot R \sin\theta \, d\varphi$$

$$= \frac{\mu_0 q\omega}{8\pi R} \sin^3\theta \, d\theta .$$

Hence

$$B_z = \frac{\mu_0 q\omega}{8\pi R} \int_0^{\pi} \sin^3\theta \, d\theta = \frac{\mu_0 q\omega}{6\pi R} .$$

As $r_p \ll R$, the magnetic induction can be taken to be uniform in the circular loop P. The magnetic flux crossing P is then

$$\phi = \pi r_p^2 B = \frac{\mu_0 r_p^2 \omega q}{6R} = \frac{\mu_0 q r_p^2 (\omega_0 + kt)}{6R} .$$

Hence

$$\oint_P \mathbf{E} \cdot d\mathbf{l} = -\frac{d\phi}{dt} = \frac{\mu_0 k r_p^2 q}{6R} .$$

2036

An isolated conducting sphere of radius R is charged to potential V and rotated about a diameter at angular speed ω.

(a) Find the magnetic induction $\mathbf{B}$ at the center of the sphere.

(b) What is the magnetic dipole moment of the rotating sphere?

(*UC, Berkeley*)

Solution:

(a) Using the answer to Problem **2035**, the magnetic field at the spherical center is

$$\mathbf{B} = \frac{\mu_0 \omega Q}{6\pi R} \mathbf{e}_z ,$$

where Q is the total charge of the sphere and $\mathbf{e}_z$ is a unit vector along the axis of rotation. From $V = \frac{Q}{4\pi\varepsilon_0 R}$, we get $Q = 4\pi\varepsilon_0 RV$. Hence

$$B = \frac{2}{3}\varepsilon_0 \mu_0 \omega V .$$

(b) Referring to Problem **2032**, the magnetic dipole moment of the sphere is

$$\mathbf{m} = \frac{4\pi}{3} R^4 \sigma \omega \mathbf{e}_z ,$$

where σ is the surface charge density

$$\sigma = \frac{Q}{4\pi R^2} = \frac{\varepsilon_0 V}{R} .$$

Hence

$$\mathbf{m} = \frac{4}{3}\pi\varepsilon_0 R^3 \omega V \mathbf{e}_z .$$

2037

A charge Q is uniformly distributed over the surface of a sphere of radius r_0. The material inside and outside the sphere has the properties of the vacuum.

(a) Calculate the electrostatic energy in all space.

(b) Calculate the force per unit area on the surface of the sphere due to the presence of the charge. For $Q = 1$ coulomb and $r_0 = 1$ cm, give a numerical answer.

(c) The sphere rotates around an axis through a diameter with constant angular velocity ω. Calculate the magnetic field at the center of the sphere.

(UC, Berkeley)

Solution:

(a) $W = \int \frac{1}{2}\varepsilon_0 E^2 dV = \frac{\varepsilon_0}{2}\left(\frac{Q}{4\pi\varepsilon_0}\right)^2 \int_{r_0}^{\infty} \frac{1}{r^2} \cdot 4\pi r^2 dr = \frac{Q^2}{8\pi\varepsilon_0 r_0}.$

(b) The surface charge density is $\sigma = \frac{Q}{4\pi r_0^2}$ and the electric field outside the sphere is

$$\mathbf{E} = \frac{Q}{4\pi\varepsilon_0 r_0^2}\mathbf{e}_r .$$

Using the answer to Problem **1021**, the electric force per unit area on the outer surface is

$$\mathbf{f} = \frac{\sigma \mathbf{E}}{2}\bigg|_{r=r_0} = \frac{Q^2}{32\pi^2\varepsilon_0 r_0^4}\mathbf{e}_r .$$

With the given data, we have

$$f = 3.6 \times 10^{12} \text{ N/cm}^2 .$$

(c) Using the answer to Problem **2035**, we have

$$\mathbf{B} = \frac{\mu_0 q \omega}{6 \pi r_0} \mathbf{e}_z$$

where $\mathbf{e}_z$ is a unit vector along the axis of rotation.

2038

A long hollow right circular cylinder made of iron of permeability μ is placed with its axis perpendicular to an initially uniform magnetic flux density $\mathbf{B}$. Assume that B_0 is small enough so that it does not saturate the iron, and that the permeability μ is a constant in the field range of our interest.

(a) Sketch the magnetic field lines in the entire region before and after the cylinder is placed in the field.

(b) Let the inner and outer radii of the cylinder be b and a respectively. Derive an expression for $\mathbf{B}$ inside the cylinder. Note: In cylindrical coordinates we have

$$\nabla^2 = \frac{1}{r} \frac{\partial}{\partial r} \left(r \frac{\partial}{\partial r} \right) + \frac{1}{r^2} \frac{\partial^2}{\partial \theta^2} + \frac{\partial^2}{\partial z^2} .$$

(*Columbia*)

Solution:

(a) The magnetic field is uniform before the cylinder is introduced and the field lines are as shown in Fig. 2.26. After the cylinder is placed in the field, the magnetic field will be distorted and the field lines are as shown in Fig. 2.27.

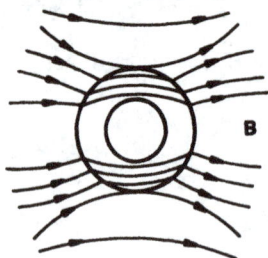

Fig. 2.26   Fig. 2.27

(b) We introduce a magnetic scalar potential ϕ which satisfies $\mathbf{H} = -\nabla\phi$. As there is no free current we have $\nabla^2\phi = 0$. In cylindrical co-ordinates (r, θ, z), where the z-axis is along the axis of the cylinder, the potential satisfies

$$\left\{\frac{1}{r}\frac{\partial}{\partial r}\left(r\frac{\partial}{\partial r}\right) + \frac{1}{r^2}\frac{\partial^2}{\partial \theta^2} + \frac{\partial^2}{\partial z^2}\right\}\phi = 0. \tag{1}$$

Because of axial symmetry we have $\frac{\partial^2\phi}{\partial z^2} = 0$. Let

$$\phi(r, \theta) = R(r)S(\theta).$$

The equation can be written as

$$\frac{1}{R}\left(r^2\frac{d^2R}{dr^2} + r\frac{dR}{dr}\right) = -\frac{1}{S}\frac{d^2S}{d\theta^2} = \text{constant} = m^2,$$

say, since varying r does not affect the expression involving S. This leads to the general solution

$$\phi = \sum_{m=1}^{\infty}(c_m r^m + d_m r^{-m})(g_m \cos m\theta + h_m \sin m\theta). \tag{2}$$

By symmetry $\phi(r, \theta) = \phi(r, -\theta)$, so that the sine functions are to be eliminated by putting $h_m = 0$. Divide the space into three parts as shown in Fig. 2.28 and write the general solutions for them as

$$\phi_i = \sum_{m=1}^{\infty}(c_{im}r^m + d_{im}r^{-m})\cos m\theta, \quad (i = 1, 2, 3)$$

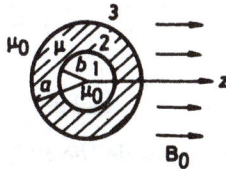

Fig. 2.28

At large distances from the cylinder, $\frac{-\partial\phi_3}{\partial z} = H_0$, or $\phi_3 = -H_0 z = -\frac{B_0}{\mu_0}r\cos\theta$. Comparing the coefficients of $\cos m\theta$ we have

$$c_{31} = -\frac{B_0}{\mu_0}, \quad C_{3m} = d_{3m} = 0 \quad (m \neq 1).$$

Hence

$$\phi_3 = -\left(\frac{B_0 r}{\mu_0} - \frac{d_{31}}{r}\right)\cos\theta.$$

We also require ϕ_1 to be finite for $r \to 0$. Hence $d_{1m} = 0$ for all m and

$$\phi_1 = \sum_{m=1}^{\infty} c_{1m} r^m \cos m\theta.$$

Next consider the boundary conditions at $r = a$ and b. We have

$$\mu_0 \frac{\partial \phi_3}{\partial r} = \mu \frac{\partial \phi_2}{\partial r}\bigg|_{r=a},$$

$$\mu_0 \frac{\partial \phi_1}{\partial r} = \mu \frac{\partial \phi_2}{\partial r}\bigg|_{r=b},$$

$$\frac{\partial \phi_3}{\partial \theta} = \frac{\partial \phi_2}{\partial \theta}\bigg|_{r=a},$$

$$\frac{\partial \phi_1}{\partial \theta} = \frac{\partial \phi_2}{\partial \theta}\bigg|_{r=b}.$$

These together give $c_{1m} = c_{2m} = d_{2m} = 0$ for $m \neq 1$ and the simultaneous equations

$$\begin{cases} B_0 + \frac{\mu_0 d_{31}}{a^2} = \mu\left(-c_{21} + \frac{d_{21}}{a^2}\right), \\ -B_0 + \frac{\mu_0 d_{31}}{a^2} = \mu_0\left(c_{21} + \frac{d_{21}}{a^2}\right), \\ \mu\left(c_{21} - \frac{d_{21}}{b^2}\right) = \mu_0 c_{11}, \\ c_{21} + \frac{d_{21}}{b^2} = c_{11}. \end{cases}$$

Solving for c_{11} we have

$$c_{11} = \frac{4\pi a^2 B_0}{b^2(\mu - \mu_0)^2 - a^2(\mu + \mu_0)^2},$$

giving

$$\phi_1 = c_{11} r \cos\theta.$$

The magnetic field intensity inside the cylinder is

$$\begin{aligned} \mathbf{H}_1 = -\nabla\phi_1 &= -\frac{\partial \phi_1}{\partial r}\mathbf{e}_r - \frac{1}{r}\frac{\partial \phi}{\partial \theta}\mathbf{e}_\theta \\ &= -c_{11}(\cos\theta\,\mathbf{e}_r - \sin\theta\,\mathbf{e}_\theta) \\ &= -c_{11}\mathbf{e}_z \\ &= \frac{4\mu a^2}{a^2(\mu + \mu_0)^2 - b^2(\mu - \mu_0)^2}\mathbf{B}_0. \end{aligned}$$

If $\mu \gg \mu_0$, the magnetic field becomes

$$\mathbf{H}_1 = \frac{4}{\mu}\frac{\mathbf{B}_0}{1 - (\frac{b}{a})^2} .$$

Obviously, the greater the value of μ, the stronger is the magnetic shielding.

2. ELECTROMAGNETIC INDUCTION (2039–2063)

2039

A uniform cylindrical coil in vacuum has $r_1 = 1$ m, $l_1 = 1$ m and 100 turns. Coaxial and at the center of this coil is a smaller coil of $r_2 = 10$ cm, $l_2 = 10$ cm and 10 turns. Calculate the mutual inductance of the two coils.

(*Columbia*)

Solution:

Suppose current I_1 passes through the outer coil, then the magnetic induction produced by it is

$$B_1 = \mu_0 \frac{N_1 I_1}{l_1} .$$

As $r_2 \ll r_1, l_2 \ll l_1$, we may consider the magnetic field B_1 as uniform across the inner coil. Thus the magnetic flux crossing the inner coil is

$$\psi_{12} = N_2 B_1 S_2 = \mu_0 \frac{N_1 N_2 I_1}{l_1} \pi r_2^2 ,$$

which gives the mutual inductance of the two coils as

$$M = \frac{\psi_{12}}{I_1} = \mu_0 \frac{N_1 N_2}{l_1} \pi r_2^2 \approx 39.5 \ \mu\text{H} .$$

2040

A circular wire loop of radius R is rotating uniformly with angular velocity ω about a diameter PQ as shown in Fig. 2.29. At its center, and lying along this diameter, is a small magnet of total magnetic moment M. What is the induced emf between the point P (or Q) and a point on the loop mid-way between P and Q?

(*Columbia*)

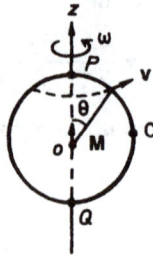

Fig. 2.29

Solution:

At a point distance r from the spherical center the magnetic field established by the small magnet is

$$\mathbf{B} = \frac{\mu_0}{4\pi}\left[\frac{3(\mathbf{M}\cdot\mathbf{r})\mathbf{r}}{r^5} - \frac{\mathbf{M}}{r^3}\right], \quad \mathbf{M} = M\mathbf{e}_z.$$

Let C be the mid-point of arc $\widehat{PQ}$. The velocity of a point on arc $\widehat{PC}$ is $\mathbf{v} = \omega R\sin\theta\,\mathbf{e}_\varphi$. The induced emf between the points P and C along arc $\widehat{PQ}$ is given by

$$\varepsilon_{PC} = \int_P^C (\mathbf{v}\times\mathbf{B})\cdot d\mathbf{l}, \quad \text{with} \quad d\mathbf{l} = Rd\theta\mathbf{e}_\theta.$$

As $\mathbf{e}_z = \cos\theta\mathbf{e}_r - \sin\theta\mathbf{e}_\theta$, $\mathbf{e}_\varphi\times\mathbf{e}_r = \mathbf{e}_\theta, \mathbf{e}_\varphi\times\mathbf{e}_\theta = -\mathbf{e}_r$, we have

$$\mathbf{v}\times\mathbf{B} = \frac{\mu_0\omega RM\sin\theta}{4\pi R^3}(2\cos\theta\mathbf{e}_\theta + \sin\theta\mathbf{e}_r)$$

and

$$\varepsilon_{PC} = \frac{\mu_0 M\omega}{4\pi R}\int_0^{\frac{\pi}{2}} 2\cos\theta\sin\theta d\theta = \frac{\mu_0 M\omega}{4\pi R}.$$

2041

Two infinite parallel wires separated by a distance d carry equal currents I in opposite directions, with I increasing at the rate $\frac{dI}{dt}$. A square loop of wire of length d on a side lies in the plane of the wires at a distance d from one of the parallel wires, as illustrated in Fig. 2.30.

(a) Find the emf induced in the square loop.

(b) Is the induced current clockwise or counterclockwise? Justify your answer.

(Wisconsin)

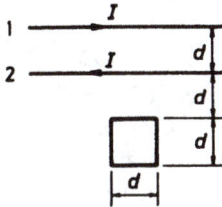

Fig. 2.30

Solution:

(a) The magnetic field produced by an infinite straight wire carrying current I at a point distance r from the wire is given by Ampère's circuital law as

$$B = \frac{\mu_0 I}{2\pi r},$$

its direction being perpendicular to the wire. Thus the magnetic flux crossing the loop due to the wire farther away from the loop is

$$\phi_1 = \int_{2d}^{3d} \frac{\mu_0 I d}{2\pi r} dr = \frac{\mu_0 I d}{2\pi} \ln \frac{3}{2}$$

directing into the page. The other wire, which is nearer the loop, gives rise to the magnetic flux

$$\phi_2 = \int_{d}^{2d} \frac{\mu_0 I d}{2\pi r} dr = \frac{\mu_0 I d}{2\pi} \ln 2$$

pointing out from the page. Hence the total flux is

$$\phi = \phi_2 - \phi_1 = \frac{\mu_0 I d}{2\pi} \ln \frac{4}{3}$$

pointing out from the page. The emf induced in the square loop is therefore

$$\varepsilon = -\frac{d\phi}{dt} = -\frac{\mu_0 d}{2\pi} \ln \left(\frac{4}{3} \right) \frac{dI}{dt}.$$

(b) The magnetic field produced by the induced current tends to oppose the change of the magnetic flux, so that this field will direct into the page. Then by the right-hand rule the induced current is clockwise as seen from the above.

2042

In Fig. 2.31 two conductors of infinite length carry a current I. They are parallel and separated by a distance $2a$. A circular conducting ring of radius a in the plane of the parallel wires lies between the two straight conductors and is insulated from them. Find the coefficient of mutual inductance between the circular conductor and the two straight conductors.

(*UC, Berkeley*)

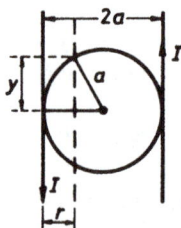

Fig. 2.31

Solution:

The magnetic field at a point between the two conductors at distance r from one conductor is

$$\mathbf{B}(r) = \frac{\mu_0 I}{2\pi} \left(\frac{1}{r} + \frac{1}{2a-r} \right) \mathbf{e}_\theta .$$

So the magnetic flux crossing the area of the ring is given by

$$\phi = \int \mathbf{B} \cdot d\mathbf{S} = 2 \int_0^a B(r) \cdot 2y \, dr$$

$$= 2 \int_0^a \frac{\mu_0 I}{2\pi} \left(\frac{1}{r} + \frac{1}{2a-r} \right) \cdot 2\sqrt{a^2 - (a-r)^2} \, dr .$$

Let $x = a - r$ and integrate:

$$\phi = 2 \int_0^a \frac{\mu_0 I}{\pi} \left(\frac{1}{a-x} + \frac{1}{a+x} \right) \sqrt{a^2 - x^2} \, dx$$

$$= \frac{4\mu_0 I a}{\pi} \arcsin \frac{x}{a} \Big|_0^a = 2\mu_0 I a .$$

Hence the coefficient of mutual inductance is

$$M = \frac{\phi}{I} = 2\mu_0 a .$$

2043

As shown in Fig. 2.32, an infinite wire carries a current I in the $+z$ direction. A rectangular loop of wire of side l is conneted to a voltmeter and moves with velocity u radially away from the wire. Indicate which terminal (a or b) of the voltmeter is positive. Calculate the reading on the voltmeter in terms of the distances r_1, r_2 and l.

(Wisconsin)

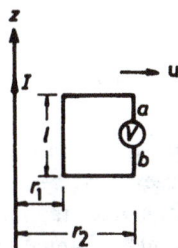

Fig. 2.32

Solution:

The magnetic field at a point of radial distance r is

$$B = \frac{\mu_0 I}{2\pi r} ,$$

and its direction is perpendicular to and pointing into the paper. The induced emf in the rectangular loop (i.e. reading of the voltmeter) is

$$V = \oint (\mathbf{u} \times \mathbf{B}) \cdot d\mathbf{l} = \frac{\mu_0 I u l}{2\pi} \left(\frac{1}{r_1} - \frac{1}{r_2} \right)$$

if we integrate in the clockwise sense. Note that $\mathbf{u} \times \mathbf{B}$ is in the $+z$ direction. As $V > 0$, terminal a is positive.

2044

A long uniform but laminated cylindrical iron core of radius $= 0.1$ m is uniformly wound with wire which excites a uniform flux density in it of magnitude $B(t) = \frac{1}{\pi}\sin(400t)$ Wb/m^2.

(a) What is the voltage per turn on the wire coil?

(b) What is the vector potential due to this core, $A(r)$, at points where $r > 0.1$ m?

(c) What is $\mathbf{B}(r,t)$ due to this core for points where $r > 0.1$ m?

(d) What is $\mathbf{A}(r,t)$ due to this core for $r < 0.1$ m?

(*Wisconsin*)

Solution:

(a) The voltage per turn on the wire coil must just balance the induced emf, i.e.,

$$V = -\varepsilon = \frac{d\phi}{dt} = \pi R^2 \frac{dB}{dt}$$
$$= 400R^2 \cos(400t) = 4\cos(400t) \ \text{V}.$$

(b) Consider a circular path C of radius $r > 0.1$ m with axis along the axis of the iron core. By symmetry we see that the magnitude of $\mathbf{A}(r)$ is the same everywhere on the circle and its direction is always tangential (in the same direction as the current). Using $\nabla \times \mathbf{A} = \mathbf{B}$ in Stokes' theorem we have

$$\oint_C \mathbf{A} \cdot d\mathbf{l} = \int_S \mathbf{B} \cdot d\mathbf{S}.$$

As $\mathbf{B}$ vanishes outside a long solenoid and is uniform inside, the right-hand side is

$$\int_S \mathbf{B} \cdot d\mathbf{S} = B \cdot \pi R^2 = 0.01 \sin(400t).$$

Stokes' theorem then reduces to $2\pi r A(r,t) = 0.01 \sin(400t)$, giving

$$A(t,r) = \frac{1}{200\pi r} \sin(400t) \ \text{Wb/m}.$$

(c) The magnetic field due to the core is zero for $r > 0.1$ m. (Strictly speaking, there is a very small magnetic field outside the solenoid tangential to the circle in (b) of magnitude $\frac{\mu_0 I}{2\pi r}$, I being the current in the core. This is however usually negligible.)

(d) For $r < 0.1$ m, Stokes' theorem

$$2\pi r A(r,t) = \pi r^2 B = \pi r^2 \cdot \frac{1}{\pi} \sin(400t)$$

gives

$$A(r,t) = \frac{r}{2\pi} \sin(400t) \text{ Wb/m}.$$

2045

An iron ring of radius 10 cm and of cross sectional area 12 cm^2 is evenly wound with 1200 turns of insulated wire. There is an air gap in the ring of length 1 mm. The permeability of the iron is 700 and is assumed independent of the field; the phenomenon of hysteresis is ignored.

(a) Calculate the magnetic field in the gap when a current of 1 amp passes through the coil.

(b) Calculate the self-inductance of the coil (with this core).

(UC, Berkeley)

Solution:

Using the results of Problem 2029 we have

(a) $B = \dfrac{\mu_0 N I}{\frac{2\pi R - d}{\mu_r} + d} = 0.795$ T.

(b) $L = \dfrac{BAN}{I} = 1.14$ H.

2046

Two single-turn circular loops are mounted as shown in Fig. 2.33. Find the mutual inductance between the two coils assuming $b \ll a$.

(Wisconsin)

Solution:

As $b \ll a$, the magnetic field at the small loop created by the large loop can be considered approximately as the magnetic field on the axis of the large loop, namely

$$B = \frac{\mu_0 a^2 I}{2(a^2 + c^2)^{3/2}},$$

where I is the current in the large loop. Hence the magnetic flux crossing the small loop is

$$\psi_{12} = \frac{\mu_0 a^2 I}{2(a^2 + c^2)^{3/2}} \cdot \pi b^2$$

and the mutual inductance is

$$M_{12} = \frac{\psi_{12}}{I} = \frac{\pi \mu_0 a^2 b^2}{2(a^2 + c^2)^{3/2}} \, .$$

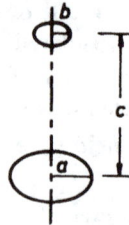

Fig. 2.33

2047

A closely wound search coil has an area of 4 cm², 160 turns and a resistance of 50 Ω. It is connected to a ballistic galvanometer whose resistance is 30 Ω. When the coil rotates quickly from a position parallel to a uniform magnetic field to one perpendicular, the galvanometer indicates a charge of 4×10^{-5} C. What is the flux density of the magnetic field?

(Wisconsin)

Solution:

Suppose the coil rotates from a position parallel to the uniform magnetic field to one perpendicular in time Δt. Since Δt is very short, we have

$$\varepsilon = \frac{\Delta \phi}{\Delta t} = i(R + r) \, .$$

As $q = i \Delta t$, the increase of the magnetic flux is

$$\Delta \phi = q(R + r) = BAN \, ,$$

since the coil is now perpendicular to the field. Hence the magnetic flux density is

$$B = \frac{(R+r)q}{AN} = \frac{(50+30) \times (4 \times 10^{-5})}{4 \times 10^{-4} \times 160}$$
$$= 0.05 \text{ T} = 50 \text{ Gs}.$$

2048

Two coaxial circular turns of wire of radii a and b are separated by a distance x and carry currents i_a and i_b respectively. Assume $a \gg b$.

(a) What is the mutual inductance?

(b) What is the force between the currents?

(*Wisconsin*)

Solution:

(a) As in Problem 2046, the mutual inductance is

$$M_{12} = \frac{\pi\mu_0 a^2 b^2}{2(a^2 + x^2)^{3/2}}.$$

(b) Consider the small coil b as a magnetic dipole of moment $m_b = \pi b^2 i_b$. The force on it is given by

$$F = m_b \left| \frac{\partial B_a}{\partial x} \right| = \frac{\pi\mu_0 a^2 b^2 i_a i_b}{2} \cdot \frac{3x}{(a^2 + b^2)^{5/2}}.$$

If the currents in the two coils are in the same direction, the force will be an attraction. If the directions of the currents are opposite, the force will be a repulsion.

2049

A d.c. electromagnet is to be constructed by winding a coil of N turns tightly on an iron yoke shaped like a doughnut with a small slab sliced out to form the gap as in Fig. 2.34. The radii for the doughnut are a and b and the width of the gap is W. The permeability μ for the iron can be assumed constant and large. A wire of radius r and resistivity ρ is to be used for the coil. The completed magnet will be operated by placing the coil across a d.c. power supply of voltage V. For simplicity, assume that $b/a \gg 1$ and $a/r \gg 1$. Derive expressions for the following quantities:

(a) The steady state value for the magnetic field in the gap.

(b) The steady state value for the power consumed in the coil.

(c) The time constant governing the response of the current in the coil to an abrupt change in V.

(UC, Berkeley)

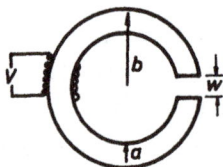

Fig. 2.34

Solution:

(a) In the steady state, as $\nabla \cdot \mathbf{B} = 0$ and the cross section of the iron yoke is the same everywhere, B must be a constant in the yoke. Applying $\oint \mathbf{H} \cdot d\mathbf{l} = NI$ to the doughnut, we have

$$\frac{B}{\mu}(2\pi b - W) + \frac{B}{\mu_0}W = NI,$$

giving

$$B = \frac{NI\mu_0\mu}{\mu_0(2\pi b - W) + \mu W} \approx \frac{NI\mu_0\mu}{\mu_0 2\pi b + \mu W}.$$

As

$$I = \frac{V}{R} = \frac{V}{\rho\frac{N2\pi a}{\pi r^2}} = \frac{Vr^2}{2a\rho N},$$

the steady state value for the magnetic field in the gap is

$$B = \frac{\mu_0\mu Vr^2}{2a\rho(2\pi b\mu_0 + W\mu)}.$$

(b) The steady state value for the power consumed in the coil is

$$P = IV = \frac{V^2r^2}{2a\rho N}.$$

(c) The self-inductance of the coil is

$$L = \frac{NB\pi a^2}{I} = \frac{N^2 \mu_0 \mu \pi a^2}{\mu_0 2\pi b + \mu W},$$

so the time constant governing the response of the current in the coil to an abrupt change in V is

$$\tau = \frac{L}{R} = \frac{N^2 \mu_0 \mu \pi a^2}{\mu_0 2\pi b + \mu W} \bigg/ \frac{\rho N 2\pi a}{\pi r^2} = \frac{N \mu_0 \mu a \pi r^2}{2\rho(\mu_0 2\pi b + \mu W)}.$$

2050

A very long solenoid of n turns per unit length carries a current which increases uniformly with time, $i = Kt$.

(a) Calculate the magnetic field inside the solenoid at time t (neglect retardation).

(b) Calculate the electric field inside the solenoid.

(c) Consider a cylinder of length l and radius equal to that of the solenoid and coaxial with the solenoid. Find the rate at which energy flows into the volume enclosed by this cylinder and show that it is equal to $\frac{d}{dt}(\frac{1}{2}lLi^2)$, where L is the self-inductance per unit length of the solenoid.

(*UC, Berkeley*)

Solution:

Use cylindrical coordinates (r, θ, z) with the z-axis along the axis of the solenoid.

(a) Applying Ampère's circuital law $\oint \mathbf{H} \cdot d\mathbf{l} = i$ to a rectangle with the long sides parallel to the z-axis, one inside and one outside the solenoid, we obtain $H = ni$, or

$$\mathbf{B} = \mu_0 n K t \mathbf{e}_z.$$

(b) Maxwell's equation $\nabla \times \mathbf{E} = -\dot{\mathbf{B}}$ gives

$$\frac{1}{r}\left[\frac{\partial}{\partial r}(r E_\theta) - \frac{\partial E_r}{\partial \theta_r}\right] = -\mu_0 n K.$$

Noting that by symmetry E does not depend on θ and integrating, we have

$$\mathbf{E} = -\frac{\mu_0 n K r}{2}\mathbf{e}_\theta.$$

(c) The Poynting vector is

$$\mathbf{N} = \mathbf{E} \times \mathbf{H} = -\frac{\mu_0 n^2 K^2}{2} r t \mathbf{e}_r \ .$$

So energy flows into the cylinder along the radial directions. The energy flowing in per unit time is then

$$\frac{dW}{dt} = 2\pi r l N = \mu_0 V n^2 K^2 t \ ,$$

where V is the volume of the cylinder. The self-inductance per unit length of the solenoid is

$$L = \frac{n B \pi r^2}{i} = \mu_0 n \pi r^2 \ .$$

Hence

$$\frac{d}{dt}\left(\frac{1}{2} l L i^2\right) = \mu_0 V n^2 K^2 t = \frac{dW}{dt} \ .$$

2051

Consider a rectangular loop of wire, of width a and length b, rotating with an angular velocity ω about the axis PQ and lying in a uniform, time dependent magnetic field $B = B_0 \sin \omega t$ perpendicular to the plane of the loop at $t = 0$ (see Fig. 2.35). Find the emf induced in the loop, and show that it alternates at twice the frequency $f = \frac{\omega}{2\pi}$.

(Columbia)

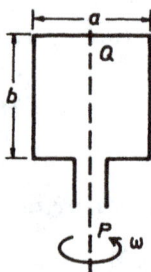

Fig. 2.35

Solution:

The magnetic flux crossing the loop is

$$\phi = \mathbf{B} \cdot \mathbf{S} = B_0 ab \sin(\omega t)\cos(\omega t) = \frac{1}{2} B_0 ab \sin(2\omega t) \ .$$

So the induced emf is

$$\varepsilon = -\frac{d\phi}{dt} = -B_0 ab\omega \cos(2\omega t).$$

Its alternating frequency is $\frac{2\omega}{2\pi} = 2 \cdot \frac{\omega}{2\pi} = 2f$.

2052

A rectangular coil of dimensions a and b and resistance R moves with constant velocity v into a magnetic field **B** as shown in Fig. 2.36. Derive an expression for the vector force on the coil in terms of the given parameters.

(*Wisconsin*)

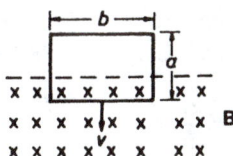

Fig. 2.36

Solution:

As it starts to cut across the magnetic field lines, an emf is induced in the coil of magnitude

$$\varepsilon = -\int \mathbf{B} \times \mathbf{v} \cdot d\mathbf{l} = -Bvb$$

and produces a current of

$$I = \frac{\varepsilon}{R} = -\frac{Bvb}{R}.$$

The minus sign indicates that the current flows counterclockwise. The force on the coil is

$$F = \left| \int I d\mathbf{l} \times \mathbf{B} \right| = IbB = -\frac{vb^2 B^2}{R}.$$

The direction of this force is opposite to **v**. That is, the force opposes the motion which tends to increase the cutting of the magnetic field lines.

2053

A constant force F is applied to a sliding wire of mass m. The wire starts from rest. The wire moves through a region of constant magnetic field B. Assume that the sliding contacts are frictionless and that the self-inductance of the loop can be ignored.

(a) Calculate the velocity of the wire as a function of time.

(b) Calculate the current through the resistor R as a function of time. What is the direction of the current?

(Wisconsin)

Solution:

(a) As the wire moves through the uniform magnetic field an emf $\varepsilon = Blv$ will be induced in it, where l is the length of the wire in the field and v is its speed. This causes a current to flow in the wire of magnitude $I = \varepsilon/R$, R being the resistance of the wire, because of which a magnetic force $|Id\mathbf{l} \times \mathbf{B}| = IlB$ acts on the wire. This force opposes the motion of the wire. Thus the equation of the motion of the wire is

$$m\frac{dv}{dt} = F - \frac{B^2l^2}{R}v \ .$$

Solving it we have

$$v(t) = \frac{RF}{B^2l^2} + C\exp\left(-\frac{B^2l^2}{mR}t\right) \ .$$

As $v = 0$ at $t = 0$, we find $C = -\frac{RF}{B^2l^2}$. Hence

$$v(t) = \frac{FR}{B^2l^2}\left[1 - \exp\left(-\frac{B^2l^2}{mR}t\right)\right] \ .$$

(b) The current is

$$I(t) = \frac{Blv(t)}{R} = \frac{F}{Bl}\left[1 - \exp\left(-\frac{B^2l^2}{mR}t\right)\right] \ .$$

2054

A rectangle of perfectly conducting wire having sides a and b, mass M, and self-inductance L, moves with an initial velocity v_0 in its plane, directed along its longest side, from a region of zero magnetic field into a region with a field B_0 which is uniform and perpendicular to the plane of the rectangle. Describe the motion of the rectangle as a function of time.

(Columbia)

Solution:

Taking $b > a$, the rectangle will move along side b and the equation of the motion is

$$m\frac{dv}{dt} = -B_0 a I,$$

where I is the current induced in the conducting wire given by

$$L\frac{dI}{dt} = B_0 a v.$$

The above two differential equations combine to give

$$\frac{d^2 v}{dt^2} + \omega^2 v = 0$$

with $\omega = \frac{B_0 a}{\sqrt{mL}}$. Solving this equation, we obtain the velocity of the rectangle

$$v = C_1 \sin \omega t + C_2 \cos \omega t.$$

As $v = v_0$ at $t = 0$ we get $C_2 = v_0$; and as $I = 0$ at $t = 0$, we get $C_1 = 0$. Hence

$$v = v_0 \cos \omega t.$$

The displacement of the rectangle of wire (with $s = 0$ at $t = 0$) is

$$s = \frac{v_0}{\omega} \sin \omega t.$$

2055

A rectangular loop of wire with dimensions l and w is released at $t = 0$ from rest just above a region in which the magnetic field is B_0 as shown in Fig. 2.37. The loop has resistance R, self-inductance L, and mass m. Consider the loop during the time that it has its upper edge in the zero field region.

(a) Assume that the self-inductance can be ignored but not the resistance, and find the current and velocity of the loop as functions of time.

(b) Assume that the resistance can be ignored but not the self-inductance, and find the current and velocity of the loop as functions of time.

$$(MIT)$$

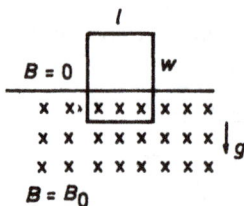

Fig. 2.37

Solution:

During the time stated above, we have

$$\varepsilon = Blv \,,$$

$$\varepsilon - L\frac{dI}{dt} = IR \,,$$

$$F = mg - BIl = m\frac{dv}{dt} \,.$$

(a) Using the results of Problem **2053**, we have

$$v = \frac{mgR}{B^2l^2}\left[1 - \exp\left(-\frac{B^2l^2}{mR}t\right)\right] \,,$$

$$I = \frac{mg}{Bl}\left[1 - \exp\left(-\frac{B^2l^2}{mR}t\right)\right] \,.$$

(b) $R = 0$. We have $L\frac{dI}{dt} = mlv$ and the equation of the motion is

$$m\frac{dv}{dt} = mg - BIl \,.$$

These give

$$\frac{d^2v}{dt} + \omega^2 v = 0 \,,$$

where $\omega^2 = \frac{B^2l^2}{mL}$. The general solution is

$$v = c_1 \cos \omega t + c_2 \sin \omega t \,.$$

As $v = 0, I = 0$ at $t = 0$, we find $c_1 = 0, c_2 = \frac{g}{\omega}$. Hence

$$v = \frac{g}{\omega} \sin \omega t,$$

$$I = \frac{mg}{Bl}(1 - \cos \omega t),$$

with $\omega = \frac{Bl}{\sqrt{mL}}$.

2056

As in Fig. 2.38 a long straight wire pointing in the y direction lies in a uniform magnetic field Be_x. The mass per unit length and resistance per unit length of the wire are ρ and λ respectively. The wire may be considered to extend to the edges of the field, where the ends are connected to one another by a massless perfect conductor which lies outside the field. Fringing effects can be neglected. If the wire is allowed to fall under the influence of gravity ($\mathbf{g} = -ge_z$), what is its terminal velocity as it falls through the magnetic field?

(*MIT*)

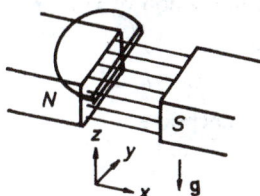

Fig. 2.38

Solution:

As the wire cuts across the lines of induction an emf is induced and produces current. Suppose the length of the wire is l and the terminal velocity is $\mathbf{v} = -ve_z$. The current so induced is given by

$$-\int_l \mathbf{B} \times \mathbf{v} \cdot d\mathbf{l} = i\lambda l.$$

Thus

$$i = -\frac{vBl}{\lambda l} = -\frac{vB}{\lambda},$$

flowing in the $-y$ direction. The magnetic force acting on the wire is

$$\mathbf{F} = \int i d\mathbf{l} \times \mathbf{B} = iBl\mathbf{e}_z = \frac{vB^2l}{\lambda}\mathbf{e}_z .$$

When the terminal velocity is reached this force is in equilibrium with the gravitation. Hence the terminal velocity of the wire is given by

$$\frac{vB^2l}{\lambda} = \rho l g ,$$

i.e.,

$$v = \frac{\rho g \lambda}{B^2}$$

or

$$\mathbf{v} = -\frac{\rho g \lambda}{B^2}\mathbf{e}_z .$$

2057

As in Fig. 2.39, what is the direction of the current in the resistor r (from A to B or from B to A) when the following operations are performed? In each case give a brief explanation of your reasoning?

(a) The switch S is closed.

(b) Coil 2 is moved closer to coil 1.

(c) The resistance R is decreased.

(*Wisconsin*)

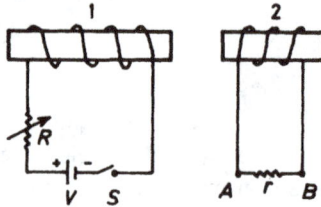

Fig. 2.39

Solution:

In all the three cases the magnetic field produced by coil 1 at coil 2 is increased. Lenz's law requires the magnetic field produced by the induced current in coil 2 to be such that as to prevent the increase of the magnetic field crossing coil 2. Applying the right-hand rule we see that the direction of the current in resistor r is from B to A.

2058

A piece of copper foil is bent into the shape as illustrated in Fig. 2.40. Assume $R = 2$ cm, $l = 10$ cm, $a = 2$ cm, $d = 0.4$ cm. Estimate the lowest resonant frequency of this structure and the inductance measured between points A and B, when the inductance is measured at a frequency much lower than the resonant frequency.

(UC, Berkeley)

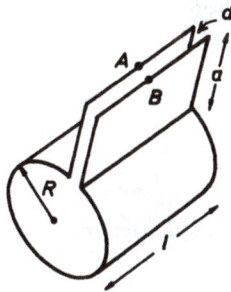

Fig. 2.40

Solution:

Consider the current in the copper foil. As $d \ll R$, we can consider the currents in two sides of the cylinder to have same phase. That is to say, the current enters from one side and leaves from the other with the same magnitude. Accordingly, the maximum wavelength is $2\pi R$. Along the axial direction, the current densities are zero at both ends of the cylinder so that the maximum half-wavelength is l, or the maximum wavelength is $2l$. As $2l \gg 2\pi R$, the maximum wavelength is $2l = 20$ cm, or the lowest resonant frequency is

$$f_0 = \frac{c}{2l} = \frac{3 \times 10^{10}}{20} = 1.5 \times 10^9 \text{ Hz}.$$

When the frequency is much lower than f_0, we can consider the current as uniformly distributed over the cylindrical surface and varying slowly with time. As a result, we are essentially dealing with a static situation. Ignoring edge effects, the magnetic induction inside the structure is

$$B = \mu_0 i = \frac{\mu_0 I}{l}.$$

The magnetic flux crossing a cross section of the structure is

$$\phi = BS = \frac{\mu_0 I}{l}(\pi R^2 + ad),$$

giving an inductance

$$L = \frac{\phi}{I} = \frac{\mu_0}{l}(\pi R^2 + ad)$$
$$= \frac{4\pi \times 10^{-7} \times (\pi \times 0.02^2 + 0.02 \times 0.004)}{0.1} = 1.68 \times 10^{-8} \text{ H}.$$

2059

A magnetized uncharged spherical conductor of radius R has an internal magnetic field given by

$$\mathbf{B(r)} = Ar_\perp^2 \hat{\mathbf{K}},$$

where A is a constant, $\hat{\mathbf{K}}$ is a constant unit vector through the center of the sphere and $r_\perp$ is the distance of the point $\mathbf{r}$ to the $\hat{\mathbf{K}}$ axis. (In a Cartesian coordinate system as in Fig. 2.41, $\hat{\mathbf{K}}$ is in the z-direction, the sphere's center is at the origin, and $r_\perp^2 = x^2 + y^2$.) The sphere is now spun (non-relativistically) about its z-axis with angular frequency ω.

(a) What electric field (in the "laboratory frame") exists in the interior of the spinning sphere?

(b) What is the electric charge distribution? (Do not calculate any surface charge.)

(c) What potential drop is measured by a stationary voltmeter (Fig. 2.42), one of whose ends is at the pole of the spinning sphere and whose other end brushes the sphere's moving equator?

(CUSPEA)

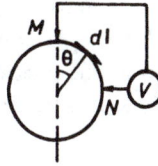

Fig. 2.41   Fig. 2.42

Solution:

(a) At the point P of radius vector $\mathbf{r}$ the magnetic field is $\mathbf{B} = Ar_\perp^2 \mathbf{e}_z$. The velocity of P is

$$\mathbf{v} = \boldsymbol{\omega} \times \mathbf{r} = \omega \mathbf{e}_z \times \mathbf{r}.$$

For a free charge q to remain stationary inside the sphere, the total force on it, $f = q(\mathbf{E} + \mathbf{v} \times \mathbf{B})$, must vanish. Thus the electric field intensity at P is

$$\mathbf{E}(\mathbf{r}) = -\mathbf{v} \times \mathbf{B} = -A\omega r_\perp^2 (\mathbf{e}_z \times \mathbf{r}) \times \mathbf{e}_z$$
$$= -A\omega(x^2 + y^2)(x\mathbf{e}_x + y\mathbf{e}_y).$$

(b) By $\nabla \cdot \mathbf{E} = \frac{\rho}{\varepsilon}$, we can get the volume charge density inside the sphere,

$$\rho = \varepsilon \nabla \cdot \mathbf{E} = -4\frac{A\omega r_\perp^2}{\varepsilon},$$

where ε is the permittivity of the conductor.

(c) To find V we integrate from N to M along a great circle of radius R (see Fig. 2.42):

$$V = -\int_M^N \mathbf{E} \cdot d\mathbf{l}.$$

In spherical coordinates $d\mathbf{l} = Rd\theta \mathbf{e}_\theta$, and for a point (x, y, z), $r_\perp = r\sin\theta$, $x\mathbf{e}_x + y\mathbf{e}_y = r_\perp(\sin\theta \mathbf{e}_r + \cos\theta \mathbf{e}_\theta)$. Thus the electric field on the surface is

$$\mathbf{E} = -A\omega R^3 \sin^3\theta(\sin\theta \mathbf{e}_r + \cos\theta \mathbf{e}_\theta),$$

giving

$$V = A\omega R^3 \int_0^{\frac{\pi}{2}} \sin^3\theta \cos\theta \, d\theta = \frac{A\omega R^3}{4}.$$

2060

Consider a square loop of wire, of side length l, lying in the x, y plane as shown in Fig. 2.43. Suppose a particle of charge q is moving with a constant velocity v, where $v \ll c$, in the xz-plane at a constant distance z_0 from the xy-plane. (Assume the particle is moving in the positive x direction.) Suppose the particle crosses the z-axis at $t = 0$. Give the induced emf in the loop as a function of time.

(*Columbia*)

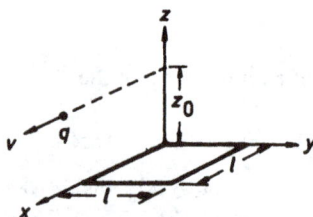

Fig. 2.43

Solution:

At time t, the position of q is $(vt, 0, z_0)$. The radius vector $\mathbf{r}$ from q to a field point (x, y, z) is $(x - vt, y, z_0)$. As $v \ll c$, the electromagnetic field due to the uniformly moving charge is

$$\mathbf{E}(\mathbf{r}, t) = \frac{q}{4\pi\varepsilon_0} \frac{\mathbf{r}}{r^3} = \frac{q}{4\pi\varepsilon_0 r^3}[(x - vt)\mathbf{i} + y\mathbf{j} + z\mathbf{k}],$$

$$\mathbf{B}(\mathbf{r}, t) = \frac{\mathbf{v}}{c^2} \times \mathbf{E} = \frac{\mu_0 q v}{4\pi r^3}(-z\mathbf{j} + y\mathbf{k}),$$

with

$$r = [(x - vt)^2 + y^2 + (z - z_0)^2]^{\frac{1}{2}}.$$

The induced emf in the loop is given by the integral

$$\varepsilon = -\int_S \frac{\partial \mathbf{B}}{\partial t} \cdot d\mathbf{S}$$

where S is the area of the loop and $dS = dxdy$. Thus

$$\varepsilon = -\frac{\mu_0 q v}{4\pi} \int_0^l dx \frac{\partial}{\partial t} \int_0^l dy \frac{y}{[(x-vt)^2 + y^2 + (z-z_0)^2]^{3/2}}$$

$$= -\frac{\mu_0 q v}{4\pi} \int_0^l dx \frac{\partial}{\partial t} \left[\frac{1}{\sqrt{(x-vt)^2 + (z-z_0)^2}} \right.$$

$$\left. - \frac{1}{\sqrt{(x-vt)^2 + l^2 + (z-z_0)^2}} \right]$$

$$= -\frac{\mu_0 q v^2}{4\pi} \int_0^l dx \left[\frac{x-vt}{[(x-vt)^2 + (z-z_0)^2]^{3/2}} \right.$$

$$\left. - \frac{x-vt}{[(x-vt)^2 + l^2 + (z-z_0)^2]^{3/2}} \right]$$

$$= -\frac{\mu_0 q v^2}{4\pi} \left\{ \frac{1}{\sqrt{v^2 t^2 + (z-z_0)^2}} - \frac{1}{\sqrt{(l-vt)^2 + (z-z_0)^2}} \right.$$

$$\left. - \frac{1}{\sqrt{(l-vt)^2 + l^2 + (z-z_0)^2}} + \frac{1}{\sqrt{v^2 t^2 + l^2 + (z-z_0)^2}} \right\}.$$

2061

A very long insulating cylinder (dielectric constant ε) of length L and radius $R (L \gg R)$ has a charge Q uniformly distributed over its outside surface. An external uniform electric field is applied perpendicular to the cylinder's axis: $\mathbf{E} = E_0 \mathbf{e}_z$ (see Fig. 2.44). Ignore edge effects.

(a) Calculate the electric potential everywhere (i.e. inside and outside the cylinder).
Now the electric field E_0 is removed and the cylinder is made to rotate with angular velocity ω.

(b) Find the magnetic field (magnitude and direction) inside the cylinder.

(c) A single-turn coil of radius $2R$ and resistance ρ is wrapped around the cylinder as shown in Fig. 2.45, and the rotation of the cylinder is slowed down linearly ($\omega(t) = \omega_0(1 - t/t_0)$) as a function of time. What is the magnitude of the induced current in the coil? In what direction does the current flow?

(d) Instead of the coil of part (c), a one-turn coil is placed through the cylinder as shown in Fig. 2.46, and the cylinder is slowed down as before. How much current will now flow?

(Princeton)

Solution:

(a) By the superposition principle the electric potential can be treated as the superposition of the potentials due to Q and $\mathbf{E}$. The potential due to Q is

$$\varphi_1(r) = \begin{cases} 0, & r < R, \\ -\frac{Q}{2\pi\epsilon_0 L} \ln \frac{r}{R}, & r > R. \end{cases}$$

Here the potential is taken to be zero at the cylinder's center.

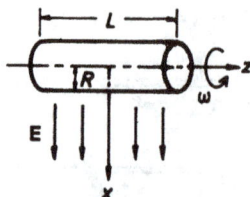

Fig. 2.44

Let φ_2 be the potential due to the uniform field $\mathbf{E}$. Then $\nabla^2 \varphi_2 = 0$ $(r \neq R)$, or in cylindrical coordinates

$$\frac{1}{r} \frac{\partial}{\partial r} \left(r \frac{\partial \varphi_2}{\partial r} \right) + \frac{1}{r^2} \frac{\partial^2 \varphi_2}{\partial \theta^2} = 0.$$

We separate the variables and obtain the general solution

$$\varphi_2 = \begin{cases} \sum_n [r^n(a_n \cos n\theta + b_n \sin n\theta) + \frac{1}{r^n}(c_n \cos n\theta + d_n \sin n\theta)], & r < R \\ \sum_n [r^n(e_n \cos n\theta + f_n \sin n\theta) + \frac{1}{r^n}(g_n \cos n\theta + h_n \sin n\theta)], & r > R. \end{cases}$$

From the boundary condition $\varphi_2 = -E_0 r \cos \theta$ for $r \to \infty$ we get $e_1 = -E_0$, $f_1 = 0$, $e_n = f_n = 0$ for $n \neq 1$. From the condition $\varphi_2 = 0$ for $r \to 0$ we get $c_n = d_n = 0$ for all n. For $r = R$, we have boundary conditions

$$\varphi_2 \Big|_{r=R-} = \varphi_2 \Big|_{r=R+}, \quad \epsilon \frac{\partial \varphi_2}{\partial r} \Big|_{r=R-} = \epsilon_0 \frac{\partial \varphi_2}{\partial r} \Big|_{r=R+},$$

which give the simultaneous equations

$$
\begin{cases}
\sum_n R^n (a_n \cos n\theta + b_n \sin n\theta) = -RE_0 \cos\theta \\
\qquad\qquad\qquad\qquad\qquad\qquad + \sum_n \frac{1}{R^n}(g_n \cos n\theta + h_n \sin n\theta) , \\
\varepsilon \sum_n n R^{n-1}(a_n \cos\theta + b_n \sin n\theta) = -\varepsilon_0 E_0 \cos\theta \\
\qquad\qquad\qquad\qquad\qquad\qquad - \varepsilon_0 \sum_n \frac{n}{R^{n+1}}(g_n \cos n\theta + h_n \sin n\theta) .
\end{cases}
$$

These have the solution

$$
a_1 = -\frac{2E_0}{\varepsilon + \varepsilon_0} , \quad g_1 = \frac{(\varepsilon - \varepsilon_0)R^2 E_0}{\varepsilon + \varepsilon_0} , \quad b_1 = h_1 = 0 ,
$$
$$
a_n = b_n = g_n = h_n = 0 \quad \text{for} \quad n \neq 1 .
$$

Thus

$$
\varphi_2 = \begin{cases}
-\frac{2E_0}{\varepsilon + \varepsilon_0} r \cos\theta , & r < R \\
-E_0 r \cos\theta + \frac{(\varepsilon - \varepsilon_0)R^2 E_0}{\varepsilon + \varepsilon_0} , & r > R .
\end{cases}
$$

Hence the total electric potential is

$$
\varphi = \varphi_1 + \varphi_2 = \begin{cases}
-\frac{2E_0}{\varepsilon + \varepsilon_0} r \cos\theta , & r < R \\
-\frac{Q}{2\pi\varepsilon_0 L} \ln\frac{r}{R} - E_0 r \cos\theta + \frac{\varepsilon - \varepsilon_0}{\varepsilon + \varepsilon_0}\frac{R^2 E_0}{r} \cos\theta , & r > R .
\end{cases}
$$

(b) With **E** removed and the cylinder rotating about its axis with angular velocity ω, a surface current of density $\frac{Q}{2\pi RL} \cdot \omega R = \frac{Q\omega}{2\pi L}$ is generated. By Ampère's circuital law we find

$$
\mathbf{B} = \frac{\mu_0 Q\omega}{2\pi L}\mathbf{e}_z
$$

for the interior of the cylinder.

(c) The magnetic flux passing through the single-turn coil shown in Fig. 2.45 is

$$
\phi = \pi R^2 B = \frac{\mu_0 Q\omega R^2}{2L} ,
$$

as there is no flux outside the cylinder. The induced emf is therefore

$$
\varepsilon = -\frac{d\phi}{dt} = \frac{\mu_0 Q R^2}{2L}\left(-\frac{d\omega}{dt}\right) = \frac{\mu_0 Q R^2 \omega_0}{2L t_0}
$$

and the induced current is

$$i = \frac{\varepsilon}{\rho} = \frac{\mu_0 Q R^2 \omega_0}{2\rho L t_0}.$$

By Lenz's law the direction of i is that of rotation.

(d) There is no magnetic flux crossing the coil shown in Fig. 2.46, so no current is induced in the coil.

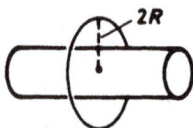   

Fig. 2.45   Fig. 2.46

2062

Consider a closed circuit of wire formed into a coil of N turns with radius a, resistance R, and self-inductance L. The coil rotates in a uniform magnetic field **B** about a diameter perpendicular to the field.

(a) Find the current in the coil as a function of θ for rotation at a constant angular velocity ω. Here $\theta(t) = \omega t$ is the angle between the plane of the coil and **B**.

(b) Find the externally applied torque required to maintain this uniform rotation. (In both parts you should assume that all transient effects have died away.)

(CUSPEA)

Solution:

(a) The emf induced in the coil is given by

$$\varepsilon = -\frac{d}{dt}\int_S \mathbf{B}\cdot d\mathbf{S}.$$

Noting that, as the vector $d\mathbf{S}$ is normal to the plane of the coil, $\mathbf{B}\cdot d\mathbf{S} = B\cos(\frac{\pi}{2} - \theta)dS$, we have, with $\theta = \omega t$,

$$\varepsilon = -\frac{d}{dt}\int_S B\sin(\omega t)dS = -\frac{d}{dt}\left[\pi a^2 N B \sin(\omega t)\right]$$
$$= -\pi a^2 \omega N B \cos(\omega t)$$
$$= -\text{Re}\left[\pi a^2 \omega N B \exp(i\omega t)\right].$$

The current in the circuit is given by

$$L\frac{dI}{dt} + IR = \varepsilon.$$

Let $I = I_0 \exp(i\omega t)$. The above gives

$$I_0 = \frac{-\pi a^2 \omega N B}{i\omega L + R} = \frac{\pi a^2 \omega N B}{\sqrt{\omega^2 L^2 + R^2}} e^{-i(\frac{\pi}{2}+\varphi)},$$

where $\varphi = \arctan(\frac{\omega L}{R})$.

Fig. 2.47

Thus we have

$$I(t) = \frac{\pi a^2 \omega N B}{\sqrt{\omega^2 L^2 + R^2}} \cos\left(\omega t - \varphi - \frac{\pi}{2}\right)$$

$$= \frac{\pi a^2 \omega N B}{\sqrt{\omega^2 L^2 + R^2}} \sin\left(\omega t - \varphi\right).$$

(b) The magnetic dipole moment of the coil is

$$\mathbf{m} = I\pi a^2 N \mathbf{n},$$

where $\mathbf{n}$ is a unit vector normal to the coil. At time t the external torque on the coil $\boldsymbol{\tau} = \mathbf{m} \times \mathbf{B}$ has magnitude

$$\tau = |\mathbf{m} \times \mathbf{B}| = I\pi a^2 N B \sin\left(\frac{\pi}{2} - \theta\right) = \frac{(\pi a^2 N B)^2 \omega}{\sqrt{R^2 + L^2\omega^2}} \cos(\omega t)\sin(\omega t - \varphi).$$

2063

You are equipped with current sources and a machine shop for constructing simple linear electric components such as coils, inductors, capacitors, and resistors. You have instruments to measure mechanical forces but no electrical meters. Devise an experiment to measure the ampere given the above equipment and your knowledge of the basic equations of electricity and magnetism.

(*Chicago*)

Solution:

Make two identical circular coils and arrange them co-axially under a pan of a balance to construct an Ampère's balance, as shown schematically in Fig. 2.48. The mutual inductance between the coils is $M_{12}(z)$. The coils are connected to the same current source. With standard weights on the pans the force F_{12} between the two coils can be measured. The mutual inductance part of the magnetic energy stored in the two coils is

$$W_{12} = M_{12}I_1 I_2 = M_{12}I^2 \,,$$

as the coils are connected to the same source. Hence the interacting force is

$$F_{12} = \frac{\partial W_{12}}{\partial z} = I^2 \frac{\partial M_{12}(z)}{\partial z} \,.$$

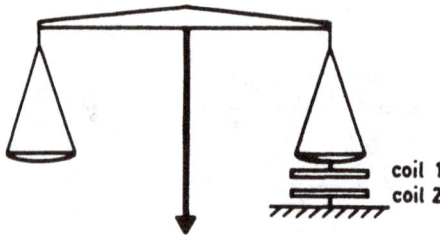

Fig. 2.48

Using the value of the force measured with the Ampère's balance and calculating $\frac{\partial M_{12}(z)}{\partial z}$, I can be determined.

Using the MKSA unit system the value of the current so determined is in amperes.

3. ACTION OF ELECTROMAGNETIC FIELD ON CURRENT-CARRYING CONDUCTORS AND CHARGED PARTICLES (2064–2090)

2064

Two parallel wires carry currents i_1 and i_2 going in the same direction.

The wires:

    (a) attract each other

    (b) repel each other

    (c) have no force on each other.

                                                                      (*CCT*)

Solution:

    The answer is (a).

2065

    Two mutually perpendicular long wires are separated a distance a and carry currents I_1 and I_2. Consider a symmetrically located segment $(-\frac{l}{2}, \frac{l}{2})$ of I_2 of length $l \ll a$ as shown in Fig. 2.49.

    (a) What are the net force and net torque on this segment?

    (b) If the wires are free to rotate about the connecting line a, what configuration will they assume? Does this correspond to a maximum or a minimum in the energy stored in the magnetic field?

                                                                      (*Wisconsin*)

Solution:

    (a) The magnetic field at a point $(0, a, z)$ produced by I_1 is

$$\mathbf{B}_1 = \frac{\mu_0 I_1}{2\pi\sqrt{a^2 + z^2}} \left[\frac{z}{\sqrt{a^2 + z^2}}\mathbf{e}_y - \frac{a}{\sqrt{a^2 + z^2}}\mathbf{e}_z \right],$$

so that a small current element $(z, z + dz)$ on I_2 will experience a force

$$d\mathbf{F}_{21} = I_2 dz \mathbf{e}_z \times \mathbf{B}_1 = I_2 dz \frac{\mu_0 I_1}{2\pi\sqrt{a^2 + z^2}} \frac{z}{\sqrt{a^2 + z^2}}(-\mathbf{e}_x)$$

$$= \frac{\mu_0 I_1 I_2 z dz}{2\pi(a^2 + z^2)}(-\mathbf{e}_x).$$

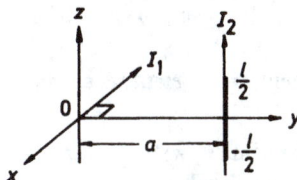

Fig. 2.49

Thus the force acting on the small segment $(-\frac{l}{2}, \frac{l}{2})$ is

$$\mathbf{F}_{21} = \int_{-l/2}^{l/2} d\mathbf{F}_{21} = 0$$

as the integrand is an odd function of z, and the torque on it is

$$\boldsymbol{\tau} = \int_{-l/2}^{l/2} z\mathbf{e}_z \times d\mathbf{F}_{21} = -\frac{\mu_0 I_1 I_2}{2\pi} \int_{-l/2}^{l/2} \frac{z^2 dz}{a^2 + z^2} (\mathbf{e}_z \times \mathbf{e}_x)$$

$$\approx -\frac{\mu_0 I_1 I_2}{2\pi a^2} \cdot \frac{1}{3} z^3 \Big|_{-l/2}^{l/2} \cdot \mathbf{e}_y = \frac{-\mu_0 I_1 I_2}{24\pi a^2} l^3 \mathbf{e}_y \, .$$

We can conclude from this that if the current I_2 is free to rotate about the connecting line a then it will finally settle in parallel with the current I_1 such that the directions of both currents I_1 and I_2 are the same. Obviously, this position corresponds to a minimum energy stored in the magnetic field.

2066

A uniform sheet of surface current of strength λ (ampères per meter in the y direction) flows eastward (in x direction) on a horizontal plane $(z = 0)$, as shown in Fig. 2.50. What are the magnitude and direction of the force on:

(a) A horizontal segment of a wire of length l, above the sheet by a distance R, carrying a current i (amperes) in a northward direction?

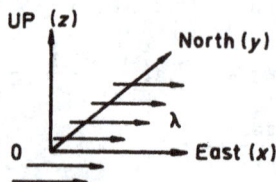

Fig. 2.50

(b) The same segment but oriented so as to carry a current in the westward direction?

(c) A loop of wire of radius r, with center above the sheet by a distance $R (r < R)$, carrying a current i whose magnetic moment is eastward?

(d) The same loop but with its magnetic moment northward?

(e) The same loop but with its magnetic moment upward?
State briefly the reason for each of your answers.

(*Wisconsin*)

Solution:

The sheet of surface current can be divided into narrow strips of width dy and each strip regarded as a current $dI = \lambda dy$. Consider a point P at $(0, 0, R)$. According to Ampère's circuital law two current strips dI_1 and dI_2 located symmetrically on two sides of the point 0 will give rise to magnetic inductions $d\mathbf{B}_1$ and $d\mathbf{B}_2$ which combine to a resultant $d\mathbf{B}$ in the $-y$ direction (see Fig. 2.51):

$$d\mathbf{B} = -\frac{2\mu_0 \lambda dy}{2\pi(y^2 + R^2)^{1/2}} \cos\theta \mathbf{e}_y ,$$

where $\cos\theta = \dfrac{R}{\sqrt{y^2 + R^2}}$.

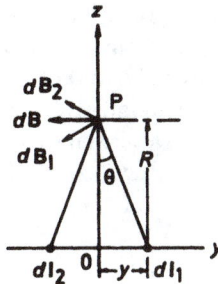

Fig. 2.51

Let $2L$ be the width of the current sheet. The total magnetic field at the point P is

$$\mathbf{B} = \int d\mathbf{B} = -\frac{\mu_0 \lambda R}{\pi} \int_0^L \frac{dy}{y^2 + R^2} \mathbf{e}_y = -\frac{\mu_0 \lambda}{\pi} \arctan\left(\frac{L}{R}\right) \mathbf{e}_y .$$

(a) The current element at P is $id\mathbf{l} = il\mathbf{e}_y$ so the force on it is

$$\mathbf{F} = il\mathbf{e}_y \times \mathbf{B} = 0 .$$

(b) The current element $id\mathbf{l} = -il\mathbf{e}_x$ and the force on it is

$$\mathbf{F} = \frac{\mu_0 il\lambda}{\pi} \arctan\left(\frac{L}{R}\right) \mathbf{e}_x \times \mathbf{e}_y = \frac{\mu_0 li\lambda}{\pi} \arctan\left(\frac{L}{R}\right) \mathbf{e}_z .$$

(c) The loop of wire carrying current i has magnetic dipole moment $\mathbf{m} = \pi r^2 i \mathbf{e}_x$. The force exerted by the magnetic field $\mathbf{B}$ on the loop is

$$\mathbf{F} = \nabla(\mathbf{m} \cdot \mathbf{B}) = \nabla(mB\mathbf{e}_x \cdot \mathbf{e}_y) = 0.$$

(d) $\mathbf{m} = \pi r^2 i \mathbf{e}_y$, and the force acting on it is

$$\mathbf{F} = \nabla(\mathbf{m} \cdot \mathbf{B}) = -\mu_0 \lambda r^2 i \frac{\partial}{\partial R}\left(\arctan\frac{L}{R}\right)\mathbf{e}_z = -\frac{\mu_0 \lambda r^2 i R^2}{R^2 + L^2}\mathbf{e}_z.$$

(e) $\mathbf{m} = \pi r^2 i \mathbf{e}_z$ and the force acting on it is

$$\mathbf{F} = \nabla(mB\mathbf{e}_z \cdot \mathbf{e}_y) = 0.$$

2067

A circular wire of radius R carries a current i electromagnetic units. A sphere of radius a $(a \ll R)$ made of paramagnetic material with permeability μ is placed with its center at the center of the circuit (see Fig. 2.52). Determine the magnetic dipole moment of the sphere resulting from the magnetic field of the current. Determine the force per unit area on the sphere.

(UC, Berkeley)

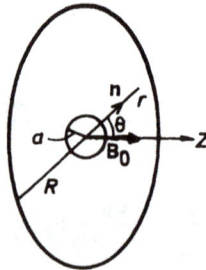

Fig. 2.52

Solution:

Take the center of the circular wire as the origin and its axis as the z-axis. The magnetic field at the origin generated by a current i in the wire is

$$\mathbf{B}_0 = \frac{\mu_0 i}{2R}\mathbf{e}_z.$$

As the radius of the small sphere $a \ll R$, we may think of the sphere as being in a uniform magnetic field $\mathbf{B}_0$ and make use of the magnetic scalar potential φ. Let φ_1 and φ_2 be the potentials outside and inside the sphere respectively. They satisfy the equations $\nabla^2\varphi_1 = \nabla^2\varphi_2 = 0$ since the inside and outside magnetizations are both uniform. We require

$$\varphi_1 \approx \varphi_0 = -\frac{B_0}{\mu_0}r\cos\theta$$

for $r \to \infty$ and φ_2 finite for $r \to 0$. Furthermore, at $r = R$ we have the boundary conditions

$$\varphi_1 = \varphi_2, \quad \mu_0\frac{\partial\varphi_1}{\partial r} = \mu\frac{\partial\varphi_2}{\partial r}.$$

Solving the Laplace's equations by separation of variables and following the procedure for solving Problem **1062** we are led to

$$\varphi_2 = -\frac{3B_0}{\mu + 2\mu_0}r\cos\theta.$$

The magnetic induction inside the sphere is then

$$\mathbf{B} = -\mu\nabla\varphi_2 = \frac{3\mu}{\mu + 2\mu_0}\mathbf{B}_0.$$

Let the magnetization of the small sphere be $\mathbf{M}$. Then as $\mathbf{B} = \mu_0(\mathbf{H}+\mathbf{M}) = \mu\mathbf{H}$ by definition we have

$$\mathbf{M} = \left(\frac{1}{\mu_0} - \frac{1}{\mu}\right)\mathbf{B} = \frac{3(\mu - \mu_0)}{\mu_0(\mu + 2\mu_0)}\mathbf{B}_0 = \frac{3(\mu - \mu_0)i}{2(\mu + 2\mu_0)R}\mathbf{e}_z.$$

The magnetic dipole moment of the sphere is then

$$\mathbf{m} = \frac{4}{3}\pi a^3\mathbf{M} = \frac{2(\mu - \mu_0)\pi a^3 i}{\mu + 2\mu_0}\mathbf{e}_z.$$

The surface current density on the sphere is given by the boundary condition

$$\boldsymbol{\alpha}_m = \mathbf{n} \times (\mathbf{H}_2 - \mathbf{H}_1).$$

As

$$\mathbf{H}_1 = \frac{\mathbf{B}_1}{\mu_0}, \quad \mathbf{H}_2 = \frac{\mathbf{B}_2}{\mu_2} - \mathbf{M}, \quad \mathbf{n} \times (\mathbf{B}_2 - \mathbf{B}_1) = 0$$

we have

$$\alpha_M = \mathbf{M} \times \mathbf{n} = \frac{3(\mu - \mu_0)i}{2(\mu + 2\mu_0)R} \sin\theta \; \mathbf{e}_\varphi .$$

Finally, the force per unit area on the sphere is

$$\mathbf{f} = \alpha_M \times \mathbf{B}_0 = \frac{3(\mu - \mu_0)i}{2(\mu + 2\mu_0)R} \sin\theta \cdot \frac{\mu_0 i}{2R}(\mathbf{e}_\varphi \times \mathbf{e}_z)$$

$$= \frac{3\mu_0(\mu - \mu_0)i^2}{4(\mu + 2\mu_0)R} \sin\theta(\cos\theta \mathbf{e}_\theta + \sin\theta \mathbf{e}_r) .$$

2068

A current loop has magnetic moment $\mathbf{m}$. The torque $\mathbf{N}$ in a magnetic field $\mathbf{B}$ is given by:

(a) $\mathbf{N} = \mathbf{m} \times \mathbf{B}$, (b) $\mathbf{N} = \mathbf{m} \cdot \mathbf{B}$, (c) $\mathbf{N} = 0$.

(CCT)

Solution:

The answer is (a).

2069

A bar magnet in the earth's field will

(a) move toward the North pole

(b) move toward the South pole

(c) experience a torque.

(CCT)

Solution:

The answer is (c).

2070

A copper penny is placed on edge in a vertical magnetic field $B = 20$ kGs. It is given a slight push to start it falling. Estimate how long it takes to fall. (Hint: The conductivity and density of Cu are $6 \times 10^5 \Omega\text{cm}^{-1}$ and 9 gcm^{-3}.)

(*Princeton*)

Solution:

Because Cu is a good conductor, the potential energy of the copper penny will be converted mainly into heat when it is falling in such a strong magnetic field. We may assume that in the falling process the magnetic torque is always in equilibrium with the gravitational torque. Let θ be the angle between the plane of the copper penny and the vertical axis. When we are considering a ring of radii r and $r + dr$, the magnetic flux crossing the area of the ring is $\phi(\theta) = \pi r^2 B \sin \theta$. The induced emf in the ring is

$$\varepsilon = \left| \frac{d\phi}{dt} \right| = \pi r^2 B \dot{\theta} \cos \theta$$

and the induced current is

$$di = \frac{\varepsilon}{R} = \frac{\pi r^2 B \dot{\theta} \cos \theta}{R} ,$$

where R is the resistance of the ring. Let h be the thickness of the penny. We then have

$$R = \frac{2\pi r}{\sigma h dr} .$$

Thus

$$di = \frac{B r \dot{\theta} \cos \theta \sigma h dr}{2} .$$

The magnetic moment of the ring is

$$dm = \pi r^2 di = \frac{\pi r^3 B \dot{\theta} \cos \theta \sigma h dr}{2} ,$$

and the magnetic torque is

$$d\tau_m = |\mathbf{dm} \times \mathbf{B}| = \frac{\pi r^2 B^2 \dot{\theta} \cos^2 \theta \sigma h dr}{2} .$$

Let the radius of the penny be r_0, then the magnetic torque on the whole copper penny is

$$\tau_m = \int d\tau_m = \int_0^{r_0} \frac{\pi B^2 \dot{\theta} \cos^2 \theta \sigma h}{2} r^3 dr = \frac{\pi r_0^4 B^2 \dot{\theta} \cos^2 \theta \sigma h}{8} .$$

The gravitational torque on the other hand is

$$\tau_g = mgr_0 \sin\theta = \pi r_0^3 \rho h g \sin\theta .$$

From $\tau_m = \tau_g$, we get

$$dt = \frac{B^2 r_0 \sigma}{8g\rho} \cdot \frac{\cos^2\theta}{\sin\theta} d\theta .$$

Suppose the penny starts falling at $\theta = \theta_0$, then the falling time will be

$$T = \int dt = \int_{\theta_0}^{\pi/2} \frac{B^2 r_0 \sigma}{8g\rho} \cdot \frac{\cos^2\theta}{\sin\theta} d\theta$$

$$= \frac{B^2 \sigma r_0}{8g\rho} \left[-\cos\theta_0 + \frac{1}{2} \ln\left(\frac{1+\cos\theta_0}{1-\cos\theta_0}\right) \right] .$$

Using the given data and taking $r_0 = 0.01$ m, $\theta_0 = 0.1$ rad., we have the estimate

$$T \approx 6.8 \text{ s} .$$

We can conclude from this that the potential energy converts mainly into heat since the time required for falling in a stong magnetic field is much longer than that when no magnetic field is present.

2071

Suppose that inside a material the following equations are valid:

$$c\nabla \times \lambda \mathbf{j} = -\mathbf{H}, \quad (\lambda \text{ constant})$$

$$\frac{\partial}{\partial t}(\lambda \mathbf{j}) = \mathbf{E},$$

rather than Ohm's law $\mathbf{j} = \sigma\mathbf{E}$. (These are known as London's equations.) Consider an infinite slab of this material of thickness $2d\,(-d < z < d)$ outside of which is a constant magnetic field parallel to the surface, $H_x = H_z = 0, H_y = H_1$ for $z < -d$ and $H_y = H_2$ for $z > d$ with $\mathbf{E} = \mathbf{D} = 0$ everywhere, as shown in Fig. 2.53. Assume that no surface currents or surface charges are present.

(a) Find $\mathbf{H}$ inside the slab.

(b) Find $\mathbf{j}$ inside the slab.

(c) Find the force per unit area on the surface of the slab.

(Princeton)

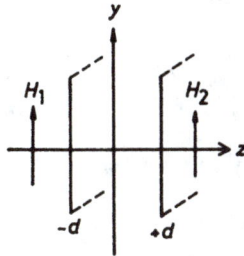

Fig. 2.53

Solution:

We use Gaussian units for this problem. In superconducting electro-dynamics there are two descriptive methods. Here we shall take the current approach, rather than treating the material as a magnetic medium. The relevant Maxwell's equations are

$$\nabla \times \mathbf{H} = \frac{4\pi}{c}\mathbf{j}, \quad \nabla \cdot \mathbf{H} = 0.$$

Since $\mathbf{E} = 0, \mathbf{j}$ is a constant current and the magnetic field is stationary. The first Maxwell's equation gives

$$\nabla \times (\nabla \times \mathbf{H}) = \frac{4\pi}{c}\nabla \times \mathbf{j},$$

i.e.,

$$\nabla(\nabla \cdot \mathbf{H}) - \nabla^2\mathbf{H} = \frac{4\pi}{c}\nabla \times \mathbf{j}.$$

Using $\nabla \cdot \mathbf{H} = 0$ and London's equations in the above we find for the material

$$\nabla^2\mathbf{H} - \frac{4\pi}{\lambda c^2}\mathbf{H} = 0.$$

From the symmetry we can assume that $\mathbf{H} = \mathbf{H}(z)$ and has only y-component, i.e., $\mathbf{H} = H_y(z)\mathbf{e}_y$. The above then becomes

$$\frac{d^2 H_y}{dz^2} - \frac{4\pi}{\lambda c^2}H_y = 0.$$

The general solution is $H_y = Ae^{-kz} + Be^{kz}$. Using the given boundary conditions we have

$$\begin{cases} H_y(d) = Ae^{-kd} + Be^{kd} = H_2, \\ H_y(-d) = Ae^{kd} + Be^{-kd} = H_1, \end{cases}$$

giving

$$A = \frac{H_1 e^{kd} - H_2 e^{-kd}}{e^{2kd} - e^{-2kd}}, \quad B = \frac{H_2 e^{kd} - H_1 e^{-kd}}{e^{2kd} - e^{-2kd}}.$$

(a) Inside the slab only the y-component is present. It is

$$H_y(z) = Ae^{-kz} + Be^{kz}$$
$$= \frac{H_2 \sinh[k(z+d)] - H_1 \sinh[k(z-d)]}{\sinh(2kd)}.$$

(b) From Maxwell's equation $\mathbf{j} = \frac{c}{4\pi} \nabla \times \mathbf{H}$ and $\mathbf{H} = H_y(z)\mathbf{e}_y$ we have $\mathbf{j} = j_x \mathbf{e}_x$ with

$$j_x = -\frac{c}{4\pi} \frac{\partial H_y}{\partial z} = -\frac{c}{4\pi} \cdot \frac{k\{H_2 \cosh[k(z+d)] - H_1 \cosh[k(z-d)]\}}{\sinh(2kd)}.$$

(c) The force on the slab is

$$\mathbf{F} = \frac{1}{c} \int \mathbf{j} \times \mathbf{H} dV$$
$$= \frac{\mathbf{e}_z}{c} \int \left(-\frac{c}{4\pi}\right) \frac{\partial H_y}{\partial z} H_y dz dS.$$

Hence the force per unit area on the surface is

$$\mathbf{f} = -\frac{\mathbf{e}_z}{4\pi} \int H_y \frac{\partial H_y}{\partial z} dz = -\frac{\mathbf{e}_z}{4\pi} \int H_y dH_y$$
$$= -\frac{1}{8\pi}(H_2^2 - H_1^2)\mathbf{e}_z.$$

2072

A long thin wire carrying a current I lies parallel to and at a distance d from a semi-infinite slab of iron, as shown in Fig. 2.54. Assuming the iron to have infinite permeability, determine the magnitude and direction of the force per unit length on the wire.

(*UC, Berkeley*)

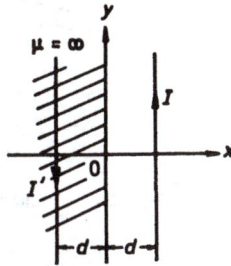

Fig. 2.54

Solution:

Use the method of images. The image current is I', located at $x = -d$ and opposite in direction to I with magnitude

$$I' = \frac{\mu - \mu_0}{\mu + \mu_0} I.$$

With $\mu \to \infty$, $I' = I$. The magnetic field at the position $x = d$ produced by I' is given by Ampère's circuital law as $B = \frac{\mu_0 I}{4\pi d} e_z$. Therefore the force per unit length on the wire is

$$\mathbf{F} = i d\mathbf{l} \times \mathbf{B} = I B e_y \times e_z = \frac{\mu_0 I^2}{4\pi d} e_x.$$

2073

An uncharged metal block has the form of a rectangular parallelepiped with sides a, b, c. The block moves with velocity v in a magnetic field of intensity $\mathbf{H}$ as shown in Fig. 2.55. What is the electric field intensity in the block and what is the electric charge density in and on the block?

(*Wisconsin*)

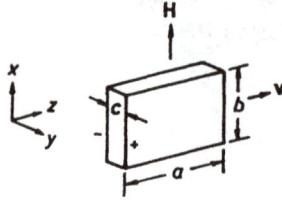

Fig. 2.55

Solution:

In equilibrium no force acts on the electrons of the metal block, i.e., $-e\mathbf{E} - e\mathbf{v} \times \mathbf{B} = 0$. Hence

$$\mathbf{E} = -\mathbf{v} \times \mathbf{B} = -\mu_0 \mathbf{v} \times \mathbf{H} = -\mu_0 v H \mathbf{e}_y .$$

In the block the charge density is

$$\rho = \varepsilon \nabla \cdot E = \varepsilon \frac{\partial}{\partial y}(-\mu_0 v H) = 0 .$$

Hence there is no charge inside the block. The surface charge density σ is given by the boundary condition

$$\sigma = \pm D_n = \pm \varepsilon_0 E_n .$$

As $\mathbf{E}$ is in the y-direction, σ occurs only for the surfaces formed by the sides a, b and has the magnitude

$$\sigma = \pm \varepsilon_0 E = \pm \varepsilon_0 \mu_0 v H ,$$

and the sign as shown in Fig. 2.55.

2074

In Fig. 2.56 an iron needle 1 cm long and 0.1 cm in diameter is placed in a uniform magnetic field of $H_0 = 1000$ Gs with its long axis along the field direction. Give an approximate formula for $\mathbf{H(r)}$ valid for distances $r \gg 1$ cm. Here r is measured from the center of the needle as origin.

(Chicago)

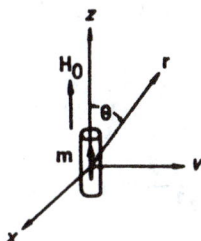

Fig. 2.56

(Note: the saturation value of **B** in iron is approximately 2000 Gs.)

Solution:

For distances $r \gg 1$ cm the iron needle can be treated as a magnetic dipole with moment **m**. Take the z-axis along the axis of the needle. Write

$$\mathbf{H}_{ext} = H_0 \mathbf{e}_z , \quad \mathbf{m} = m \mathbf{e}_z .$$

In Gaussian units the magnetic field at position **r** is approximately

$$\mathbf{H(r)} = \mathbf{H}_0 - \nabla \left(\frac{\mathbf{m} \cdot \mathbf{r}}{r^3} \right) = \mathbf{H}_0 - \frac{\mathbf{m}}{r^3} + \frac{3(\mathbf{m} \cdot \mathbf{r})\mathbf{r}}{r^5} .$$

As the magnetic field due to the iron needle is much weaker than the external field $\mathbf{H}_0, m/r^3 \ll H_0$. As the tangential component of H is continuous at the boundary, the magnetic field within the iron needle may be taken as approximately

$$\mathbf{H}_{in} = \mathbf{H}_0 .$$

The magnetization of the needle is then

$$\mathbf{M} = \frac{1}{4\pi} (\mathbf{B}_{in} - \mathbf{H}_{in}) .$$

With the volume of the needle equal to

$$V = (0.05)^2 \pi \text{ cm}^3 ,$$

$B_{in} = 2 \times 10^4$ Gs, $H_{in} \approx H_0 = 10^3$ Gs, the magnetic moment of the needle is

$$m = vM = \frac{(0.05)^2 \pi}{4\pi} \times (2 - 0.1) \times 10^4 = 11.9 \text{ Gs} \cdot \text{cm}^3 .$$

In polar coordinates the magnetic field at distances $r \gg 1$ cm has components

$$H_r = H_0 \cos\theta + 2m\frac{\cos\theta}{r^3} = \left(1000 + \frac{23.8}{r^3}\right)\cos\theta \text{ Gs},$$

$$H_\theta = -H_0 \cos\theta + \frac{m\sin\theta}{r^3} = \left(-1000 + \frac{11.9}{r^3}\right)\sin\theta \text{ Gs}.$$

2075

A charged metal sphere of mass 5 kg, radius 10 cm is moving in vacuum at 2400 m/sec. You would like to alter the direction of motion by acting on the sphere either electrostatically or magnetically within a region 1 m × 1 m × 100 m.

(a) If limited by the total stored energy (electric or magnetic) in the volume of 100 m^2, will you obtain a greater force by acting on the sphere with a magnetic field **B** or an electric field **E**?

(b) For a maximum electric field (due to its charge) of 10 kV/cm at the sphere's surface find the transverse velocity of the sphere at the end of the 100 m flight path as a function of the applied field (**B** or **E**)?

(*Princeton*)

Solution:

(a) The electric energy density is $w_e = \frac{1}{2}\epsilon_0 E^2$, and the magnetic energy density is $w_m = \frac{B^2}{2\mu_0}$. To estimate order of magnitude, we may assume the field intensity to be the same everywhere in the region under consideration. For the same energy density, $\frac{1}{2}\epsilon_0 E^2 = \frac{B^2}{2\mu_0}$, we have $\frac{E}{B} = \frac{1}{\sqrt{\mu_0\epsilon_0}} = c$ and

$$\frac{f_e}{f_m} = \frac{qE}{qvB} = \frac{c}{v} \gg 1.$$

This shows that the force exerted on the metal sphere by an electric field is much greater than that by a magnetic field for the same stored energy.

(b) The maximum electric field on the metal sphere's surfaces of $E_0 = 10$ kV/cm limits the maximum charge Q_m carried by the sphere as well as the magnitude of the applied field (**E** or **B**). If an external electric field **E** is applied the surface charge density is (see Problem 1065)

$$\sigma = \sigma_0 + 3\epsilon_0 E\cos\theta$$

where the polar axis has been taken along the direction of $\mathbf{E}$, and σ_0 is the surface charge density due to the charge Q carried by the sphere, i.e.,

$$\sigma_0 = \frac{Q}{4\pi r^2},$$

r being the radius of the sphere. The electric field on the sphere's surface is given by $E = \frac{\sigma}{\varepsilon_0}$ and the maximum electric field, E_0, occurs at $\theta = 0$. Hence

$$E_0 = \frac{\sigma_0}{\varepsilon_0} + 3E,$$

and the total charge of the sphere is

$$Q = 4\pi r^2 \sigma_0 = 4\pi \varepsilon_0 r^2 (E_0 - 3E), \quad \left(E < \frac{1}{3}E_0\right).$$

The time taken for the sphere to travel a distance l is $\Delta t = \frac{l}{v}$. The transverse acceleration is $\frac{QE}{m}$ if we assume $E_\perp = E$. Then the transverse velocity at the end of Δt is

$$v_\perp = \frac{QE}{m}\Delta t = \frac{4\pi\varepsilon_0 r^2 (E_0 - 3E) \cdot El}{mv_0}, \quad \left(E < \frac{1}{3}E_0\right).$$

If an external magnetic field is applied instead, the above needs to be modified only by substituting $v_0 B$ for E, the result being

$$v_\perp = \frac{4\pi\varepsilon_0 r^2 (E_0 - 3v_0 B) Bl}{m}, \quad \left(B < \frac{E_0}{3v_0}\right).$$

If $E \geq \frac{1}{3}E_0$ or $B \geq \frac{E_0}{3v_0}$, the charge of the metal sphere is zero and the transverse velocity is also zero. From the above results we can also show that the transverse velocity is maximum for $E = E_0/9$ or $B = E_0/9V_0$, as the case may be. It follows that

$$v_{\perp m} = \frac{8\pi\varepsilon_0 r^2 E_0^2 l}{27mv_0} = \frac{8\pi \times \frac{10^{-9}}{4\pi \times 9} \times 0.1^2 \times 10^{6\times 2} \times 100}{27 \times 5 \times 2400}$$
$$= 6.86 \times 10^{-4} \text{ m/s}$$

and the maximum transverse displacement of the sphere is

$$v_{\perp m}\Delta t = \frac{v_m l}{v} = \frac{6.86 \times 10^{-4} \times 100}{2400} = 2.86 \times 10^{-5} \text{ m},$$

which is negligible in comparison with the transverse size (1 m) of the space.

2076

Show that the force between two magnetic dipoles varies as the inverse fourth power of the distance between their centers, whatever their relative orientation in space is. Assume that the dipoles are small compared with the separation between them.

(Columbia)

Solution:

Let the magnetic moments of the two dipoles be $\mathbf{m}_1$ and $\mathbf{m}_2$. The potential produced by $\mathbf{m}_2$ at the location of $\mathbf{m}_1$ is

$$\varphi_m = \frac{1}{4\pi}\frac{\mathbf{m}_2 \cdot \mathbf{r}}{r^3},$$

where $\mathbf{r}$ is directed from $\mathbf{m}_2$ to $\mathbf{m}_1$. Because the magnetic field is $\mathbf{B} = -\mu_0 \nabla \varphi_m$, the force on $\mathbf{m}_1$, is

$$\mathbf{F}_m = \nabla(\mathbf{m}_1 \cdot \mathbf{B}) = \nabla[\mathbf{m}_1 \cdot (-\mu_0 \nabla \varphi_m)]$$

$$= -\frac{\mu_0}{4\pi}\nabla\left[\mathbf{m}_1 \cdot \nabla\left(\frac{\mathbf{m}_2 \cdot \mathbf{r}}{r^3}\right)\right]$$

$$= -\frac{\mu_0}{4\pi}\nabla\left[\mathbf{m}_1 \cdot \left(\frac{\mathbf{m}_2}{r^3} - \frac{3\mathbf{r}(\mathbf{m}_2 \cdot \mathbf{r})}{r^5}\right)\right]$$

$$= \frac{\mu_0}{4\pi}\nabla\left[\frac{3(\mathbf{m}_1 \cdot \mathbf{r})(\mathbf{m}_2 \cdot \mathbf{r})}{r^5} - \frac{\mathbf{m}_1 \cdot \mathbf{m}_2}{r^3}\right].$$

As both terms in the expression for the gradient are proportional to $\frac{1}{r^3}$, $\mathbf{F}$ will be proportional to $\frac{1}{r^4}$.

2077

A magnetic dipole $\mathbf{m}$ is moved from infinitely far away to a point on the axis of a fixed perfectly conducting (zero resistance) circular loop of radius b and self-inductance L. In its final position the dipole is oriented along the loop axis and is at a distance z from the center of the loop. Initially, when the dipole is very far away, the current in the loop is zero (Fig. 2.57).

(a) Calculate the current in the loop when the dipole is in its final position.

(b) Calculate for the same positions the force between the dipole and the loop.

(UC, Berkeley)

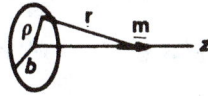

Fig. 2.57

Solution:

(a) The induced emf of the loop is given by

$$\varepsilon = -L\frac{dI}{dt} = -\frac{\partial}{\partial t}\int \mathbf{B}\cdot d\mathbf{S}\,.$$

Integrating over time we have

$$L[I(f) - I(i)] = \int [\mathbf{B}(f) - \mathbf{B}(i)]\cdot d\mathbf{S}\,.$$

Initially, when the dipole is far away,

$$I(i) = 0, \quad \mathbf{B}(i) = 0\,.$$

Writing for the final position $I = I(f), \mathbf{B} = \mathbf{B}(f)$, we have

$$LI = \int \mathbf{B}\cdot d\mathbf{S}\,.$$

Consider a point P in the plane of the loop. Use cylindrical coordinates (ρ, θ, z) such that P has radius vector $\rho\mathbf{e}_\rho$. Then the radius vector from $\mathbf{m}$ to P is $\mathbf{r} = \rho\mathbf{e}_\rho - z\mathbf{e}_z$. The magnetic induction at P due to $\mathbf{m}$ is

$$\mathbf{B} = \frac{\mu_0}{4\pi}\left[\frac{3(\mathbf{m}\cdot\mathbf{r})\mathbf{r}}{r^5} - \frac{\mathbf{m}}{r^3}\right],$$

where $\mathbf{m} = m\mathbf{e}_z$. As $d\mathbf{S} = \rho d\rho d\theta \mathbf{e}_z$ we have

$$\int \mathbf{B}\cdot d\mathbf{S} = \frac{\mu_0}{4\pi}\int\int\left(\frac{3(\mathbf{m}\cdot\mathbf{r})(\mathbf{r}\cdot\mathbf{e}_z)}{r^5} - \frac{\mathbf{m}\cdot\mathbf{e}_z}{r^3}\right)\rho d\rho d\theta$$

$$= \frac{\mu_0}{4\pi}\cdot 2\pi\int_0^b\left[\frac{3mz^2}{(\rho^2+z^2)^{5/2}} - \frac{m}{(\rho^2+z^2)^{3/2}}\right]\rho d\rho$$

$$= \frac{\mu_0 m}{2}\left[(b^2+z^2)^{-\frac{1}{2}} - z^2(b^2+z^2)^{-\frac{3}{2}}\right],$$

and the induced current in the loop is

$$I = \frac{\mu_0 m}{2L}\left[(b^2 + z^2)^{-\frac{1}{2}} - z^2(b^2 + z^2)^{-\frac{3}{2}}\right].$$

By Lenz's law the direction of flow is clockwise when looking from the location of $\mathbf{m}$ positioned as shown in Fig. 2.57.

(b) For the loop, with the current I as above, the magnetic field at a point on its axis is

$$\mathbf{B}' = -\frac{\mu_0 I}{2}\frac{b^2}{(b^2 + z^2)^{3/2}}\mathbf{e}_z = -\frac{\mu_0^2 m}{4L}\frac{b^4}{(b^2 + z^2)^3}\mathbf{e}_z.$$

The energy of the magnetic dipole $\mathbf{m}$ in the field $\mathbf{B}'$ is

$$W = \mathbf{m} \cdot \mathbf{B}'$$

and the force between the dipole and the loop is

$$F = -\frac{\partial W}{\partial z} = -\frac{3\mu_0^2 m^2 b^4 z}{2L(b^2 + z^2)^4}.$$

2078

The force on a small electric current loop of magnetic moment $\boldsymbol{\mu}$ in a magnetic field $\mathbf{B(r)}$ is given by

$$\mathbf{F} = (\boldsymbol{\mu} \times \nabla) \times \mathbf{B}.$$

On the other hand, the force on a magnetic charge dipole $\boldsymbol{\mu}$ is given by

$$\mathbf{F} = (\boldsymbol{\mu} \cdot \nabla)\mathbf{B}.$$

(a) Using vector analysis and expanding the expression for the force on a current loop, discuss in terms of local sources of the magnetic field the conditions under which the forces would be different.

(b) Propose an experiment using external electric or magnetic fields that could in principle determine whether the magnetic moment of a nucleus arises from electric current or from magnetic charge.

(UC, Berkeley)

Solution:

(a) We expand the expression for the force on a current loop:

$$\mathbf{F} = (\boldsymbol{\mu} \times \nabla) \times \mathbf{B} = \nabla(\boldsymbol{\mu} \cdot \mathbf{B}) - \boldsymbol{\mu}(\nabla \cdot \mathbf{B}).$$

The external magnetic field $\mathbf{B}(\mathbf{r})$ satisfies $\nabla \cdot \mathbf{B} = 0$ so the above equation can be written as

$$\mathbf{F} = \nabla(\boldsymbol{\mu} \cdot \mathbf{B}) = (\boldsymbol{\mu} \cdot \nabla)\mathbf{B} + \boldsymbol{\mu} \times (\nabla \times \mathbf{B}).$$

Compared with the expression for the force on a magnetic dipole, it has an additional term $\boldsymbol{\mu} \times (\nabla \times \mathbf{B})$. Thus the two forces are different unless $\nabla \times \mathbf{B} = 0$ in the loop case which would mean $\mathbf{J} = \dot{\mathbf{D}} = 0$ in the region of the loop.

(b) Take the z-axis along the direction of the magnetic moment of the nucleus and apply a magnetic field $\mathbf{B} = B(z)\mathbf{e}_z$ in this direction. According to $\mathbf{F} = (\boldsymbol{\mu} \times \nabla) \times \mathbf{B}$, the magnetic force is zero. But according to $\mathbf{F} = (\boldsymbol{\mu} \cdot \nabla)\mathbf{B}$, the force is not zero. So whether the magnetic moment arises from magnetic charge or from electric current depends on whether or not the nucleus suffers a magnetic force.

2079

A particle with charge q is traveling with velocity $\mathbf{v}$ parallel to a wire with a uniform linear charge distribution λ per unit length. The wire also carries a current I as shown in Fig. 2.58. What must the velocity be for the particle to travel in a straight line parallel to the wire, a distance r away?

(Wisconsin)

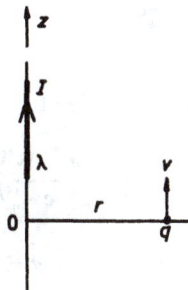

Fig. 2.58

Solution:

Consider a long cylinder of radius r with the axis along the wire. Denote its curved surface for unit length by S and the periphery of its cross section by C. Using Gauss' flux theorem and Ampère's circuital law

$$\oint_S \mathbf{E} \cdot d\mathbf{S} = \lambda/\varepsilon_0 , \quad \oint_C \mathbf{B} \cdot d\mathbf{l} = \mu_0 I ,$$

by the axial symmetry we find

$$\mathbf{E}(r) = \frac{\lambda}{2\pi\varepsilon_0 r}\mathbf{e}_r , \quad \mathbf{B}(r) = \frac{\mu_0 I}{2\pi r}\mathbf{e}_\theta$$

in cylindrical coordinates (r, θ, z) with origin 0 at the wire.

The total force acting on the particle which has velocity $\mathbf{v} = v\mathbf{e}_z$ is

$$\mathbf{F} = \mathbf{F}_e + \mathbf{F}_m = q\mathbf{E} + q\mathbf{v} \times \mathbf{B}$$
$$= \frac{q\lambda}{2\pi\varepsilon_0 r}\mathbf{e}_r + \frac{q\mu_0 I}{2\pi r}v(-\mathbf{e}_r) .$$

For the particle to maintain the motion along the z direction, this radial force must vanish, i.e.,

$$\frac{q\lambda}{2\pi\varepsilon_0 r} - \frac{q\mu_0 I}{2\pi r}v = 0 ,$$

giving

$$v = \frac{\lambda}{\varepsilon_0\mu_0 I} = \frac{\lambda c^2}{I} .$$

2080

The Lorentz force law for a particle of mass m and charge q is

$$\mathbf{F} = q\left(\mathbf{E} + \frac{\mathbf{v} \times \mathbf{B}}{c}\right) .$$

(a) Show that if the particle moves in a time-independent electric field $\mathbf{E} = -\nabla\phi(x, y, z)$ and any magnetic field, then the energy $\frac{1}{2}mv^2 + q\varphi$ is a constant.

(b) Suppose the particle moves along the x-axis in the electric field $\mathbf{E} = Ae^{-t/\tau}\mathbf{e}_x$, where A and τ are both constants. Suppose the magnetic field is zero along x axis and $x(0) = \dot{x}(0) = 0$. Find $x(t)$.

(c) In (b) is $\frac{1}{2}mv^2 - qxAe^{-t/\tau}$ a constant (indicate briefly your reasoning)?

(UC, Berkeley)

Solution:

(a) As

$$\mathbf{F} = m\dot{\mathbf{v}} = q\left(\mathbf{E} + \frac{\mathbf{v} \times \mathbf{B}}{c}\right)$$

we have

$$(m\dot{\mathbf{v}} - q\mathbf{E}) = q\frac{\mathbf{v} \times \mathbf{B}}{c}.$$

It follows that

$$\mathbf{v} \cdot (m\dot{\mathbf{v}} - q\mathbf{E}) = \mathbf{v} \cdot (\mathbf{v} \times \mathbf{B})\frac{q}{c} = 0.$$

Consider

$$\frac{d}{dt}\left[\frac{1}{2}mv^2 + q\phi\right] = m\mathbf{v} \cdot \dot{\mathbf{v}} + q\frac{d\phi}{dt} = m\mathbf{v} \cdot \dot{\mathbf{v}} + q\mathbf{v} \cdot \nabla\phi$$

$$= \mathbf{v} \cdot (m\dot{\mathbf{v}} + q\nabla\phi) = \mathbf{v} \cdot (m\dot{\mathbf{v}} - q\mathbf{E}) = 0,$$

where we have made use of

$$\frac{d\phi}{dt} = \frac{\partial\phi}{\partial x}\frac{dx}{dt} + \frac{\partial\phi}{\partial y}\frac{dy}{dt} + \frac{\partial\phi}{\partial z}\frac{dz}{dt} = \mathbf{v} \cdot \nabla\phi.$$

Hence

$$\frac{1}{2}mv^2 + q\phi = \text{Const.}$$

(b) The magnetic force $\mathbf{F}_m = q\frac{\mathbf{v} \times \mathbf{B}}{c}$ is perpendicular to $\mathbf{v}$ so that if the particle moves in the x direction the magnetic force will not affect the x-component of the motion. With $\mathbf{E}$ in the x direction the particle's motion will be confined in that direction. Newton's second law gives

$$m\ddot{x} = qE = qAe^{-t/\tau},$$

i.e.,

$$mdv = qAe^{-t/\tau}dt,$$

with

$$v(0) = 0, \quad mv = -qA\tau e^{-t/\tau} + qA\tau,$$

or

$$dx = qA\tau(1 - e^{-t/\tau})\frac{dt}{m}.$$

With $x(0) = 0$, this gives

$$x(t) = qA\tau\frac{t}{m} + \frac{qA\tau^2}{m}e^{-t/\tau} - \frac{qA\tau^2}{m}$$

$$= \frac{qA\tau}{m}[(t - \tau) + \tau e^{-t/\tau}].$$

(c) $$\frac{1}{2}mv^2 - qxAe^{-t/\tau} = \frac{1}{2}m\left[\frac{qA\tau}{m}(1 - e^{t/\tau})\right]^2$$

$$- qA\frac{qA\tau}{m}[(t - \tau) + \tau e^{-t/\tau}]e^{-t/\tau}.$$

As

$$\frac{d}{dt}\left(\frac{1}{2}mv^2 - qxAe^{-t/\tau}\right) \neq 0$$

the expression is not a constant.

2081

A point particle of mass m and magnetic dipole moment $\mathbf{M}$ moves in a circular orbit of radius R about a fixed magnetic dipole, moment $\mathbf{M_0}$, located at the center of the circle. The vectors $\mathbf{M_0}$ and $\mathbf{M}$ are antiparallel to each other and perpendicular to the plane of the orbit.

(a) Compute the velocity v of the orbiting dipole.

(b) Is the orbit stable against small perturbations? Explain. (Consider only the motion in the plane.)

(CUSPEA)

Solution:

(a) The magnetic field produced at a point of radius vector $\mathbf{r}$ from the center of the circle by the dipole of moment $\mathbf{M_0}$ is

$$\mathbf{B} = \frac{\mu_0}{4\pi}\left[\frac{3(\mathbf{M_0} \cdot \mathbf{r})\mathbf{r}}{r^5} - \frac{\mathbf{M_0}}{r^3}\right].$$

This exerts a force on the particle moving in the circular orbit of

$$\mathbf{F} = \nabla(\mathbf{M} \cdot \mathbf{B})|_{r=R} \cdot$$

Noting $\mathbf{M} \cdot \mathbf{M_0} = -MM_0, \mathbf{M_0} \cdot \mathbf{r} = \mathbf{M} \cdot \mathbf{r} = 0$, we have

$$\mathbf{F} = -\frac{3\mu_0 M M_0}{4\pi R^4} \mathbf{e}_r \cdot$$

This force acts towards the center and gives rise to the circular motion of the particle. Balancing the force with the centrifugal force,

$$m\frac{v^2}{R} = \frac{3\mu_0 M M_0}{4\pi R^4} ,$$

gives the particle velocity as

$$v = \sqrt{\frac{3\mu_0 M M_0}{4\pi m R^3}} \cdot$$

(b) The energy of the particle is

$$E = \frac{1}{2}m\left(\frac{d\mathbf{r}}{dt}\right)^2 - \frac{\mu_0 M M_0}{4\pi r^3} = \frac{1}{2}m\left(\frac{d\mathbf{r}}{dt}\right)^2 + U(r),$$

with

$$U(r) = -\frac{\mu_0 M M_0}{4\pi r^3} + \frac{L^2}{2mr^2} ,$$

where L is the conserved angular momentum and the first term is the potential energy of $\mathbf{M}$ in the magnetic field $\mathbf{B}, -\mathbf{M} \cdot \mathbf{B}$, for a circular orbit of radius r. We note that $(\frac{dU}{dr})_{r=R} = 0$ and $(\frac{d^2U}{dr^2})_{r=R} < 0$, so that $U(R)$ is a maximum and the orbit is not stable against small perturbations in r.

2082

A long solenoid of radius b and length l is wound so that the axial magnetic field is

$$\mathbf{B} = \begin{cases} B_0\mathbf{e}_z, & r < b, \\ 0, & r > b. \end{cases}$$

A particle of charge q is emitted with velocity $\mathbf{v}$ perpendicular to a central rod of radius a (see Fig. 2.59). The electric force on the particle is given by $q\mathbf{E} = f(r)\mathbf{e}_r$, where $\mathbf{e}_r \cdot \mathbf{e}_z = 0$. We assume $\mathbf{v}$ is sufficiently large so that the particle passes out of the solenoid and does so without hitting the solenoid.

(a) Find the angular momentum of the particle about the axis of the solenoid, for $r > b$.

(b) If the electric field inside the solenoid is zero before the particle leaves the rod and after the particle has gone far way, it becomes

$$\mathbf{E} = \begin{cases} -\frac{q}{2\pi\epsilon_0 r}\mathbf{e}_r, & r > a, \\ 0, & r < a, \end{cases}$$

calculate the electromagnetic field angular momentum and discuss the final state of the solenoid if the solenoid can rotate freely about its axis. Neglect end effects.

(*Wisconsin*)

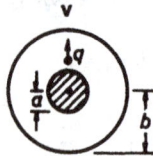

Fig. 2.59

Solution:

(a) As $\mathbf{v}$ is very large, we can consider any deviation from a straight line path to be quite small in the emitting process. Let $\mathbf{v}_\perp$ be the transverse velocity of the particle. We have

$$m\frac{d\mathbf{v}_\perp}{dt} = q\mathbf{v} \times \mathbf{B}.$$

As $\mathbf{v} \perp \mathbf{B}, dv_\perp = \frac{q}{m}B_0 v dt = \frac{q}{m}B_0 dr$ and

$$v_\perp(b) = \int_a^b \frac{q}{m}B_0 dr = \frac{q}{m}B_0(b-a).$$

At $r = b$ the angular momentum of the particle about the axis of the solenoid has magnitude

$$|\mathbf{r} \times m v_\perp(b)|_{r=b} = mbv_\perp(b) = qB_0 b(b-a).$$

and direction $-\mathbf{e}_z$. Thus the angular momentum is

$$\mathbf{J}_p = -qB_0b(b-a)\mathbf{e}_z \, .$$

For $r > b$, $\mathbf{B} = 0$ and $\mathbf{J}_p$ is considered. So $\mathbf{J}_p$ is the angular momentum of the particle about the axis for $r > b$.

(b) After the particle has gone far away from the solenoid, the momentum density of the electromagnetic field at a point within the solenoid is

$$\mathbf{g} = \frac{\mathbf{E} \times \mathbf{H}}{c^2} = \varepsilon_0 \mathbf{E} \times \mathbf{B} \, ,$$

and the angular momentum density is

$$\mathbf{j} = \mathbf{r} \times \mathbf{g} = \varepsilon_0 \mathbf{r} \times (\mathbf{E} \times \mathbf{B}) = \frac{B_0 q}{2\pi l}\mathbf{e}_z \, ,$$

which is uniform. As there is no field outside the solenoid and inside the central rod the total angular momentum of the electromagnetic field is

$$\mathbf{J}_{\text{EM}} = \pi(b^2 - a^2)l\mathbf{j} = \frac{qB_0(b^2 - a^2)}{2}\mathbf{e}_z \, .$$

Initially, $\mathbf{E} = 0$, $\mathbf{v}_\perp = 0$ and the solenoid is at rest, so the total angular momentum of the system is zero. The final angular momentum of the solenoid can be obtained from the conservation of the total angular momentum:

$$\mathbf{J}_S = -\mathbf{J}_{\text{EM}} - \mathbf{J}_p = \frac{qB_0}{2}(b-a)^2\mathbf{e}_z \, .$$

This signifies that the solenoid in the final state rotates with a constant angular speed about its central axis, the sense of rotation being related to the direction $\mathbf{e}_z$ by the right-hand screw rule.

2083

Suppose a bending magnet with poles at $x = \pm x_0$ has a field in the median plane that depends only on z, $B_x = B_x(z)$, where the origin is chosen at the center of the magnet gap. What component must exist outside the median plane? If a particle with charge e and momentum P is incident down the z-axis in the median plane, derive integral expressions for the bending angle θ and the displacement y as a function of z within the magnet. Do not evaluate the integrals.

(Wisconsin)

Solution:

Since there is no current between the magnet poles, $\nabla \times \mathbf{B} = 0$, or

$$\frac{\partial B_x}{\partial z} - \frac{\partial B_z}{\partial x} = 0 .$$

This implies that as B_x depends on z, $B_z \neq 0$, i.e. there is a z-component outside the median plane.

The kinetic energy of a charged particle moving in a magnetic field is conserved. Hence the magnitude of its velocity is a constant. Let θ be the bending angle, then

$$v_y = v \sin \theta , \quad v_z = v \cos \theta .$$

The equation of the motion of the particle in the y direction, since $v_x = 0$, is

$$m \frac{dv_y}{dt} = e B_x v_z ,$$

or

$$mv \frac{d}{dt}(\sin \theta) = e B_x v \cos \theta .$$

This gives

$$d\theta = \frac{e B_x}{m} dt = \frac{e B_x}{m} \cdot \frac{dz}{v \cos \theta} ,$$

or

$$\cos \theta d\theta = \frac{e}{P} B_x dz .$$

Suppose at $t = 0$ the particle is at the origin and its velocity is along the $+z$ direction. Then $\theta(z)|_{z=0} = 0$ and we have

$$\int_0^\theta \cos \theta d\theta = \frac{e}{P} \int_0^z B_x dz' ,$$

or

$$\theta = \sin^{-1} \left[\frac{e}{P} \int_0^z B_x dz' \right] .$$

The displacement y is given by

$$y = \int_0^t v_y dt' = \int_0^z v \sin \theta \frac{dz'}{v \cos \theta}$$

$$= \int_0^z \tan \theta dz' = \int_0^z \frac{\frac{e}{P} \int_0^{z'} B_x(z'')dz''}{[1 - (\frac{e}{P} \int_0^{z'} B_x(z'')dz'')^2]^{1/2}} dz' .$$

2084

An infinitely long wire lies along the z-axis (i.e., at $x = 0, y = 0$) and carries a current i in the $+z$ direction. A beam of hydrogen atoms is injected at the point $x = 0, y = b, z = 0$ with a velocity $\mathbf{v} = v_0 \mathbf{e}_z$. The hydrogen atoms were polarized such that their magnetic moments μ_H are all pointing in the $+x$ direction, i.e., $\mu = \mu_H \mathbf{e}_x$.

(a) What are the force and torque on these hydrogen atoms due to the magnetic field of the wire?

(b) How would your answer change if the hydrogen atoms were polarized in such a way that initially their magnetic moments point in the $+z$ direction, i.e. $\mu = \mu_H \mathbf{e}_z$.

(c) In which of the above two cases do the hydrogen atoms undergo Larmor precession? Describe the direction of the precession and calculate the precession frequency.

(*Columbia*)

Solution:

The hydrogen atoms are moving in the yz plane. In this plane the magnetic field produced by the infinitely long wire at a point distance y from the wire is

$$\mathbf{B} = -\frac{\mu_0 i}{2\pi y} \mathbf{e}_x .$$

(a) With $\mu = \mu_H \mathbf{e}_x$, the energy of such a hydrogen atom in the field $\mathbf{B}$ is

$$W = \mu \cdot \mathbf{B} = -\frac{\mu_0 \mu_H i}{2\pi y} .$$

Thus the magnetic force on the atom is

$$\mathbf{F} = \nabla W|_{y=b} = \frac{\mu_0 \mu_H i}{2\pi b^2} \mathbf{e}_y ,$$

and the torque on the atom is

$$\mathbf{L} = \mu \times \mathbf{B} = 0 .$$

(b) With $\mu = \mu_H \mathbf{e}_z$, the energy is $W = \mu \cdot \mathbf{B} = 0$. So the force exerted is $\mathbf{F} = 0$ and the torque is

$$\mathbf{L} = \mu \times \mathbf{B}|_{y=b} = -\frac{\mu_0 \mu_H i}{2\pi b} \mathbf{e}_y .$$

(c) In case (b), because the atom is exerted by a torque, Larmor precession will take place. The angular momentum of the atom, M, and its magnetic moment are related by

$$\mu_{\mathrm{H}} = g\frac{e}{2m}M \,,$$

where g is the Landé factor. The rate of change of the angular momentum is equal to the torque acting on the atom,

$$\frac{dM}{dt} = \mathbf{L} \,.$$

The magnitude of M does not change, but L will give rise to a precession of **M** about **B**, called the Larmor precession, of frequency ω given by

$$\left|\frac{d\mathbf{M}}{dt}\right| = M\omega \,,$$

or

$$\omega = \frac{L}{M} = \frac{\mu_0\mu_{\mathrm{H}}i}{2\pi bM} = \frac{eg\mu_0 i}{4\pi bm} \,.$$

The precession is anti-clockwise if viewed from the side of positive x.

2085

A uniformly magnetized iron sphere of radius R is suspended from the ceiling of a large evacuated metal chamber by an insulating thread. The north pole of the magnet is up and south pole is down. The sphere is charged to a potential of 3,000 volts relative to the walls of the chamber.

(a) Does this static system have angular momentum?

(b) Electrons are injected radially into the chamber along a polar axis and partially neutralize the charge on the sphere. What happens to the sphere?

(*UC, Berkeley*)

Solution:

Use coordinates as shown in Fig. 2.60.

(a) This system has an angular momentum.

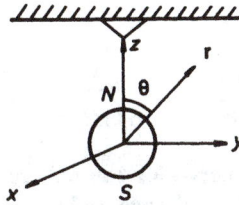

Fig. 2.60

(b) Let **m** be the magnetic moment of the sphere. The magnetic field at a point **r** outside the sphere is

$$\mathbf{B} = \frac{\mu_0}{4\pi}\left[\frac{3(\mathbf{m}\cdot\mathbf{r})\mathbf{r}}{r^5} - \frac{\mathbf{m}}{r^3}\right].$$

Suppose the sphere carries a charge Q. As the sphere is a conductor, the electric field inside is zero and the electric field outside is

$$\mathbf{E} = \frac{Q}{4\pi\varepsilon_0 r^3}\mathbf{r}.$$

Therefore the electromagnetic momentum density in the space outside the sphere, as $\mathbf{m} = m\mathbf{e}_z = m(\cos\theta\,\mathbf{e}_r - \sin\theta\,\mathbf{e}_\theta)$ in spherical coordinates, is

$$\mathbf{g} = \varepsilon_0\mathbf{E}\times\mathbf{B} = \frac{Q\mu_0 m\sin\theta}{16\pi^2 r^5}\mathbf{e}_\varphi,$$

and the angular momentum density is

$$\mathbf{j} = \mathbf{r}\times\mathbf{g} = -\frac{\mu_0 mQ\sin\theta}{16\pi^2 r^4}\mathbf{e}_\theta.$$

Because of symmetry the total angular momentum has only the z-component, which is obtained by the integration of the z component of $\mathbf{j}$:

$$J = \int_0^{2\pi}\int_0^\pi\int_R^\infty \frac{\mu_0 mQ\sin^3\theta}{16\pi^2 r^2}\,d\varphi\,d\theta\,dr = \frac{\mu_0 mQ}{6\pi R} = \frac{2mV}{3c^2},$$

where V is the voltage of the sphere. As the electrons are being injected radially on the sphere, the charge Q decreases, causing the electromagnetic angular momentum to decrease also. However, because of the conservation of total angular momentum, the sphere will rotate about the polar axis, with the sense of rotation determined by the right-handed screw rule.

The rotating angular velocity is

$$\omega = -\frac{\Delta I}{I} = -\frac{\mu_0 m \Delta Q}{6\pi RI} = -\frac{2m\Delta V}{3c^2 I},$$

where I is the rotational inertia of the sphere about the polar axis, and $\Delta Q, \Delta V$ are the changes of its charge and potential respectively, which are both negative.

2086

A cylinder of length L and radius R carries a uniform current I parallel to its axis, as in Fig. 2.61.

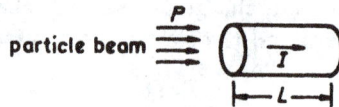

Fig. 2.61

(a) Find the direction and magnitude of the magnetic field everywhere inside the cylinder. (Ignore end effects.)

(b) A beam of particles, each with momentum P parallel to the cylinder axis and each with positive charge q, impinges on its end from the left. Show that after passing through the cylinder the particle beam is focused to a point. (Make a "thin lens" approximation by assuming that the cylinder is much shorter than the focal length. Neglect the slowing down and scattering of the beam particles by the material of the cylinder.) Compute the focal length.

(*CUSPEA*)

Solution:

(a) Use cylindrical coordinates (r, φ, z) with the z-axis along the cylindrical axis. The magnetic field at a point distance r from the axis is given by Ampère's circuital law to be

$$\mathbf{B} = \frac{\mu_0 Ir}{2\pi R^2}\mathbf{e}_\varphi .$$

(b) The magnetic force acting on a particle of the beam is

$$\mathbf{F} = q\mathbf{v} \times \mathbf{B} = -qvBe_r \, .$$

On account of this force the particle will receive a radial momentum towards the axis after traversing the cylinder of

$$P_r = q \int vB dt = \frac{qvBL}{v} = \frac{\mu_0 qIL}{2\pi R^2} r \, .$$

If we neglect the slowing down of the beam particles through the cylinder and use thin lens approximation, the axial momentum of a beam particle is still P after it comes out of the cylinder. The combination of P and P_r will make the particle cross the cylindrical axis at a point M, as shown in Fig. 2.62. From the diagram we find the relation

$$\frac{P_r}{P} = \frac{r}{d} \, .$$

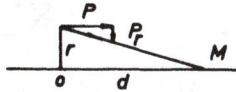

Fig. 2.62

Thus the focal length is

$$d = \frac{Pr}{P_r} = \frac{2\pi R^2 P}{\mu_0 eIL}$$

and is independent of r. Hence all the particles will be focused at the point M.

2087

A dipole electromagnet has rectangular pole faces in horizontal planes with length l and width w. The main component of the magnetic field $\mathbf{B}$ is vertical. A parallel beam of particles, each with velocity v, mass m, and charge q, enters the magnet with the velocity v parallel to the horizontal plane but at an angle φ with the center line of the magnet. The vertical size of the beam is comparable to the gap of the magnet. The particles leave the magnet at an angle $-\varphi$ with the center line of the magnet, having been bent an angle of 2φ (see Fig. 2.63 and Fig. 2.64). Show that the fringe field of the magnet will have a vertically focusing effect on the beam. Calculate the approximate focal length.

(Columbia)

Solution:

As the pole area of the dipole electromagnet is limited, the magnetic field has fringe lines as shown in Fig. 2.63. If the y-component of the fringe field is neglected, the fringe field will only have z- and x-components. From $\nabla \times \mathbf{B} = 0$, we have

$$\frac{\partial B_x}{\partial z} = \frac{\partial B_z}{\partial x} \, .$$

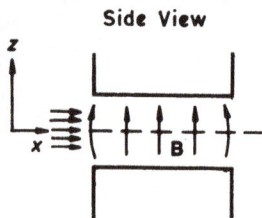

Fig. 2.63

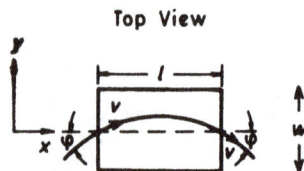

Fig. 2.64

Suppose the extent of the fringe field is b. At the entrance of the electromagnet B_z increases from 0 to B in a distance b. To first approximation the above relation gives $B_{x,\text{in}} = \frac{B}{b} z$. Whereas, at the exit B_z decreases from B to 0 and one has $B_{x,\text{out}} = -\frac{B}{b} z$. The velocities of the particles at the entrance and the exit are $\mathbf{v} = v \cos\varphi \mathbf{e}_x + v \sin\varphi \mathbf{e}_y$ and $\mathbf{v} = v \cos\varphi \mathbf{e}_x - v \sin\varphi \mathbf{e}_y$ respectively. Thus at both the entrance and the exit the particles will be acted on by a force, which is along the z direction and near the center line of the magnet, of

$$F_z = -\frac{qvBz \sin\varphi}{b} \, .$$

This force gives a vertical momentum to the particles. The time taken for the particles to pass through the fringe width b is

$$\Delta t = \frac{b}{v \cos\varphi} \, .$$

Hence, the vertical momentum is approximately

$$P_z = -F_z \Delta t = -qBz \tan\varphi \, .$$

As P_z is negative for $+z$ and positive for $-z$, it will have a vertical focusing effect on the particles. The momentum of the particles in the xy plane is $P = mv$. Letting the focal length be f (from the extrance), we have

$$\frac{2|P_z|}{P} = \frac{|z|}{f},$$

giving

$$f = \frac{mv}{2qB\tan\varphi}.$$

The equation of the motion of a particle between the poles of the magnet is

$$m\frac{dv_y}{dt} = qv_x B$$

or

$$\frac{d\varphi}{dt} = \frac{qB}{m},$$

as $v_x = v\cos\varphi \approx v, v_y = v\sin\varphi \approx v\varphi$. If the deflecting angle φ is small, we can take the time elapse in traversing the distance $\frac{l}{2}$ to be

$$t = \frac{l}{2}/v,$$

and have approximately

$$\varphi \approx \frac{qBl}{2mv}.$$

Substituting it in the expression for focal length and taking $\tan\varphi \approx \varphi$, we have

$$f = \frac{m^2 v^2}{q^2 B^2 l}.$$

2088

A dipole magnet with rectangular pole faces, magnetic field B_0 and dimensions as shown in Fig. 2.65 and 2.66 has been constructed. We introduce a coordinate system with x-axis parallel to the magnetic field and y- and z-axes parallel to the edges of the pole faces. Choose the $x = 0$ plane so that it lies midway between the pole faces. Suppose that a particle of charge q and momentum P parallel to the z-axis is projected into the magnet, entering the region between the pole faces at a height x above the $x = 0$ mid-plane.

(a) What is the approximate angular deflection θ_y in the yz plane after passing through the magnet? (Assume $P \gg qBL$.)

(b) Show that the angular deflection in the xz plane after passing through the magnet is given approximately by $\theta_x \approx \theta_y^2 x/L$, where θ_y is the deflection found in (a). (Hint: This deflection is caused by the fringe field acting on the particle as it exits the magnet.)

(c) Is the effect found in (b) to focus or defocus off-axis particles?

(Columbia)

Side View

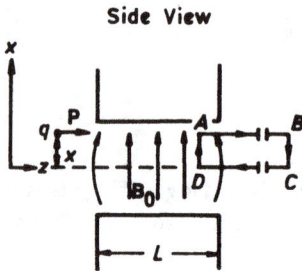

Fig. 2.65

Top View

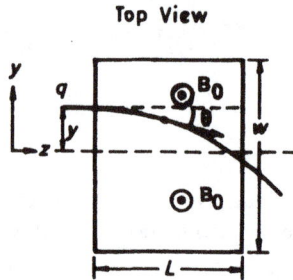

Fig. 2.66

Solution:

The magnetic force acting on the particle has components

$$Fx = -q(v_y B_z - v_z B_y),$$
$$Fy = -q(v_z B_x - v_x B_z),$$
$$Fz = -q(v_x B_y - v_y B_x).$$

Note that as indicated in Figs. 2.65 and 2.66, a left-handed coordinate system is used here.

(a) As $\mathbf{B} = B_0 \mathbf{e}_x$ and is uniform between the pole faces, the equation of the transverse motion of the particle is

$$m\frac{dv_y}{dt} = -qv_z B_0.$$

Since the speed v does not change in a magnetic field, we have $v_y = -v\sin\theta_1, v_z = v\cos\theta_1$, where θ_1 is the deflecting angle in yz plane. As $v_y \approx -v\theta_1, v_z = v$, and $P = mv$ =constant, the above becomes

$$\frac{d\theta_1}{dt} = \frac{qvB_0}{P},$$

or

$$d\theta_1 = \frac{qvB_0}{P}dt = \frac{qvB_0}{Pv_z}dz,$$

i.e.,

$$\cos\theta_1 d\theta_1 = \frac{qB_0}{P}dz.$$

Integrating

$$\int_0^{\theta_y} \cos\theta_1 d\theta_1 = \int_0^L \frac{qB_0}{P}dz$$

we find

$$\sin\theta_y = \frac{qB_0L}{P}.$$

As $P \gg qB_0L$, we have approximately

$$\theta_y \approx \frac{qB_0L}{P}.$$

(b) To take account of the fringe effect, we can assume $B_y \approx 0$ and a small B_z in addition to the main field $B_0\mathbf{e}_z$. The equation of the vertical motion of the particle is

$$m\frac{dv_x}{dt} = -qv_yB_z.$$

As $v_y \approx -v\theta_y$, $v_x \approx -v\theta_2$, $v_z \approx v$, $dz \approx vdt$, the above equation becomes

$$\frac{d\theta_2}{dt} = -\frac{qv\theta_y}{P}B_z.$$

From (a) we have $P \approx \frac{qB_0L}{\theta_y}$. Thus

$$d\theta_2 = -\frac{\theta_y^2}{LB_0}B_z\,dz,$$

and the angular deflection in xz plane is

$$\theta_x = \int_0^{\theta_x} d\theta_2 = -\frac{\theta_y^2}{LB_0}\int_{z_0}^{\infty} B_z\,dz.$$

At the exit of the magnet, $B_x \approx B_0$. We choose the closed path ABCD shown in Fig. 2.65 for the integral

$$\oint \mathbf{B}\cdot d\mathbf{l} = 0.$$

Integrating segment by segment:

$$\int_A^B B_z\,dz = \int_{z_0}^\infty B_z\,dz, \quad \int_B^C B_x\,dx \approx 0,$$

$$\int_C^D B_z\,dz = 0, \quad \int_D^A B_x\,dx = xB_0.$$

Note that we have taken the points B, C at infinity and used the fact that $B_z = 0$ for the mid-plane. Hence

$$\int_{z_0}^\infty B_z\,dz = -xB_0,$$

and

$$\theta_x = \frac{\theta_y^2}{L}x.$$

(c) As $\theta_x \gtrless 0$ for $x \gtrless 0$, the particle will always deflect to the middle of the magnet. Hence the effect found in (b) focuses the particles, the focal length being

$$f = v\left|\frac{x}{v_x}\right| = \frac{x}{\theta_x} = \frac{L}{\theta_y^2} = \frac{P^2}{q^2 B_0^2 L}.$$

2089

When a dilute suspension of diamagnetically anisotropic cylindrical particles is placed on a uniform magnetic field **H**, it is observed that the particles align with their long axes parallel to the field lines. The particles are cylindrically symmetric and they have magnetic susceptibility tensor components characterized by

$$\chi_x = \chi_y < \chi_z < 0.$$

Assume that the suspending fluid has a negligible magnetic susceptibility.

(a) The z-axis of the particle initially makes an angle of θ with the magnetic field. What is the magnetic energy of orientation?

(b) What is the torque on the particle in part (a)?

(c) The tendency toward alignment will be counteracted by Brownian motion. When the particle rotates in the fluid it experiences a viscous torque of magnitude $\zeta\dot\theta$ where

$$\zeta = 10^{-10} \text{ gcm}^2 \text{ sec}^{-1}.$$

The moment of inertia of a particle is $I = 10^{-15} \text{gcm}^2$. If the particles are initially aligned by the magnetic field, estimate the root-mean-square angle $\Delta\theta_{\text{rms}}$ by which the molecular axes will have deviated from the alignment direction in a time $t = 10\,\text{sec}$ after the magnetic field has been turned off. The temperature of the suspension is $T = 300$ K.

(Princeton)

Solution:

(a) In the Cartesian coordinates (x, y, z) attached to a particle, the magnetic field can be expressed as

$$\mathbf{B} = B \sin\theta \cos\varphi \mathbf{e}_x + B \sin\theta \sin\varphi \mathbf{e}_y + B \cos\theta \mathbf{e}_z \,.$$

As $|\chi_x|, |\chi_y|$ and $|\chi_z|$ are generally much smaller than 1, the magnetic field inside the particle (a small cylinder) may be taken to be $\mathbf{B}$ also. The magnetization is given by

$$\mathbf{M} = \chi \cdot \mathbf{H} = \chi \cdot \frac{\mathbf{B}}{\mu_0} \,.$$

Let the volume of the small cylinder be V, then the energy of orientation is

$$E = \mathbf{m} \cdot \mathbf{B} = \mathbf{B} \cdot \left(\chi \cdot \frac{\mathbf{B}}{\mu_0} V \right)$$

$$= \frac{V}{\mu_0} (B_x, B_y, B_z) \begin{pmatrix} \chi_x & & \\ & \chi_y & \\ & & \chi_z \end{pmatrix} \begin{pmatrix} B_x \\ B_y \\ B_z \end{pmatrix}$$

$$= \frac{V}{\mu_0} (\chi_x B_x^2 + \chi_y B_y^2 + \chi_z B_z^2)$$

$$= \frac{V}{\mu_0} [\chi_x (B_x^2 + B_y^2) + \chi_z B_z^2]$$

$$= \frac{V}{\mu_0} [\chi_x B^2 \sin^2\theta + x_z B^2 \cos^2\theta] \,.$$

(b) The torque on the particle is

$$\tau = -\frac{\partial E}{\partial\theta} = -\frac{V}{\mu_0} [\chi_x \cdot 2\sin\theta\cos\theta + \chi_z \cdot 2\cos\theta(-\sin\theta)] B^2$$

$$= \frac{B^2 V}{\mu_0} (\chi_z - \chi_x) \sin 2\theta \,.$$

(c) The rotation of the particle satisfies the equation

$$I\frac{d^2\theta}{dt^2} = -\zeta\frac{d\theta}{dt} + F(t),$$

where $F(t)$ is the random force acting on the particle. Noting that

$$\frac{d^2\theta^2}{dt^2} = 2\dot{\theta}^2 + 2\theta\ddot{\theta},$$

we have

$$I\left(\frac{1}{2}\frac{d^2\theta^2}{dt^2} - \dot{\theta}^2\right) = -\zeta\theta\frac{d\theta}{dt} + \theta F(t),$$

or

$$\frac{1}{2}I\frac{d^2\theta^2}{dt^2} - I\dot{\theta}^2 = -\frac{1}{2}\zeta\frac{d\theta^2}{dt} + \theta F(t).$$

Averaging over the particles, $\overline{\theta F(t)} = 0$, and we have

$$\frac{1}{2}I\frac{d^2\overline{\theta^2}}{dt^2} + \frac{1}{2}\zeta\frac{\overline{\theta^2}}{dt} - I\overline{\dot{\theta}^2} = 0.$$

The principle of equipartition of energy gives

$$\frac{1}{2}I\overline{\dot{\theta}^2} = \frac{1}{2}kT,$$

so the above becomes

$$\frac{d^2\overline{\theta^2}}{dt^2} + \frac{\zeta}{I}\frac{d\overline{\theta^2}}{dt} - \frac{2kT}{I} = 0.$$

This has solution

$$\overline{\theta^2} = \frac{2kT}{\zeta} + Ce^{-\frac{\zeta}{I}t},$$

where C is a constant to be determined. Note that, as θ is not restricted to the zone $[0, \pi]$, we must take the number of turns rotated by the particle about the magnetic field into consideration.

To estimate $\Delta\theta_{rms}$, let $\overline{\theta^2} = 0$ at $t = 0$, then $C = 0$. Hence $\overline{\theta^2} = \frac{2kT}{\zeta}t$ at time t. Numerically

$$\overline{\theta^2} = \frac{2 \times 1.38 \times 10^{-23} \times 300}{10^{-17}} \times 10 = 8.28 \times 10^{-3},$$

or

$$\Delta\theta_{rms} \equiv \sqrt{\overline{\theta^2}} = 0.091 \text{ rad. } = 5.2 \text{ deg}.$$

2090

An electron is introduced in a region of uniform electric and magnetic fields at right angles to each other (let us say $\mathbf{E} = E\mathbf{e}_x, \mathbf{B} = B\mathbf{e}_z$).

(a) For what initial velocity will the electron move with constant velocity (both the direction and the magnitude of velocity are constant)?

(b) Consider a beam of electrons of arbitrary velocity distribution simultaneously injected into a plane normal to the electric field. Is there a time at which all the electrons are in this plane again?

(*Columbia*)

Solution:

(a) If the electron moves with constant velocity, the total force acting on it must be zero, i.e.,

$$\mathbf{F}_B = -\mathbf{F}_E = -eE\mathbf{e}_x .$$

As $\mathbf{F}_B = -e\mathbf{v} \times \mathbf{B} = -eB\mathbf{v} \times \mathbf{e}_z$, we have

$$\mathbf{v} = -\left(\frac{E}{B}\right)\mathbf{e}_y .$$

(b) Suppose all electrons are in the YOZ plane at $t = 0$. Consider an electron with initial position $(0, y_0, z_0)$ and initial velocity (v_{0x}, v_{0y}, v_{0z}). The equations of its motion are then

$$m\frac{dv_x}{dt} = -e(E + Bv_y) , \tag{1}$$

$$m\frac{dv_y}{dt} = ev_x B , \tag{2}$$

$$m\frac{dv_z}{dt} = 0 . \tag{3}$$

Let $v_+ = v_x + iv_y$, then Eqs. (1) and (2) combine to give

$$m\frac{dv_+}{dt} = -eE + ieBv_+$$

with solution

$$v_+ = ce^{i\omega t} - i\frac{E}{B}$$

where $\omega = \frac{eB}{m}$. The intial conditions give

$$c = v_{0x} + i\left(v_{0y} + \frac{E}{B}\right) .$$

Hence

$$v_+ = \left[v_{0x} \cos \omega t - \left(v_{0y} + \frac{E}{B} \right) \sin \omega t \right]$$
$$+ i \left[v_{0x} \sin \omega t + \left(v_{0y} + \frac{E}{B} \right) \cos \omega t - \frac{E}{B} \right],$$

from which we obtain

$$v_x(t) = v_{0x} \cos \omega t - \left(v_{0y} + \frac{E}{B} \right) \sin \omega t,$$

$$v_y(t) = v_{0x} \sin \omega t + \left(v_{0y} + \frac{E}{B} \right) \cos \omega t - \frac{E}{B},$$

$$v_z(t) = v_{0z}.$$

Integrating the above expressions we have

$$x(t) = \frac{v_{0x}}{\omega} \sin \omega t + \frac{1}{\omega} \left(v_{0y} + \frac{E}{B} \right) \cos \omega t - \frac{1}{\omega} \left(v_{0y} + \frac{E}{B} \right),$$

$$y(t) = -\frac{v_{0x}}{\omega} \cos \omega t + \frac{1}{\omega} \left(v_{0y} + \frac{E}{B} \right) \sin \omega t - \frac{E}{B} t + \frac{v_{0x}}{\omega} + y_0,$$

$$z(t) = z_0 + v_{0z} t.$$

For $x(t) = 0$ we require that $t = \frac{2\pi n}{\omega}$ $(n = 1, 2, \dots)$. Hence all the electrons will be in the YOZ plane again at times $\frac{2\pi n}{\omega}$.

4. MISCELLANEOUS APPLICATIONS (2091–2119)

2091

Figure 2.67 shows a simplified electron lens consisting of a circular loop (of radius a) of wire carrying current I. For $\rho \ll a$ the vector potential is approximately given by

$$A_\theta = \frac{\pi I a^2 \rho}{(a^2 + z^2)^{3/2}}.$$

(a) Write down the Lagrangian in cylindrical coordinates (ρ, θ, z) and the Hamiltonian for a particle of charge q moving in this field.

(b) Show that the canonical momentum p_θ vanishes for the orbit shown and find an expression for $\dot\theta$.

In parts (c) and (d) it is useful to make a simplifying approximation that the magnetic force is most important when the particle is in the vicinity of the lens (impulse approximation). Since ρ is small we can assume that $\rho \approx b$ and $\dot z \approx u$ are nearly constant in the interaction region.

(c) Calculate the impulsive change in the radial momentum as the particle passes through the lens. Then show that the loop acts like a thin lens

$$\frac{1}{l_0} + \frac{1}{l_i} = \frac{1}{f},$$

where

$$f = \frac{8a}{3\pi} \left(\frac{muc}{\pi qI} \right)^2 .$$

(d) Show that the image rotates by an angle $\theta = -4\sqrt{\frac{2a}{3\pi f}}$ in passing through the lens.

(*Wisconsin*)

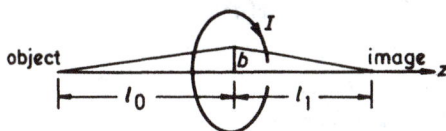

Fig. 2.67

Solution:

(a) The Lagrangian of a charge q in an electromagnetic field is

$$L = T - V = \frac{1}{2}mv^2 - q\left(\varphi - \frac{\mathbf{v} \cdot \mathbf{A}}{c}\right),$$

where v is the velocity of the charge of mass m, φ is the scalar potential, and $\mathbf{A}$ is the vector potential. As

$$\mathbf{v} = \dot\rho\mathbf{e}_\rho + \rho\dot\theta\mathbf{e}_\theta + \dot z\mathbf{e}_z,$$

$$\mathbf{A} = \frac{I\pi a^2\rho}{(a^2 + z^2)^{3/2}}\mathbf{e}_\theta, \quad \varphi = 0,$$

we have

$$L = \frac{m}{2}(\dot{\rho}^2 + \rho^2\dot{\theta}^2 + \dot{z}^2) + \frac{I\pi a^2 q\rho^2}{c(a^2 + z^2)^{3/2}}\dot{\theta}.$$

Hence the components of the canonical momentum:

$$P_\rho = \frac{\partial L}{\partial \dot{\rho}} = m\dot{\rho},$$

$$P_\theta = \frac{\partial L}{\partial \dot{\theta}} = m\rho^2\dot{\theta} + \frac{I\pi a^2 q\rho^2}{c(a^2 + z^2)^{3/2}},$$

$$P_z = \frac{\partial L}{\partial \dot{z}} = m\dot{z}.$$

The Hamiltonian is then

$$H = P_\rho\dot{\rho} + P_\theta\dot{\theta} + P_z\dot{z} - L$$
$$= \frac{1}{2m}\left[P_\rho^2 + \frac{1}{\rho^2}\left(P_\theta - \frac{I\pi a^2 q\rho^2}{c(a^2 + z^2)^{3/2}}\right)^2 + P_z^2\right].$$

(b) Using Hamilton's canonical equation $\dot{P}_\theta = -\frac{\partial H}{\partial \theta}$, we obtain $\dot{P}_\theta = 0$, i.e.,

$$P_\theta = m\rho^2\dot{\theta} + \frac{I\pi a^2 q\rho^2}{c(a^2 + z^2)^{3/2}} = \text{const}.$$

Initially when the particle was far away from the lens it was traveling along the axis of the circular loop ($\rho = 0$) with $v_\theta = 0$. Since P_θ is a constant of the motion, we have $P_\theta = 0$, giving

$$\dot{\theta} = -\frac{I\pi a^2 q}{mc(a^2 + z^2)^{3/2}}.$$

(c) Another Hamilton's canonical equation $\dot{P}_\rho = -\frac{\partial H}{\partial \rho}$, with $P_\theta = 0$, gives

$$\dot{P}_\rho = -\frac{I^2\pi^2 a^4 q^2\rho}{mc^2(a^2 + z^2)^3},$$

or

$$dP_\rho = -\frac{I^2\pi^2 a^4 q^2\rho}{mc^2(a^2 + z^2)^3\dot{z}}dz.$$

Since $\rho \simeq b$ and $\dot{z} \simeq u$ are nearly constant in the interaction region, the change in the radial momentum is

$$\Delta P_\rho \approx -\frac{I^2\pi^2 a^4 q^2 b}{mc^2 u}\int_{-\infty}^{\infty}\frac{1}{(a^2 + z^2)^3}dz = -\frac{3\pi b}{8mau}\left(\frac{Iq\pi}{c}\right)^2.$$

Consider the orbit shown in Fig. 2.67. We have $\frac{\dot{\rho}_0}{u} = \frac{b}{l_0}$ at the object point and $-\frac{\dot{\rho}_i}{u} = \frac{b}{l_i}$ at the image point of the lens. Hence

$$\frac{b}{l_0} + \frac{b}{l_i} = \frac{1}{u}(\dot{\rho}_0 - \dot{\rho}_i) = -\frac{\Delta P_\rho}{mu} = \frac{3\pi b}{8a}\left(\frac{Iq\pi}{muc}\right)^2,$$

which can be written as

$$\frac{1}{l_0} + \frac{1}{l_i} = \frac{1}{f}$$

with

$$f = \frac{8a}{3\pi}\left(\frac{muc}{Iq\pi}\right)^2.$$

(d) The expression for $\dot{\theta}$ can be written as

$$d\theta = -\frac{I\pi a^2 q dz}{mc(a^2 + z^2)^{3/2}u}.$$

Hence passing through the lens the image will have rotated with respect to the object by an angle

$$\Delta\theta = -\frac{I\pi a^2 q}{mcu}\int_{-\infty}^{\infty}\frac{1}{(a^2 + z^2)^{3/2}}dz = -\frac{2I\pi q}{mcu} = -4\sqrt{\frac{2a}{3\pi f}}.$$

2092

In Fig. 2.68 a block of semiconductor (conductivity $= \sigma$) has its bottom face ($z = 0$) attached to a metal plate (its conductivity $\sigma \to \infty$) which is held at potential $\phi = 0$. A wire carrying current J is attached to the center of the top face ($z = c$). The sides ($x = 0, x = a, y = 0, y = b$) are insulated and the top is insulated except for the wire. Assume that the charge density is $\rho = 0$ and $\epsilon = \mu = 1$ inside the block.

(a) Write down the equations satisfied by the potential inside the box and the general solution for the potential.

(b) Write down the boundary conditions for all faces and express the arbitrary constants in the solution from (a) in terms of the given quantities.

(Princeton)

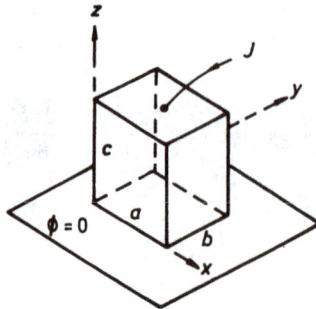

Fig. 2.68

Solution:

(a) Inside the box $\phi(x, y, z)$ satisfies Laplace's equation

$$\nabla^2 \phi(z, y, z) = 0.$$

Separate the variables by writing

$$\phi(x, y, z) = X(x)Y(y)Z(z).$$

Laplace's equations then becomes

$$\frac{1}{X}\frac{d^2 X}{dx^2} + \frac{1}{Y}\frac{d^2 Y}{dy^2} + \frac{1}{Z}\frac{d^2 Z}{dz^2} = 0. \tag{2}$$

Each of the three terms on the left-hand side depends on one variable only and must thus be equal to a constant:

$$\frac{1}{X}\frac{d^2 X}{dx^2} = -\alpha^2.$$

$$\frac{1}{Y}\frac{d^2 Y}{dy^2} = -\beta^2,$$

$$\frac{1}{Z}\frac{d^2 Z}{dz^2} = \gamma^2,$$

where $\gamma^2 = \alpha^2 + \beta^2$. The solutions of these equations are

$$X = A\cos\alpha x + B\sin\alpha x,$$
$$Y = C\cos\beta y + D\sin\beta y,$$
$$Z = Ee^{\gamma z} + Fe^{-\gamma z},$$

where A, B, C, D, E, F are constants. Hence

$$\phi(x,y,z) = (A\cos\alpha x + B\sin\alpha x)(C\cos\beta y + D\sin\beta y)$$
$$\cdot [E\exp(\sqrt{\alpha^2 + \beta^2}z) + F\exp(-\sqrt{\alpha^2 + \beta^2}z)].$$

(b) The boundary condition $\mathbf{E}_t = 0$ for the four vertical surfaces gives

$$\left.\frac{\partial\phi}{\partial x}\right|_{x=0,a} = \left.\frac{\partial\phi}{\partial y}\right|_{y=0,b} = 0. \tag{1}$$

For the top face $(z = c)$, Ohm's law gives

$$\left.\frac{\partial\phi}{\partial z}\right|_{z=c} = -E_z = -\frac{J}{\sigma}\delta\left(x - \frac{a}{2}, y - \frac{b}{2}\right). \tag{2}$$

The bottom face $(z = 0)$ has zero potential, thus

$$\phi(x,y,0) = 0. \tag{3}$$

The conditions (1) require

$$B = D = 0,$$

$$\alpha = \alpha_m = \frac{m\pi}{a}, \ \beta = \beta_n = \frac{n\pi}{b}, \gamma = \gamma_{mn} = \pi\sqrt{\left(\frac{m}{a}\right)^2 + \left(\frac{n}{b}\right)^2},$$

where m, n are positive integers. (Negative integers only repeat the solution). Equation (3) requires $F = -E$. Thus for a given set of integers m, n we have

$$\phi_{mn}(x,y,z) = A_{mn}\cos(\alpha_m x)\cos(\beta_n y)\sinh(\gamma_{mn}z).$$

Hence the general solution is

$$\phi(x,y,z) = \sum_{m,n=1}^{\infty} \phi_{mn}(x,y,z)$$

$$= \sum_{m,n=1}^{\infty} A_{mn}\cos(\alpha_m x)\cos(\beta_n y)\sinh(\gamma_{mn}z).$$

Substituting this in (2) we have

$$\sum_{m,n=1}^{\infty} A_{mn}\cos(\alpha_m x)\cos(\beta_n y)\gamma_{mn}\cosh(\gamma_{mn}c)$$

$$= \frac{J}{\sigma}\delta\left(x - \frac{a}{2}, y - \frac{b}{2}\right).$$

Multiplying both sides by $\cos(\gamma_m x)\cos(\beta_n y)$ and integrating over the top surface we have

$$A_{mn} = \frac{4J}{ab\sigma\gamma_{mn}\cosh(\gamma_{mn}c)}\int_0^a dx \int_0^b \delta\left(x-\frac{a}{2}, y-\frac{b}{2}\right)$$
$$\cdot \cos(\alpha_m x)\cos(\beta_n y)dy$$
$$= \frac{4J}{ab\sigma\gamma_{mn}\cosh(\gamma_{mn}c)}\cos\frac{m\pi}{2}\cos\frac{n\pi}{2}.$$

Note that $A_{mn} \neq 0$ only for m and n both being positive even numbers.

2093

A magnetic dipole of moment **m** is placed in a magnetic lens whose field components are given by

$$B_x = \alpha(x^2 - y^2), \quad B_y = -2\alpha xy, \quad B_z = 0,$$

where z is the axis of the lens and α is a constant. (This is called a sextupole field.)

(a) What are the components of the force on the dipole?

(b) Could one or more such lenses be used to focus a beam of neutral particles possessing a magnetic dipole moment? Give the reasons for your answer.

(*UC, Berkeley*)

Solution:

(a) As **m** is a constant vector, the force exerted by the external magnetic field on **m** is

$$\mathbf{F} = \nabla(\mathbf{m} \cdot \mathbf{B}).$$

Thus we have

$$F_x = 2\alpha(m_x x - m_y y),$$
$$F_y = -2\alpha(m_x y + m_y x),$$
$$F_z = 0,$$

where we have written $\mathbf{m} = m_x \mathbf{e}_x + m_y \mathbf{e}_y + m_z \mathbf{e}_z$.

(b) If $\mathbf{m} = m\mathbf{e}_y$, we have

$$F_x = -2\alpha my, \quad F_y = -2\alpha mx.$$

This force is opposite to the displacement of the dipole from the axis. Hence we can use the sextupole lens to focus a beam of neutral particles with magnetic moment. If $\mathbf{m} = m\mathbf{e}_x$, we have

$$F_x = 2\alpha m x, \quad F_y = -2\alpha m y.$$

Then the lens is diverging in x-direction but converging in y-direction. Hence to focus the beam, we need a pair of sextupole lenses, with the phase angles of the sextupole fields differing by π, i.e., the field of the second sextupole is

$$B_x = -\alpha(x^2 - y^2), \quad B_y = 2\alpha xy.$$

The force exerted by the second sextupole is

$$F_x = -2\alpha m x, \quad F_y = 2\alpha m y,$$

so that a converging power is obtained in x-direction as well.

2094

A charged particle enters a uniform static magnetic field $\mathbf{B}$ moving with a nonrelativistic velocity $\mathbf{v}_0$ which is inclined at an angle α to the direction of $\mathbf{B}$.

(a) What is the rate of emission of radiation?

(b) What is the condition on $\mathbf{v}_0$ that the radiation be dominantly of one multipolarity?

(c) If a uniform static electric field $\mathbf{E}$ is added parallel to $\mathbf{B}$, how large must it be to double the previous rate of radiation?

(*Wisconsin*)

Solution:

(a) The radiation emitted per unit time by an accelerating nonrelativistic particle of charge q with velocity $v \ll c$ is approximately

$$P = \frac{2q^2}{3c^3} \dot{v}^2.$$

in Gaussian units. The equation of the motion of the particle in the magnetic field $\mathbf{B}$ is

$$m_0 \dot{\mathbf{v}}_0 = \frac{q}{c}(\mathbf{v}_0 \times \mathbf{B}),$$

giving

$$\dot{v}_0^2 = \frac{q^2}{m_0^2 c^2} v_0^2 B^2 \sin^2 \alpha \,,$$

where α is the angle between $\mathbf{v}_0$ and $\mathbf{B}$. The rate of emission of radiation is then

$$P = \frac{2q^4}{3m_0^2 c^5} B^2 v_0^2 \sin^2 \alpha \ \text{erg/s} \,.$$

(b) The radiation emitted by a charged particle moving in a magnetic field $\mathbf{B}$ is known as cyclotron radiation and has the form of the radiation of a Hertzian dipole. The particle executes Larmor precession perpendicular to $\mathbf{B}$ with angular frequency $\omega_0 = \frac{qB}{m_0 c}$. Actually there are also weaker radiations of higher harmonic frequencies $2\omega_0, 3\omega_0, \ldots$. However, if $v_0 \ll c$ is satisfied, the dipole radiation is the main component and the others may be neglected.

(c) When a uniform static field $\mathbf{E}$ is added parallel to $\mathbf{B}$, the equation of the particle's motion becomes

$$m_0 \dot{\mathbf{v}} = \frac{q}{c} (\mathbf{v}_0 \times \mathbf{B}) + q\mathbf{E} \,,$$

or

$$\dot{\mathbf{v}} = \dot{\mathbf{v}}_\perp + \dot{\mathbf{v}}_{//} = \frac{q}{m_0 c} (\mathbf{v}_0 \times \mathbf{B}) + \frac{q}{m_0} \mathbf{E} \,,$$

$\dot{\mathbf{v}}_\perp$ and $\dot{\mathbf{v}}_{//}$ being components of the particle's acceleration perpendicular and parallel to the electric field respectively. To double the radiation power in (a) $\dot{v}^2$ is to be doubled. Writing the above equation as

$$\dot{v}^2 = \dot{v}_\perp^2 + \dot{v}_{//}^2 = \frac{q^2}{m_0^2 c^2} v_0^2 B^2 \sin^2 \alpha + \frac{q^2}{m_0^2} E^2$$

since as $\mathbf{E}$ is parallel to $\mathbf{B}$, $\mathbf{E}$ is perpendicular to $\mathbf{v}_0 \times \mathbf{B}$, we see that to obtain $\dot{v}^2 = 2\dot{v}_0^2$ we require

$$E = \frac{v_0}{c} B \sin \alpha \,.$$

2095

A circular loop of wire, radius r, weighing m kilograms, carries a steady current I amperes. It is constrained to have its axis perpendicular to a large planar sheet of a perfect conductor. It is free to move vertically, and its instantaneous height is x meters. It is moving at a speed v in the y direction with $v \ll c$.

(a) What is the boundary condition on the magnetic field **B** at the planar conducting sheet?

(b) Draw and describe algebraically a single image current that, combined with the real current, exactly reproduces the magnetic field in the region above the plane.

(c) Find the approximate equilibrium height x and frequency of small vertical oscillations for a value of the current such that $x \ll r$.

(*Princeton*)

Solution:

(a) The normal component of **B**, which is continuous across a boundary, is zero on the conducting surface: $B_n = 0$.

(b) As shown in Fig. 2.69, the image of the current is a current loop symmetric with respect to the surface of the conducting plane, but with an opposite direction of flow. The magnetic field above the planar conductor is the superposition of the magnetic fields produced by the two currents which satisfies the boundary condition $B_n = 0$.

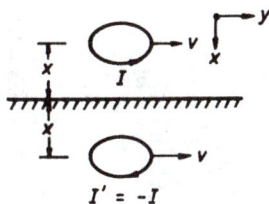

Fig. 2.69

(c) Consider a current element Idl of the real current loop. As $x \ll r$, we may consider the image current as an infinite straight line current. Then the current element Idl will experience an upward force of magnitude

$$dF = I|dl \times \mathbf{B}| = I\left|dl\frac{\mu_0(-I)}{2\pi(2x)}\right| = \frac{\mu_0 I^2 dl}{4\pi x}.$$

The force on the entire current loop is

$$F = \frac{\mu_0 I^2}{4\pi x} \cdot 2\pi r = \frac{\mu_0 I^2}{2}\frac{r}{x}.$$

At the equilibrium height this force equals the downward gravity:

$$\frac{\mu_0 I^2 r}{2x} = mg \, ,$$

giving

$$x = \frac{\mu_0 I^2 r}{2mg} \, .$$

Suppose the loop is displaced a small distance δ vertically from the equilibrium height x, i.e., $x \to x + \delta$, $\delta \ll x$. The equation of the motion of the loop in the vertical direction is

$$-m\ddot{\delta} = mg - \frac{\mu_0 I^2 r}{2(x + \delta)} \simeq mg - \frac{\mu_0 I^2 r}{2x}\left(1 - \frac{\delta}{x}\right) \, .$$

Noting that $mg = \frac{\mu_0 I^2 r}{2x}$, we get

$$\ddot{\delta} + \frac{\mu_0 I^2 r}{2mx^2}\delta = 0 \, .$$

This shows that the vertical motion is harmonic with angular frequency

$$\omega_0 = \sqrt{\frac{\mu_0 I^2 r}{2mx^2}} = \frac{g}{I}\sqrt{\frac{2m}{\mu_0 r}} \, .$$

2096

We assume the existence of magnetic charge related to the magnetic field by the local reaction

$$\nabla \cdot \mathbf{B} = \mu_0 \rho_m \, .$$

(a) Using the divergence theorem, obtain the magnetic field of a point magnetic charge at the origin.

(b) In the absence of magnetic charge, the curl of the electric field is given by Faraday's law

$$\nabla \times \mathbf{E} = -\frac{\partial \mathbf{B}}{\partial t} \, .$$

Show that this law is incompatible with a magnetic charge density that is a function of time.

(c) Assuming that magnetic charge is conserved, derive the local relation between the magnetic charge current density $\mathbf{J}_m$ and the magnetic density ρ_m.

(d) Modify Faraday's law as given in part (b) to obtain a law consistent with the presence of a magnetic charge density that is a function of position and time. Demonstrate the consistency of the modified law.

(UC, Berkeley)

Solution:

(a) Consider a spherical surface S of radius r at the origin. As $\nabla \cdot \mathbf{B} = \mu_0 \rho_m$ the divergence theorem gives

$$\int_V \nabla \cdot \mathbf{B}\, dV = \oint_S \mathbf{B} \cdot d\mathbf{S} = 4\pi r^2 B(r) = \mu_0 q_m \,.$$

Hence

$$\mathbf{B}(r) = \frac{\mu_0 q_m}{4\pi r^2} \mathbf{e}_r \,.$$

(b)

$$\frac{\partial}{\partial t} \nabla \cdot \mathbf{B} = \nabla \cdot \frac{\partial \mathbf{B}}{\partial t} = -\nabla \cdot (\nabla \times \mathbf{E}) = 0 \,,$$

since $\nabla \cdot (\nabla \times \mathbf{E}) = 0$ identically. On the other hand,

$$\frac{\partial}{\partial t} \nabla \cdot \mathbf{B} = \mu_0 \frac{\partial \rho_m}{\partial t} \,.$$

Thus Faraday's law is incompatible with a time-varying magnetic charge density.

(c) The conservation of magnetic charge can be expressed as

$$\frac{\partial}{\partial t} \int_V \rho_m\, dV = -\oint_S \mathbf{J}_m \cdot d\mathbf{S} = -\int_V \nabla \cdot \mathbf{J}_m\, dV \,.$$

As V is arbitrary we must have

$$\frac{\partial \rho_m}{\partial t} + \nabla \cdot \mathbf{J}_m = 0 \,.$$

This is the continuity equation for magnetic charge.

(d) If we modify Faraday's law to

$$\nabla \times \mathbf{E} = -\mu_0 \mathbf{J}_m - \frac{\partial \mathbf{B}}{\partial t}$$

and take divergence on both sides, we shall obtain

$$-\mu_0 \nabla \cdot \mathbf{J}_m - \frac{\partial}{\partial t} \nabla \cdot \mathbf{B} = -\mu_0 \left(\nabla \cdot \mathbf{J}_m + \frac{\partial \rho_m}{\partial t} \right) = 0 \,.$$

Hence

$$\frac{\partial}{\partial t} \nabla \cdot \mathbf{B} = -\mu_0 \nabla \cdot \mathbf{J}_m = \mu_0 \frac{\partial \rho_m}{\partial t} \,,$$

consistent with the second equation of (b).

2097

(a) Suppose that isolated magnetic charges (magnetic monopoles) exist. Rewrite Maxwell's equations including contributions from a magnetic charge density ρ_m and a magnetic current density $\mathbf{j}_m$. Assume that, except for the sources, the fields are in vacuum.

(b) Alvarez and colleagues looked for magnetic monopoles in matter by making pieces of matter go a number of successive times through a coil of n turns. If the coil has a resistance R, and we assume that the magnetic charges are moved slowly enough to make the effect of its inductance small, calculate how much charge Q flows through the coil after N circuits of a monopole q_m.

(c) Suppose that the coil is made superconducting so that its resistance is zero, and only its inductance L limits the current induced in it. Assuming that initially the current in the coil is zero, calculate how much current it carries after N circuits of the monopole.

(CUSPEA)

Solution:

(a) Use the analysis of Problem 2096. When electric charge density ρ, electric current density $\mathbf{j}$, magnetic charge density ρ_m, and magnetic current density $\mathbf{j}_m$ are all present in vacuum, Maxwell's equations (in Gaussian

units) are

$$\nabla \cdot \mathbf{E} = 4\pi\rho,$$

$$\nabla \cdot \mathbf{B} = 4\pi\rho_m,$$

$$\nabla \times \mathbf{E} = -\frac{1}{c}\frac{\partial \mathbf{B}}{\partial t} - \frac{4\pi}{c}\mathbf{j}_m,$$

$$\nabla \times \mathbf{B} = \frac{1}{c}\frac{\partial \mathbf{E}}{\partial t} + \frac{4\pi\mathbf{j}}{c},$$

where c is the velocity of light in vacuum.

(b) As shown in Fig. 2.70, we take one of the turns as the closed loop l in Stokes' theorem and let the area surrounded by l be S. Then, using Stokes' theorem and the third equation above, we have

$$\oint_l \mathbf{E} \cdot d\mathbf{l} = \int_S \nabla \times \mathbf{E} \cdot d\mathbf{S} = -\frac{1}{c}\frac{\partial}{\partial t}\int_S \mathbf{B} \cdot d\mathbf{S} - \frac{4\pi}{c}\int_S \mathbf{j}_m \cdot d\mathbf{S}.$$

Fig. 2.70

Letting I_m be the magnetic current in the coil, we have

$$I_m = \int_S \mathbf{j}_m \cdot d\mathbf{S}.$$

Letting V be the potential across the coil and I the electric current flowing through it, we have

$$\oint_l \mathbf{E} \cdot d\mathbf{l} = V = IR.$$

The magnetic flux crossing the coil is $\phi = \int_S \mathbf{B} \cdot d\mathbf{S}$, and the induced emf in the coil is

$$\varepsilon = -\frac{1}{c}\frac{\partial \phi}{\partial t}.$$

Combining the above we have the circuit equation

$$IR = \varepsilon - \frac{4\pi}{c}I_m.$$

If the inductance can be neglected, $\varepsilon = 0$ and

$$IR = -\frac{4\pi}{c} I_m .$$

From $I = \frac{dQ}{dt}$, $I_m = \frac{dq_m}{dt}$, this reduces to

$$R\frac{dQ}{dt} = -\frac{4\pi}{c}\frac{dq_m}{dt} .$$

Integrating leads to

$$Q = -\frac{4\pi q_m}{RC} .$$

After q_m goes N times through the coil of n turns, the total charge flowing through the coil is

$$q = -\frac{4\pi N n q_m}{RC} .$$

(c) If the resistance is negligible, i.e., $R = 0$, while the inductance L is not, we have $\varepsilon = -L\frac{dI}{dt}$. The circuit equation now gives

$$-L\frac{dI}{dt} = \frac{4\pi}{c} N n \frac{dq_m}{dt} .$$

Integrating we get

$$I = -\frac{4\pi N n q_m}{LC} .$$

2098

In Fig. 2.71, the cylindrical cavity is symmetric about its long axis. For the purposes of this problem, it can be approximated as a coaxial cable (which has inductance and capacitance) shorted at one end and connected to a parallel plate disk capacitor at the other.

(a) Derive an expression for the lowest resonant frequency of the cavity. Neglect end and edge effects ($h \gg r_2, d \ll r_1$).

(b) Find the direction and radial-dependence of the Poynting vector N in the regions near points A and B.

(Princeton)

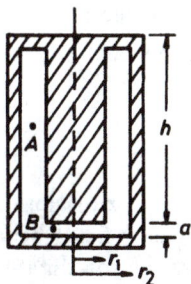

Fig. 2.71

Solution:

(a) To find the inductance and capacitance per unit length of the coaxial cable, we suppose that the inside and outside conductors respectively carry currents I and $-I$ and uniform linear charges λ and $-\lambda$. Use cylindrical coordinates (r, θ, z) with the z-axis along the axis of the cable. Let the direction of flow of the current in the inner conductor be along the $+z$-direction. From Problem **2022** the inductance and capacitance per unit length of the coaxial cable are

$$L = \frac{\mu_0 h}{2\pi} \ln \frac{r_2}{r_1}, \quad C = \frac{2\pi \varepsilon_0 h}{\ln \frac{r_2}{r_1}}.$$

The capacitance of the parallel-plate condenser connected to the coaxial cable is

$$C_0 = \frac{\pi \varepsilon_0 r_1^2}{d}.$$

Hence the lowest resonant angular frequency of the cavity is

$$\omega_0 = \frac{1}{L(C + C_0)} = \frac{2dc^2}{h(2dh + r_1^2 \ln \frac{r_2}{r_1})}.$$

(b) At point $A, r_1 < r < r_2, \mathbf{E}(r) \sim \frac{\mathbf{e}_r}{r}, \mathbf{B}(r) \sim \frac{\mathbf{e}_\theta}{r}$, so $\mathbf{N} \sim \frac{1}{r^2}\mathbf{e}_z$. At point $B, 0 < r < r_1, \mathbf{E}(r) \sim -\mathbf{e}_z, \mathbf{B}(r) \sim r\mathbf{e}_\theta$, so $\mathbf{N} \sim r\mathbf{e}_r$.

2099

An electromagnetic wave can propagate between two long parallel metal plates with **E** and **B** perpendicular to each other and to the direction of propagation. Show that the characteristic impedance $Z_0 = \sqrt{L/C}$

is $\sqrt{\frac{\mu_0}{\varepsilon_0}} \cdot \frac{s}{w}$, where L and C are the inductance and capacitance per unit length, s is the plate separation, and w is the plate width. Use the long wavelength approximation.

(Wisconsin)

Solution:

In the long wavelength approximation, $\lambda \gg w, \lambda \gg s$, and we can consider the electric and magnetic fields between the two metal plates as approximately stationary. Use the coordinate system as shown in Fig. 2.72 with the z-axis along the direction of propagation. Since the electric and magnetic fields are perpendicular to the z-axis and are zero in the metal plates, the continuity of $\mathbf{E}_t$ gives $E_y = 0$, while the continuity of B_n gives $B_x = 0$.

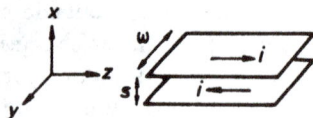

Fig. 2.72

Suppose the two plates carry currents $+i$ and $-i$. The magnetic field between the plates is given by the boundary condition $\mathbf{n} \times \mathbf{H} = \mathbf{I}_l$, where $\mathbf{I}_l$ is the current per unit width of the conductor, to be

$$\mathbf{B} = -\frac{\mu_0 i}{w}\mathbf{e}_y .$$

The inductance per unit length of the plates is obtained by considering the flux crossing a rectangle of unit length and width s parallel to the z-axis to be

$$L = \frac{Bs}{i} = \frac{\mu_0 s}{w} .$$

Let the surface charge density of the two metal plates be σ and $-\sigma$. The electric field between the plates is

$$\mathbf{E} = \frac{\sigma}{\varepsilon_0}\mathbf{e}_x ,$$

and the potential difference between the plates is

$$V = Es = \frac{\sigma s}{\varepsilon_0} .$$

Hence the capacitance per unit length is

$$C = \frac{\sigma w}{V} = \frac{\varepsilon_0 w}{s}.$$

Therefore the characteristic impedance per unit length of the plates is

$$Z = \sqrt{\frac{L}{C}} = \sqrt{\frac{\mu_0 s}{w} \Big/ \frac{\varepsilon_0 w}{s}} = \sqrt{\frac{\mu_0}{\varepsilon_0}} \frac{s}{w}.$$

2100

Reluctance in a magnetic circuit is analogous to:

(a) resistance in a direct current circuit

(b) volume of water in a hydraulic circuit

(c) voltage in an alternating current circuit.

(*CCT*)

Solution:

The answer is (a).

2101

The permeability of a paramagnetic substance is:

(a) slightly less than that of vacuum

(b) slightly more than that of vacuum

(c) much more than that of vacuum

(*CCT*)

Solution:

The answer is (b).

2102

Magnetic field is increasing through a copper plate. The eddy currents:

(a) help the field increase

(b) slow down the increase

(c) do nothing

(*CCT*)

Solution:

The answer is (b).

2103

A golden ring is placed on edge between the poles of a large magnet. The bottom of the ring is prevented from slipping by two fixed pins. It is disturbed from the vertical by 0.1 rad and begins to fall over. The magnetic field is 10^4 gauss, the major and minor radii of the ring are 1 cm and 1 mm respectively (see Fig. 2.73), the conductivity of gold is 4×10^{17} s^{-1} and the density of gold is 19.3 g/cm^3.

(a) Does the potential energy released by the fall go mainly into kinetic energy or into raising the temperature of the ring? Show your reasoning (order of magnitude analysis only for this part).

(b) Neglecting the smaller effect calculate the time of the fall. (Hint: $\int_{0.1}^{\frac{\pi}{2}} \frac{\cos^2 \theta \, d\theta}{\sin \theta} = 2.00$)

(MIT)

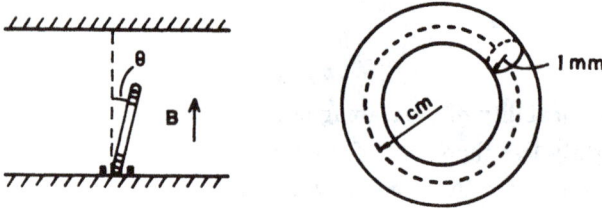

Fig. 2.73

Solution:

(a) Let the time of the fall be T. In the process of falling, potential energy is converted into thermal energy W_t and kinetic energy W_k given by an order of magnitude analysis to be roughly (in Gaussian units)

$$W_t \sim I^2 R T \approx \left(\frac{\phi}{cTR}\right)^2 RT = \frac{B^2(\pi r_1^2)^2}{c^2 RT}$$

$$= \frac{mB^2(\pi r_1^2)^2}{c^2 T \cdot \left(\frac{2\pi r_1}{\sigma \pi r_2^2}\right) \cdot (\rho \cdot 2\pi r_1 \cdot \pi r_2^2)} = \frac{mr_1^2}{T} \cdot \frac{\sigma B^2}{4\rho c^2},$$

$$W_k \sim \frac{1}{2} I \omega^2 \approx \frac{1}{2} \cdot \frac{3}{2} m r_1^2 \left(\frac{\pi}{2T}\right)^2 = \frac{3\pi^2 m r_1^2}{16 T^2},$$

where

r_1, r_2 = major and minor radii of the ring respectively,
ϕ = magnetic flux crossing the ring $\approx B\pi r_1^2$,
ρ = density of gold,
σ = conductivity of gold,
R = resistance of the ring = $\frac{2\pi r_1}{\sigma \pi r_2^2}$,
m = mass of the ring = $\rho 2\pi r_1 \cdot \pi r_2^2$,
c = velocity of light in vacuum,
ω = angular velocity of fall $\approx \frac{\pi}{2T}$.

Putting

$$T_g = \sqrt{\frac{r_1}{g}} = \sqrt{\frac{1}{980}} = 3.2 \times 10^{-2} \text{ s},$$

$$T_B = \frac{4\rho c^2}{\sigma B^2} = \frac{4 \times 19.3 \times 9 \times 10^{20}}{4 \times 10^{17} \times 10^8} = 1.74 \times 10^{-3} \text{ s},$$

we can write the above as

$$W_t \sim mgr_1 \cdot \frac{T_g^2}{T \cdot T_B}, \quad W_k \sim mgr_1 \cdot \frac{3\pi^2}{16}\left(\frac{T_g}{T}\right)^2.$$

The energy balance gives

$$mgr_1 = W_t + W_k,$$

or

$$T^2 - \left(\frac{T_g^2}{T_B}\right)T - \frac{3\pi^2}{16}T_g^2 = 0.$$

Solving for T we have

$$T = \frac{T_g}{2}\left[\frac{T_g}{T_B} + \sqrt{\left(\frac{T_g}{T_B}\right)^2 + \frac{3\pi^2}{4}}\right].$$

As $\left(\frac{T_g}{T_B}\right)^2 \gg \frac{3\pi^2}{4}, T \approx \frac{T_g^2}{T_B}$. Hence

$$W_t \sim mgr_1, \quad W_k \sim mgr_1 \cdot \frac{3\pi^2}{16}\left(\frac{T_B}{T_g}\right) \ll mgr_1.$$

It follows that the potential energy released by the fall goes mainly into raising the temperature of the ring.

(b) We neglect the kinetic energy of the ring. That is, we assume that the potential energy is changed entirely into thermal energy. Then the gravitational torque and magnetic torque on the ring approximately balance each other.

The magnetic flux crossing the ring is

$$\phi(\theta) = B\pi r_1^2 \sin\theta .$$

The induced emf is

$$\varepsilon = \frac{1}{c}\left|\frac{d\phi}{dt}\right| = \frac{1}{c}B\pi r_1^2 \cos\theta\dot{\theta} ,$$

giving the induced electric current as

$$i = \frac{\varepsilon}{R} = B\pi r_1^2 \cos\theta \cdot \frac{\dot{\theta}}{cR}$$

and the magnetic moment of the ring as

$$m = \frac{i\pi r_1^2}{c} = \frac{B(\pi r_1^2)^2 \cos\theta\dot{\theta}}{c^2 R} .$$

Thus the magnetic torque on the ring is

$$\tau_m = |\mathbf{m} \times \mathbf{B}| = \frac{(B\pi r_1^2 \cos\theta)^2\dot{\theta}}{c^2 R} .$$

The gravitational torque on the ring is $\tau_g = mgr_1 \sin\theta$. Therefore

$$\tau_m = \tau_g ,$$

or

$$\frac{(B\pi r_1^2 \cos\theta)^2\dot{\theta}}{c^2 R} = mgr_1 \sin\theta ,$$

giving

$$dt = \frac{\sigma B^2 r_1 \cos^2\theta d\theta}{4\rho g c^2 \sin\theta} .$$

Integrating we find

$$T = \frac{\sigma B^2 r_1}{4\rho g c^2} \int_{0.1}^{\frac{\pi}{2}} \frac{\cos^2\theta d\theta}{\sin\theta} = \frac{\sigma B^2 r_1}{4\rho g c^2} \cdot 2$$

$$= 2\frac{T_g^2}{T_B} = 2 \times \frac{(3.2 \times 10^{-2})^2}{1.74 \times 10^{-3}} = 1.2 \text{ s} .$$

2104

A particle with given charge, mass and angular momentum moves in a circular orbit.

(a) Starting from the fundamental laws of electrodynamics, find the static part of the magnetic field generated at distances large compared with the size of the loop.

(b) What magnetic charge distribution would generate the same field?

(*UC, Berkeley*)

Solution:

(a) Let the charge, mass and angular momentum of the particle be q, m, L respectively. Use cylindrical coordinates (R, θ, z) with the z-axis along the axis of the circular orbit and the origin at its center. As we are interested in the steady component of the field, we can consider the charge orbiting the circle as a steady current loop. The vector potential at a point of radius vector $\mathbf{R}$ from the origin is

$$\mathbf{A}(\mathbf{R}) = \frac{\mu_0}{4\pi} \int \frac{\mathbf{J}(\mathbf{r}')}{r} dV',$$

where $r = |\mathbf{R} - \mathbf{r}'|$. For large distances take the approximation $r = R(1 - \frac{\mathbf{R} \cdot \mathbf{r}'}{R^2})$ and write $\mathbf{J}(\mathbf{r}')dV' = I d\mathbf{r}'$. Then

$$\mathbf{A}(\mathbf{R}) \approx \frac{\mu_0 I}{4\pi R} \oint \left(1 + \frac{\mathbf{R} \cdot \mathbf{r}'}{R^2} + \dots \right) d\mathbf{r}',$$

$$\approx \frac{\mu_0 I}{4\pi R^3} \oint (\mathbf{R} \cdot \mathbf{r}') d\mathbf{r}',$$

integrating over the circular orbit. Write

$$(\mathbf{R} \cdot \mathbf{r}')d\mathbf{r}' = \frac{1}{2}[(\mathbf{R} \cdot \mathbf{r}')d\mathbf{r}' - (\mathbf{R} \cdot d\mathbf{r}')\mathbf{r}']$$

$$+ \frac{1}{2}[(\mathbf{R} \cdot \mathbf{r}')d\mathbf{r}' + (\mathbf{R} \cdot d\mathbf{r}')\mathbf{r}'].$$

The symmetric part gives rise to an electric quadrupole field and will not be considered. The antisymmetric part can be written as

$$\frac{1}{2}(\mathbf{r}' \times d\mathbf{r}') \times \mathbf{R}.$$

Hence, considering only the magnetic dipole field we have

$$\mathbf{A}(\mathbf{R}) = \frac{\mu_0}{4\pi} \left[\frac{I}{2} \oint \mathbf{r}' \times d\mathbf{r}' \right] \times \frac{\mathbf{R}}{R^3},$$

$$= \frac{\mu_0}{4\pi} I\pi r^2 \mathbf{e}_z \times \frac{\mathbf{R}}{R^3} = \frac{\mu_0}{4\pi} \mathbf{M} \times \frac{\mathbf{R}}{R^3},$$

where $\mathbf{M} = I\pi r^2 \mathbf{e}_z$ is the magnetic dipole moment of the loop. From $\mathbf{B} = \nabla \times \mathbf{A}$ we have

$$\mathbf{B} = \frac{\mu_0}{4\pi} \nabla \times \left(\mathbf{M} \times \frac{\mathbf{R}}{R^3} \right) = -\frac{\mu_0}{4\pi} (\mathbf{M} \cdot \nabla) \frac{\mathbf{R}}{R^3}$$

$$= \frac{\mu_0 q L}{8\pi m} \left(\frac{3\cos\theta}{R^3} \mathbf{e}_R - \frac{\mathbf{e}_z}{R^3} \right) = \frac{\mu_0 q L}{8\pi m R^3} (3\cos\theta \mathbf{e}_R - \mathbf{e}_z),$$

where we have used

$$I = \frac{dq}{dt} = \frac{dq}{dl} \frac{dl}{dt} = \frac{qv}{2\pi r}$$

and

$$M = I\pi r^2 = \frac{qvr}{2} = \frac{qL}{2m}.$$

(b) A magnetic dipole layer can generate the same field if we consider distances far away from the source. Let the magnetic dipole moment be $\mathbf{p}_m$, then the scalar magnetic potential far away is

$$\varphi_m = \frac{1}{4\pi} \frac{\mathbf{p}_m \cdot \mathbf{R}}{R^3},$$

giving

$$\mathbf{B} = \mu_0 \mathbf{H} = -\mu_0 \nabla \varphi_m = -\frac{\mu_0}{4\pi} \nabla \frac{\mathbf{p}_m \cdot \mathbf{R}}{R^3} = -\frac{\mu_0}{4\pi} (\mathbf{p}_m \cdot \nabla) \frac{\mathbf{R}}{R^3},$$

which is the same as the expression for $\mathbf{B}$ in (a) with $\mathbf{p}_m = \mathbf{M}$.

2105

A conducting loop of area A and total resistance R is suspended by a torsion spring of constant k in a uniform magnetic field $\mathbf{B} = B\mathbf{e}_y$. The loop is in the yz plane at equilibrium and can rotate about the z-axis with moment of inertia I as shown in Fig. 2.74(a). The loop is displaced by a small angle θ from equilibrium and released. Assume the torsion spring is non-conducting and neglect self-inductance of the loop.

(a) What is the equation of motion for the loop in terms of the given parameters?

(b) Sketch the motion and label all relevant time scales for the case when R is large.

(MIT)

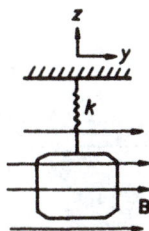

Fig. 2.74(a)

Solution:

(a) when the angle between the plane of the loop and the magnetic field is α, the magnetic flux passing through the loop is $\phi = BA \sin \alpha$. The induced emf and current are given by

$$\varepsilon = -\frac{d\phi}{dt} = -BA \cos \alpha \, \dot{\alpha}, \quad i = \frac{\varepsilon}{R} = -\frac{BA\dot{\alpha} \cos \alpha}{R}.$$

The magnetic moment of the loop is

$$m = iA = -\frac{BA^2 \cos \alpha}{R}\dot{\alpha}.$$

Thus the magnetic torque on the loop is

$$\tau_m = |\mathbf{m} \times \mathbf{B}| = -\frac{B^2 A^2 \cos^2 \alpha}{R}\dot{\alpha}.$$

Besides, the torsion spring also provides a twisting torque $k\alpha$. Both torques will resist the rotation of the loop. Thus one has

$$I\ddot{\alpha} + \frac{B^2 A^2 \cos^2 \alpha}{R}\dot{\alpha} + k\alpha = 0.$$

As $\alpha \ll \theta$ and θ is itself small, we have $\cos^2 \alpha \approx 1$ and

$$I\ddot{\alpha} + \frac{B^2 A^2}{R}\dot{\alpha} + k\alpha = 0.$$

Let $\alpha = e^{Ct}$ and obtain the characteristic equation

$$IC^2 + \frac{B^2 A^2}{R}C + k = 0.$$

The solution is

$$C = \frac{-\frac{B^2 A^2}{R} \pm \sqrt{\left(\frac{B^2 A^2}{R}\right)^2 - 4Ik}}{2I} = -\frac{B^2 A^2}{2IR} \pm j\sqrt{\frac{k}{I} - \left(\frac{B^2 A^2}{2IR}\right)^2}.$$

Defining

$$\beta = \frac{B^2 A^2}{2IR}, \quad \gamma = \sqrt{-\left(\frac{B^2 A^2}{2IR}\right)^2 + \frac{k}{I}} = \sqrt{-\beta^2 + \frac{k}{I}},$$

we have two solutions

$$C_1 = -\beta + j\gamma, \quad C_2 = -\beta - j\gamma.$$

The general solution of the equation of motion is therefore

$$\alpha = e^{-\beta t}[A_1 \cos \gamma t + A_2 \sin \gamma t].$$

Since $\alpha|_{t=0} = \theta, \dot{\alpha}|_{t=0} = 0$, we find

$$A_1 = \theta, \quad A_2 = \frac{\beta}{\gamma}A_1 = \frac{\beta}{\gamma}\theta.$$

Hence the rotational oscillation of the loop is described by

$$\alpha(t) = \theta e^{-\beta t}\left[\cos \gamma t + \frac{\beta}{\gamma}\sin \gamma t\right].$$

Note that for the motion to be oscillatory, we require that $k > \beta^2 I$, which was assumed to be the case.

(b) If R is large, $\beta \ll \gamma$ and we have

$$\alpha(t) \approx \theta e^{-\beta t} \cos \gamma t .$$

The motion is harmonic with exponentially attenuating amplitude, as shown in Fig. 2.74(b).

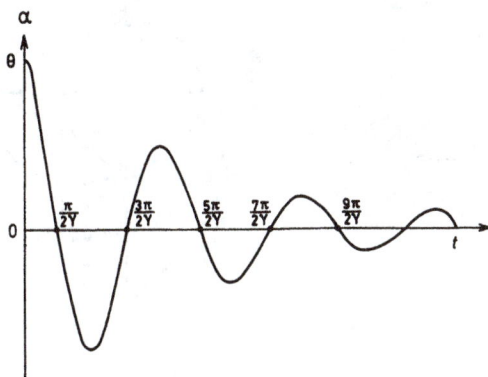

Fig. 2.74(b)

2106

Figure 2.75 shows two long parallel wires carrying equal and opposite steady currents I and separated by a distance $2a$.

(a) Find an expression for the magnetic field strength at a point in the median plane (i.e. xz plane in Fig. 2.75) lying a distance z from the plane containing the wires.

(b) Find the ratio of the field gradient dB_z/dz to the field strength B.

(c) Show qualitatively that the above "two-wire field" may be produced by cylindrical pole pieces of circular cross sections which coincide with the appropriate equipotentials. Further, give arguments to show that the analogous electric field and field gradient may be produced by equivalent circular pipes with the current I being replaced by q, the charge per unit length on the pipes.

(d) Consult the diagram 2.76 which gives specific dimensions and which represents two long pipes of circular cross section carrying equal and opposite charges q (per cm). Given that the field is $E = 8000$ V/cm at the position $z = a = 0.5$ cm, calculate the value of q and the potential difference between the two pipes.

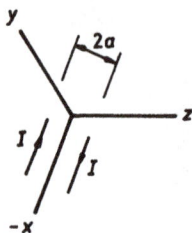

Fig. 2.75

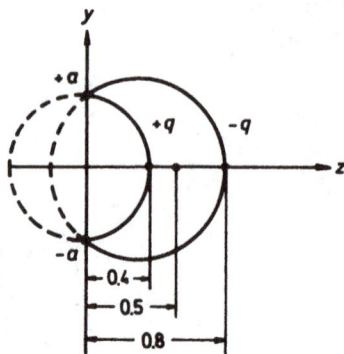

Fig. 2.76

Solution:

(a) Suppose the long wires carrying currents $+I$ and $-I$ cross the y-axis at $+a$ and $-a$ respectively. Consider an arbitrary point P and without loss of generality we can take the yz plane to contain P. Let the distances of P from the y- and z-axes be z and y respectively, and its distances from the wires be r_1 and r_2 as shown in Fig. 2.77. Ampère's circuital law gives the magnetic inductions $\mathbf{B}_1$ and $\mathbf{B}_2$ at P due to $+I$ and $-I$ respectively with magnitudes

$$B_1 = \frac{\mu_0 I}{2\pi r_1}, \quad B_2 = \frac{\mu_0 I}{2\pi r_2},$$

where $r_1 = [z^2 + (a-y)^2]^{\frac{1}{2}}$ and $r_2 = [z^2 + (a+y)^2]^{\frac{1}{2}}$, and directions as shown in Fig. 2.77. The total induction at P, $\mathbf{B} = \mathbf{B}_1 + \mathbf{B}_2$, then has components

$$B_x = 0,$$
$$B_y = -B_1 \sin\theta_1 + B_2 \sin\theta_2$$
$$= \frac{\mu_0 I}{2\pi}\left(-\frac{z}{r_1^2} + \frac{z}{r_2^2}\right) = -\frac{2\mu_0 I a y z}{\pi r_1^2 r_2^2},$$
$$B_z = -B_1 \cos\theta_1 - B_2 \cos\theta_2$$
$$= -\frac{\mu_0 I}{2\pi}\left(\frac{a-y}{r_1^2} + \frac{a+y}{r_2^2}\right).$$

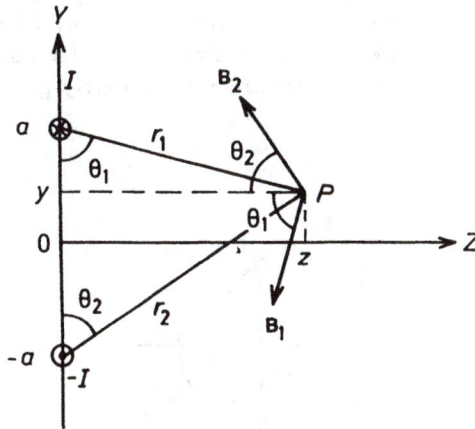

Fig. 2.77

For a point in the xz plane and distance z from the y-axis, i.e., at coordinates $(0, 0, z)$, the above reduces to

$$\mathbf{B} = -\frac{\mu_0 I}{\pi} \frac{a}{(z^2 + a^2)} \mathbf{e}_z . \tag{1}$$

(b) Equation (1) gives

$$\frac{dB_z}{dz} = \frac{2\mu_0 I}{\pi} \frac{az}{(z^2 + a^2)^2} .$$

Hence

$$\frac{dB_z}{dz} \Big/ B_z = -\frac{2z}{z^2 + a^2} .$$

(c) The magnetic lines of force are parallel to the yz plane and are given by

$$\frac{dy}{B_y} = \frac{dz}{B_z} .$$

They have mirror symmetry with respect to the xz plane, as shown by the dashed curves in Fig. 2.78. If we define the scalar magnetic potential ϕ_m by $\mathbf{H} = -\nabla \phi_m$, then the equipotentials are cylindrical surfaces everywhere perpendicular to the lines of force. Their intercepts in the yz plane are shown as solid curves in the figure. Hence if the two wires carrying currents were replaced by a cylindrical permanent magnet piece with the two side surfaces ($+z$ and $-z$) coinciding with the equipotential surfaces, the same magnetic lines of force would be obtained, since for an iron magnet of $\mu \to \infty$ the magnetic lines of force are approximately perpendicular to the surface of the magnetic poles.

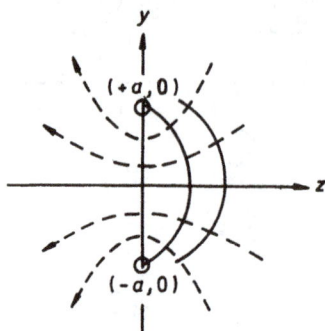

Fig. 2.78

Carrying the idea of magnetic charges further, we have

$$\nabla \cdot \mathbf{H} = -\nabla^2 \phi_m = \rho_m \,,$$

where ρ_m is the magnetic charge density. Then applying the divergence theorem we have

$$\oint_S \mathbf{H} \cdot d\mathbf{S} = \int_V \nabla \cdot \mathbf{H} dV = q_m$$

where q_m is the magnetic charge enclosed by S, showing that $\mathbf{H}$ is analogous to $\mathbf{D}$ in electrostatics. Applying the integral to a uniform cylinder, we have

$$\oint_C \mathbf{H} \cdot d\mathbf{l} = \lambda_m \,,$$

where C is the circumference of a cross section of the cylinder and λ_m is the magnetic charge per unit length. Comparing with Ampère's circuital law, $\oint_C \mathbf{H} \cdot d\mathbf{l} = I$, we have the equivalence

$$I \leftrightarrow \lambda_m \,.$$

Proceeding further the analogy between electric and magnetic fields we suppose that a metal pipe of the same cross section is used, instead of the cylindrical magnet piece, with charges $\pm \lambda$ per length on the side surfaces

$\mp z$. Then an electrostatic field distribution is produced similar to the lines of force of **B** above. With the substitution $\mathbf{H} \to \mathbf{D}, I \to \lambda$, the relations in (a) and (b) are still valid.

(d) By analogy, Eq. (1) gives

$$E_z = -\frac{q}{4\pi\varepsilon_0}\frac{a}{(z^2 + a^2)}.$$

With $a = z = 5 \times 10^{-3}$m, $E_z = 8 \times 10^5$ V/m, we have

$$q = 8.90 \times 10^{-7}\ \text{C}.$$

The potential difference between the two cylindrical sides carrying opposite charges q per unit length is

$$\Delta V = \frac{qa}{4\pi\varepsilon_0}\int_{z_1}^{z_2}\frac{dz}{z^2 + a^2} = \frac{q}{4\pi\varepsilon_0}\arctan\left(\frac{z}{a}\right)\Big|_{z_1}^{z_2} = 2.7 \times 10^3\ \text{V}$$

with $z_1 = 4 \times 10^{-3}$ m, $z_2 = 8 \times 10^{-3}$ m.

2107

In a measurement of e/m for electron using a Thomson type apparatus, i.e., crossed electric and magnetic fields in a cathode-ray tube, it is noted that if the accelerating potential difference is sufficiently large, the ratio e/m becomes one-half as large as the accepted value. Take $e/m_0 = 1.8 \times 10^{11}$ C/kg.

(a) Draw a simple sketch of the apparatus used and give a brief explanation of how it is supposed to function.

(b) Find the accelerating potential difference V which causes e/m to be one-half its accepted value. Take $c = 3 \times 10^8$ m/sec.

(SUNY, Buffalo)

Solution:

(a) A Thomson type apparatus is shown schematically in Fig. 2.79, where V_1 is the accelerating voltage and V_2 is the deflecting voltage.

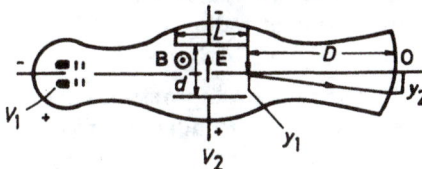

Fig. 2.79

With the addition of a magnetic field **B** as shown, the electromagnetic field has the action of a velocity-filter. With given values of V_1 and V_2, we adjust the magnitude of **B** so that the electrons strike the center O of the screen. At this time the velocity of the electron is $v = E/B$ (since $eE = evB$). Afterward the magnetic field **B** is turned off and the displacement y_2 of the electrons on the screen is measured. The ratio e/m is calculated as follows:

$$y_1 = \frac{1}{2} \cdot \frac{eE}{m}\left(\frac{L}{v}\right)^2 ,$$

$$y_2 = \frac{D + \frac{L}{2}}{L/2} y_1 = \frac{eE}{mv^2}\left(\frac{L^2}{2} + LD\right) = \frac{e}{m} \cdot \frac{dB^2}{V_2}\left(\frac{L^2}{2} + LD\right),$$

giving

$$e/m = \frac{V_2 y_2}{dB^2(\frac{L^2}{2} + LD)} .$$

(b) When the accelerating voltage is very large, relativistic effects must be considered. From energy conversation

$$eV_1 + m_0 c^2 = mc^2 ,$$

we find

$$V_1 = \left(\frac{m}{e} - \frac{m_0}{e}\right)c^2 .$$

As $\frac{e}{m} = \frac{1}{2}\frac{e}{m_0}$, the accelerating voltage is

$$V_1 = \frac{m_0 c^2}{e} = \frac{9 \times 10^{16}}{1.8 \times 10^{11}} = 5 \times 10^5 \text{ V} .$$

2108

The betatron accelerates particles through the emf induced by an increasing magnetic field within the particle's orbit. Let $\bar{B}_1$ be the average field within the particle orbit of radius R, and let B_2 be the field at the orbit (see Fig. 2.80).

(a) What must be the relationship between $\bar{B}_1$ and B_2 if the particle is to remain in the orbit at radius R independent of its energy?

(b) Does the above relationship hold at relativistic energies? Explain.

(MIT)

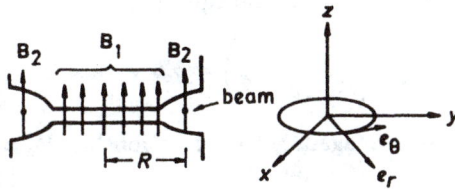

Fig. 2.80

Solution:

(a) Suppose the magnetic field is oriented in the z direction, i.e., $\mathbf{B}_2 = B_2\mathbf{e}_z$. From $\nabla \times \mathbf{E} = -\frac{\partial \mathbf{B}}{\partial t}$, where $\frac{\partial B}{\partial t} > 0$, we see that the electric field is along the $-\mathbf{e}_\theta$ direction and has axial symmetry. Then from

$$\oint_C \mathbf{E} \cdot d\mathbf{l} = -\int_S \dot{\mathbf{B}} \cdot d\mathbf{S}$$

we have

$$2\pi R E = -\frac{\partial}{\partial t}\int \mathbf{B}_1 \cdot d\mathbf{S}.$$

The average magnetic field is

$$\bar{B}_1 = \frac{\int \mathbf{B}_1 \cdot d\mathbf{S}}{\pi R^2}.$$

Hence

$$E = -\frac{R}{2}\frac{d\bar{B}_1}{dt}.$$

If the effect of radiation damping is negligible the equation of the motion of the particle is

$$\frac{d(m\mathbf{v})}{dt} = q\mathbf{E} + q\mathbf{v} \times \mathbf{B}_2.$$

In cylindrical coordinates, this is equivalent to two equations:

$$\frac{mv^2}{R} = qvB_2 \qquad \text{in} \quad \mathbf{e}_r \quad \text{direction},$$

$$\frac{d(mv)}{dt} = -qE = \frac{qR}{2}\frac{d\bar{B}_1}{dt} \qquad \text{in} \quad \mathbf{e}_\theta \quad \text{direction}.$$

The last equation can be integrated to give $mv = \frac{1}{2}qR\bar{B}_1$ assuming $v = 0$, $\bar{B}_1 = 0$ at $t = 0$. Thus $B_2 = \frac{mv}{Rq}$, $\bar{B}_1 = \frac{2mv}{Rq}$. Hence we require $B_2 = \bar{B}_1/2$.

(b) For the relativistic case, the equation of motion for the particle is

$$\frac{d}{dt}\left(\frac{m\mathbf{v}}{\sqrt{1 - v^2/c^2}}\right) = q\mathbf{E} + q\mathbf{v} \times \mathbf{B}.$$

By a similar analysis, we again get the relationship $B_2 = \bar{B}_1/2$.

2109

(a) Calculate the electric polarization vector **P** and also the surface and volume bound charge densities in a long dielectric cylinder spinning at an angular velocity ω about its axis in a uniform magnetic field **B** which is parallel to the axis.

(b) A doughnut-shaped solenoid winding has dimensions $R = 1$ meter, diameter of the loop $= 10$ cm, and the number of windings $= 1000$. If a current of 10 amperes runs through the wire, what is the magnitude and the direction of the force on one loop?

(c) Find the radiation pressure on a mirror 1 meter away from a 70 watt bulb. Assume normal incidence.

(d) A plane electromagnetic wave is normally incident on a perfect conductor (superconductor). Find the reflected **E** and **B** fields, the surface charge and current densities in terms of the incoming fields.

(e) Two charges q and $-q$ are brought from infinity to a distance d from a conducting plane and a distance r from each other. Find the work done in the process by the external force which moved the charges. Give both magnitude and sign.

(*UC, Berkeley*)

Solution:

(a) The constitutive equation for electric fields in a dielectric medium moving with velocity **v** in a magnetic field **B** is

$$\mathbf{D} = k\varepsilon_0\mathbf{E} + \varepsilon_0(k - 1)\mathbf{v} \times \mathbf{B},$$

where k is its relative dielectric constant. For a point distance r from the axis of rotation, $\mathbf{v} = \omega \times \mathbf{r}$ and $\mathbf{v} \times \mathbf{B} = (\omega \cdot \mathbf{B})\mathbf{r} - (\mathbf{r} \cdot \mathbf{B})\omega = \omega B\mathbf{r}$ as $\mathbf{r}$ is perpendicular to **B**. As there are no free charges, Gauss' flux theorem $\oint \mathbf{D} \cdot d\mathbf{S} = 0$ gives $\mathbf{D} = 0$. Then from $\mathbf{D} = \varepsilon_0\mathbf{E} + \mathbf{P}$ we get

$$\mathbf{P} = -\varepsilon_0\mathbf{E} = \varepsilon_0\left(1 - \frac{1}{k}\right)\omega B\mathbf{r}.$$

Hence the volume bound charge density is

$$\rho' = -\nabla \cdot \mathbf{P} = -\frac{1}{r}\frac{\partial}{\partial r}(rP_r) = -2\varepsilon_0\left(1 - \frac{1}{k}\right)\omega B$$

and the surface bound charge density is

$$\sigma' = P_r = \varepsilon_0\left(1 - \frac{1}{k}\right)\omega Ba,$$

as $r = a$ for the cylinder's surface.

(b) By symmetry and using Ampère's circuital law, we obtain the magnetic induction in a doughnut-shaped solenoid:

$$B = \frac{\mu_0 NI}{2\pi r},$$

where r is the distance from the center of the doughnut. Consider a small section of length dl of the solenoid. This section contains $\frac{N}{2\pi R}dl$ turns of the winding, where R is the radius of the doughnut. Take as current element a segment of this section which subtends an angle $d\theta$ at the axis of the solenoid:

$$\Delta I = \frac{NIdl}{2\pi R}\rho d\theta,$$

where θ is the angle made by the radius from the axis to the segment and the line from the axis to center of the doughnut and ρ is the radius of a loop of winding. The magnetic force on the current element is in the radial direction and has magnitude

$$dF = \Delta I \cdot \frac{B}{2} = \frac{NI\rho}{4\pi R}Bd\theta dl$$
$$= \frac{\mu_0 N^2 I^2 \rho}{8\pi^2 Rr}d\theta dl,$$

where $B/2$ is used, instead of B, because the magnetic field established by the current element itself has to be taken out from the total field. Note that dF is perpendicular to the surface of the solenoid and only its component $dF \cdot \cos\theta$ along the line from the axis to the center of the doughnut is not canceled out with another element at $2\pi - \theta$. As

$$r = R + \rho\cos\theta,$$

we have the total force on the doughnut

$$F = \int \cos\theta \, dF$$

$$= \frac{\mu_0 N^2 I^2}{8\pi^2 R} \int_0^{2\pi R} dl \int_0^{2\pi} \frac{\rho \cos\theta}{R + \rho \cos\theta} d\theta$$

$$= \frac{\mu_0 N^2 I^2}{4\pi} \int_0^{2\pi} \left(1 - \frac{R}{R + \rho \cos\theta}\right) d\theta$$

$$= \frac{\mu_0 N^2 I^2}{4\pi} \int_0^{2\pi} \left[1 - \left(1 + \frac{\rho}{R}\cos\theta\right)^{-1}\right] d\theta$$

$$= \frac{\mu_0 N^2 I^2}{2} \left\{1 - \left[1 - \left(\frac{\rho}{R}\right)^2\right]^{-\frac{1}{2}}\right\}$$

$$= \frac{4\pi \times 10^{-7} \times 1000^2 \times 10^2}{2} \left[1 - \frac{1}{\sqrt{1 - 0.05^2}}\right]$$

$$= -0.079 \text{ N}$$

Hence, the force on one loop is

$$\frac{F}{N} = -\frac{0.079}{1000} = -7.9 \times 10^{-5} \text{ N}$$

and points to the center of the doughnut.

(c) The electromagnetic field momentum incident on the mirror per unit time per unit area is $\frac{W}{4\pi d^2 c}$, where W is the wattage of the bulb and d is the distance of the mirror from the bulb. Suppose the mirror reflects totally. The change of momentum occurring on the mirror per unit time per unit area is the pressure

$$p = \frac{2W}{4\pi d^2 c} = \frac{2 \times 70}{4\pi \times 1^2 \times 3 \times 10^8} = 3.7 \times 10^{-8} \text{ N/m}^2 .$$

(d) Let $\mathbf{E}_0$ and $\mathbf{B}_0$ be incoming electromagnetic field vectors and let $\mathbf{E}'$ and $\mathbf{B}'$ be the reflected fields. Applying the boundary relation $\mathbf{n} \times (\mathbf{E}_2 - \mathbf{E}_1) = 0$ to the surface of the conductor we obtain $\mathbf{E}' + \mathbf{E}_0 = 0$, or $\mathbf{E}' = -\mathbf{E}_0$, since $\mathbf{E}_0$ and $\mathbf{E}'$ are both tangential to the boundary. For a plane electromagnetic wave we have

$$\mathbf{B}' = \frac{1}{\omega}\mathbf{k}' \times \mathbf{E}' = \frac{1}{\omega}(-\mathbf{k}_0) \times (-\mathbf{E}_0) = \mathbf{B}_0 .$$

For the conductor the surface charge density is $\sigma = 0$ and the surface current density is

$$\mathbf{i} = \mathbf{n} \times (\mathbf{H}' + \mathbf{H}_0) = 2\mathbf{n} \times \mathbf{H}_0 = -2(\mathbf{k}_0 \times \mathbf{H})/k_0$$
$$= 2\varepsilon_0\omega_0\mathbf{E}_0/k_0 = 2\varepsilon_0 c\mathbf{E}_0 .$$

(e) The work done by the external force can be considered in three steps:

1. Point charge q is brought from infinity to a distance d from the conducting plane. When the distance between q and the conducting plane is z, the (attractive) force on q is given by the method of images to be

$$F = -\frac{q^2}{4\pi\varepsilon_0(2z)^2} .$$

In this step the external force does work

$$W_1 = -\int_\infty^d F\,dz = -\frac{q^2}{16\pi\varepsilon_0 d} .$$

Note that the first minus sign applies because F and dz are in opposite directions.

2. Point charge $-q$ is brought from infinity to a distance d from the conducting plane, but far away from charge q. The work done in this process by the external force is exactly the same as in step 1:

$$W_2 = W_1 = -\frac{q^2}{16\pi\varepsilon_0 d} .$$

3. The charge $-q$ is moved to a distance r from q keeping its distance from the conducting plane constant at d. When the charge $-q$ is at distance x from q, the horizontal component of the (attractive) force on $-q$ is given by the method of images to be

$$F = -\frac{q^2}{4\pi\varepsilon_0 x^2} + \frac{q^2 x}{4\pi\varepsilon_0(x^2 + 4d^2)^{3/2}} .$$

In this step the work done by the external force is

$$W_3 = -\int_\infty^r F\,dx = \int_\infty^r \frac{q^2}{4\pi\varepsilon_0 x^2}\,dx - \int_\infty^r \frac{q^2 x}{4\pi\varepsilon_0(x^2 + 4d^2)^{3/2}}\,dx$$
$$= -\frac{q}{4\pi\varepsilon_0 r} + \frac{q^2}{4\pi\varepsilon_0(r^2 + 4d^2)^{1/2}} .$$

Hence the total work done by the external force is

$$W = W_1 + W_2 + W_3$$
$$= -\frac{q^2}{4\pi\varepsilon_0}\left[\frac{1}{r} + \frac{1}{2d} - \frac{1}{(r^2 + 4d^2)^{1/2}}\right].$$

We can also solve the problem by considering the electrostatic energy of the system. The potential at the position of q is

$$\varphi_1 = \frac{q}{4\pi\varepsilon_0}\left(-\frac{1}{r} - \frac{1}{2d} + \frac{1}{\sqrt{r^2 + 4d^2}}\right)$$

and that at $-q$ is

$$\varphi_2 = \frac{q}{4\pi\varepsilon_0}\left(\frac{1}{r} + \frac{1}{2d} - \frac{1}{\sqrt{r^2 + 4d^2}}\right),$$

again using the method of images. The electrostatic energy of the system is given by $W_e = \frac{1}{2}\Sigma q\varphi$. Taking the potential on the conducting surface to be zero, we find the work done by the external force to be

$$W = W_e = -\frac{q^2}{4\pi\varepsilon_0}\left(\frac{1}{r} + \frac{1}{2d} - \frac{1}{\sqrt{r^2 + 4d^2}}\right).$$

2110

A Hall probe with dimensions as shown in Fig. 2.81 has conductivity σ and carries charge density ρ. The probe is placed in an unknown magnetic field B oriented along the $+y$ direction. An external potential V_{ext} is applied to two ends producing an electric field in the $+z$ direction. Between which pair of ends is the equilibrium Hall voltage V_{Hall} observed? Derive an expression for B in terms of $V_{Hall}, V_{ext}, \sigma, \rho$ and the dimensions of the probe.

(*Wisconsin*)

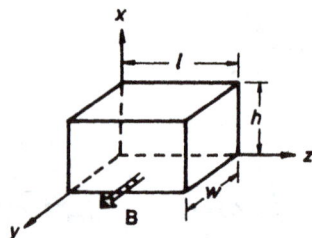

Fig. 2.81

Solution:

The Hall voltage is between the planes $x = 0$ and $x = h$. For equilibrium we have

$$q E_{\text{Hall}} = q B v .$$

As

$$E_{\text{Hall}} = \frac{V_{\text{Hall}}}{h} , \quad v = \frac{j}{\rho} = \frac{\sigma E_{\text{ext}}}{\rho} = \frac{\sigma}{\rho} \cdot \frac{V_{\text{ext}}}{l} ,$$

the above gives

$$\frac{V_{\text{Hall}}}{h} = B \frac{\sigma}{\rho} \cdot \frac{V_{\text{ext}}}{l} ,$$

or

$$B = \frac{V_{\text{Hall}}}{V_{\text{ext}}} \cdot \frac{\rho l}{\sigma h} .$$

2111

A uniform magnetic field is applied perpendicular to the flow of a current in a conductor as shown in Fig. 2.82. The Lorentz force on the charged carriers will deflect the carriers across the sample to develop a potential, the Hall voltage, which is perpendicular to both the directions of the current I_y and magnetic field B_z. Thus the total electric field can be expressed as

$$\mathbf{E} = \frac{\mathbf{j}}{\sigma} + R_{\text{H}} \mathbf{j} \times \mathbf{B} ,$$

where R_{H} is the Hall coefficient, σ the conductivity, and $\mathbf{j}$ the current density.

(a) For the case of a single type of carrier, show that R_{H} gives the sign of the charge of the carrier and the carrier density.

(b) Describe an experimental method determining R_{H} for a sample at room temperature. Draw a diagram based on Fig. 2.82 which shows all the electrical connections (and contacts with the sample) which are required, including circuits and measuring instruments to determine the true Hall voltage (its magnitude and polarity).

(c) Prepare a table of all the parameters which must be measured with the B-field on or off. State the units in which each parameter is measured.

(d) How do you compensate experimentally for rectifying effects which may exist at the electrical contacts with the sample?

(e) The sample (a semiconductor) is found to have a negative value for R_H at room temperature. Describe the charged carriers.

(f) At liquid nitrogen temperature the R_H of this sample reverses to become positive. How do you explain the results for room and low temperatures under the simplifying assumptions that: (1) all the charged carriers of one type have the same drift velocity, and (2) we neglect the fact that most semiconductors have two distinct overlapping bands?

(Chicago)

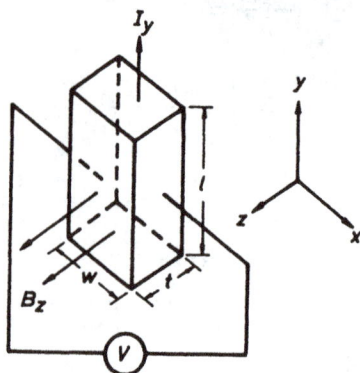

Fig. 2.82

Solution:

(a) Let the charge of a carrier be q and its drift velocity be $\mathbf{v}$, then in equilibrium

$$q\mathbf{E}_\perp + q\mathbf{v} \times \mathbf{B} = 0.$$

As $\mathbf{j} = nq\mathbf{v}$, n being the carrier density, we have

$$\mathbf{E}_\perp = -\frac{1}{qn}\mathbf{j} \times \mathbf{B}.$$

But we also have

$$\mathbf{E}_\perp = R_H\mathbf{j} \times \mathbf{B},$$

hence

$$R_H = -\frac{1}{qn}.$$

Thus R_H gives the sign of the charge and the charge density of the carriers.

(b) An experimental arrangement for determining R_H is shown in Fig. 2.82. The magnitude and polarity of the Hall voltage V can be measured using a voltmeter with high internal resistance. The Hall electric field is given by $E_\perp = V/w$. Accordingly

$$R_H = \frac{E_\perp}{jB_z} = \frac{Vt}{jwtB_z} = \frac{Vt}{I_y B_z} .$$

I_y can be measured with an ammeter, B_z can be determined using a sample of known Hall coefficient.

(c) All the parameters to be measured are listed below:

$$\text{Parameter}: \quad B_z \quad I_y \quad t \quad V$$
$$\text{Unit}: \quad \text{T} \quad \text{A} \quad \text{m} \quad \text{V}$$

(d) Repeat the experiment for two different sets of I_y and B_z. We have

$$R_H = \frac{(V_1 - V_0)t}{I_{y_1} B_{z_1}} = \frac{(V_2 - V_0)t}{I_{y_2} B_{z_2}} ,$$

where V_0 is the contact potential difference caused by rectifying effects and can be determined from the above expressions to be

$$V_0 = \frac{V_2 I_{y_1} B_{z_1} - V_1 I_{y_2} B_{z_2}}{I_{y_1} B_{z_1} - I_{y_2} B_{z_2}} .$$

Once V_0 is determined, it can be compensated for.

(e) As R_H is negative the carriers of the sample have positive charge. Hence the sample is a p type semiconductor.

(f) At liquid nitrogen temperature the concentration of the holes relating to the main atoms is greatly reduced mainly because of the eigen electrons and holes. The concentrations of the eigen electrons and holes are equal, but because of the greater mobility of the electrons their Hall effect exceeds that of the holes. As a result, the R_H of the sample reverses sign to become positive.

2112

The Hall effect has to do with:

(a) the deflections of equipotential lines in a material carrying a current in a magnetic field,

(b) rotation of the plane of polarization of light going through a transparent solid,

(c) the space charge in electron flow in a vacuum.

(CCT)

Solution:

The answer is (a).

2113

(a) Prove that in a stationary plasma of ohmic conductivity σ and permeability $\mu = 1$ the magnetic field **B** satisfies the equation

$$\frac{\partial \mathbf{B}}{\partial t} = D\nabla^2 \mathbf{B},$$

where $D = c^2/4\pi\sigma$.

(b) If the plasma is in motion with velocity **v**, prove that the above equation is replaced by

$$\frac{\partial \mathbf{B}}{\partial t} = \nabla \times (\mathbf{v} \times \mathbf{B}) + D\nabla^2 \mathbf{B}.$$

(c) At $t = 0$ a stationary plasma contains a magnetic field

$$\mathbf{B} = B(x)\mathbf{e}_z,$$
$$B(x) = \begin{cases} B_0, & |x| < L \\ 0, & |x| > L, \end{cases}$$

where B_0 is a constant. Determine the time evolution of the field assuming that the plasma remains stationary.

(d) The average conductivity of the earth is roughly equal to that of copper, i.e., $\sigma \sim 10^{15}\text{s}^{-1}$. Can the earth's magnetic field be a primordial field which has survived since the formation of the solar system, about 5×10^9 years?

(MIT)

Solution:

(a) If a plasma is stationary and its displacement current can be neglected, the electromagnetic field inside the plasma satisfies the following

Maxwell's equations (on Gaussian units)

$$\nabla \cdot \mathbf{D} = 4\pi \rho_f, \qquad \nabla \times \mathbf{E} = -\frac{1}{c}\frac{\partial \mathbf{B}}{\partial t},$$

$$\nabla \cdot \mathbf{B} = 0, \qquad \nabla \times \mathbf{B} = \frac{4\pi}{c}\mathbf{j}_f,$$

and if the plasma is ohmic we have also

$$\mathbf{j}_f = \sigma \mathbf{E}.$$

Thus

$$\nabla \times \mathbf{B} = \frac{4\pi}{c}\sigma \mathbf{E},$$

$$\nabla \times (\nabla \times \mathbf{B}) = \nabla(\nabla \cdot \mathbf{B}) - \nabla^2 \mathbf{B} = -\nabla^2 \mathbf{B} = \frac{4\pi}{c}\sigma \nabla \times \mathbf{E} = -\frac{4\pi\sigma}{c^2}\frac{\partial \mathbf{B}}{\partial t},$$

or

$$\frac{\partial \mathbf{B}}{\partial t} = D\nabla^2 \mathbf{B} \tag{1}$$

with $D = \frac{c^2}{4\pi\sigma}$. The equation (1) is a diffusion equation.

(b) If the velocity of the plasma is not zero, we have

$$\mathbf{j}_f = \sigma\left(\mathbf{E} + \frac{1}{c}\mathbf{v} \times \mathbf{B}\right).$$

In the nonrelativistic approximation $v \ll c$, use of the above Maxwell's equation gives

$$\nabla \times \mathbf{B} = \frac{4\pi\sigma}{c}\left(\mathbf{E} + \frac{1}{c}\mathbf{v} \times \mathbf{B}\right).$$

Taking curl of both sides gives

$$\frac{\partial \mathbf{B}}{\partial t} = D\nabla^2 \mathbf{B} + \nabla \times (\mathbf{v} \times \mathbf{B}). \tag{2}$$

(c) For a stationary plasma the magnetic field is determined by (1). From the initial condition we see that (1) can be reduced to the one-dimensional diffusion equation

$$\frac{\partial B_z(x,t)}{\partial t} = D\frac{\partial^2 B_z(x,t)}{\partial x^2}.$$

We separate the variables by letting $B_z(x,t) = X(x)T(t)$ and obtain

$$\frac{1}{DT}\frac{dT}{dt} = \frac{1}{X}\frac{d^2X}{dx^2} = -\omega^2$$

with solutions

$$T(t) = Ae^{-\omega^2 Dt}, \quad X(x) = Ce^{i\omega x}.$$

Hence

$$B_z(x,t,\omega) = A(\omega)e^{-\omega^2 Dt}e^{i\omega x}.$$

As ω is arbitrary the general solution is

$$B_z(x,t) = \int_{-\infty}^{\infty} A(\omega)e^{-\omega^2 Dt}e^{i\omega x}d\omega.$$

For $t = 0$, the above reduces to

$$B_z(x) = \int_{-\infty}^{\infty} A(\omega)e^{i\omega x}d\omega,$$

and by Fourier transform we obtain

$$A(\omega) = \frac{1}{2\pi}\int_{-\infty}^{\infty} B_z(\xi)e^{-i\omega\xi}d\xi.$$

Hence

$$B_z(x,t) = \int_{-\infty}^{\infty} B_z(\xi)\left[\frac{1}{2\pi}\int_{-\infty}^{\infty} e^{-\omega^2 Dt}e^{i\omega(x-\xi)}d\omega\right]d\xi,$$

where the definite integral inside the brackets can be evaluated,

$$\int_{-\infty}^{\infty} e^{-\omega^2 Dt}e^{i\omega(x-\xi)}d\omega = \sqrt{\frac{\pi}{Dt}}e^{-(x-\xi)^2/4Dt},$$

and $B(\xi)$ is given by the initial condition

$$B_z(\xi) = \begin{cases} B_0, & \text{for} \quad |\xi| \le L \\ 0, & \text{for} \quad |\xi| > L. \end{cases}$$

Therefore, the time evolution of the field is given by

$$B_z(x,t) = \frac{B_0}{\sqrt{4\pi Dt}}\int_{-\frac{L}{2}}^{\frac{L}{2}} e^{-\frac{(x-\xi)^2}{4Dt}}d\xi. \tag{3}$$

(d) It is not possible that the earth's magnetic field is a primordial field which has survived the formation of the solar system, about 5×10^9 years ago, as B_0 would have fast disappeared by diffusion. A semi-quantitative proof is given below.

As the conductivity of the earth is approximately $\sigma \approx 10^{15}$ s^{-1}, the diffusion coefficient of the earth's magnetic field is

$$D = \frac{c^2}{4\pi\sigma} \approx 10^5 \text{ cm}^2/\text{s}$$

and $Dt \approx 10^{22}$ for $t = 5 \times 10^9$ years $= 1.5 \times 10^{17}$s. The linear dimension of the earth is $L \simeq 10^9$ cm. Thus the exponent in (3) is approximately

$$\frac{(x-\xi)^2}{4Dt} \approx \frac{L^2}{4Dt} \sim 10^{-4},$$

giving

$$e^{-(x-\xi^2)/4Dt} \approx 1$$

and

$$\int_{-\frac{L}{2}}^{\frac{L}{2}} e^{-\frac{(x-\xi)^2}{4Dt}} d\xi \approx L.$$

Hence

$$B_z(x,t) = \frac{LB_0}{\sqrt{4\pi Dt}} \approx 10^{-4} B_0.$$

This shows that the present earth's magnetic field would be only $10^{-4} B_0$ if it has arisen from the primordial field B_0. With the present earth's magnetic field of ~ 1 Gs, the primordial field would have been $B_0 \sim 10^4$ Gs. This value is much higher than the magnetic fields in the plasmas of the various celestial bodies.

2114

A model for an electron consists of a shell of charge distributed uniformly on the surface of a sphere of radius a. The electron moves with velocity $v \ll c$ (see Fig. 2.83).

(a) What are **E** and **B** at a point (r, θ) outside the sphere?

(b) Find the value of a such that the total momentum carried by the field is just equal to the mechanical momentum mv, v being the electron's speed.

(c) Use the value of a to calculate the energy in the field of the moving charge and compare it with the rest-mass energy and kinetic energy.

(*Wisconsin*)

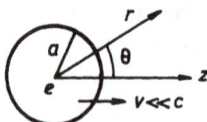

Fig. 2.83

Solution:

(a) In the rest frame Σ' of the electron, the electromagnetic field at a point of radius vector $\mathbf{r}'$ from it is, in Gaussian units,

$$\mathbf{E}' = \frac{e\mathbf{r}'}{r'^3}, \quad \mathbf{B}' = 0.$$

In the laboratory frame Σ, by Lorentz transformation (with $v \ll c$) the field is

$$\mathbf{E} = \mathbf{E}' - \frac{\mathbf{v}}{c} \times \mathbf{B}' = \mathbf{E}',$$

$$\mathbf{B} = \mathbf{B}' + \frac{\mathbf{v}}{c} \times \mathbf{E}' = \frac{\mathbf{v}}{c} \times \mathbf{E}'.$$

The field point has coordinates (r, θ) in Σ as shown in Fig. 2.83. As $v \ll c$, we have $r' \approx r$ and

$$\mathbf{E} = \mathbf{E}' = \frac{e\mathbf{r}'}{r'^3} \simeq \frac{e\mathbf{r}}{r^3},$$

$$\mathbf{B} = \frac{\mathbf{v}}{c} \times \mathbf{E}' \simeq \frac{e}{c} \cdot \frac{\mathbf{v} \times \mathbf{r}}{r^3},$$

with magnitudes

$$E = \frac{e}{r^2}, \quad B = \frac{ev \sin \theta}{cr^2}.$$

(b) The momentum density of the field is

$$\mathbf{g} = \frac{\mathbf{N}}{c^2} = \frac{1}{4\pi c}(\mathbf{E} \times \mathbf{B}).$$

Substituting in $\mathbf{E}$ and $\mathbf{B}$ yields

$$\mathbf{g} = \frac{e^2}{4\pi c^2} \frac{\mathbf{r} \times (\mathbf{v} \times \mathbf{r})}{r^6} = \frac{e^2}{4\pi c^2 r^5}(\mathbf{v}r - v\mathbf{r}\cos\theta).$$

Hence the momentum of the electromagnetic field of the electron is

$$\mathbf{P} = \iiint_\infty \mathbf{g}dV = \mathbf{e}_z \int_0^{2\pi} d\varphi \int_0^\pi d\theta \int_a^\infty dr \frac{e^2 v(1 - \cos^2\theta)}{4\pi c^2 r^4} r^2 \sin\theta$$

$$= \frac{2e^2 v}{3c^2 a}\mathbf{e}_z = \frac{2e^2}{3c^2 a}\mathbf{v}.$$

Note that in the integrand above the component of **g** perpendicular to **v** will cancel out on integration; only the component parallel to **v** needs to be considered.

If the electromagnetic field momentum of the electron is equal to its mechanical momentum, $m\mathbf{v}$, i.e., $\frac{2e^2}{3c^2 a}\mathbf{v} = m\mathbf{v}$, then

$$a = \frac{2e^2}{3mc^2} = \frac{2}{3} \times 2.82 \times 10^{-5}\,\text{Å} = 1.88 \times 10^{-5}\,\text{Å}.$$

(c) The energy of the field of the electron is

$$W = \iiint_\infty \frac{1}{8\pi}(E^2 + B^2)dV = \iiint_\infty \frac{1}{8\pi}\left[\frac{e^2}{r^4} + \frac{e^2 v^2 \sin^2\theta}{c^2 r^4}\right]dV$$

$$= \frac{e^2}{8\pi} \int_0^{2\pi} d\varphi \int_0^\pi d\theta \int_a^\infty r^2 \sin\theta \left[\frac{1 + \frac{v^2}{c^2}\sin^2\theta}{r^4}\right]dr$$

$$= \frac{e^2}{2a}\left(1 + \frac{2}{3}\frac{v^2}{c^2}\right) = \frac{3mc^2}{4}\left[1 + \frac{2}{3}\frac{v^2}{c^2}\right] = \frac{3}{4}mc^2 + \frac{1}{2}mv^2.$$

It follows that for $v \ll c$, $\frac{1}{2}mv^2 \ll W \lesssim mc^2$.

2115

A beam of Na atoms (ground state $^2S_{1/2}$), polarized in the $+x$ direction, is sent in the z direction through a region in which there is a magnetic field in the $+y$ direction. Describe the form of the beam downstream from the magnetic field region (both its spatial structure and polarization), assuming the field has a large gradient in the y-direction.

(Wisconsin)

Solution:

Na atoms polarized in the $+x$ direction have the probabilities of one-half in eigenstate $S_y = +\frac{\hbar}{2}$ and one-half in eigenstate $S_y = -\frac{\hbar}{2}$. Under the action of a magnetic field $\mathbf{B} = B\mathbf{e}_y$ with $\frac{dB}{dy} > 0$, the Na atoms of $S_y = +\frac{\hbar}{2}$ will deflect to the $-y$ direction, while those of $S_y = -\frac{\hbar}{2}$ will deflect to the $+y$ direction. Thus, going through the magnetic field, the Na atoms will split into two beams with directions of polarization $-y$ and $+y$. (Since $\Delta E = -\mathbf{m} \cdot \mathbf{B} = -\frac{e}{mc}(-\mathbf{S}) \cdot \mathbf{B} = \frac{e}{mc} S_y B$, and $\frac{dB}{dy} > 0$, Na atoms of $S_y = \frac{\hbar}{2}$ deflect to the $-y$ direction and Na atoms of $S_y = -\frac{\hbar}{2}$ deflect to the $+y$ direction.)

2116

In a frame S there is a uniform electromagnetic field

$$\mathbf{E} = 3A\mathbf{e}_x, \quad \mathbf{B} = 5A\mathbf{e}_z$$

(in Gaussian units). An ion of rest mass m_0 and charge q is released from rest at $(0, b, 0)$. What time elapses before it returns to the y-axis?

(SUNY, Buffalo)

Solution:

The Lorentz force equation

$$m\ddot{\mathbf{r}} = q\left(\mathbf{E} + \frac{1}{c}\dot{\mathbf{r}} \times \mathbf{B}\right)$$

has component equations

$$\begin{cases} m_0\ddot{x} = 3Aq + \frac{5Aq}{c}\dot{y}, & (1) \\ m_0\ddot{y} = -\frac{5Aq}{c}\dot{x}, & (2) \\ m_0\ddot{z} = 0. & (3) \end{cases}$$

Integrating (3) and noting $z|_{t=0} = 0, \dot{z}|_{t=0} = 0$, we have $z = 0$. Integrating (2) and using $x|_{t=0} = 0, \dot{y}|_{t=0} = 0$, we find

$$\dot{y} = -\frac{5Aq}{m_0 c}x. \tag{4}$$

Use of (4) in (1) gives

$$\ddot{x} + \left(\frac{5Aq}{m_0 c}\right)^2\left(x - \frac{3m_0 c^2}{25Aq}\right) = 0.$$

With $\dot{x}|_{t=0} = 0$ we get

$$x = \frac{3m_0 c^2}{25Aq}(1 - \cos \omega t),$$

where

$$\omega = \frac{5Aq}{m_0 c}.$$

Note that $x = 0$ at $t = \frac{2n\pi}{\omega}$. Let $n = 1$, then

$$t = \frac{2\pi}{\omega} = \frac{2\pi m_0 c}{5Aq}.$$

This is the time that elapses before the ion returns to the y-axis.

2117

A magnetic field can suppress the flow of current in a diode. Consider a uniform magnetic field $\mathbf{B} = (0, 0, B_0)$ filling the gap between two infinite conductors in the yz plane. The cathode is located at $x = 0$ and the anode at $x = d$. A positive potential V_0 is applied to the anode.

Electrons leave the cathode with zero initial velocity and their charge density causes the electric field to be non-uniform:

$$\mathbf{E} = \left(-\frac{\partial \phi}{\partial x}, 0, 0 \right).$$

(a) Under steady state conditions what quantities are constants of the electron's motion?

(b) Determine the strength of the magnetic field required to reflect the electrons before they reach the anode.

(MIT)

Solution:

When gravity is neglected, the motion of an electron is described by (assuming $v \ll c$)

$$\begin{cases} m\frac{dv_x}{dt} = -e\left(-\frac{\partial \phi}{\partial x} + v_y B_0\right), & (1) \\ m\frac{dv_y}{dt} = ev_x B_0, & (2) \\ m\frac{dv_z}{dt} = 0. & (3) \end{cases}$$

(a) Integrating (3) we have

$$v_z(t) = v_z(t = 0) = 0,$$
$$z(t) = z(t = 0) = \text{const.}$$

Hence the coordinate and speed of the electron in the z direction are the constants of the motion, in particular $v_z = 0$.

(b) The work done by the electric field in moving an electron from cathode to anode is

$$W = \int_0^d -e \cdot \left(-\frac{\partial \phi}{\partial x} \right) dx = eV_0,$$

since the magnetic field does no work. When the electron reaches the anode, the magnitude of its velocity $\mathbf{v}_f = v_{fx}\mathbf{i} + v_{fy}\mathbf{j}$ can be obtained by equating the kinetic energy of the electron to the work done by the electric field:

$$\frac{1}{2}mv_f^2 = eV_0,$$

giving

$$v_f = \sqrt{\frac{2eV_0}{m}}.$$

If the electrons are not to reach the anode, we require that

$$v_{fx} = 0, \quad v_{fy} = \sqrt{\frac{2eV_0}{m}}.$$

Writing (2) as

$$m\frac{dv_y}{dt} = eB_0\frac{dx}{dt}$$

and integrating both sides, noting $v_y = 0, x = 0$, at $t = 0$ we obtain

$$m\sqrt{\frac{2eV_0}{m}} = eB_0d,$$

giving

$$B_0 = \sqrt{\frac{2mV_0}{ed^2}}.$$

Therefore the induction of the magnetic field must be greater than $\sqrt{\frac{2mV_0}{ed^2}}$ for the electrons to reflect back before reaching the anode.

2118

Specialized bacteria can be found living in quite unattractive places, such as oil and sewage disposal plants. A bacterium that lives in an absolutely dark and essentially homogeneous soup faces a serious navigational problem if he must sometimes rise for oxygen and then descend for an important part of his dinner. Which way is up?

One class of bacteria has solved the problem by incorporating an iron oxide magnet inside its cell. Discuss the following questions quantitatively making clear the nature of the necessarily rough approximations used.

(a) Why not sense the pressure gradient in the fluid instead of using a magnet?

(b) Estimate the minimum magnetic moment that could be used to line up the bacterium.

(c) Asssuming 10^{-4} cm for the length of the magnetic needle, estimate its minimum diameter.

(d) Why is a needle better than a spherical magnet?

(Princeton)

Solution:

(a) The pressure gradient in a fluid (buoyancy) can cause a baterium to rise or descend, depending on its specific weight relative to the fluid. On the other hand, a magnet inside a bacterium can cause it to rise or fall, depending on the relative orientation between the magnetic moment and geomagnetic field. Because of the random thermal motion (Brownian movement) this relative orientation is randomly changed so that a bacterium can both rise for oxygen and descend for food. Actually such small magnets of the bacteria string together to form large magnets of moment **m**. In the inhomogeneous geomagnetic field the force causing such a magnet to rise or fall is given by

$$F_z = \mathbf{m} \cdot \frac{\partial \mathbf{B}}{\partial z},$$

where **B** is the earth's magnetic induction. If we represent the height above the earth's surface by z, then $\frac{\partial}{\partial z}|\mathbf{B}|$ is negative as we go up. F_z can be pointing up or down depending on the orientation of **m** relative to $\frac{\partial \mathbf{B}}{\partial z}$.

(b) In Gaussian units the interaction energy between two magnetic dipoles of moments $\mathbf{m}_1$ and $\mathbf{m}_2$ is

$$W = -\mathbf{m}_1 \cdot \left[\frac{\mathbf{m}_2}{r^3} - \frac{3(\mathbf{m}_2 \cdot \mathbf{r})}{r^5}\mathbf{r} \right]$$

where **r** is the radius vector from $\mathbf{m}_2$ to $\mathbf{m}_1$.

For the magnets inside two adjacent bacteria which line up end to end, $m_1 = m_2 = m, m_1//m_2, r = d = 10^{-4}$ cm, and the interaction energy is

$$W = \frac{2m^2}{d^3}.$$

The energy of the Brownian movement of bacteria is $\sim kT$, where k is Boltzmann's constant and T is the absolute temperature. This movement tends to destroy the ordering arrangement of the bacteria. Hence for the linear arrangement of the magnets of the bacteria to be possible we require that

$$\frac{2m^2}{d^3} \gtrsim kT,$$

giving the minimum magnetic moment inside a bacterium as

$$|\mathbf{m}_{min}| \gtrsim \sqrt{kTd^3}.$$

(c) Let r, M and d be the radius of the cross-section, the magnetization and the length of the magnetic needle respectively, then its magnetic dipole moment is

$$m = \pi r^2 dM.$$

Combining with the result in (b), we get

$$r \gtrsim \left(\frac{\sqrt{kTd}}{\pi M}\right)^{\frac{1}{2}}.$$

Take T to be the room temperature, $T \sim 300$ K. The saturation magnetization of a ferromagnet at this temperature is $M \sim 1.7 \times 10^3$ Gs. The above equation then gives

$$r \simeq 0.6 \times 10^{-6} \text{ cm}.$$

(d) Needle-shaped magnets are better than spherical ones because they can be more easily lined up.

2119

As a model to describe the electrodynamical properties of a pulsar we consider a sphere of radius R which rotates like a rigid body with angular velocity ω about a fixed axis. The charge and current distributions are thus symmetric with respect to this axis (and with respect to the normal midplane of the pulsar). The net charge of the sphere is zero. In the vacuum outside the pulsar the magnetic field is that of a magnetic dipole **m** parallel to the axis of rotation. The magnetic field in the inside is consistent with the outside field, but otherwise arbitrary.

(a) The magnitudes of the electric and magnetic forces on charged particles inside the pulsar are very large compared with all other forces. Since the charged particles are assumed to share in the rotational motion of the pulsar it follows that, to a good approximation, $\mathbf{E} = -\mathbf{v} \times \mathbf{B}$, where $\mathbf{v} = \omega \times \mathbf{r}$ is the local velocity, everywhere inside the pulsar. Imposing this condition at points just inside the surface of the pulsar, show that at such points

$$E_\theta = -\frac{\mu_0 m \omega \sin\theta \cos\theta}{2\pi R^2} ,$$

where θ is the polar angle with respect to the axis.

(b) On the basis of the above result, find the electrostatic potential everywhere outside the sphere.

(c) Show that the equation $\mathbf{E} = -(\omega \times \mathbf{r}) \times \mathbf{B}$ does not hold immediately outside the pulsar.

(*UC, Berkeley*)

Solution:

(a) As shown in Fig. 2.84, at a point P just inside the surface of the sphere ($r \approx R$), the magnetic field generated by the dipole **m** is

$$\mathbf{B} = B_r \mathbf{e}_r + B_\theta \mathbf{e}_\theta$$
$$= \frac{\mu_0}{4\pi} \left(\frac{m \cos\theta}{R^3} \mathbf{e}_r + \frac{m \sin\theta}{R^3} \mathbf{e}_\theta \right).$$

So the electric field at the point P is

$$\mathbf{E} = -\mathbf{v} \times \mathbf{B} = -(\omega \times \mathbf{r}) \times \mathbf{B}|_{r \approx R}$$
$$= \frac{\mu_0}{4\pi} \left(\frac{m\omega \sin^2\theta}{R^2} \mathbf{e}_r - \frac{2m\omega \cos\theta \sin\theta}{R^2} \mathbf{e}_\theta \right),$$

with the θ-component

$$E_\theta = -\frac{\mu_0 m \omega \sin\theta \cos\theta}{2\pi R^2} .$$

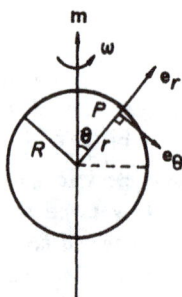

Fig. 2.84

(b) Taking the potential on the equator as reference level the induced potential at a point of latitude $\alpha = \frac{\pi}{2} - \theta$ is

$$V = -\int_{\frac{\pi}{2}}^{\frac{\pi}{2}-\alpha} E_\theta R d\theta$$

$$= \int_{\frac{\pi}{2}}^{\frac{\pi}{2}-\alpha} \frac{\mu_0 m\omega \sin\theta d(\sin\theta)}{2\pi R} = \frac{\mu_0 m\omega \sin^2\theta}{4\pi R}\bigg|_{\frac{\pi}{2}}^{\frac{\pi}{2}-\alpha}$$

$$= \frac{\mu_0 m\omega}{4\pi R}(\cos^2\alpha - 1) = \frac{B_P \omega R^2}{2}(\cos^2\alpha - 1),$$

where $B_P = \frac{\mu_0 m}{2\pi R^3}$ is the induction of the magnetic dipole at the north pole.

(c) The equation $\mathbf{E} = -\mathbf{v} \times \mathbf{B}$ does not hold outside the sphere. The reason is as follows. This equation follows from the transformation of the electromagnetic field (with $v \ll c$),

$$\mathbf{E}' = \mathbf{E} + \mathbf{v} \times \mathbf{B},$$

where $\mathbf{E}$ and $\mathbf{B}$ are the fields at a point just inside the surface of the pulsar as measured by an observer in the rest reference frame K fixed at a distant star, $\mathbf{E}'$ is the electric field observed in the moving frame K' fixed with respect to the surface of the pulsar, and $\mathbf{v} = \omega \times \mathbf{R}$ is the velocity of K' with respect to K. Since in K' the surface layer of the pulsar is equivalent to a stationary conductor, $\mathbf{E}' = 0$ (otherwise $\mathbf{j}' = \sigma\mathbf{E}' \neq 0$, i.e., the observer in K' would see a current). Thus we have $\mathbf{E} = -\mathbf{v} \times \mathbf{B} = -(\omega \times \mathbf{R}) \times \mathbf{B}$. But at points just outside the surface of the pulsar the requirement $\mathbf{E}' = 0$ is not needed. Also the charge density on the surface is generally not zero, so that the boundary condition would not give $\mathbf{E}' = 0$ on points just outside the surface; hence $\mathbf{E} \neq -\mathbf{v} \times \mathbf{B}$ outside the sphere.

PART 3

CIRCUIT ANALYSIS

1. BASIC CIRCUIT ANALYSIS (3001-3026)

3001

Suppose the input voltages V_1, V_2, and V_3 in the circuit of Fig. 3.1 can assume values of either 0 or 1 (0 means ground). There are thus 8 possible combinations of input voltage. Compute V_{out} for each of these possibilities.

(UC, Berkeley)

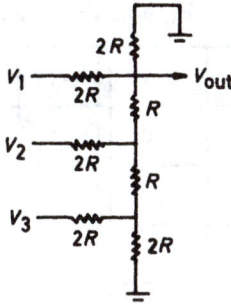

Fig. 3.1

Solution:

The circuit in Fig. 3.1 can be redrawn as that in Fig. 3.2.

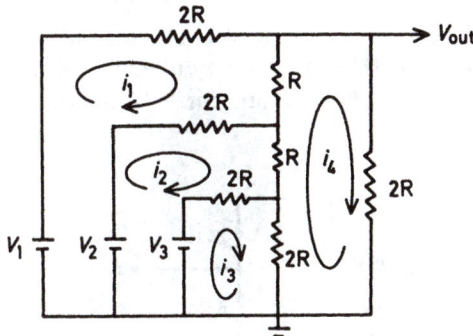

Fig. 3.2

Let the currents flowing in the component circuits be as shown. By Kirchhoff's laws we have

$$V_3 = [2(i_3 - i_2) + 2(i_3 - i_4)]R,$$
$$V_2 - V_3 = [2(i_2 - i_1) + (i_2 - i_4) + 2(i_2 - i_3)]R,$$

315

$$V_1 - V_2 = [2i_1 + (i_1 - i_4) + 2(i_1 - i_2)]R,$$
$$0 = 2(i_4 - i_3) + (i_4 - i_2) + (i_4 - i_1) + 2i_4 .$$

After solving for i_4 we obtain

$$V_{\text{out}} = 2i_4 R = \frac{V_1}{3} + \frac{V_2}{6} + \frac{V_3}{12} .$$

V_{out} for various values of V_1, V_2, and V_3 are shown in the table below.

V_1	V_2	V_3	V_{out}	V_1	V_2	V_3	V_{out}
0	0	0	0	1	0	0	$\frac{1}{3}$
0	0	1	$\frac{1}{12}$	1	0	1	$\frac{5}{12}$
0	1	0	$\frac{1}{6}$	1	1	0	$\frac{1}{2}$
0	1	1	$\frac{1}{4}$	1	1	1	$\frac{7}{12}$

3002

The current–voltage characteristic of the output terminals A, B (Fig. 3.3) is the same as that of a battery of emf ε_0 and internal resistance r. Find ε_0 and r and the short-circuit current provided by the battery.

(*Wisconsin*)

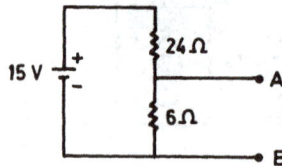

Fig. 3.3

Solution:

According to Thévenin's theorem, the equivalent emf is the potential across AB when the output current is zero, i.e., the open-circuit voltage:

$$\varepsilon_0 = V_{\text{AB}} = \frac{6}{24 + 6} \times 15 = 3 \text{ V}$$

The equivalent internal resistance is the resistance when the battery is shorted, i.e., the parallel combination of the resistances:

$$r = \frac{6 \times 24}{6 + 24} = 4.8 \ \Omega \ .$$

Then the short-circuit current provided by the battery is

$$I = \frac{\varepsilon_0}{r} = \frac{3 \ \mathrm{V}}{4.8 \ \Omega} = 0.625 \ \mathrm{A} \ .$$

3003

Any linear dc network (a load R is connected between the two arbitrary points A and B of the network) is equivalent to a series circuit consisting of a battery of emf V and a resistance r, as shown in Fig. 3.4.

(a) Calculate V and r of the circuit in Fig. 3.5.

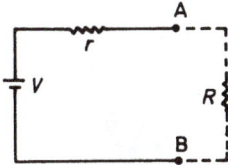

Fig. 3.4

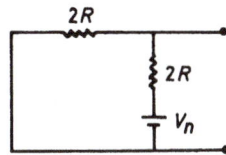

Fig. 3.5

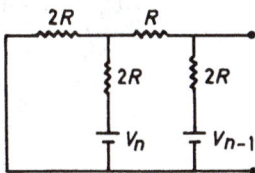

Fig. 3.6

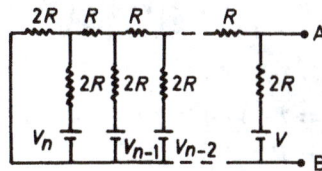

Fig. 3.7

(b) Calculate V and r of the circuit in Fig. 3.6.

(c) Calculate V and r of the circuit in Fig. 3.7.

(Hint: Use mathematical induction)

(*Chicago*)

Solution:

(a) We find by Thévenin's theorem that

$$V = \frac{2R}{2R + 2R} \cdot V_n = \frac{1}{2} V_n \,,$$

$$r = \frac{2R \times 2R}{2R + 2R} = R \,.$$

(b) Using the result of (a) for the circuit in Fig. 3.6, we obtain a simpler circuit shown in Fig. 3.8. Thévenin's theorem then gives

$$V = \frac{1}{2}\left(\frac{V_n}{2}\right) + \frac{1}{2} V_{n-1}$$

$$= \frac{1}{2}\left(V_{n-1} + \frac{1}{2} V_n\right) \,,$$

$$r = \frac{2R \times 2R}{2R + 2R} = R \,.$$

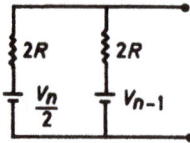

Fig. 3.8

(c) By induction we have

$$V = \frac{1}{2}\left\{ V_1 + \frac{1}{2}\left[V_2 + \frac{1}{2}\left(V_3 + \cdots + \underbrace{\frac{1}{2}\left(V_{n-1} + \frac{1}{2} V_n \right) \cdots \right) }_{n-1} \right] \right\}$$

$$= 2^{-1} V_1 + 2^{-2} V_2 + \cdots + 2^{-n} V_n \,,$$

$$r = R \,.$$

3004

Four one-microfarad capacitors are connected in parallel, charged to 200 volts and discharged through a 5 mm length of fine copper wire. This wire has a resistance of 4 ohms per meter and a mass of about 0.045 gram per meter. Would you expect the wire to melt? Why?

(Columbia)

Solution:

The relevant data are

total capacitance $C = 4 \times 1 = 4 \ \mu F$,

energy stored in the capacitance

$$E = \frac{1}{2} \ CV^2 = \frac{1}{2} \times 4 \times 10^{-6} \times 200^2 = 0.08 \, \text{J} \ ,$$

resistance of copper wire $R = 4 \times 5 \times 10^{-3} = 0.02 \ \Omega$,

mass of copper wire $m = 0.045 \times 5 \times 10^{-3} = 0.225$ mg,

melting point of copper $t = 1356°C$,

specific heat of copper $c = 0.091$ cal/g.$°C$.

If the copper wire is initially at room temperature ($t = 25°C$), the heat needed to bring it to melting point is

$$Q = cm\Delta t = 0.091 \times 0.225 \times 10^{-3} \times (1356 - 25)$$
$$= 0.027 \ \text{cal} \ = \ 0.11 \ \text{J} \ .$$

As $Q > E$ the copper wire will not melt.

3005

As in Fig. 3.9, switch S is closed and a steady dc current $I = V/R$ is established in a simple LR series circuit. Now switch S is suddenly opened. What happens to the energy $\frac{1}{2} \ LI^2$ which was stored in the circuit when the current I was present?

(Wisconsin)

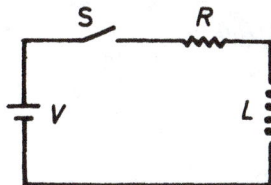

Fig. 3.9

Solution:

When switch S is suddenly opened, the energy $\frac{1}{2}LI^2$ will be radiated in the form of electromagnetic waves.

3006

(a) The capacitor in the circuit in Fig. 3.10 is made from two flat square metal plates of length L on a side and separated by a distance d. What is the capacitance?

(b) Show that if any electrical energy is stored in C, it is entirely dissipated in R after the switch is closed.

(Wisconsin)

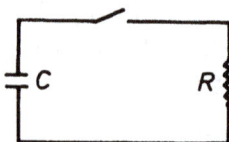

Fig. 3.10

Solution:

(a) The capacitor has capacitance $C = \frac{\varepsilon S}{d}$. As $\varepsilon = \varepsilon_0$ for air, $C = \varepsilon_0 L^2/d$.

(b) Let V_0 be the voltage across the plates of the capacitor initially. The energy stored is then $W_C = \frac{1}{2}CV_0^2$. After the switch is closed at $t = 0$, one has

$$V_C(t) = V_0 e^{-t/RC} ,$$

$$i(t) = -C\,\frac{dV_C(t)}{dt} = \frac{V_0}{R}\,e^{-t/RC} .$$

The energy dissipated in the resistance is

$$W_R = \int_0^\infty i^2(t)R\,dt = \frac{V_0^2}{R}\int_0^\infty e^{-2t/RC}\,dt = \frac{1}{2}CV_0^2 .$$

Hence

$$W_R = W_C ,$$

which implies that the energy stored in the capacitor is entirely dissipated in the resistance R.

3007

(a) Given the following infinite network (Fig. 3.11):

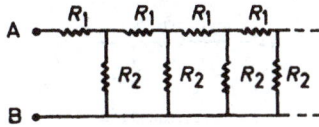

Fig. 3.11

Find the input resistance, i.e., the equivalent resistance between terminals A and B.

(b) Figure 3.12 shows two resistors in parallel, with values R_1 and R_2. The current I_0 divides somehow between them. Show that the condition $I_0 = I_1 + I_2$ together with the requirement of minimum power dissipation leads to the same current values that we would calculate by ordinary circuit formulae.

(SUNY, Buffalo)

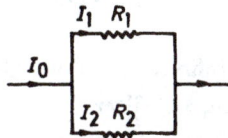

Fig. 3.12

Solution:

(a) Let the total resistance of the infinite network be R. After removing the resistances of the first section, the remaining circuit is still an infinite network which is equivalent to the original one. Its equivalent circuit is shown in Fig. 3.13 and has total resistance

$$R = R_1 + \frac{RR_2}{R + R_2} \ .$$

This gives a quadratic equation in R

$$R^2 - R_1 R - R_1 R_2 = 0 \ .$$

The positive root

$$R = \frac{R_1}{2} + \frac{\sqrt{R_1^2 + 4R_1 R_2}}{2}$$

gives the equivalent resistance.

(b) As $I_0 = I_1 + I_2$, the power dissipation is
$$P = I_1^2 R_1 + I_2^2 R_2 = I_1^2 R_1 + (I_0 - I_1)^2 R_2 .$$

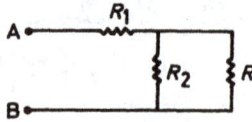

Fig. 3.13

To minimize, put $\frac{dP}{dI_1} = 0$, which gives $2I_1 R_1 - 2(I_0 - I_1)R_2 = 0$, or
$$\frac{I_1}{I_2} = \frac{I_1}{I_0 - I_1} = \frac{R_2}{R_1} .$$
This is the formula one usually uses.

3008

The frequency response of a single low-pass filter (RC–circuit) can be compensated ideally:

(a) exactly only by an infinite series of RC–filters

(b) exactly only by using LC–filters

(c) exactly by a single high-pass (RC) filter.

(*CCT*)

Solution:

The answer is (c).

3009

A square voltage pulse (Fig. 3.15) is applied to terminal A in the circuit shown in Fig. 3.14. What signal appears at B?

(*Wisconsin*)

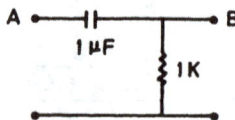

Fig. 3.14

Solution:

The time constant of this circuit is

$$\tau = RC = 1 \times 10^3 \times 1 \times 10^{-6} = 10^{-3} \text{ s} = 1 \text{ ms} .$$

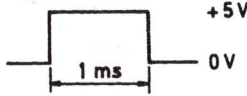

Fig. 3.15

The voltages at A and B are

$$V_A = 5u(t) - 5u(t-1) ,$$
$$V_B = 5e^{-t/\tau}u(t) - 5e^{-(t-1)/\tau}u(t-1)$$
$$= 5e^{-t}u(t) - 5e^{-(t-1)}u(t-1) ,$$

where t is in ms. The time curve of V_B is shown in Fig. 3.16.

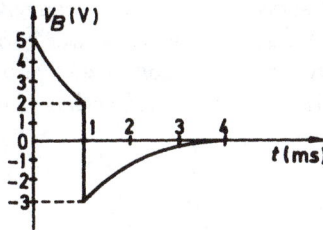

Fig. 3.16

3010

Calculate the energy in the 3 μF capacitor in Fig. 3.17.

(*Wisconsin*)

Solution:

The voltage across the two ends of the capacitors in series is

$$V = \left| \frac{(1.5\|1)}{1.4 + (1.5\|1)} \cdot 4 - 2 \right| = 0.8 \text{ V} .$$

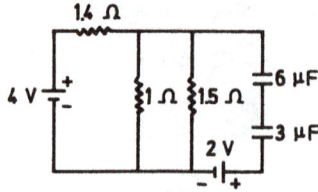

Fig. 3.17

The voltage across the two ends of the 3 μF capacitor is $\frac{6}{3+6} \times 0.8 = 0.53$ V. So the energy stored in the 3 μF capacitor is

$$E = \frac{1}{2} \times 3 \times 10^{-6} \times 0.53^2 = 0.42 \times 10^{-6} \text{ J} .$$

3011

The diagram 3.18 shows a circuit of 2 capacitors and 2 ideal diodes driven by a voltage generator. The generator produces a steady square wave of amplitude V, symmetrical around zero potential, shown at point a in the circuit. Sketch the waveforms and assign values to the voltage levels at points b and c in the circuit.

(*Wisconsin*)

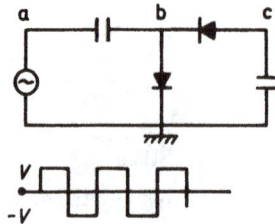

Fig. 3.18

Solution:

The resistance of an ideal diode is 0 in the positive direction and ∞ in the negative direction. Figure 3.19 gives the equivalent circuits corresponding to the positive and negative voltages at point a. We shall assume that the voltage generator is always working and the circuit has already entered a steady state.

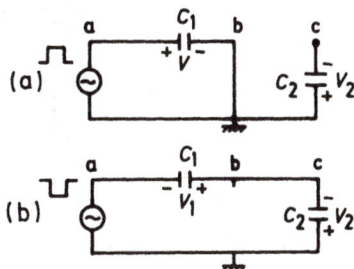

Fig. 3.19

Suppose that during a negative pulse the voltage drops across C_1 and C_2 are V_1 and V_2 respectively with the directions as shown in Fig. 3.19(b). The points a, b and c have potentials

$$V_a = -V = -V_1 - V_2 \,,$$
$$V_b = V_c = -V_2 \,.$$

Now the potential at a jumps to $+V$. The voltage drop across C_2 remains at V_2 as it is unable to discharge (see Fig. 3.19(a)), while that across C_1 is changed to $+V$. We have

$$V_a = V, \qquad V_b = 0, \qquad V_c = -V_2$$

Then the potential at a jumps again to $-V$. We have

$$V_a = -V = V - V_2 \,,$$

giving

$$V_2 = 2V, \qquad V_1 = -V \,,$$

and

$$V_b = V_c = -V_2 = -2V \,.$$

Combining the above we have

$$V_1 = -V, \qquad V_2 = 2V \,,$$

$$V_b = \begin{cases} 0 & \text{when } V_a = V \\ -2V & \text{when } V_a = -V \end{cases}$$
$$V_c = -2V \text{ at all times}.$$

The waveforms at points a, b and c are shown in Fig. 3.20.

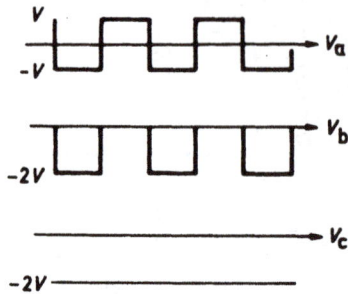

Fig. 3.20

3012

In the circuit shown in Fig. 3.21, the capacitors are initially charged to a voltage V_0. At $t = 0$ the switch is closed. Derive an expression for the voltage at point A at a later time t.

(UC, Berkeley)

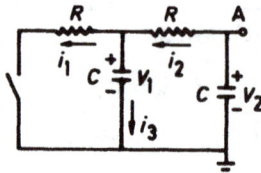

Fig. 3.21

Solution:

Suppose at time t the voltage drops across the two capacitors are V_1, V_2 and the currents in the three branches are i_1, i_2, i_3 as shown in Fig. 3.21.

By Kirchhoff's laws and the capacitor equation we have

$$\begin{cases} i_1 R + i_2 R - V_2 = 0, & (1) \\ i_1 R - V_1 = 0, & (2) \\ i_1 - i_2 + i_3 = 0; & (3) \end{cases}$$

$$\begin{cases} i_2 = -C\dfrac{dV_2}{dt}\,, & (4) \\[2mm] i_3 = C\dfrac{dV_1}{dt}\,. & (5) \end{cases}$$

Equations (2) and (5) give

$$i_3 = RC\frac{di_1}{dt}\,.$$

This and Eqs. (3) and (4) give

$$i_1 + C\frac{dV_2}{dt} + RC\frac{di_1}{dt} = 0\,. \qquad (6)$$

From Eqs. (1) and (4) one has

$$i_1 = \frac{1}{R}V_2 + C\frac{dV_2}{dt}\,.$$

Substituting it into (6), we obtain

$$\frac{d^2V_2}{dt^2} + \frac{3}{RC}\frac{dV_2}{dt} + \frac{1}{R^2C^2}V_2 = 0\,.$$

Solving this equation we have

$$V_2 = Ae^{-\frac{3+\sqrt{5}}{2}t/RC} + Be^{-\frac{3-\sqrt{5}}{2}t/RC}\,,$$

and hence

$$V_1 = i_1 R = V_2 + RC\frac{dV_2}{dt}$$
$$= -\frac{1+\sqrt{5}}{2}Ae^{-\frac{3+\sqrt{5}}{2}t/RC} - \frac{1-\sqrt{5}}{2}Be^{-\frac{3-\sqrt{5}}{2}t/RC}\,.$$

Using the initial condition that at $t = 0$

$$V_1(0) = V_2(0) = \pm V_0\,,$$

we obtain

$$V_A = V_2 = \pm\frac{5-3\sqrt{5}}{10}V_0 e^{-\frac{3+\sqrt{5}}{2}t/RC} \pm \frac{5+3\sqrt{5}}{10}V_0 e^{-\frac{3-\sqrt{5}}{2}t/RC}$$
$$\approx \pm(1.17e^{-0.38t/RC} - 0.17e^{-2.62t/RC})V_0\,.$$

3013

A network is composed of two loops and three branches. The first branch contains a battery (of emf ε and internal resistance R_1) and an open switch S. The second branch contains a resistor of resistance R_2 and an uncharged capacitor of capacitance C. The third branch is only a resistor of resistance R_3 (see Fig. 3.22).

(a) The switch is closed at $t = 0$. Calculate the charge q on C as a function of time t, for $t \geq 0$.

(b) Repeat the above, but with an initial charge Q_0 on C.

(*SUNY, Buffalo*)

Solution:

Let the currents in the three branches be I, I_1, and I_2 as shown in Fig. 3.22 and the charge on C be q at a time $t > 0$. We have by Kirchhoff's laws

$$\varepsilon = I R_1 + I_1 R_3 ,$$

$$\varepsilon = I R_1 + I_2 R_2 + \frac{q}{C} ,$$

$$I = I_1 + I_2 .$$

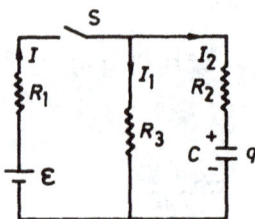

Fig. 3.22

As $\frac{dq}{dt} = I_2$, the above give

$$\frac{dq}{dt} = -Aq + B ,$$

where

$$A = \frac{R_1 + R_2}{(R_1 R_2 + R_2 R_3 + R_3 R_1)C} , \qquad B = \frac{\varepsilon R_3}{R_1 R_2 + R_2 R_3 + R_3 R_1} .$$

Solving for q we have

$$q = de^{-At} + \frac{B}{A},$$

with d to be determined by the initial conditions.

(a) If $q(0) = 0$, then $d = -\frac{B}{A}$, and

$$q = \frac{B}{A}(1 - e^{-At}) = \frac{\varepsilon R_3}{R_1 + R_2}\left(1 - \exp\left\{-\frac{R_1 + R_2}{(R_1 R_2 + R_2 R_3 + R_3 R_1)C}t\right\}\right).$$

(b) If $q(0) = Q_0$, then $d = Q_0 - \frac{B}{A}$, and

$$q = \frac{B}{A} + \left(Q_0 - \frac{B}{A}\right)e^{-At}$$

$$= \frac{\varepsilon R_3}{R_1 + R_2} + \left(Q_0 - \frac{\varepsilon R_3}{R_1 + R_2}\right)\exp\left\{-\frac{R_1 + R_2}{(R_1 R_2 + R_2 R_3 + R_3 R_1)C}t\right\}.$$

3014

In the circuit shown in Fig. 3.23, the resistance of L is negligible and initially the switch is open and the current is zero. Find the quantity of heat dissipated in the resistance R_2 when the switch is closed and remains closed for a long time. Also, find the heat dissipated in R_2 when the switch, after being closed for a long time, is opened and remains open for a long time. (Notice the circuit diagram and the list of values for V, R_1, R_2, and L_0.)

(*UC, Berkeley*)

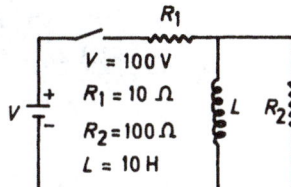

R_1

$V = 100$ V
$R_1 = 10\ \Omega$
V
$R_2 = 100\ \Omega$
$L = 10$ H

L R_2

Fig. 3.23

Solution:

Consider a resistance R and an inductance L in series with a battery of emf ε. We have

$$\varepsilon - L\frac{dI}{dt} = IR \ ,$$

or

$$\frac{-R\,dI}{\varepsilon - IR} = -R\frac{dt}{L} \ .$$

Integrating we have

$$\ln\left[\varepsilon - I(t)R\right] = -\frac{t}{\tau} + K \ ,$$

where $\tau = \frac{L}{R}$ and K is a constant. Let $I = I(0)$ at $t = 0$ and $I = I(\infty)$ for $t \to \infty$. Then

$$K = \ln\left[\varepsilon - I(0)R\right], \qquad I(\infty) = \frac{\varepsilon}{R} \ ,$$

and the solution can be written as

$$I(t) = I(\infty) + [I(0) - I(\infty)]e^{-\frac{t}{\tau}} \ .$$

Now consider the circuit in Fig. 3.23.

(1) When the switch is just closed, we have

$$I_{R_2}(0) = \frac{V}{R_1 + R_2} = 0.91 \text{ A} \ .$$

After it remains closed for a long time, we have

$$I_{R_2}(\infty) = 0 \ ,$$

since in the steady state the entire current will pass through L which has negligible resistance.

As the time constant of the circuit is

$$\tau = \frac{L}{R_1 \| R_2} = 1.1 \text{ s} \ ,$$

we have

$$I_{R_2}(t) = I_{R_2}(\infty) + [I_{R_2}(0) - I_{R_2}(\infty)]e^{-t/\tau}$$
$$= 0.91e^{-0.91t} \text{ A} \ ,$$

$$W_{R_2} = \int_0^\infty I_{R_2}^2(t)R_2 dt = \int_0^\infty 0.91^2 e^{-1.82t} \times 100 dt$$
$$= 45.5 \text{ J}.$$

(2) When the switch is just opened, we have

$$I_L(0) = \frac{V}{R_1} = 10 \text{ A}.$$

The energy stored in the inductance L at this time will be totally dissipated in the resistance R_2. Thus the heat dissipated in R_2 is

$$W_{R_2} = \frac{1}{2} L I_L^2(0) = \frac{1}{2} \times 10 \times 100 = 500 \text{ J}.$$

3015

The switch S in Fig. 3.24 has been opened for a long time. At time $t = 0$, S is closed. Calculate the current I_L through the inductor as a function of the time.

(*Wisconsin*)

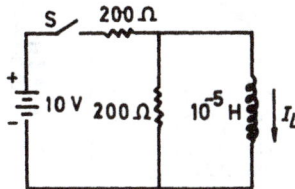

Fig. 3.24

Solution:

Assume the inductor has negligible resistance. Then at $t = 0$ and $t = \infty$,

$$I_L(0) = 0,$$
$$I_L(\infty) = \frac{10}{200} = 0.05 \text{ A}.$$

The equivalent resistance as seen from the ends of L is

$$R = 200 \| 200 = 100 \ \Omega,$$

giving the time constant as

$$\tau = \frac{L}{R} = \frac{10^{-5}}{100} = 10^{-7} \text{ s} .$$

At time t, the current passing through L is

$$I_L(t) = I_L(\infty) + (I_L(0) - I_L(\infty))e^{-\frac{t}{\tau}}$$
$$= 0.05(1 - e^{-10^7 t}) \text{ A} .$$

3016

Refer to Fig. 3.25.

(a) The switch has been in position A for a long time. The emf's are dc. What are the currents (magnitude and direction) in ε_1, R_1, R_2 and L?

(b) The switch is suddenly moved to position B. Just after the switching, what are currents in ε_2, R_1, R_2 and L?

(c) After a long time in position B, what are the currents in ε_2, R_1, R_2 and L?

(Wisconsin)

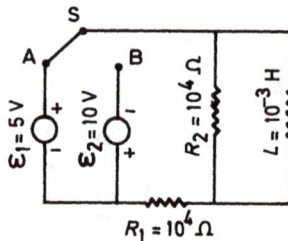

Fig. 3.25

Solution:

(a) After the switch has been in position A for a long time, L corresponds to a shorting. Then one finds that

$$I_{R_2} = 0,$$
$$I_{R_1} = \frac{\varepsilon_1}{R_1} = \frac{5}{10^4} = 0.5 \text{ mA, leftward};$$
$$I_{\varepsilon_1} = I_{R_1} = 0.5 \text{ mA, upward};$$
$$I_L = I_{\varepsilon_1} = 0.5 \text{ mA, downward}.$$

(b) When the switch is suddenly moved to position B, I_L holds constant instantaneously, namely, $I_L = 0.5$ mA and flows downward. Let the currents through R_1 and R_2 be I_{R_1} and I_{R_2} and their directions be rightward and upward respectively. Now we have

$$\begin{cases} I_{R_1} + 0.5 \times 10^{-3} = I_{R_2}, \\ I_{R_1}R_1 + I_{R_2}R_2 = (I_{R_1} + I_{R_2}) \times 10^4 = \varepsilon_2 = 10. \end{cases}$$

Solving these equations we have

$$I_{R_1} = 0.25 \text{ mA, rightward}; \qquad I_{R_2} = 0.75 \text{ mA, upward};$$

$$I_{\varepsilon_2} = 0.25 \text{ mA, downward}; \qquad I_L = 0.5 \text{ mA, downward}.$$

(c) Using the results of (a) but replacing $\varepsilon = 5$ V by $\varepsilon = -10$ V, we have

$$I_{R_1} = \frac{\varepsilon_2}{R_1} = 1 \text{ mA, rightward}; \qquad I_{R_2} = 0;$$

$$I_L = 1 \text{ mA, upward}; \qquad I_{\varepsilon_2} = 1 \text{ mA, downward}.$$

3017

As shown in Fig. 3.26, the switch has been in position A for a long time. At $t = 0$ it is suddenly moved to position B. Immediately after contact with B:

(a) What is the current through the inductor L?

(b) What is the time rate of change of the current through R?

(c) What is the potential of point B (with respect to ground)?

(d) What is the time rate of change of the potential difference across L?

(e) Between $t = 0$ and $t = 0.1$ s, what total energy is dissipated in R?

(*Wisconsin*)

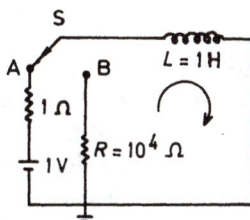

Fig. 3.26

Solution:

(a) Because the current through an inductor cannot be changed suddenly, we still have

$$i_L(0) = \frac{1}{1} = 1 \text{ A}.$$

(b) As $-L\frac{di_L}{dt} = i_L R$,

$$\left.\frac{di_L}{dt}\right|_{t=0} = -i_L(0)\frac{R}{L} = 1 \times \frac{10^4}{1} = -10^4 \text{ A/s}.$$

(c) $v_B(0) = -i_L(0)R = -1 \times 10^4 = -10^4$ V.

(d) As $v_L = v_B = i_L R$,

$$\left.\frac{dv_L}{dt}\right|_{t=0} = \left.\frac{di_L}{dt}\right|_{t=0} R = -i_L(0)\frac{R^2}{L}$$

$$= -1 \times \frac{(10^4)^2}{1} = -10^8 \text{ V/s}.$$

(e) As $W_L = \frac{1}{2}Li_L^2(t)$,

$$i_L(t) = i_L(0)e^{-\frac{R}{L}t} = e^{-10^4 t} \text{ A},$$

the total energy dissipated in R from $t = 0$ to $t = 0.1$ s is

$$W_R = \frac{1}{2}Li_L^2(0) - \frac{1}{2}Li_L^2(0.1)$$

$$= \frac{1}{2} \times 1 \times (1)^1 - \frac{1}{2} \times 1 \times e^{-2\times10^4\times0.1} = 0.5 \text{ J}.$$

3018

The pulsed voltage source in the circuit shown in Fig. 3.27 has negligible impedance. It outputs a one-volt pulse whose duration is 10^{-6} seconds. The resistance in the circuit is changed from 10^3 ohms to 10^4 ohms and to 10^5 ohms. You can assume the scope input is properly compensated so that it does not load the circuit being inspected. Sketch the oscilloscope waveforms when $R = 10^3$ ohms, 10^4 ohms, and 10^5 ohms.

(*Wisconsin*)

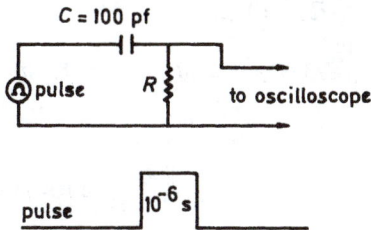

Fig. 3.27

Solution:

The output of the pulsed voltage source is $u(t) - u(t - 1)$ V, where t is in μs. The step-response of the CR circuit is $u(t)e^{-t/RC}$ with RC in μs. So the output of the CR circuit is

$$v_0 = u(t)e^{-t/RC} - u(t-1)e^{-(t-1)/RC} \text{ V} .$$

The oscilloscope waveforms are as sketched in Fig. 3.28 and Fig. 3.29.

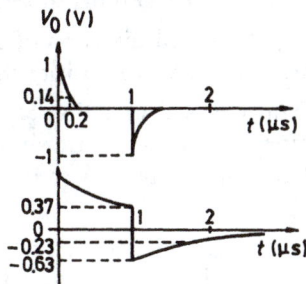

Fig. 3.28

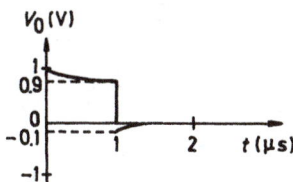

Fig. 3.29

(1) $R = 10^3 \ \Omega, \qquad RC = 10^{-1} \mu s,$

$$v_0 = u(t)e^{-10t} - u(t-1)e^{-10(t-1)} \text{ V} .$$

(2) $R = 10^4\ \Omega$, $RC = 1\ \mu s$,

$$v_0 = u(t)e^{-t} - u(t-1)e^{-t+1}\ \text{V}\ .$$

(3) $R = 10^5\ \Omega$, $RC = 10\ \mu s$,

$$v_0 = u(t)e^{-0.1t} - u(t-1)e^{-0.1(t-1)}\ \text{V}\ .$$

In all the above t is in μs.

3019

Switch S is thrown to position A as shown in Fig. 3.30.

(a) Find the magnitude and direction ("up" or "down" along page) of the currents in R_1, R_2, and R_3, after the switch has been in position A for several seconds.
Now the switch is thrown to position B (open position).

(b) What are the magnitude and direction of the currents in R_1, R_2, and R_3 just after the switch is thrown to position B?

(c) What are the magnitude and direction of the currents in R_1, R_2, and R_3 one-half second after the switch is thrown from A to B?
One second after the switch is thrown from A to B, it is finally thrown from B to C.

(d) What are the magnitude and direction of the currents in R_2, R_3, R_4, and R_5 just after the switch is thrown from B to C?

(*Wisconsin*)

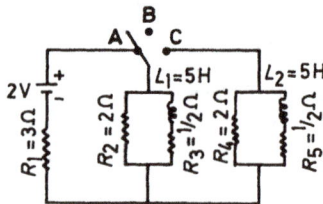

Fig. 3.30

Solution:

Let the currents in R_1, R_2, R_3 be i_1, i_2, i_3 respectively.

(a) When the switch is thrown to position A, we have instantaneously

$$i_1(0) = i_2(0) = \frac{2}{R_1 + R_2} = \frac{2}{3 + 2} = 0.4 \text{ A} ,$$

$$i_3(0) = 0 ,$$

$$i_1(\infty) = \frac{2}{R_1 + R_2 \| R_3} = 0.59 \text{ A} .$$

After the switch is in A for some time, we have

$$i_2(\infty) = \frac{R_3}{R_2 + R_3} i_1(\infty) = 0.12 \text{ A} ,$$

$$i_3(\infty) = \frac{R_2}{R_2 + R_3} i_1(\infty) = 0.47 \text{ A} .$$

As seen from the ends of L_1 the resistance in the circuit is

$$R = R_3 + R_1 \| R_2 = 1.7 \ \Omega ,$$

and the time constant is

$$\tau = L_1/R = \frac{5}{1.7} = \frac{1}{0.34} \text{ s} .$$

Using $i(t) = i(\infty) + [i(0) - i(\infty)]e^{-t/\tau}$, (see Problem **3014**), we have

$$i_1(t) = 0.59 - 0.19e^{-0.34t} \text{ A, the direction is upward,}$$

$$i_2(t) = 0.12 + 0.28e^{-0.34t} \text{ A, the direction is downward,}$$

$$i_3(t) = 0.47(1 - e^{-0.34t}) \text{ A, the direction is downward.}$$

(b) After the switch has been in A for several seconds, we can consider $e^{-0.34t} \approx 0$. From the rule that the current in an inductor cannot be changed abruptly, at the instant the switch is thrown to B we have

$$i_3(0) = 0.47 \text{ A, downward} ,$$

and so

$$i_1(0) = 0 ,$$
$$i_2(0) = 0.47 \text{ A, upward} .$$

(c) As the circuit is open,

$$i_1(0.5) = 0 .$$

For the inductor part, the time constant is

$$\tau = \frac{L}{R_2 + R_3} = \frac{5}{2 + 0.5} = 2 \text{ s} .$$

Using $i(0)$ obtained in (b) and

$$i_1(\infty) = i_2(\infty) = i_3(\infty) = 0 ,$$

we have

$$i_2(t) = 0.47e^{-0.5t} \text{ A, upward};$$

$$i_3(t) = 0.47e^{-0.5t} \text{ A, downward} .$$

Hence for $t = 0.5$ s

$$i_2(0.5) = 0.37 \text{ A, upward};$$

$$i_3(0.5) = 0.37 \text{ A, downward} .$$

(d) We denote by $1\pm$ the instants just after and before $t = 1$ s. We have $i_3(1-) = 0.47e^{-0.5} = 0.29$ A, flowing downward. As the current in an inductor cannot be changed suddenly, we have

$$i_5(1+) = 0, \qquad i_3(1+) = 0.29 \text{ A, downward} .$$

For the rest of the circuit, we have

$$i_2(1+) + i_4(1+) = 0.29 \text{ A} ,$$
$$2i_2(1+) = 2i_4(1+) .$$

Hence $i_2(1+) = i_4(1+) = 0.145$ A, upward.

3020

A source of current $i_0 \sin \omega t$, with i_0 a constant, is connected to the circuit shown in Fig. 3.31. The frequency ω is controllable. The inductances L_1 and L_2 and capacitances C_1 and C_2 are all lossless. A lossless voltmeter reading peak sine-wave voltage is connected between A and B. The product $L_2C_2 > L_1C_1$.

(a) Find an approximate value for the reading V on the voltmeter when ω is very small but not zero.

(b) The same, for ω very large but not infinite.

(c) Sketch qualitatively the entire curve of voltmeter reading versus ω, identifying and explaining each distinctive feature.

(d) Find an expression for the voltmeter reading valid for the entire range of ω.

(*Princeton*)

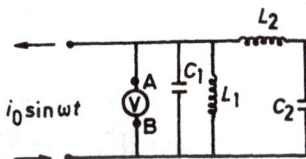

Fig. 3.31

Solution:

(a) The impedance of an inductor is $j\omega L$ and the impedance of a capacitor is $\frac{1}{j\omega C}$. For ω very small, the currents passing through the capacitors may be neglected and the equivalent circuit is as shown in Fig. 3.32.

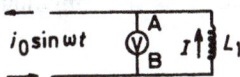

Fig. 3.32

We thus have

$$V_{BA} = IZ = j\omega L_1 I \; ,$$

where $I = i_0 e^{j\omega t}$. As ac meters usually read the rms values, we have

$$V_{meter} = \frac{1}{\sqrt{2}} i_0 \omega L_1 \; .$$

(b) For ω very large, neglect the currents passing through the inductors and the equivalent circuit is as shown in Fig. 3.33. We have

$$V_{BA} = IZ = -\frac{jI}{\omega C_1} \; .$$

and

$$V_{\text{meter}} = \frac{i_0}{\sqrt{2}\omega C_1} \ .$$

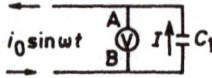

Fig. 3.33

(c) Let $\omega_1 = \frac{1}{\sqrt{L_1 C_1}}$, $\omega_2 = \frac{1}{\sqrt{L_2 C_2}}$. As $L_2 C_2 > L_1 C_1$, $\omega_1 > \omega_2$. The voltmeter reading versus ω is as shown in Fig. 3.34. The system is net inductive when ω is in the region $(0, \omega_2)$, and net capacitive when ω is in the region (ω_1, ∞). Resonance occurs at the characteristic angular frequencies ω_1 and ω_2.

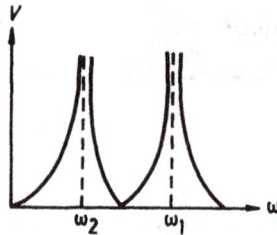

Fig. 3.34

(d) The total impedance L is the combination of two impedances L_1, L_2 in parallel:

$$Z = \frac{Z_1 Z_2}{Z_1 + Z_2} \ ,$$

where

$$Z_1 = \frac{-jL_1}{C_1(\omega L_1 - \frac{1}{\omega C_1})} \ , \qquad Z_2 = j\left(\omega L_2 - \frac{1}{\omega C_2}\right) .$$

Thus

$$Z = \frac{j}{\frac{1}{\omega L_1} - \omega C_1 + \frac{1}{\omega L_2 - \frac{1}{\omega C_2}}} \ .$$

Hence the voltmeter reading is

$$V_{\text{BA}} = \frac{i_0}{\sqrt{2}\left|\frac{1}{\omega L_1} - \omega C_1 + \frac{1}{\omega L_2 - \frac{1}{\omega C_2}}\right|} \ .$$

Note that this reduces to the results in (a) and (b) for ω very small and very large.

3021

For the circuit shown in Fig. 3.35, the coupling coefficient of mutual inductance for the two coils L_1 and L_2 is unity.

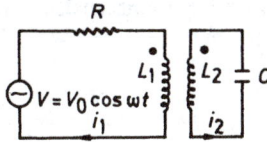

Fig. 3.35

(a) Find the instantaneous current $i(t)$ the oscillator must deliver as a function of its frequency.

(b) What is the average power supplied by the oscillator as a function of frequency?

(c) What is the current when the oscillator frequency equals the resonant frequency of the secondary circuit?

(d) What is the phase angle of the input current with respect to the driving voltage as the oscillator frequency approaches the resonant frequency of the secondary circuit?

(*UC, Berkeley*)

Solution:

(a) Let the currents of the primary and secondary circuits be I_1 and I_2 respectively. We have

$$\dot{V} = \dot{I}_1 R + L_1 \ddot{I}_1 + M \ddot{I}_2 \ ,$$

$$0 = L_2 \ddot{I}_2 + M \ddot{I}_1 + \frac{I_2}{C} \ .$$

Solving for $I_1 \sim \exp(j\omega t)$, we have

$$I_1 = \frac{V}{R + j\left(\omega L_1 + \frac{\omega^3 M^2}{\frac{1}{C} - \omega^2 L_2}\right)}$$

$$= \frac{V_0 e^{-j\varphi}}{\sqrt{R^2 + \left(\omega L_1 + \frac{\omega^3 M^2}{\frac{1}{C} - \omega^2 L_2}\right)^2}} \ ,$$

where

$$\varphi = \arctan\left(\frac{\omega L_1 + \frac{\omega^3 M^2}{\frac{1}{C}-\omega^2 L_2}}{R}\right)$$

is the phase angle of the input current with respect to the driving voltage. Applying the given conditions $L_1 = L_2 = M = L$, say, we have

$$I_1 = \frac{V_0 e^{-j\varphi}}{\sqrt{R^2 + \left(\frac{\omega L}{1-\omega^2 LC}\right)^2}},$$

$$\varphi = \arctan\left(\frac{\omega L/R}{1 - \omega^2 LC}\right),$$

or, taking the real part,

$$i_1(t) = \frac{V_0}{Z} \cos(\omega t - \varphi),$$

with

$$Z = \sqrt{R^2 + \left(\frac{\omega L}{1 - \omega^2 LC}\right)^2}.$$

(b)

$$p(t) = V(t)i_1(t) = \frac{V_0^2}{Z} \cos(\omega t - \varphi) \cos \omega t.$$

Averaging over one cycle we have

$$P = \bar{p} = \frac{V_0^2}{Z}\overline{\cos(\omega t - \varphi) \cos \omega t} = \frac{V_0^2}{2Z} \cos \varphi$$

$$= \frac{R}{2Z^2} V_0^2 = \frac{RV_0^2/2}{R^2 + \left(\frac{\omega L}{1-\omega^2 LC}\right)^2}.$$

(c) When $\omega = \frac{1}{\sqrt{LC}}$, $Z = +\infty$, and $i_1(t) = 0$.

(d) When $\omega \to \frac{1}{\sqrt{LC}}$, $\tan \varphi = \infty$, and $\varphi = \frac{\pi}{2}$.

3022

In the electrical circuit shown in Fig. 3.36, ω, R_1, R_2 and L are fixed; C and M (the mutual inductance between the identical inductors L) can be varied. Find values of M and C which maximize the power dissipated in resistor R_2. What is the maximum power?

You may assume, if needed, $R_2 > R_1$, $\omega L/R_2 > 10$.

(*Princeton*)

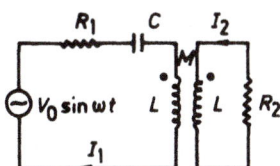

Fig. 3.36

Solution:

Assuming that the primary and secondary currents are directed as in Fig. 3.36, we have the circuit equations

$$\dot{V}_0 = R_1 \dot{I}_1 + \dot{I}_1 \left[\frac{1}{j\omega C} + j\omega L \right] + j\omega M \dot{I}_2 ,$$

$$0 = \dot{I}_2 R_2 + j\omega L \dot{I}_2 + j\omega M \dot{I}_1 .$$

The above simultaneous equations have solution

$$I_2 = \frac{j\omega M C V_0}{C[\omega^2(L^2 - M^2) - R_1 R_2] - L + j\left[\frac{R_2}{\omega} - \omega L C (R_1 + R_2)\right]} .$$

As $P_2 = \frac{1}{2}|I_2|^2 R_2$, when $|I_2|$ is maximized P_2 is maximized also. We have

$$|I_2| = \frac{\omega V_0}{\left\{ \left[\frac{1}{M}(\omega^2 L^2 - R_1 R_2 - L/C) - \omega^2 M \right]^2 + \left[\frac{1}{MC} \left(\frac{R_2}{\omega} - \omega L C (R_1 + R_2) \right) \right]^2 \right\}^{\frac{1}{2}}} .$$

As the numerator is fixed and the denominator is the square root of the sum of two squared terms, when the two squared terms are minimum at the same time $|I_2|$ will achieve its maximum. The minimum of the second squared term is zero, for which we require

$$C = \frac{R_2}{\omega^2 L (R_1 + R_2)} ,$$

giving

$$|I_2| = \frac{\omega V_0}{\omega^2 M + \frac{R_1 R_2 + R_1 \omega^2 L^2/R_2}{M}} \ .$$

Minimizing the above denominator, we require

$$M = \sqrt{\frac{R_1 R_2}{\omega^2} + \frac{R_1}{R_2} L^2} \ .$$

Hence, for

$$M = \sqrt{\frac{R_1 R_2}{\omega^2} + \frac{R_1}{R_2} L^2}, \qquad C = \frac{R_2}{\omega^2 L (R_1 + R_2)}$$

P_2 is maximum, having the value

$$P_2 = \frac{1}{2}|I_2|^2 R_2 = \frac{V_0^2}{8 R_1 \left(1 + \frac{\omega^2 L^2}{R_2^2}\right)} \ .$$

Supposing $\omega L/R_2 > 10$, we obtain

$$P_2 = \frac{V_0^2 R_2^2}{8 \omega^2 L^2 R_1}$$

as the maximum power dissipated in R_2.

3023

In Fig. 3.37 the capacitor is originally charged to a potential difference V. The transformer is ideal: no winding resistance, no losses. At $t = 0$ the switch is closed. Assume that the inductive impedances of the windings are very large compared with R_p and R_s. Calculate:

(a) The initial primary current.

(b) The initial secondary current.

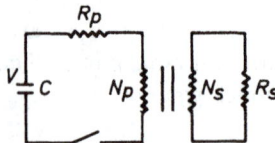

Fig. 3.37

(c) The time for the voltage V to fall to e^{-1} of its original value.

(d) The total energy which is finally dissipated in R_s.

(Wisconsin)

Solution:

As the transformer is ideal,

$$N_p/N_s = V_p/V_s , \qquad N_p/N_s = I_s/I_p .$$

(a) The equivalent resistance in the primary circuit due to the resistance R_s in the secondary circuit is

$$R'_s = \left(\frac{N_p}{N_s}\right)^2 R_s .$$

Hence the time constant of the primary circuit is

$$\tau = (R_p + R'_s)C ,$$

and the voltage drop across C is

$$V_C = Ve^{-t/\tau} .$$

The primary current is then

$$i_p = -C\frac{dV_C}{dt} = -CVe^{-t/\tau}\left(-\frac{1}{\tau}\right) = \frac{V}{R_p + R'_s}e^{-t/\tau} .$$

Initially, the primary current is

$$i_p(0) = \frac{V}{R_p + R'_s} = \frac{V}{R_p + \left(\frac{N_p}{N_s}\right)^2 R_s} .$$

(b)

$$i_s(0) = i_p(0)\frac{N_p}{N_s} = \frac{N_p N_s V}{N_s^2 R_p + N_p^2 R_s} .$$

(c) For V_C to fall to $V_C = e^{-1}V$, the time required is

$$t = \tau = \left[R_p + \left(\frac{N_p}{N_s}\right)^2 R_s\right]C .$$

(d) As

$$i_s = i_p \frac{N_p}{N_s} = \frac{N_p N_s V}{N_s^2 R_p + N_p^2 R_s} e^{-t/\tau} ,$$

the energy dissipated in R_s is

$$W_{RS} = \int_0^\infty i_s^2 R_s \, dt = \left(\frac{N_p N_s V}{N_s^2 R_p + N_p^2 R_s}\right)^2 R_s \int_0^\infty e^{-2t/\tau} dt$$

$$= \left(\frac{N_p N_s V}{N_s^2 R_p + N_p^2 R_s}\right)^2 R_s \cdot \frac{1}{2}\left[R_p + \left(\frac{N_p}{N_s}\right)^2 R_s\right]C$$

$$= \frac{1}{2} \frac{N_p^2 V^2}{N_s^2 R_p + N_p^2 R_s} R_s C .$$

3024

Show that for a given frequency the circuit in Fig. 3.38 can be made to "fake" the circuit in Fig. 3.39 to any desired accuracy by an appropriate choice of R and C. ("Fake" means that if $V_0 = IZ_R$ in one circuit and $V_0 = IZ_L$ in the other, then Z_L can be chosen such that $Z_L/Z_R = e^{i\theta}$ with θ arbitrarily small.) Calculate values of R and C that would fake a mutual inductance $M = 1$ mH at 200 Hz with $\theta < 0.01$.

(*UC, Berkeley*)

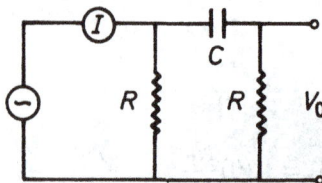

Fig. 3.38

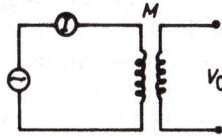

Fig. 3.39

Solution:

For the circuit in Fig. 3.38, we have

$$\dot{V}_0 = \dot{I} \cdot \left[\left(R + \frac{1}{j\omega C} \right) \| R \right] \cdot \frac{R}{R + \frac{1}{j\omega C}}$$

$$= \dot{I} \cdot \frac{R^2}{2R + \frac{1}{j\omega C}} = \dot{I} \frac{R^2}{\sqrt{4R^2 + \frac{1}{\omega^2 C^2}}} \angle \arctan\left(1/2\omega RC\right) .$$

For the circuit in Fig. 3.39, we have

$$V_0 = j\omega M \dot{I} = \dot{I} M \omega \angle \pi/2 .$$

For the former to "fake" the latter, we require

$$\begin{cases} \dfrac{R^2}{\sqrt{4R^2 + \frac{1}{\omega^2 C^2}}} = M\omega , & (1) \\[3mm] \dfrac{\pi}{2} - \arctan\left(\frac{1}{2R\omega C}\right) = \theta . & (2) \end{cases}$$

Equation (2) gives $\omega RC = \frac{1}{2}\tan\theta$. With $\theta = 0.01$, $\omega RC = 0.005$. Equation (1) then gives

$$R = M\omega \sqrt{4 + \frac{1}{(\omega RC)^2}} = 10^{-3} \times 2\pi \times 200 \times \sqrt{4 + \frac{1}{0.005^2}} \approx 251 \ \Omega ,$$

and hence

$$C = \frac{0.005}{\omega R} = \frac{0.005}{2\pi \times 200 \times 251} \approx 1.6 \times 10^{-8} \ \text{F} = 0.016 \ \mu\text{F} .$$

3025

A two-terminal "black box" is known to contain a lossless inductor L, a lossless capacitor C, and a resistor R. When a 1.5 volt battery is connected to the box, a current of 1.5 milliamperes flows. When an ac voltage of 1.0 volt (rms) at a frequency of 60 Hz is connected, a current of 0.01 ampere (rms) flows. As the ac frequency is increased while the applied voltage is maintained constant, the current is found to go through a maximum exceeding 100 amperes at $f = 1000$ Hz. What is the circuit inside the box, and what are the values of R, L, and C?

(Princeton)

Solution:

When a dc voltage is connected to the box a finite current flows. Since both C and L are lossless, this shows that R must be in parallel with C or with both L, C. At resonance a large current of 100 A is observed for an ac rms voltage of 1 V. This large resonance is not possible if L and C are in parallel, whatever the connection of R. The only possible circuit is then the one shown in Fig. 3.40 with L, C in series. Since a dc voltage of 1.5 V gives rise to a current of 1.5 mA, we have

$$R = \frac{V}{I} = \frac{1.5}{1.5 \times 10^{-3}} = 10^3 \ \Omega \ .$$

The impedance for the circuit in Fig. 3.40 is

$$Z = \frac{1}{\frac{1}{R} + \frac{1}{j(\omega L - \frac{1}{\omega C})}} = \frac{1}{\frac{1}{R} + \frac{1}{j\omega L(1 - \frac{\omega_0^2}{\omega^2})}} \ ,$$

giving

$$L = \left[\frac{1}{\frac{1}{Z^2} - \frac{1}{R^2}} \cdot \frac{1}{\omega^2 \left(1 - \frac{\omega_0^2}{\omega^2}\right)^2} \right]^{-1/2} \ ,$$

where $\omega_0^2 = \frac{1}{LC}$.

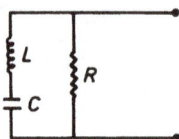

Fig. 3.40

The resonance occurs at $\omega_0 = 2000\pi$ rad/s. At $\omega = 120\pi$ rad/s, $V_{rms} = 1$ V gives $I_{rms} = \frac{1}{100}$ A, corresponding to

$$|Z| = \frac{V_{rms}}{I_{rms}} = 100 \ \Omega = \frac{R}{10} \ ,$$

at

$$\frac{\omega_0}{\omega} = \frac{50}{3}.$$

Hence

$$L \approx \frac{\omega}{\omega_0^2} |Z| = \frac{60 \times 2\pi}{(1000 \times 2\pi)^2} \times 100 = 0.95 \text{ mH},$$

$$C = \frac{1}{L\omega_0^2} = \frac{1}{\omega|Z|} = \frac{1}{60 \times 2\pi \times 100} \approx 27 \text{ } \mu\text{F}.$$

3026

In Fig. 3.41 a box contains linear resistances, copper wires and dry cells connected in an unspecified way, with two wires as output terminals A, B. If a resistance $R = 10 \text{ }\Omega$ is connected to A, B, it is found to dissipate 2.5 watts. If a resistance $R = 90 \text{ }\Omega$ is connected to A, B, it is found to dissipate 0.9 watt.

Fig. 3.41

(a) What power will be dissipated in a 30 Ω resistor connected to A, B (Fig. 3.42a)?

(b) What power will be dissipated in a resistance $R_1 = 10 \text{ }\Omega$ in series with a 5 Volt dry cell when connected to A, B (Fig. 3.42b)?

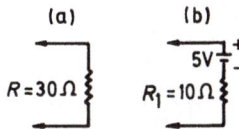

Fig. 3.42

(c) Is your answer to (b) unique? Explain.

(*UC, Berkeley*)

Solution:

Using Thévinin's theorem, we can treat the box as what is shown in Fig. 3.43. When $R = 10 \text{ }\Omega$, $P = \frac{V_R^2}{R} = 2.5$ W, giving $V_R = 5$ V. When

$R = 90 \ \Omega$, $P = 0.9$ W, giving $V_R = 9$ V. Therefore we have

$$\begin{cases} \varepsilon \cdot \frac{10}{10+R_s} = 5, \\ \varepsilon \cdot \frac{90}{90+R_s} = 9, \end{cases} \quad \text{giving} \quad \begin{cases} \varepsilon = 10 \text{ V}, \\ R_s = 10 \ \Omega. \end{cases}$$

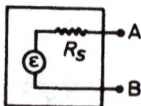

Fig. 3.43

(a) When $R = 30 \ \Omega$, we have

$$V_R = \frac{30}{30 + 10} \times 10 = 7.5 \text{ V}, \qquad P = V_R^2/R = 1.875 \text{ W} .$$

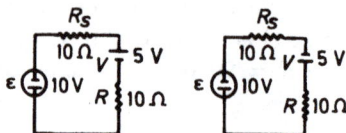

Fig. 3.44

(b) If the resistance $R = 10 \ \Omega$ is in series with a dry cell of $\varepsilon' = 5$ V, we will have

$$V_R = \frac{10}{10 + 10} (\varepsilon \pm \varepsilon') = \begin{cases} 2.5 \text{ V} \\ 7.5 \text{ V}, \end{cases}$$

$$P = V_R^2/R = \begin{cases} 0.625 \text{ W}, \\ 5.625 \text{ W}. \end{cases}$$

(c) As two polarities are possible for the connection of the 5 Volt dry cell, two different answers are obtained.

2. ELECTRIC AND MAGNETIC CIRCUITS (3027–3044)

3027

A solenoid having 100 uniformly spaced windings is 2 cm in diameter and 10 cm in length. Find the inductance of the coil.

$$\left(\mu_0 = 4\pi \times 10^{-7} \frac{\text{T} \cdot \text{m}}{\text{A}} \right)$$

(*Wisconsin*)

Solution:

Neglecting edge effects, the magnetic field in the solenoid is uniform everywhere. From Ampère's circuital law $\oint \mathbf{B} \cdot d\mathbf{l} = \mu_0 I$, we find the magnetic field induction inside the solenoid as $B = \mu_0 n I$, where $n = \frac{N}{l}$ is the turn density of the solenoid. The total magnetic flux crossing the coil is $\psi = NBA$. The inductance of the coil is given by the definition

$$L = \frac{\psi}{I} = \frac{N \mu_0 N I A}{lI} = \frac{N^2 \mu_0 A}{l} \ .$$

With $A = \pi \times 10^{-4}$ m^2, we have

$$L = \frac{100^2 \times 4\pi \times 10^{-7} \times \pi \times 10^{-4}}{0.1} = 3.95 \times 10^{-5} \text{ H} \ .$$

3028

A circuit contains a ring solenoid (torus) of 20 cm radius, 5 cm^2 cross-section and 10^4 turns. It encloses iron of permeability 1000 and has a resistance of 10 Ω. Find the time for the current to decay to e^{-1} of its initial value if the circuit is abruptly shorted.

(UC, Berkeley)

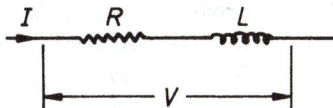

Fig. 3.45

Solution:

The equivalent circuit is shown in Fig. 3.45, for which

$$V = IR + L\frac{dI}{dt} \ ,$$

with

$$V|_{t<0} = V_0 = I_0 R , \qquad V|_{t \geq 0} = 0 \ .$$

Thus for $t \geq 0$

$$-\frac{dt}{\tau} = \frac{dI}{I} \ ,$$

where $\tau = \frac{L}{R}$. Hence

$$I = I_0 e^{-t/\tau} = \frac{V_0}{R} e^{-t/\tau} .$$

The self-inductance of the torus is

$$L = \mu\mu_0 \frac{N^2 A}{2\pi R}$$
$$= \frac{10^{4\times 2} \times 4\pi \times 10^{-7} \times 10^3 \times 5 \times 10^{-4}}{2\pi \times 20 \times 10^{-2}} = 50 \text{ H} .$$

For $I = I_0 e^{-1}$, $t = \tau = \frac{L}{R} = \frac{50}{10} = 5$ s.

3029

A circular loop of wire is placed between the pole faces of an electro-magnet with its plane parallel to the pole faces. The loop has radius a, total resistance R, and self-inductance L. If the magnet is then turned on, producing a **B** field uniform across the area of the loop, what is the total electric charge q that flows past any point on the loop?

(Wisconsin)

Solution:

When the magnetic flux crossing the circular loop changes an emf ε will be induced producing an induced current i. Besides, a self-inductance emf $L \frac{di}{dt}$ is produced as well. Thus we have

$$\varepsilon + iR + L \frac{di}{dt} = 0 ,$$

with

$$\varepsilon = -\frac{d\phi}{dt} , \quad i = \frac{dq}{dt} , \quad i(\infty) = 0 , \quad i(0) = 0 .$$

The circuit equations can be written as

$$-d\phi + Rdq + Ldi = 0 .$$

Integrating over t from 0 to ∞ then gives

$$-\Delta\phi + Rq = 0$$

as $\Delta i = 0$. Hence

$$q = \frac{\Delta \phi}{R} = \frac{B \pi a^2}{R} .$$

This shows that L has no effect on the value of q. It only leads to a slower decay of i.

3030

A solenoid has an air core of length 0.5 m, cross section 1 cm^2, and 1000 turns. Neglecting end effects, what is the self-inductance? A secondary winding wrapped around the center of the solenoid has 100 turns. What is the mutual inductance? A constant current of 1 A flows in the secondary winding and the solenoid is connected to a load of 10^3 ohms. The constant current is suddenly stopped. How much charge flows through the resistance?

(Wisconsin)

Solution:

Let the current in the winding of the solenoid be i. The magnetic induction inside the solenoid is then $B = \mu_0 n i$ with direction along the axis, n being the number of turns per unit length of the winding.

The total magnetic flux linkage is

$$\psi = N\phi = NBS = N^2 \mu_0 Si/l .$$

Hence the self-inductance is

$$L = \frac{\psi}{i} = N^2 \mu_0 S/l$$

$$\approx \frac{1000^2 \times 4\pi \times 10^{-7} \times 10^{-4}}{1/2} = 2.513 \times 10^{-4} \text{ H} .$$

The total magnetic flux linkage in the secondary winding produced by the currrent i is $\psi' = N'\phi$, giving the mutual inductance as

$$M = \frac{\psi'}{i} = \frac{NN'\mu_o S}{l} = 2.513 \times 10^{-5} \text{ H} .$$

Because of the magnetic flux linkage $\psi' = MI$, I being the current in the secondary, an emf will be induced in the solenoid when the constant current I in the secondary is suddenly stopped. Kirchhoff's law gives for the induced current i in the solenoid

$$-\frac{d\psi'}{dt} = Ri + L\frac{di}{dt} ,$$

or

$$-d\psi' = Ri\,dt + L\,di = R\,dq + L\,di .$$

Integrating over t from $t = 0$ to $t = \infty$ gives $-\Delta\psi' = Rq$, since $i(0) = i(\infty) = 0$. Thus the total charge passing through the resistance is

$$q = \frac{-\Delta\psi'}{R} = \frac{MI}{R} = \frac{2.513 \times 10^{-5} \times 1}{10^3} = 2.76 \times 10^{-7} \text{ C} .$$

3031

As in Fig. 3.46, G is a ballistic galvanometer (i.e., one whose deflection θ is proportional to the charge Q which quickly flows through it). The coil L as shown is initially in a magnetic field $B_0 = 0$. Switch S is then closed, current $I = 1$ amp flows, and G deflects $\theta_1 = 0.5$ radian and returns to rest. Then the coil is quickly moved into a magnetic field B_2, and G is observed to deflect $\theta_2 = 1$ radian. What is the field B_2 (in any specified units)?

(*UC, Berkeley*)

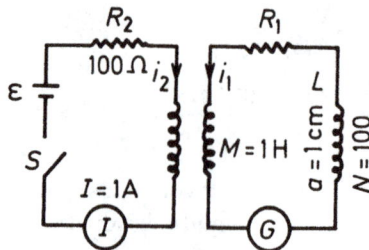

Fig. 3.46

Solution:

The direction of $\mathbf{B}_2$ is illustrated in Fig. 3.47.

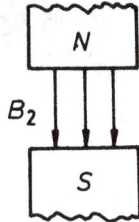

Fig. 3.47

Let the self-inductance of the coil L be L_1, then

$$\varepsilon_1 = M\frac{di_2}{dt} + L_1\frac{di_1}{dt} \; ,$$

or

$$\frac{dq}{dt} = i_1 = \frac{\varepsilon_1}{R_1} = \frac{M}{R_1}\frac{di_2}{dt} + \frac{L_1}{R_1}\frac{di_1}{dt} \; ,$$

with

$$i_2(0) = 0 \; , \quad i_2(\infty) = 1 \; \text{A} \; , \quad i_1(0) = i_1(\infty) = 0 \; .$$

Integrating the circuit equation we obtain

$$q_1 = \int_0^\infty \frac{M}{R_1}di_2 + \int_0^\infty \frac{L_1}{R_1}di_1 = \frac{M}{R_1}i_2(\infty) \; .$$

When the coil is moved into the magnetic field $\mathbf{B_2}$, its induced emf is

$$\varepsilon_2 = -\frac{d\psi}{dt} \; ,$$

with

$$\psi_1(\infty) = -NB_2S \; , \quad \psi_1(0) = 0 \; .$$

Thus

$$\frac{dq_2}{dt} = i_2 = \frac{\varepsilon_2}{R_1} = \frac{-1}{R_1}\frac{d\psi_1}{dt} \; ,$$

giving

$$q_2 = \frac{NB_2\pi a^2}{R_1} \; .$$

As $q \propto \theta$, we have

$$\frac{\theta_1}{\theta_2} = \frac{q_1}{q_2} = \frac{Mi_2(\infty)}{NB_2\pi a^2} \; ,$$

or

$$B_2 = \frac{\theta_2 M i_2(\infty)}{\theta_1 N \pi a^2} = \frac{1 \times 1 \times 1}{0.5 \times 100 \times \pi \times 10^{-4}} = 63.4 \text{ T} .$$

3032

Two perfectly conducting disks of radius a are separated by a distance h ($h \ll a$). A solid cylinder of radius b, length h and volume resistivity ρ fills the center portion of the gap between the disks (see Fig. 3.48). The disks have been connected to a battery for an infinite time.

Fig. 3.48

(a) Calculate the electric field everywhere in the gap as a function of time after the battery has been disconnected from the capacitor. Neglect edge effects and inductance.

(b) Calculate **B** everywhere within the gap as a function of time and distance r from the axis of the disks.

(c) Calculate the Poynting vector in the space between the plates. Explain qualitatively its direction at $r = a$ and at $r = b$.

(d) Show, by a detailed calculation for the special case $a = b$, that the conservation of energy theorem (Poynting's theorem) is satisfied in the volume bounded by the plates and $r = a$.

(*UC, Berkeley*)

Solution:

(a) Let the upper plate carry charge $+Q$ and the lower plate carry charge $-Q$ at time t. Due to the continuity of the tangential component of electric intensity across an interface, we have

$$\mathbf{E} = \frac{Q}{\pi a^2 \varepsilon_0} \mathbf{e}_z \text{ at } z < h .$$

For $r \leq b$, $\mathbf{j} = \sigma \mathbf{E}$, where $\sigma = \frac{1}{p}$, giving

$$I = j\pi b^2 = \sigma E \pi b^2 .$$

Thus

$$\sigma \frac{Q}{\pi a^2 \varepsilon_0} \pi b^2 = -\frac{dQ}{dt} ,$$

or

$$Q = Q_0 e^{-t/\tau} ,$$

where $\tau = \frac{a^2 \varepsilon_0 \rho}{b^2}$. Hence

$$\mathbf{E} = E_0 e^{-t/\tau} \mathbf{e}_z = \frac{V_0}{h} e^{-t/\tau} \mathbf{e}_z .$$

(b) Applying Ampère's circuital law $\oint \mathbf{B} \cdot d\mathbf{l} = \mu_0 I$ to a coaxial circle of radius $r < b$ on a cross section of the solid cylinder:

$$\oint \mathbf{B} \cdot d\mathbf{r} = \mu_0 \iint \mathbf{j} \cdot d\mathbf{S} ,$$

one has

$$B \cdot 2\pi r = \mu_0 j \pi r^2 ,$$

or

$$\mathbf{B} = \frac{\mu_0 j r}{2} \mathbf{e}_\theta = \frac{\mu_0 r V_0}{2\rho h} e^{-t/\tau} \mathbf{e}_\theta , \qquad (r < b)$$

where $\mathbf{e}_\theta$ is a unit vector tangential to the circle. For $b < r < a$, the circuital law

$$\oint \mathbf{B} \cdot d\mathbf{r} = j\pi b^2 \mu_0$$

gives

$$\mathbf{B} = \frac{\mu_0 b^2 V_0}{2r\rho h} e^{-t/\tau} \mathbf{e}_\theta .$$

(c) For $0 < r < b$ and between the conducting plates, the Poynting vector is $\mathbf{S} = \mathbf{E} \times \mathbf{H} = \frac{\mathbf{E} \times \mathbf{B}}{\mu_0} = \frac{V_0}{h} e^{-t/\tau} \cdot \frac{r V_0}{2\rho h} e^{-t/\tau} (\mathbf{e}_z \times \mathbf{e}_\theta) = -\frac{r}{2\rho} (\frac{V_0}{h})^2 e^{-2t/\tau} \mathbf{e}_r$. For $b < r < a$, we have

$$\mathbf{S} = \frac{V_0}{h} e^{-t/\tau} \cdot \frac{b^2 V_0}{2r\rho h} e^{-t/\tau} \mathbf{e}_r = -\frac{1}{2r\rho} \left(\frac{V_0 b}{h} \right)^2 e^{-2t/\tau} \mathbf{e}_r .$$

The directions of S at $r = a$ and $r = b$ are both given by $-e_r$, i.e., the electromagnetic energy flows radially inwards into the solid cylinder between the plates (in the ideal case). This energy provides for the loss of energy due to Joule heating in the solid cylinder where a current flows. This can be seen as follows. For $b < r < a$ the inward energy flow per unit time is $\frac{1}{2r\rho}(\frac{V_0 b}{h})^2 \cdot 2\pi r h e^{-2t/\tau} = \frac{\pi}{\rho h}(V_0 b)^2 e^{-2t/\tau}$, independent of r. But for $r < b$, the power in-flow is $\frac{r}{2\rho}(\frac{V_0}{h})^2 \cdot 2\pi r h e^{-2t/\tau} = \frac{\pi}{\rho h}(V_0 r)^2 e^{-2t/\tau}$, decreasing as r decreases.

(d) For $a = b$ the power flowing into the cylinder is

$$P_1 = \iint_{r=a} \mathbf{S} \cdot d\mathbf{A} = -\frac{\pi}{\rho h}(V_0 a)^2 e^{-2t/\tau} .$$

The loss of power due to Joule heating in the cylinder is

$$P_2 = I^2 R = \left[\pi a^2 \cdot \frac{1}{\rho}\frac{V_0}{h}e^{-t/\tau}\right]^2 \cdot \frac{\rho h}{\pi a^2}$$

$$= \frac{\pi}{\rho h}(V_0 a)^2 e^{-2t/\tau} .$$

Thus $P_1 + P_2 = 0$, and the conservation law of energy is satisfied.

3033

In the diagram 3.49, the two coils are wound on iron cores in the same direction. Indicate whether the current flow in resistor r is to the right, or to the left, and give a reason for your answer in each of the following cases:

(a) Switch S is opened.

(b) Resistor R is decreased.

(c) An iron bar is placed alongside the two coils.

(d) Coil A is pulled away from coil B.

(*Wisconsin*)

Solution:

The direction of the current in coil A with the switch S closed is shown in Fig. 3.49. According to the right-handed screw rule, the magnetic field produced points to the left. If the current is steady, there is no current in r.

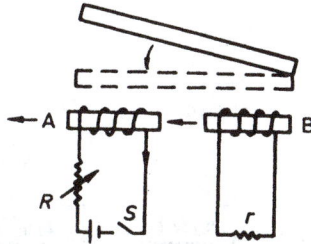

Fig. 3.49

(a) When the switch S is opened, the magnetic field pointing to the left decreases. Lenz's law requires that a magnetic field pointing to the left is induced in coil B. Hence the induced current in the resistance r flows from right to left.

(b) If R is decreased, the current flowing through coil A is increased. Then the magnetic flux, which points to the left, piercing B will also increase. Lenz's law requires the induced current in the resistance r to flow from left to right.

(c) An iron bar placed alongside the coils will increase the original field. So the current in r is from left to right.

(d) When coil A is pulled away from coil B, the magnetic flux piercing B will decrease, so that the induced current in r will flow from right to left.

3034

A solenoid is designed to generate a magnetic field over a large volume. Its dimensions are as follows: length = 2 meters, radius = 0.1 meter, number of turns = 1000. (Edge effects should be neglected.)

(a) Calculate the self-inductance of the solenoid in Henrys.

(b) What is the magnetic field (in Webers/m^2) produced on the axis of the solenoid by a current of 2000 Amperes?

(c) What is the stored energy when the solenoid is operated with this current?

(d) The total resistance of the solenoid is 0.1 ohm. Derive the equation describing the transient current as a function of time immediately after connecting the solenoid to a 20 Volt power supply. What is the time constant of the circuit?

(*Princeton*)

Solution:

(a) Suppose the solenoid carries a current I. The magnetic induction inside it is

$$B = \mu_0 n I = \mu_0 N I / l ,$$

and the magnetic flux linkage is

$$\psi = NBS = N\frac{\mu_0 N I}{l} \cdot \pi r^2 = \frac{I \mu_0 N^2 \pi r^2}{l} .$$

Hence the self-inductance is

$$L = \frac{\psi}{I} = \frac{\mu_0 N^2 \pi r^2}{l} = \frac{4\pi \times 10^{-7} \times 1000^2 \times \pi \times 0.1^2}{2}$$
$$= 1.97 \times 10^{-2} \text{ H} .$$

(b)

$$B = \frac{\mu_0 N I}{l} = \frac{4\pi \times 10^{-7} \times 1000 \times 2000}{2} = 1.26 \text{ Wb/m}^2 .$$

(c)

$$W_m = \frac{L}{2} I^2 = \frac{1.97 \times 10^{-2} \times 2000^2}{2} = 3.94 \times 10^4 \text{ J} .$$

(d) The circuit equation is

$$\varepsilon = iR + L\frac{di}{dt} ,$$

giving

$$i = \frac{\varepsilon}{R}(1 - e^{-t/\tau}) = i(\infty)(1 - e^{-t/\tau}) ,$$

where $\tau = \frac{L}{R} = 0.197$ s is the time constant of the circuit. As

$$\varepsilon = 20 \text{ V}, \quad R = 0.1 \ \Omega, \quad L = 1.97 \times 10^{-2} \text{ H} ,$$

we have

$$i(t) = 200(1 - e^{-st}) \text{ A} .$$

3035

The electrical circuit shown in Fig. 3.50 consists of two large parallel plates. Plate B is grounded except for a small section (the detector). A sinusoidal voltage of frequency ω is applied to plate A.

(a) For what value of ω is V_0 (the amplitude of V_{out}) a maximum?

(b) With ω fixed, plate A is moved left and right. Make a sketch of V_0 as a function of position. Indicate the points at which the edge of plate A passes the detector.

(c) Suppose A is held at a fixed potential. How is the resulting electrostatic field related to the function sketched in part (b)? Explain.

(*Wisconsin*)

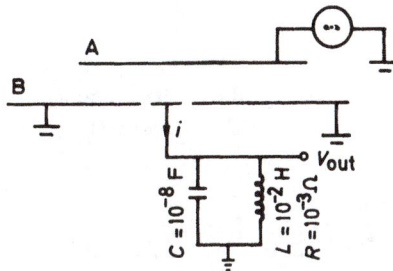

Fig. 3.50

Solution:

(a) Resonance will take place when

$$\omega = \frac{1}{\sqrt{LC}} = \frac{1}{\sqrt{10^{-2} \times 10^{-8}}} = 10^5 \text{ rad/s} .$$

At this time the equivalent impedance for the parallel circuit is maximum. Hence V_0 will also be maximum.

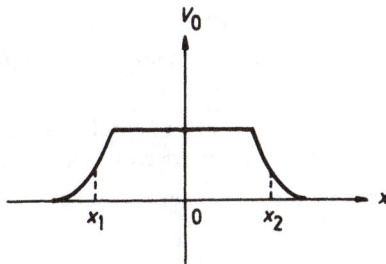

Fig. 3.51

(b) V_0 as a function of position is shown in Fig. 3.51, where x is the horizontal distance of the middle of A from the middle of the detector, and x_1 and x_2 correspond to the right edge and the left edge respectively of A passing by the middle of the detector.

(c) The variation of the electrostatic field with x is similar to that of the function sketched in part (b). The field intensity decreases near the edges of the plate A.

Since the magnitiude of V_0 reflects the amount of charge carried by plate A and the detector through $Q = CV_0$, the electrostatic field intensity is large where V_0 is large. The plate B is larger than A so that the movement of the latter can be ignored.

3036

In the circuit shown in Fig. 3.52, the capacitor has circular plates of radius r_0 separated by a distance d. Between the plates there is a vacuum. At $t = 0$, when there is a charge Q_0 on the capacitor, the switch is closed.

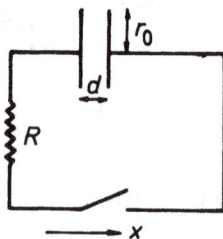

Fig. 3.52

(a) For $t \geq 0$, the electric field between the plates is approximately $E(t) = E_0 e^{-t/\tau} \mathbf{i}$. Find E_0 and τ (if you cannot find them, take them as given constants and go on to part (b)).

(b) Mention some approximations and idealizations made when deriving the form of $\mathbf{E}$ given in (a).

(c) Find the magnetic field between the plates for $t > 0$. You may use idealizations and approximations similar to those in (b).

(d) What is the electromagnetic energy density in the vacuum region between the plates?

(e) Consider a small cylindrical portion of the vacuum region between the plates (see Fig. 3.53). Suppose it has radius r_1, length l and is centered.

Using (a), (c), and the Poynting vector compute the total energy which flows through the surface of the small cylinder during the time $0 < t < \infty$.

(*UC, Berkeley*)

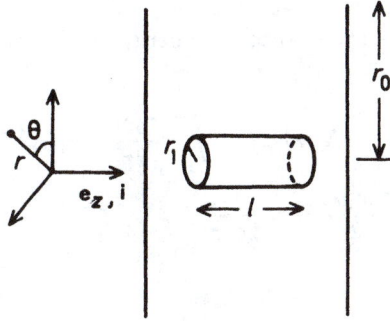

Fig. 3.53

Solution:

(a) Since $\frac{Q}{C} = iR = -\frac{dQ}{dt}R$, we have $Q = Q_0 e^{-t/\tau}$ with $\tau = RC$. As $E = \frac{\sigma}{\varepsilon_0}$, we have $E = \frac{Q}{\pi r^2 \varepsilon_0} e^{-t/\tau}$. Comparing this with $E = E_0 e^{-t/\tau}$, we find

$$E_0 = \frac{Q}{\pi r^2 \varepsilon_0}, \qquad \tau = RC = \frac{R\varepsilon_0 \pi r_0^2}{d} .$$

(b) To find **E** for case (a), we have assumed that the charge Q is uniformly distributed over the plates at any time and the edge effects may be neglected. These approximations are good if $d \ll r_0$.

(c) By symmetry and Maxwell's integral equation

$$\oint \mathbf{H} \cdot d\mathbf{l} = \iint_S \frac{\partial \mathbf{D}}{\partial t} \cdot d\mathbf{S} ,$$

where

$$\mathbf{D} = \varepsilon_0 \mathbf{E} ,$$

we find

$$\mathbf{H} = -\frac{\varepsilon_0 r E}{2\tau} \mathbf{e}_\theta , \qquad \mathbf{B} = -\frac{\mu_0 \varepsilon_0 r E}{2\tau} \mathbf{e}_\theta ,$$

taking approximations similar to those stated in (b).

(d)

$$U = \frac{\varepsilon_0}{2} E^2 + \frac{1}{2\mu_0} B^2 = \frac{1}{2} \varepsilon_0 E^2 \left[1 + \frac{\mu_0 \varepsilon_0 r^2}{4\tau^2} \right] .$$

(e) The Poynting vector of the electromagnetic field is

$$\mathbf{N} = \mathbf{E} \times \mathbf{H} = E\mathbf{e}_z \times \left(-\frac{\varepsilon_0 r}{2\tau}E\right)\mathbf{e}_\theta = \frac{\varepsilon_0 r E^2}{2\tau}\mathbf{e}_r .$$

Thus during the time $t = 0$ to ∞ the energy flowing through the cylinder's surface is

$$W = \int_0^\infty N 2\pi r l\, dt = \int_0^\infty \frac{\varepsilon_0 r_1}{2\tau} 2\pi r_1 l E e^{-2t/\tau}\, dt$$

$$= \frac{\varepsilon_0 r_1^2 \pi l}{2} E = \frac{r_1^2 l Q^2}{2\varepsilon_0 \pi r_0^4} .$$

3037

A resonant circuit consists of a parallel-plate capacitor C and an inductor of N turns wound on a toroid. All linear dimensions of the capacitor and inductor are reduced by a factor 10, while the number of turns on the toroid remains constant.

(a) By what factor is the capacitance changed?

(b) By what factor is the inductance changed?

(c) By what factor is the resonant frequency of the resonant circuit changed?

(*Wisconsin*)

Solution:

(a) The capacitance is $C \propto \frac{S}{d}$, then $C_f = \frac{1}{10}C_i$.

(b) The inductance is $L \propto N^2 S/l$, hence $L_f = \frac{1}{10}L_i$.

(c) The resonant frequency is $\omega \propto \frac{1}{\sqrt{LC}}$, hence $\omega_f = 10\omega_i$.

3038

You have n storage cells, each with internal resistance R_i and output voltage V. The cells are grouped in sets of k series-connected cells each. The n/k sets are connected in parallel to a load-resistance R. Find the k which maximizes the power in R. How much is the power?

(*Wisconsin*)

Solution:

For each set the voltage is kV and internal resistance is kR_i. After the n/k sets are connected in parallel, the total voltage is still kV, but the total internal resistance becomes $\frac{kR_i}{n/k} = \frac{k^2 R_i}{n}$. The power in R will be maximum when the load-resistance R matches the internal resistance, i.e., $R = \frac{k^2 R_i}{n}$. Hence $k = \sqrt{\frac{nR}{R_i}}$ for maximum power, which has the value

$$P_{max} = \left(\frac{kV}{2R}\right)^2 R = \frac{k^2 V^2}{4R} = \frac{nV^2}{4R_i} \; .$$

3039

When a capacitor is being discharged:

(a) the energy originally stored in the capacitor can be completely transferred to another capacitor;

(b) the original charge decreases exponentially with time;

(c) an inductor must be used.

(*CCT*)

Solution:

The answer is (b).

3040

If $L =$ inductance and $R =$ resistance, what units does $\frac{L}{R}$ have?

(a) sec   (b) sec^{-1} (c) amperes

(*CCT*)

Solution:

The answer is (a).

3041

Two inductances L_1 and L_2 are placed in parallel far apart. The inductance of both is

(a) $L_1 + L_2$ (b) $\frac{L_1 L_2}{L_1 + L_2}$ (c) $(L_1 + L_2)\frac{L_1}{L_2}$

(*CCT*)

Solution:

The answer is (b).

3042

An alternating current generator with a resistance of 10 ohms and no reactance is coupled to a load of 1000 ohms by an ideal transformer. To deliver maximum power to the load, what turn ratio should the transformer have?

(a) 10   (b) 100   (c) 1000

(CCT)

Solution:

The answer is (a).

3043

An electrical circuit made up of a capacitor and an inductor in series can act as an oscillator because:

(a) there is always resistance in the wires;

(b) voltage and current are out of phase with each other;

(c) voltage and current are in phase with each other.

(CCT)

Solution:

The answer is (b).

3044

The force in the x-direction between two coils carrying currents i_1 and i_2 in terms of the mutual inductance M is given by

(a) $i_1 \frac{di_2}{dx} M$ (b) $i_1 i_2 \frac{dM}{dx}$ (c) $i_1 i_2 \frac{d^2 M}{dx^2}$.

(CCT)

Solution:

The answer is (b).

3. ANALOG CIRCUITS (3045–3057)

3045

In order to obtain the Zener effect, the Zener diode has to be:
(a) reverse biased   (b) forward biased   (c) connected to ac.

(CCT)

Solution:

The answer is (a).

3046

A transistor amplifier in a "grounded base" configuration has the following characteristics:

(a) low input impedance

(b) high current gain

(c) low output impedance.

(CCT)

Solution:

For a transistor amplifier in a grounded base configuration we have

input impedance $r_i = R_e \| \frac{r_{be}}{1+\beta}$, which is small,

current gain $A_i = \frac{\beta}{1+\beta} \cdot \frac{R_e}{R_c+r_i} \cdot \frac{R_e}{R_e+R_L} < 1$,

output impedance $r_0 \approx R_c$,

where R_e, R_c and R_L are the resistances of the ejector, collector and load respectively. Hence answer (a) is correct.

3047

It is possible to measure the impedance of a coaxial cable

(a) with an ohmmeter across the cable

(b) making use of the reflection properties of terminations

(c) by measuring the attenuation of signals through the cable.

(CCT)

Solution:

When the impedances of the terminals of the cable match, no reflection occurs. This method may be used to measure the impedance of a coaxial cable. Hence answer (b) is correct.

3048

The transmission of high frequencies in a coaxial cable is determined by:

(a) the impedance

(b) $\frac{1}{\sqrt{LC}}$, with L and C the distributed inductance and capacitance

(c) dielectric losses and skin-effect.

(*CCT*)

Solution:

The answer is (b).

3049

The high frequency limit of a transistor is determined by

(a) the increase of noise figure with frequency

(b) type of circuit (grounded base/emitter/collector)

(c) mechanical dimensions of active zones and drift velocity of charge carriers.

(*CCT*)

Solution:

The answer is (c).

3050

A Si transistor with $\beta = 100$ is used in the amplifier circuit shown in Fig. 3.54. Fill in the information requested. You may assume that for the frequencies involved $\frac{1}{\omega C}$ is negligible and that the emf source providing V_{in} has a negligible internal impedance.

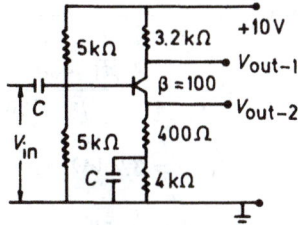

Fig. 3.54

$I_B =$ _____ $I_C =$ _____

$I_E =$ _____ $V_C =$ _____

$V_E =$ _____ $V_B =$ _____

Sign of 10 V supply = _____
$R_{in} =$ _____
Small signal gain at output 1 = _____
Small signal gain at output 2 = _____
$R_{out1} =$ _____ $R_{out2} =$ _____

(*Wisconsin*)

Solution:

The values are calculated below:

$$V_B = \frac{5}{5+5} \times 10 = 5 \text{ V},$$

$$V_E = 5 - 0.6 = 4.4 \text{ V},$$

$$I_E = \frac{V_E}{R_E} = 1 \text{ mA},$$

$$I_C = \frac{\beta}{1+\beta} I_E \approx 0.99 \text{ mA},$$

$$I_B = \frac{I_E}{1+\beta} \approx 0.01 \text{ mA},$$

$$V_C = 10 - I_C \cdot 3.2 \text{ k}\Omega = 10 - 0.99 \times 3.2 \approx 6.8 \text{ V},$$

$$R_{in} = 5\text{k}||5\text{k}|| [r_{be} + (1+\beta) \cdot 400] \approx 2.4 \text{ k}\Omega,$$

where $r_{be} = 300 + (1+\beta)\dfrac{26}{I_C} \approx 3 \text{ k}\Omega,$

small signal gain at output $1 = -\dfrac{\beta R_C}{r_{be} + (1 + \beta) \cdot 400} \approx -7.4$,

small signal gain at output $2 = \dfrac{(1 + \beta) \cdot 400}{r_{be} + (1 + \beta) \cdot 400} \approx 0.93$,

$$R_{out1} = 3.2 \text{ k}\Omega ,$$

$$R_{out2} = 400 \| \left(\frac{r_{be} + 5\|5}{1 + \beta} \right) \approx 48 \ \Omega .$$

3051

Calculate $A_F = V_0/V_i$, the amplification of the circuit with feedback shown in Fig. 3.55. $A_0 = V_0/V$, the amplification without feedback, is large and negative. The input resistance to A_0 is much greater than R_1 and R_2 and the output resistance is much less than R_1 and R_2. Discuss the dependence of A_F on A_0.

(*Wisconsin*)

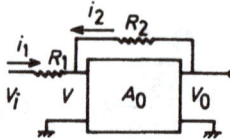

Fig. 3.55

Solution:

As A_0 is large and the input resistance is much greater than R_1 and R_2, while the output resistance is much less than R_1 and R_2, we can consider the circuit with feedback as an ideal amplifier.

Taking $i_1 = -i_2$, then

$$V_i - V = -\frac{R_1}{R_2}(V_0 - V) ,$$

or

$$\frac{V_i}{V_0} - \frac{V}{V_0} = -\frac{R_1}{R_2} \left(1 - \frac{V}{V_0} \right) .$$

Putting

$$A_F = V_0/V_i, \qquad A_0 = V_0/V,$$

the above becomes

$$\frac{1}{A_F} = -\frac{R_1}{R_2} + \left(1 + \frac{R_1}{R_2}\right)\frac{1}{A_0},$$

giving

$$A_F = \frac{1}{-\frac{R_1}{R_2} + \left(1 + \frac{R_1}{R_2}\right)\frac{1}{A_0}}.$$

As A_0 is large, $A_F \approx -\frac{R_2}{R_1}$. It follows that A_F is independent of A_0 but is determined by R_1/R_2. Hence the amplification is stable.

3052

The amplifier in the circuit shown in Fig. 3.56 is an operational amplifier with a large gain (say gain = 50,000). The input signal V_{in} is sinusoidal with an angular frequency ω in the middle of the amplifier's bandwidth. Find an expression for the phase angle ϕ between the input and output voltages as measured with respect to ground. Assume that the values of R_1 and R_2 are within an order of magnitude of each other. Note the non-inverting input to the amplifier is grounded.

(*Wisconsin*)

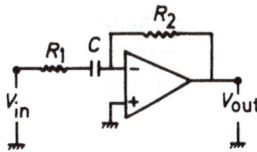

Fig. 3.56

Solution:

The amplifier may be considered as an ideal operational amplifier with "virtually grounded" inverting input. Then

$$\frac{V_{in} - 0}{R_1 + \frac{1}{j\omega C}} = \frac{0 - V_{out}}{R_2},$$

or

$$\frac{V_{out}}{V_{in}} = -\frac{R_2}{R_1 + \frac{1}{j\omega C}} \; .$$

The phase difference between the input and output voltages is

$$\phi = \pi - \arctan\left(\frac{-\frac{1}{\omega C}}{R_1}\right) = \pi + \arctan\left(\frac{1}{\omega C R_1}\right) \; .$$

3053

A very high-gain differential input amplifier is connected in the negative feedback op-amp configuration shown in the digram 3.57.

The output impedance may be considered negligibly small. The open-loop gain A may be considered "infinitely large" in this application, but the amplifier saturates abruptly when V_{out} reaches ± 10 volts.

(a) Write the ideal operational expression relating $V_{out}(t)$ to $V_{in}(t)$.

(b) What is the input impedance at terminals (J_1, J_2)?

(c) A two-volt step input is applied for V_{in} as shown in the first graph. Copy the second graph on your answer sheet and sketch in the output response.

(*Wisconsin*)

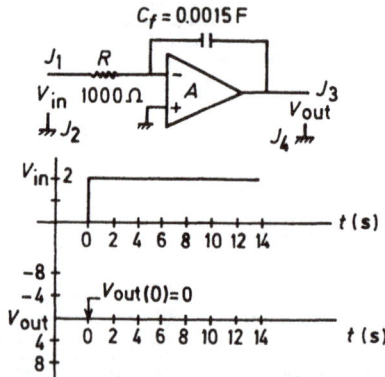

Fig. 3.57

Solution:

(a) The circuit is that of an opposite-phase integrator made up of an ideal amplifier. We have

$$V_{\text{out}} = -\frac{1}{C_f} \int_0^t \frac{V_{\text{in}}}{R} dt + V_0(0), \qquad (|V_{\text{out}}| < 10 \text{ V})$$

If $V_0(0) = 0$ at the initial time, then

$$V_{\text{out}} = -\frac{1}{C_f} \int_0^t \frac{V_{\text{in}}}{R} dt .$$

(b) If the input voltage at terminals (J_1, J_2) is sinusoidal with frequency ω, the input impedance across the terminals is $R + \frac{1}{j\omega C_f}$.

(c) If a two-volt step input is applied for V_{in} and $V_{\text{out}} = 0$ at the initial time, we have

$$V_{\text{out}} = -\frac{1}{RC_f} \int_0^t V_{\text{in}} dt = -\frac{V_{\text{in}}}{RC_f} \cdot t .$$

As the amplifier saturates abruptly at $V_{\text{out}} = \pm 10$ V, the saturated time is

$$t = \frac{-V_{\text{out}}}{V_{\text{in}}} RC_f = \frac{10}{2} \times 1000 \times 0.0015 = 7.5 \text{ s} .$$

Hence

$$V_{\text{out}} = \begin{cases} 0 & t \leq 0, \\ -\frac{4}{3} t & 0 < t \leq 7.5 \text{ s}, \\ -10 & t > 7.5 \text{ s}. \end{cases}$$

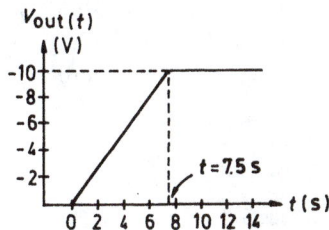

Fig. 3.58

The output response is shown in Fig. 3.58.

3054

Consider the operational amplifier circuit shown in **Fig. 3.59.**

(a) Is this an example of positive or negative feedback?

(b) Show that the circuit functions as an operational integrator. (State any assumptions necessary.)

(c) Indicate a circuit using the same components which will perform operational differentiation.

(Wisconsin)

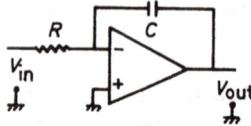

Fig. 3.59

Solution:

(a) Negative feedback.

(b) From Fig. 3.59, we find

$$\begin{cases} i = C \dfrac{dV_{out}}{dt}, \\ Ri = -V_{in}, \end{cases}$$

giving

$$V_{out} = -\frac{1}{RC} \int_0^t V_{in}\,dt + V_0 .$$

Thus the circuit is an operational integrator.
The above calculation is based on the following assumptions:

(1) the open-loop input impedance is infinite,

(2) the open-loop voltage gain is infinite.

(c) The corresponding operational differentiating circuit is shown in Fig. 3.60.

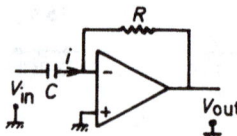

Fig. 3.60

3055

The circuit shown in Fig. 3.61 is a relaxation oscillator built from an ideal (infinite open-loop gain, infinite input impedance) differential amplifier. The amplifier saturates at an output of ± 10 V.

(a) Calculate the frequency of oscillation for the component values given.

(b) Sketch the waveforms at the inverting input (A), the non-inverting input (B) and the output (C).

(*Wisconsin*)

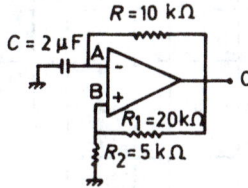

Fig. 3.61

Solution:

(a) This is a relaxation oscillator having positive feedback shunted by R_1 and R_2 and discharged through an RC circuit. When stability is reached, the output is a rectangular wave with amplitude equal to the saturated voltage. Let $V_C = +10$ V. The potential at point B is $V_B = \frac{R_2}{R_1+R_2} \times V_C = 2$ V. The capacitor C is charged through R, and V_A will increase from -2 V to $+2$ V. When V_A is higher than V_B, V_C will decrease to -10 V and C will discharge through R. When V_A is lower than the potential at B, which is now $\frac{R_2}{R_1+R_2} \times (-10) = -2$ V, V_C will again increase to $+10$ V. So following each charging the circuit relaxes back to the starting point, i.e., relaxation oscillation occurs. The charging of the capacitor follows

$$V = V_0 \left[1 - \exp(-t/RC)\right], \quad \text{or} \quad t = RC \ln\left(\frac{V_0}{V_0 - V}\right) .$$

The charging time T from V_1 to V_2 is

$$T = RC \ln \frac{V_0 - V_1}{V_0 - V_2} .$$

The charging time is given by $V_0 = 10$ V, $V_1 = -2$ V, $V_2 = 2$ V, i.e., $T_1 = RC \ln \frac{10+2}{10-2} = 8.1$ ms; the discharging time is given by $V_0 = -10$ V, $V_1 = 2$ V, $V_2 = -2$ V, i.e., $T_2 = RC \ln \frac{-10-2}{-10+2} = 8.1$ ms.

Hence the oscillation frequency is $\frac{1}{T_1+T_2} = 61.6$ Hz.

(b) The waveforms of V_A, V_B, and V_C are shown in Fig. 3.62.

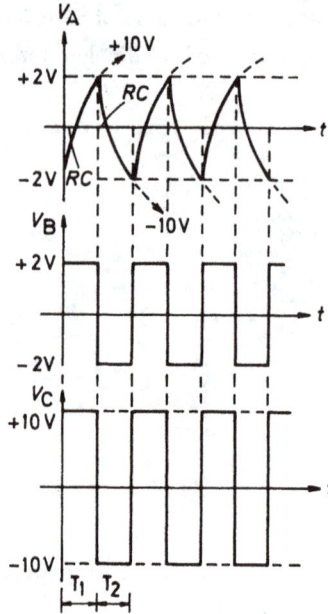

Fig. 3.62

3056

An analog computer circuit, as shown in Fig. 3.63, is made using high gain operational amplifiers. What differential equation does the analog computer solve? If the analog computation is started by simultaneously opening switches S_1 and S_2, what are the initial conditions appropriate for the solution to the differential equation?

(*Wisconsin*)

Solution:

Let the output voltage be V_0. As the operational amplifiers can be considered ideal, we have the following equations:

$$\text{at point 1:} \quad \frac{v_1}{R} = -C\,\frac{dv_2}{dt}, \tag{1}$$

at point 2: $\quad \dfrac{v_2}{R} = -C\,\dfrac{dv_0}{dt}\,,$ $\hspace{2cm}$ (2)

at point 3: $\quad \dfrac{v_0}{\frac{R}{2}} + \dfrac{v_2}{3R} = -\dfrac{v_1}{R}\,.$ $\hspace{1.5cm}$ (3)

(1) and (2) give $\quad \dfrac{v_1}{R} = RC^2\,\dfrac{d^2 v_0}{dt^2}\,,$ $\hspace{1.5cm}$ (4)

(2) and (3) give $\quad \dfrac{2v_0}{R} - \dfrac{C}{3}\,\dfrac{dv_0}{dt} = -\dfrac{v_1}{R}\,.$ $\hspace{1cm}$ (5)

Then (4) and (5) give

$$\frac{d^2 v_0}{dt^2} - \frac{1}{3}\frac{dv_0}{dt} + 2v_0 = 0\,,$$

taking $RC = 1$. This is the differential equation that can be solved by the analog computer.

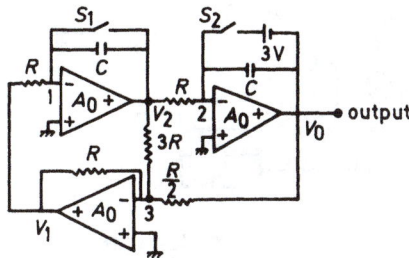

Fig. 3.63

Initially when S_1 and S_2 are just opened we have

$$v_0(0) = -3 \text{ V}\,,$$
$$v_1(0) = v_2(0) = 0\,,$$
$$C\,\frac{dv_0}{dt}\bigg|_{t=0} = 0\,,$$

so the initial conditions are

$$\begin{cases} v_0 = -3 \\[2mm] \dfrac{dv_0}{dt}\bigg|_{t=0} = 0\,. \end{cases}$$

3057

Design an analog computer circuit using operational amplifiers that will produce in the steady state a voltage $V(t)$ that is a solution to the equation

$$\frac{d^2V}{dt^2} + 10\frac{dV}{dt} - \frac{1}{3}V = 6\sin \omega t .$$

(*Wisconsin*)

Solution:

The equation can be written as

$$\frac{d^2v}{dt^2} = -10\frac{dv}{dt} + \frac{1}{3}v + 6\sin \omega t .$$

The block diagram of the design is shown in Fig. 3.64 and the circuit diagram in Fig. 3.65.

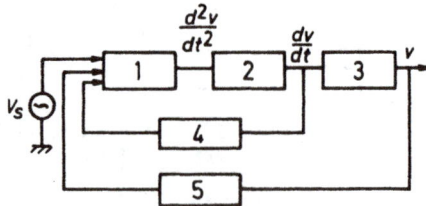

1   - addometer
2,3 - integrator
4,5 - ratiometer

Fig. 3.64

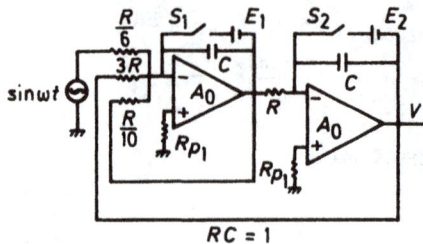

$RC = 1$

Fig. 3.65

Note when switches S_1 and S_2 are closed at the initial time, the voltage of the source is $V_s = \sin \omega t$.

4. DIGITAL CIRCUITS (3058–3065)

3058

What is the direct application in standard NIM electronics of the De Morgan relation $\overline{A \cap B} = \overline{A} \cup \overline{B}$?

(a) Transformation of an "OR" unit into an "AND" unit

(b) Inversion of signals

(c) Realization of an "EXCLUSIVE OR".

(*CCT*)

Solution:

The answer is (c). The truth table of this problem is given below:

$$\overline{A \cdot B} = \overline{A} + \overline{B}$$

A	B	output
1	1	0
1	0	1
0	1	1
0	0	1

3059

A digital system can be completely fabricated using:

(a) AND and OR gates only

(b) all NOR gates or all NAND gates

(c) neither of the above.

(*CCT*)

Solution:

The answer is (b). A NOR or a NAND gate can be used as a NON gate. Using the De Morgan law, we can translate an OR into an AND,

and an AND into an OR. Therefore, NOR gates or NAND gates alone are sufficient to fabricate a complete digital system.

3060

In Fig. 3.66 the 4 basic logic gate symbols are shown.

(a) Match them to the negative logic equivalents on the right.

(b) Write the truth table for each.

(c) Name the logic function.

(Wisconsin)

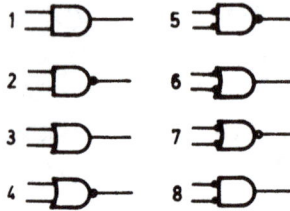

Fig. 3.66

Solution:

(a) Refer to Fig. 3.66 and denote the output by Q. The outputs of the gates are given below:

gate	output
1	$Q = A \cdot B$
2	$Q = \overline{A \cdot B} = \overline{A} + \overline{B}$
3	$Q = A + B$
4	$Q = \overline{A + B} = \overline{A} \cdot \overline{B}$
5	$Q = \overline{\overline{A} \cdot \overline{B}} = A + B$
6	$Q = \overline{A} + \overline{B}$
7	$Q = \overline{\overline{A} + \overline{B}} = A \cdot B$
8	$Q = \overline{A} \cdot \overline{B}$

It is seen that the equivalences are: 1 and 7, 2 and 6, 3 and 5, 4 and 8.

(b) The truth table for each pair of gates is given below:

1 and 7

A	B	Q
0	0	0
0	1	0
1	0	0
1	1	1

2 and 6

A	B	Q
0	0	1
0	1	1
1	0	1
1	1	0

3 and 5

A	B	Q
0	0	0
0	1	1
1	0	1
1	1	1

4 and 8

A	B	Q
0	0	1
0	1	0
1	0	0
1	1	0

(c) The logic function for each gate is given below:

$$
\begin{aligned}
&1. \quad Q = A \cdot B, \quad \text{“AND”;} \\
&2. \quad Q = \overline{A} + \overline{B}, \quad \text{“NOR”;} \\
&3. \quad Q = A + B, \quad \text{“OR”;} \\
&4. \quad Q = \overline{A} \cdot \overline{B}, \quad \text{“NAND”;} \\
&5. \quad Q = A + B, \quad \text{“AND”;} \\
&6. \quad Q = \overline{A} + \overline{B}, \quad \text{“NOR”;} \\
&7. \quad Q = A \cdot B, \quad \text{“AND”;} \\
&8. \quad Q = \overline{A} \cdot \overline{B}, \quad \text{“NAND”.}
\end{aligned}
$$

3061

Inside of the programming counter in a microprocessor there is:

(a) the address of the instruction

(b) the address of the data

(c) the sentence's number of the program.

(*Wisconsin*)

Solution:

The answer is (a).

3062

A Schmitt trigger has a dead time

(a) smaller than the pulse width
(b) about equal to the pulse width
(c) larger than the pulse width.

(*CCT*)

Solution:

The answer is (b).

3063

Refer to Fig. 3.67.

(a) Is Q_2 saturated? Justify your answer.

(b) What is the base-emitter voltage of Q_1?

(c) When this monostable circuit is triggered how long will Q_2 be off?

(d) How can this circuit be triggered? Show the triggering circuit and the waveform.

(*Wisconsin*)

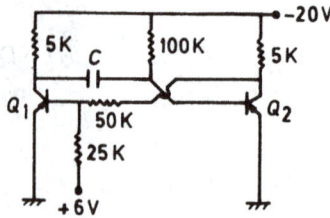

Fig. 3.67

Solution:

(a) In the circuit for Q_2, $\beta = \frac{I_c}{I_b} = \frac{100\ K}{5\ K} = 20$. Since in a practical circuit β is always much larger than 20, Q_2 is saturated.

(b) As Q_2 is saturated, $V_c(Q_2) = -0.3$ V. Hence

$$V_b(Q_1) = 6 - \frac{6+0.3}{25+50} \times 25 = 3.9 \text{ V} .$$

Thus the base-emitter voltage of Q_1 is 3.9 V.

(c) The monostable pulse width is

$$\Delta t = RC \ln 2 = 100 \times 10^3 \times 100 \times 10^{-12} \times 0.7$$
$$= 7 \times 10^{-6} \text{ s} = 7 \ \mu\text{s} ,$$

during which Q_2 is off.

(d) The triggering circuit is shown in Fig. 3.68 and the waveforms are shown in Fig. 3.69.

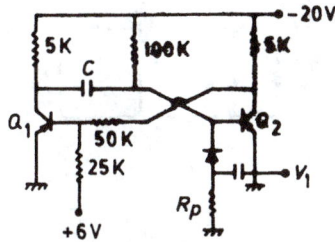

Fig. 3.68

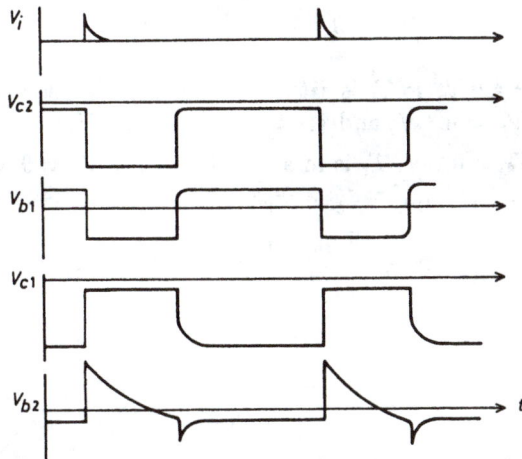

Fig. 3.69

3064

In Fig. 3.70 the circuit is a "typical" TTL totom pole output circuit. You should assume that all the solid state devices are silicon unless you specifically state otherwise. Give the voltages requested within 0.1 volt for the two cases below.

Case 1: $V_A = 4.0$ volts, give V_B, V_C, and V_E.

Case 2: $V_A = 0.2$ volts, give V_B, V_C, V_D, and V_E.

(*Wisconsin*)

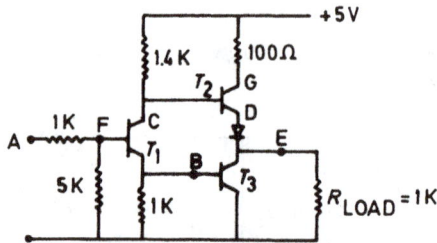

Fig. 3.70

Solution:

As all the solid state devices are silicon, the saturation voltages are

$$V_{be} = 0.7 \text{ V}, \qquad V_{ce} = 0.3 \text{ V}.$$

Case 1: $V_A = 4.0$ V, so T_1 is saturated. Then T_3 is also saturated, so that $V_B = 0.7$ V, $V_E = 0.3$ V, and $V_C = V_B + 0.3 = 1.0$ V.

Case 2: As $V_A = 0.2$ V, T_1 is in a cutoff state, so $V_B = 0$ and T_3 is also in a cutoff state. For T_2, $\beta = \frac{I_c}{I_b} = \frac{1400}{100} = 14$ so that T_2 is saturated. Thus

$$V_C = 5 \text{ V}, \quad V_D = 5 - 0.7 = 4.3 \text{ V}, \quad V_E = V_D - 0.7 = 3.6 \text{ V}.$$

3065

A register in a microprocessor is used to

(a) store a group of related binary digits

(b) provide random access data memory

(c) store a single bit of binary information.

(CCT)

Solution:

The answer is (a).

5. NUCLEAR ELECTRONICS (3066–3082)

3066

A coaxial transmission line has an impedance of 50 Ω which changes suddenly to 100 Ω. What is the sign of the pulse that returns from an initial positive pulse?

(a) none   (b) positive   (c) negative.

(CCT)

Solution:

The answer is (b).

3067

A positive pulse is sent into a transmission line which is short-circuited at the other end. The pulse reflected back:

(a) does not exist(= 0)

(b) is positive

(c) is negative.

(CCT)

Solution:

The answer is (c).

3068

What is the mechanism of discharge propagation in a self-quenched Geiger counter?

(a) Emission of secondary electrons from the cathode by UV-quanta.

(b) Ionization of the gas near the anode by UV-quanta.

(c) Production of metastable states and subsequent deexcitation.

(*CCT*)

Solution:

The answer is (c).

3069

For low noise charge-sensitive amplifier, FET-imput stages are preferred over bipolar transistors because:

(a) they have negligible parallel noise

(b) they are faster

(c) they have negligible series noise.

(*CCT*)

Solution:

The answer is (a).

3070

Using comparable technology, which ADC-type has the lowest value for the conversion time divided by the range, t_c/A, with t_c = conversion time and $A = 2^n$ with n = number of bits?

(a) flash ADC

(b) successive approximation converter

(c) Wilkinson converter.

(*CCT*)

Solution:

The answer is (a).

3071

A "derandomizer" is a circuit which consists of:

(a) trigger circuit

(b) FIFO memories

(c) phase locked loop.

(CCT)

Solution:

The answer is (c).

3072

A discriminator with a tunnel diode can be built with a threshold as low as:

(a) 1 mV   (b) 10 mV   (c) 100 mV.

(CCT)

Solution:

The answer is (c).

3073

Pulses with subnanosecond rise time and a few hundred volts amplitude can be produced using:

(a) avalance transistor

(b) thyratrons

(c) mechanical switches.

(CCT)

Solution:

The answer is (a).

3074

The square-box in Fig. 3.71 represents an unknown linear lumped-constant passive network. The source of emf at the left is assumed to have zero internal impedance.

It is known that if the input emf $e_i(t)$ is a step function, i.e.,

$$e_i(t) = \begin{cases} 0 & t \leq 0, \\ A & t > 0, \end{cases}$$

then the open-circuit (no-load) output voltage $e_0(t)$ is of the form

$$e_0(t) = \begin{cases} 0, & t \leq 0, \\ \frac{1}{2} A[1 - \exp(-t/\tau)], & t > 0, \end{cases}$$

where the constant τ has the value $\tau = 1.2 \times 10^{-4}$ s.

Find the open-circuit (no load) output voltage $e_0(t)$ when the input is given by

$$e_i(t) = 4\cos(\omega t) \text{ volts },$$

where ω corresponds to the frequency 1500 cycles/sec.

(UC, Berkeley)

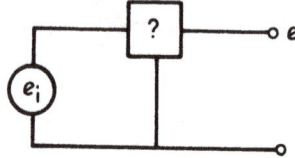

Fig. 3.71

Solution:

We first use the Laplace transform to find the transmission function $H(s)$ of the network in the frequency domain. The Laplace transform of the equation $e_i(t) = A \cdot U(t)$ is $E_i(s) = A/S$. Similarly, the Laplace transform of the output $e_0(t)$ is

$$E_0(s) = \frac{1}{2} A \left[\frac{1}{S} - \frac{1}{S + 1/\tau} \right].$$

Hence the transmission function is

$$H(s) = \frac{E_0(s)}{E_i(s)} = \frac{\frac{1}{2\tau}}{s + \frac{1}{\tau}}.$$

The Laplace transform of the new input $e_i(t) = 4\cos(\omega t)$ is

$$E_i(s) = \frac{4s}{\omega^2 + s^2},$$

giving the output as

$$E_0(s) = E_i(s) \cdot H_i(s) = \frac{4s}{\omega^2 + s^2} \times \frac{\frac{1}{2\tau}}{s + \frac{1}{\tau}}$$

$$= \frac{2}{\tau} \left[\frac{1}{s + \frac{1}{\tau}} \cdot \frac{-\frac{1}{\tau}}{\omega^2 + (\frac{1}{\tau})^2} + \frac{\frac{1}{2}}{s - i\omega} \times \frac{1}{i\omega + \frac{1}{\tau}} + \frac{\frac{1}{2}}{s + i\omega} \times \frac{1}{i\omega + \frac{1}{\tau}} \right],$$

where $\omega\tau = 2\pi \times 1500 \times 1.2 \times 10^{-4} \approx 1$. The reverse transformation of $E_0(s)$ gives the open-circuit output voltage

$$V_0(t) = \frac{2}{\tau}\left[\frac{-\frac{1}{\tau}}{\omega^2 + (\frac{1}{\tau})^2}e^{-\frac{t}{\tau}} + \frac{\frac{1}{\tau}\cos(\omega t) + \omega\sin(\omega t)}{\omega^2 + (\frac{1}{\tau})^2}\right]$$

$$\simeq -e^{-t/\tau} + \cos(\omega t) + \sin(\omega t) .$$

3075

To describe the propagation of a signal down a coaxial cable, we can think of the cable as a series of inductors, resistors and capacitors, as in Fig. 3.72(a). Thus, the cable is assigned an inductance, capacitance and resistance per unit length called L, C and R respectively. Radiation can be neglected.

(a) Show that the current in the cable, $I(x,t)$, obeys

$$\frac{\partial^2 I}{\partial x^2} = LC\frac{\partial^2 I}{\partial t^2} + RC\frac{\partial L}{\partial t} .$$

(b) Derive analogous equations for the voltage $V(x,t)$ and charge per unit length $\rho(x,t)$.

(c) What is the energy density (energy per unit length) on the cable? What is the energy flux? What is the rate of energy dissipation per unit length?

(d) Suppose that a semi-infinite length ($x \geq 0$) of this cable is coupled at $x = 0$ to an oscillator with frequency $\omega > 0$ so that

$$V(0,t) = \text{Re}(V_0 e^{i\omega t}) .$$

After the transients have decayed find the current $I(x,t)$. In the limit $R/L\omega \ll 1$ find the attenuation length and propagation speed of the signal.

(*MIT*)

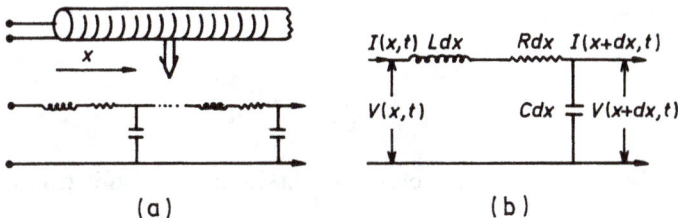

(a)   (b)

Fig. 3.72

Solution:

(a) From Fig. 3.72b, we have

$$\begin{cases} V(x,t) = V(x+dx,t) + RI\{t,x)dx + Ldx\,\frac{\partial I(t,x)}{\partial t} \\ I(x,t) = \frac{\partial V(x+dx,t)}{\partial t}Cdx + I\{x+dx,t), \end{cases}$$

or

$$\begin{cases} -\frac{\partial V}{\partial x} = IR + L\frac{\partial I}{\partial t}, \\ -\frac{\partial I}{\partial x} = C\frac{\partial V}{\partial t}. \end{cases}$$

Eliminating V we have

$$\frac{\partial^2 I}{\partial x^2} = -C\frac{\partial}{\partial x}\left(\frac{\partial V}{\partial t}\right) = -C\frac{\partial}{\partial t}\left(\frac{\partial V}{\partial x}\right)$$

$$= +C\frac{\partial}{\partial t}\left(IR + L\frac{\partial I}{\partial t}\right)$$

$$= RC\frac{\partial I}{\partial t} + LC\frac{\partial^2 I}{\partial t^2}.$$

(b) Similarly, eliminating I we have

$$\frac{\partial^2 V}{\partial x^2} = -R\frac{\partial I}{\partial x} - L\frac{\partial}{\partial x}\left(\frac{\partial I}{\partial t}\right)$$

$$= RC\frac{\partial V}{\partial t} - L\frac{\partial}{\partial t}\left(\frac{\partial I}{\partial x}\right)$$

$$= RC\frac{\partial V}{\partial t} + LC\frac{\partial^2 V}{\partial t^2}.$$

As

$$\rho dx = Cdx \cdot V, \qquad V = \rho/C,$$

the above then gives

$$\frac{\partial^2 \rho}{\partial x^2} = RC\frac{\partial \rho}{\partial t} + LC\frac{\partial^2 \rho}{\partial t^2}.$$

(c) The energy and rate of energy dissipation per unit length are respectively

$$W = \frac{1}{2}LI^2 + \frac{1}{2}CV^2, \qquad P = I^2 R.$$

The energy flux is

$$\mathbf{S} = IV\mathbf{e}_x .$$

(d) As the wave is sinusoidal, let

$$V = V_0 \exp[i(kx - \omega t)] .$$

Substitution in the differential equation for V gives

$$k^2 = LC\omega^2 + iRC\omega .$$

Since k is complex, putting $k = K + i\lambda$ and equating the real and imaginary parts separately we have

$$K^2 - \lambda^2 = LC\omega^2 ,$$
$$2K\lambda = RC\omega .$$

Solving these we obtain

$$K^2 = \frac{1}{2}\left(\sqrt{L^2C^2\omega^4 + R^2C^2\omega^2} + LC\omega^2\right) ,$$
$$\lambda^2 = \frac{1}{2}\left(\sqrt{L^2C^2\omega^4 + R^2C^2\omega^2} - LC\omega^2\right) .$$

As V is sinusoidal, so is I. Hence the equation $-\frac{\partial I}{\partial x} = C\frac{\partial V}{\partial t}$ gives

$$I = \frac{\omega C}{k}V = I_0 e^{-\lambda x} \exp\left[i(Kx - \omega t + \varphi_0)\right] ,$$

where

$$I_0 = \frac{C\omega V_0}{\sqrt{K^2 + \lambda^2}} , \qquad \varphi_0 = \arctan\left(\frac{K}{\lambda}\right) .$$

Actually $I(x,t) = \text{Re}\, I = I_0 e^{-\lambda x} \cos(Kx - \omega t + \varphi_0)$. In the limit $R/L\omega \ll 1$

$$K^2 = LC\omega^2 , \quad \lambda^2 = \frac{R^2C}{4L} .$$

So the attenuation length is

$$\frac{1}{\lambda} = \frac{2\sqrt{L/C}}{R} ,$$

and the propagation speed is

$$v = \frac{\omega}{K} = \frac{1}{\sqrt{LC}} \cdot$$

3076

A pulse generator of negligible internal impedance sends a pulse for which $V = 0$ at $t < 0$ and $t > 5$ μs, and $V = 1$ volt for $0 < t < 5$ μs into a lossless coaxial cable of characteristic impedance 20 ohms. The cable has a length equivalent to a delay of 1 μs and the end opposite the generator is open-circuited. Calculate (taking into account reflections at both ends of the cable) the form of the voltage-pulse at the open-circuited end of the cable for the time interval $t = 0$ to $t = 12$ μs. How much energy is supplied by the generator to the cable?

(Columbia)

Solution:

The reflection coefficient ρ is given by $\rho = \frac{Z_l - Z_0}{Z_l + Z_0}$, where $Z_0 = 20$ Ω. At the generator end, $Z_l = 0$ and

$$\rho_i = \frac{-20}{20} = -1 .$$

At the open-circuit end, $Z_l = \infty$ and

$$\rho_f = \frac{\infty - 20}{\infty + 20} = +1 .$$

Let the voltages at time t at the generator and open-circuit ends be v_t and V_t respectively. Then

$$v_t = v_{in} + \rho_i \rho_f v_{t-2}$$
$$= v_{in} - v_{t-2} ,$$
$$V_t = v_{t-1} + \rho_f v_{t-1} = 2v_{t-1} ,$$

where

$$v_{in} = 0 \quad \text{for} \quad t < 0 \quad \text{and} \quad t > 5 \ \mu s$$
$$= 1 \text{ V} \quad \text{for} \quad t = 1 - 5 \ \mu s .$$

Hence we have

t (μs)	0	1	2	3	4	5	6	7	8	9	10	11	12	13	14	...
V_t (V)	0	2	2	0	0	2	0	−2	0	2	0	−2	0	2	0	...

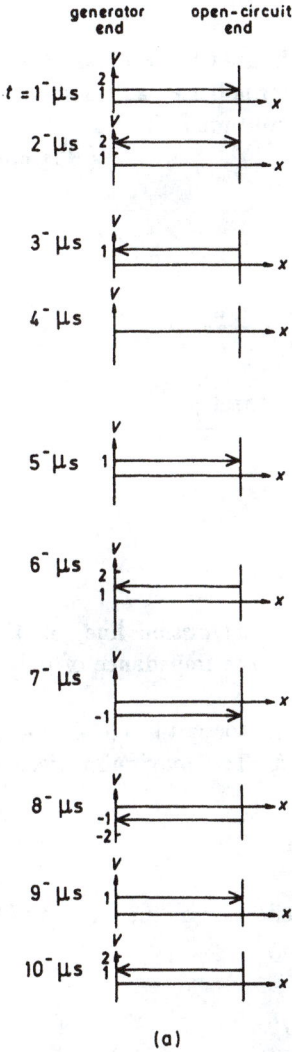

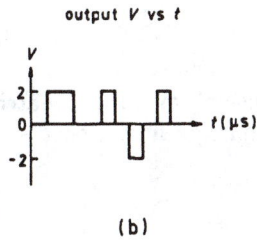

Fig. 3.73

The corresponding waveforms along the line are given in Fig. 3.73(a) for times just before each second and the output voltage as a function of time is given in Fig. 3.73(b).

3077

The emitter follower shown in Fig. 3.74 is used to drive fast negative pulses down a 50 Ω coaxial cable. If the emitter is biased at +3 volts, V_{out} is observed to saturate at -0.15 volt pulse amplitude. Why?

(*Wisconsin*)

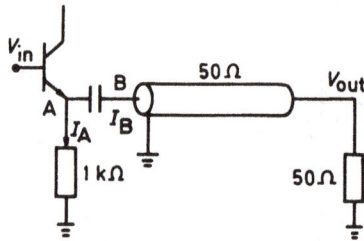

Fig. 3.74

Solution:

As the characteristic impedance of the transmission line, 50 Ω, is matched by the impedance at the output end, the impedance of point B with respect to earth is $R_B = 50$ Ω.

When a negative pulse is input, the transistor is turned off and the capacitor will be discharged through point A. The maximum discharge current is

$$I_A = \frac{3}{1000} = 3 \text{ mA} .$$

Because of impedance matching, there is no reflection at the far end of the transmission line. Hence

$$I_B = I_A = 3 \text{ mA} ,$$

and $V_{out} = -3$ mA $\times$ 50 $\Omega = -0.15$ V at saturation.

3078

A coaxial transmission line has a characteristic impedance of 100 ohms. A wave travels with a velocity of 2.5×10^8 m/s on the transmission line.

(a) What is the capacitance per meter and the inductance per meter?

(b) A voltage pulse of 15 V magnitude and 10^{-8} s duration is propagating on the cable. What is the current in the pulse?

(c) What is the energy carried in the pulse?

(d) If the pulse encounters another pulse of the opposite voltage magnitude but going in the opposite direction, what happens to the energy at the moment the two pulses cross so that the voltage everywhere is zero?

(*Wisconsin*)

Solution:

As $v = \frac{1}{\sqrt{LC}}$, $Z_c = \sqrt{L/C}$, we have

$$C = \frac{1}{vZ_c} = \frac{1}{2.5 \times 10^8 \times 100} = 4 \times 10^{-11} \text{ F/m} = 40 \text{ pF/m} ,$$

$$L = \frac{Z_c}{v} = \frac{100}{2.5 \times 10^8} = 0.4 \text{ mH/m} .$$

(b) The magnitude of the current in the pulse is

$$I_0 = \frac{v}{Z_c} = \frac{15}{100} = 0.15 \text{ A} .$$

(c) The energy carried in the pulse is distributed over the coaxial transmission line in the form of electric and magnetic fields. The line length is

$$l = vt = 2.5 \times 10^8 \times 10^{-8} = 2.5 \text{ m} ,$$

so the field energies are

$$W_e = \frac{1}{2}(Cl) \cdot V^2 = \frac{1}{2}(4 \times 10^{-11} \times 2.5) \times 15^2 = 1.125 \times 10^{-8} \text{ J} ,$$

$$W_m = \frac{1}{2}(Ll)I^2 = \frac{1}{2} \times 4 \times 10^{-7} \times 2.5 \times 0.15^2 = 1.125 \times 10^{-8} \text{ J} ,$$

giving

$$W = W_e + W_m = 2.25 \text{ J} .$$

(d) When the two pulses encounter each other, their voltages cancel out and the currents add up, giving $V' = 0$, $I' = 2I = 0.3$ A. Where there is no encounter, $V = 15$ V, $I = 0.15$ A. Where the pulses encounter electric energy is converted into magnetic energy. The more encounters occur, the more conversion of energy will take place.

3079

A lossless coaxial electrical cable transmission line is fed a step function voltage $V = 0$ for $t < 0$, $V = 1$ volt for $t > 0$. The far end of the line is an open circuit and a signal takes 10 μs to traverse the line.

(a) Calculate the voltage vs time for $t = 0$ to 100 μs at the open circuit end.

(b) Repeat for an input pulse $V = 1$ volt for $0 \le t \le 40$ μs, $V = 0$ otherwise.

(*Columbia*)

Solution:

Suppose the input end is matched, then the coefficient of reflection is

$$K = \begin{cases} 0 & \text{at input end,} \\ 1 & \text{at open circuit end.} \end{cases}$$

At the open circuit end,

$$\begin{aligned} V(t) &= V_i(t - 10) + K V_i(t - 10) \\ &= 2V_i(t - 10) . \end{aligned}$$

(a) As

$$\begin{aligned} V_i(t - 10) &= 0 \quad \text{for} \quad t < 10 \ \mu\text{s} , \\ &= 1 \text{ V for} \quad t > 10 \ \mu\text{s} , \end{aligned}$$

$V(t)$ is as shown in Fig. 3.75(a).

(b) As

$$\begin{aligned} V_i(t - 10) &= 0 \quad \text{for} \quad t < 10 \ \mu\text{s} , \\ &= 1 \text{ V} \quad \text{for} \quad 10 \le t \le 50 \ \mu\text{s} , \\ &= 0 \quad \text{for} \quad t > 50 \ \mu\text{s} , \end{aligned}$$

$V(t)$ is as shown in Fig. 3.75(b).

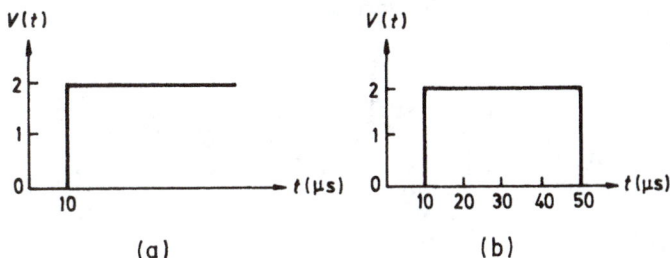

Fig. 3.75

3080

In Fig. 3.76(a) the transistor at A is normally ON so that the potential at A is normally very close to 0 V. Descibe and explain what you would see on an oscilloscope at points A and B if the transistor is turned OFF within a time < 1 ns. (Assume that the 5 volt supply has a low ac impedance to ground.)

(*Wisconsin*)

Solution:

The arrangements in Figs. 3.76(a) and (b) are equivalent so that at the input end $Z_s = 80\ \Omega$, $V_s = 4$ V. The reflection coefficients at the two ends are

$$K_{\rm B} = \frac{Z_s - Z_0}{Z_s + Z_0} = \frac{80 - 240}{80 + 240} = -0.5\,,$$

$$K_{\rm A\,ON} = \frac{Z_H - Z_0}{Z_H + Z_0} \approx \frac{0 - 240}{0 + 240} = -1\,,$$

$$K_{\rm A\,OFF} = \frac{\infty - 240}{\infty + 240} = 1\,.$$

Take $t = 0$ at the instant the transistor is turned off. When the transistor is ON, the voltage at B due to the source is

$$4 \times \frac{240}{240 + 80} = 3\ \text{V}\,.$$

Because of reflection at A,

$$V_{\rm B}(0^-) = 3 - 3 = 0\,.$$

We also have $V_A(0^-) = 0$. The waveform at $t < 0$ is as shown in Fig. 3.76(c).

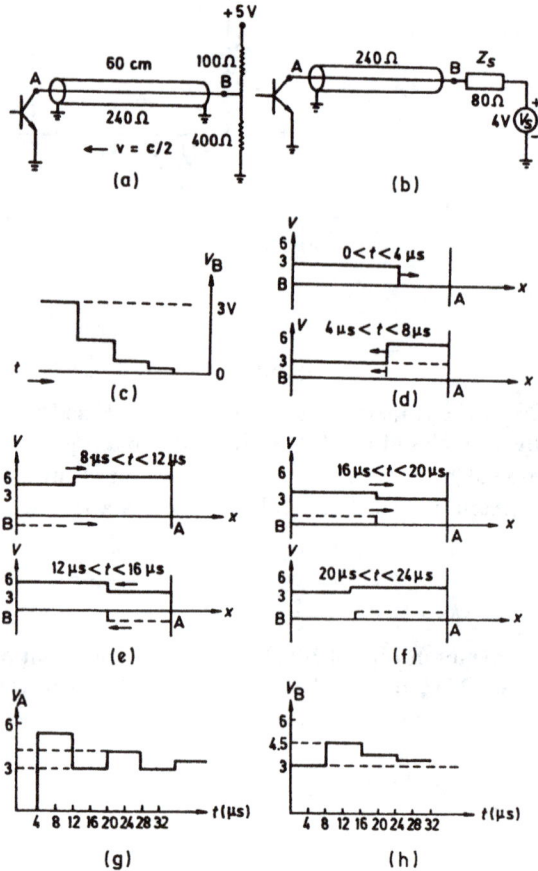

Fig. 3.76

When $t > 0$, the transistor is turned off and the circuit is open at point A. At that instant $K_{A\,OFF} = 1$, $V_B(O^+) = 3$ V. This is equivalent to a jump pulse of 3 V being input through point B at $t = 0$. Thereafter the voltage waveforms are as given in Figs. (d)–(f), where a single-pass transmission is taken to be 4 μs, the dotted lines denote reflected waves and the solid lines denote the sum of forward and reflected waves. Hence the voltage waveforms at points A and B as seen on an oscilloscope are as shown in Figs. (g) and (h).

3081

(a) In order to make a "charge sensitive amplifier", one can connect a capacitance across an ideal inverting amplifier as indicated by Fig. 3.77(a).

The triangular symbol represents an ideal inverting amplifier with the characteristics: input impedance $\gg 1$, output impedance $\ll 1$, gain $\gg 1$, and output voltage $V_{\text{out}} = -(\text{gain } G) \times (\text{input voltage } V_{\text{in}})$. Compute the output voltage as a function of the input charge.

(b) It is common practice when interconnecting electronic equipment for handling short-pulsed electrical signals to use coaxial cable terminated in its characteristic impedance. For what reason might one terminate the input end, the output end or both ends of such a coaxial cable?

(c) The following circuit (Fig. 3.77(b)) is used to generate a short, high-voltage pulse. How does it work? What is the shape, amplitude and duration of the output pulse?

(*Princeton*)

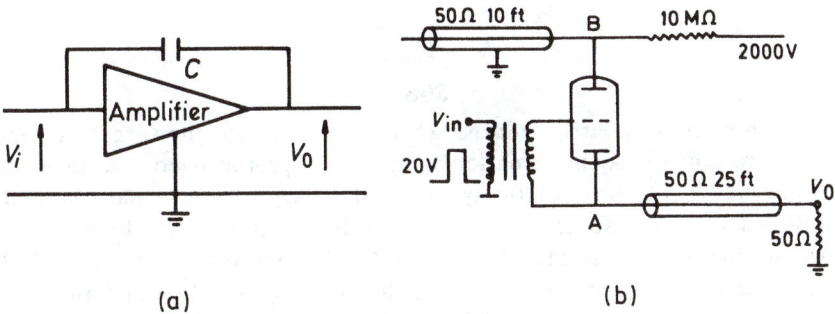

(a)   (b)

Fig. 3.77

Solution:

(a) As

$$V_0 = -GV_i , \qquad C(V_i - V_0) = Q ,$$

we have

$$V_0 = -\frac{G}{1+G} \cdot \frac{Q}{C} \approx -\frac{Q}{C} ,$$

since $G \gg 1$.

(b) Let Z_0 and Z_l be the characteristic and load impedances respectively. The reflection coefficient is

$$\rho = \frac{Z_l - Z_0}{Z_l + Z_0} .$$

Thus reflection normally takes place at the end of the delay line unless $\rho = 0$, i.e., $Z_0 = Z_l$, and the line is said to be matched. In order that the signal is not disturbed by the reflection, the ends of the line must be matched.

(c) When a positive pulse is applied to the input end V_{in}, the thyratron conducts and the potential at point A will be the same as at point B so that a potential drop of 2000 V is produced, generating a negative high-voltage pulse at the output end V_0. The width of the pulse is determined by the upper delay line in the open circuit to be

$$t_w = 2\tau = \frac{2 \times 10 \times 30.48}{3 \times 10^{10}} \approx 20 \ \mu s \ .$$

The amplitude of the output pulse is given by the voltage drop across the matching resistance of the lower delay line to be

$$\frac{2000 \times 50}{50 + 50} = 1000 \ V \ .$$

3082

The pions that are produced when protons strike the target at Fermilab are not all moving parallel to the initial proton beam. A focusing device, called a "horn", (actually two of them are used as a pair) is used to deflect the pions so as to cause them to move more closely towards the proton beam direction. This device (Fig. 3.78(a)) consists of an inner cylindrical conductor along which a current flows in one direction and an outer cylindrical conductor along which the current returns. Between these two surfaces there is produced a toroidal magnetic field that deflects the mesons that pass through this region.

(a) At first, calculate the approximate inductance of this horn using the dimensions shown in the figure. The current of charged pions and protons is negligible compared with the current in the conductors.

(b) The current is provided by a capacitor bank ($C = 2400 \ \mu F$), that is discharged (at an appropriate time before the pulse of protons strikes the target) into a transmission line that connects the two horns. The total inductance of both horns and the transmission line is 3.8×10^{-6} henries as in Fig. 3.78(b). In the circuit the charged voltage of the capacitor is $V_0 = 14$ kV and the resistance is $R = 8.5 \times 10^{-3}$ ohm. How many seconds after the switch is thrown does it take for the current to reach its maximum value?

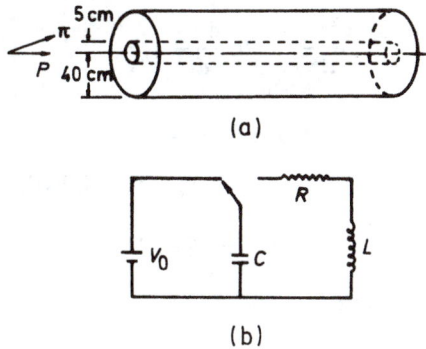

(a)

(b)

Fig. 3.78

(c) What is the maximum current in amps?

(d) At this time, what is the value of the magnetic field at a distance of 15 cm from the axis?

(e) By what angle would a 100 GeV/c meson be deflected if it traversed 2 meters of one horn's magnetic field at very nearly this radius of 15 cm?

(UC, Berkeley)

Solution:

(a) The magnetic induction at a point between the cylinders distance r from the axis is in the e_θ direction and has magnitude

$$B = \frac{\mu_0 I}{2\pi r} \ ,$$

I being the current in the inner conductor. The magnetic flux crossing a longitudinal cross section of a unit length of the horn is

$$\phi = 2 \int_{0.05}^{0.4} B\,dr \approx \frac{\mu_0 I}{\pi} \int_{0.05}^{0.4} \frac{1}{r}dr = 8.3 \times 10^{-7} \cdot I \ .$$

Hence the inductance is approximately

$$L = \frac{\phi}{I} \approx 8.3 \times 10^{-7} \text{ H} \ .$$

(b) Let the current of the RCL loop be $i(t)$. We have

$$u_C + u_L + u_R = 0$$

with

$$I = C\frac{du_C}{dt}\ , \quad u_R = RC\frac{du_C}{dt}\ , \quad u_L = L\frac{dI}{dt} = LC\frac{d^2 u_C}{dt^2}\ ,$$

i.e.,

$$LC\frac{d^2 u_C}{dt^2} + RC\frac{du_C}{dt} + u_C = 0 ,$$

and the initial condition

$$u_C(0) = V_0 .$$

To solve the equation for u_C, let $u_C = u_0 e^{-i\omega t}$. Substituting, we have

$$\omega = -i\alpha \pm \omega_d ,$$

where $\omega_d = \sqrt{\omega_0^2 - \alpha^2}$ with

$$\omega_0 = \frac{1}{\sqrt{LC}} ,$$

$$\alpha = \frac{R}{2L} .$$

Thus

$$u_C = u_0 e^{-\alpha t \pm i\omega_d t} ,$$

or

$$I = Cu_0(-\alpha \pm i\omega_d)e^{-\alpha t \pm i\omega_d t} .$$

With the data given, we have

$$\alpha = \frac{R}{2L} = 1.118 \times 10^3 \text{ s}^{-1} ,$$

$$\omega_0 = \frac{1}{\sqrt{LC}} = 1.047 \times 10^5 \text{ s}^{-1} ,$$

so that $\omega_0 \gg \alpha$ and $\omega_d \approx \omega_0$. Hence the current in the loop is

$$\begin{aligned}
I(t) &\approx \text{Re}[\mp iC\omega_0 V_0 e^{-\alpha t} e^{\pm i\omega_0 t}] \\
&= -C\omega_0 V_0 e^{-\alpha t} \sin(\omega_0 t) \\
&= -3.52 \times 10^6 e^{-1118t} \sin(1.047 \times 10^5 t) .
\end{aligned}$$

For maximum $I(t)$,

$$\frac{dI(t)}{dt} = 0 ,$$

i.e.,

$$\tan(\omega_0 t) \approx \omega_0 t = \frac{\omega_0}{\alpha} ,$$

giving

$$t = \frac{1}{\alpha} \ .$$

Therefore the current is maximum at $t = 8.94 \times 10^{-4}$ s.

(c)

$$I_{max} = 3.52 \times 10^6 e^{-1118 \times 8.94 \times 10^{-4}} \sin(1.047 \times 10^5 \times 8.94 \times 10^{-4})$$
$$= 1.29 \times 10^6 \text{ A} \ .$$

(d) At $r = 15$ cm, we have

$$B = \frac{\mu_0 I}{2\pi r} = \frac{4\pi \times 10^{-7} \times 1.29 \times 10^6}{2\pi \times 0.15} = 1.72 \text{ wbm}^{-2} \ .$$

(e) As $p = \frac{m_0 v}{\sqrt{1 - \frac{v^2}{c^2}}} = 100$ GeV/c, we have

$$\frac{m_0 v c}{\sqrt{1 - \frac{v^2}{c^2}}} = 10^{11} \text{ eV} \gg m_0 c^2 = 1.4 \times 10^8 \text{ eV}$$

for the meson so that we can take its speed to be

$$v \approx c \ .$$

The deflecting force is

$$F = eBv = 1.6 \times 10^{-19} \times 0.41 \times 3 \times 10^8 = 2.0 \times 10^{-11} \text{ N} \ ,$$

so the deflected transverse distance is

$$x = \frac{1}{2} a t^2 = \frac{1}{2} \frac{F}{m} \left(\frac{1}{c}\right)^2$$
$$= \frac{1}{2} \times \frac{2.0 \times 10^{-11}}{100 \times 10^9 \times 1.6 \times 10^{-19}/c^2} \cdot \frac{4}{c^2} = 0.0025 \text{ m}$$
$$= 2.5 \text{ mm} \ .$$

The angle of deflection is

$$\theta = \arctan\left(\frac{0.0025}{2}\right) = 0.0013 \text{ rad} \ .$$

6. MISCELLANEOUS PROBLEMS (3083–3090)

3083

Consider the circuit shown in Fig. 3.79(a).

(a) When $V_{in} = \text{Re}\{V_0 e^{i\omega t}\}$, find an expression for the complex V_{out}.

(b) Under what condition is the ratio V_{out}/V_{in} independent of ω? (It may be useful to recall Thévenin's thereom.)

(c) If V_{in} consists of a single "rectangular" pulse as shown in Fig. 3.79(b), sketch V_{out} (as a function of t) when the condition mentioned in (b) is satisfied.

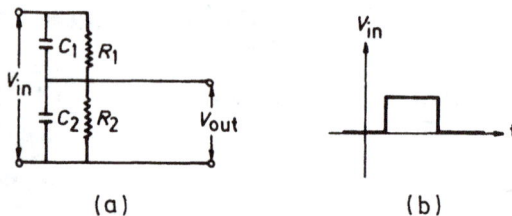

(a)   (b)

Fig. 3.79

(d) For a "rectangular" pulse V_{in} in (c), qualitatively sketch $V_{out}(t)$ when the condition mentioned in (b) is not satisfied.

(CUSPEA)

Solution:

(a) According to Thévenin's theorem, we can use two equivalent circuits to replace the capacitive and resistance networks as shown in Figs. 3.80(a) and (b). Connecting their output ends together we obtain the total equivalent circuit shown in Fig. 3.80(c) or Fig. 3.80(d).

For the circuit Fig. 3.80(d), using Kirchhoff's law we have

$$I\left[\frac{R_1 R_2}{R_1 + R_2} + \frac{1}{j\omega(C_1 + C_2)}\right] = \left[\frac{R_2}{R_1 + R_2} - \frac{C_1}{C_1 + C_2}\right]V_{in} ,$$

giving

$$I = \frac{j\omega[R_2(C_1 + C_2) - C_1(R_1 + R_2)]}{R_1 + R_2 + j\omega(C_1 + C_2)R_1 R_2} V_{in} .$$

The output voltage is

$$V_{out} = \frac{I}{j\omega(C_1 + C_2)} + \frac{C_1}{C_1 + C_2} V_{in} ,$$

hence we have the ratio

$$\frac{V_{\text{out}}}{V_{\text{in}}} = \frac{C_1}{C_1 + C_2} + \frac{R_2(C_1 + C_2) - C_1(R_1 + R_2)}{(C_1 + C_2)[R_1 + R_2 + j\omega R_1 R_2(C_1 + C_2)]} ,$$

giving V_{out} in terms of V_{in}.

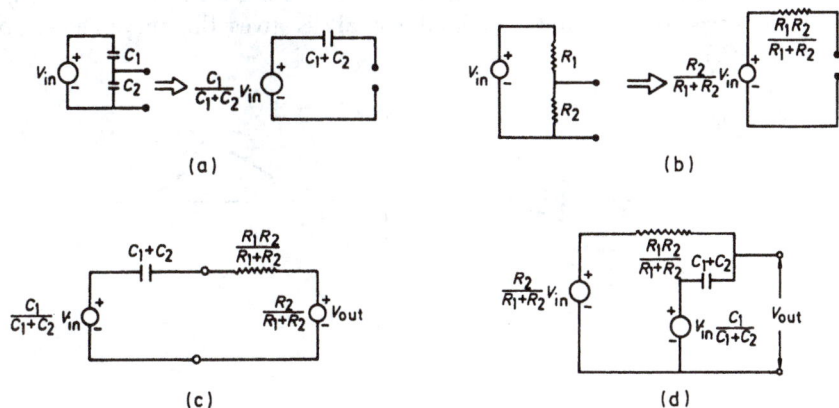

(a)

(b)

(c)

(d)

Fig. 3.80

(b) In the equivalent circuit Fig. 3.80(d), if the two sources of voltage are the same, there will be no current flowing, i.e. $I = 0$, giving

$$\frac{R_2}{R_1 + R_2} = \frac{C_1}{C_1 + C_2} ,$$

or $R_1 C_1 = R_2 C_2$. Then

$$\frac{V_{\text{out}}}{V_{\text{in}}} = \frac{C_1}{C_1 + C_2} = \frac{R_2}{R_1 + R_2} .$$

This ratio is independent of ω. Hence $R_1 C_1 = R_2 C_2$ is the necessary condition for $V_{\text{out}}/V_{\text{in}}$ to be independent of ω.

(c) When $R_1 C_1 = R_2 C_2$ is satisfied, $V_{\text{out}} = \frac{C_1}{C_1 + C_2} V_{\text{in}}$ for all frequencies. This is shown in Fig. 3.81.

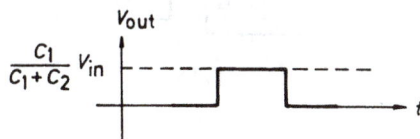

Fig. 3.81

(d) When the condition mentioned in (b) is not satisfied, $\frac{C_1}{C_1+C_2} - \frac{R_2}{R_1+R_2} = \frac{C_1R_1-C_2R_2}{(C_1+C_2)(R_1+R_2)}$. First consider the case $R_1C_1 > R_2C_2$. The attenuation in the capacitive voltage divider is less than in the resistive voltage divider. Hence when the rectangular pulse passes through the circuit, the former takes priority immediately; thereafter the output relaxes to that given by the latter. The variation of V_{out} with t is shown in Fig. 3.82(a). For the case $R_1C_1 < R_2C_2$, a similar analysis gives the curve shown in Fig. 3.82(b).

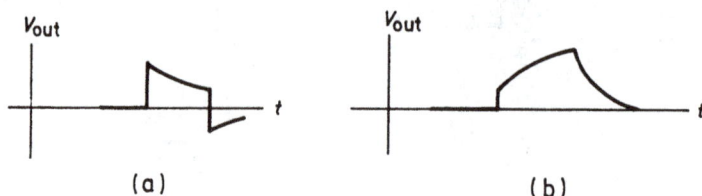

Fig. 3.82

3084

An electric circuit consists of two resistors (resistances R_1 and R_2), a single condenser (capacitor C) and a variable voltage source V joined together as shown in Fig. 3.83.

(a) When $V(t) = V_0 \cos \omega t$, what is the amplitude of the voltage drop across R_1?

(b) When $V(t)$ is a very sharp pulse at $t = 0$, we approximate $V(t) = A\delta(t)$. What is the time history of the potential drop across R_1?

(CUSPEA)

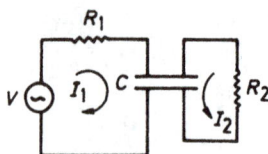

Fig. 3.83

Solution:

(a) Let the complex voltage be

$$\tilde{V} = V_0 e^{-i\omega t} \ .$$

Kirchhoff's equations for loops 1 and 2 are respectively

$$\tilde{V} = \tilde{I}_1 R_1 + \frac{1}{i\omega C}(\tilde{I}_1 + \tilde{I}_2) \ , \tag{1}$$

$$O = \tilde{I}_2 R_2 + \frac{1}{i\omega C}(\tilde{I}_1 + \tilde{I}_2) \ . \tag{2}$$

Eq. (2) gives

$$\left(R_2 - \frac{i}{\omega C}\right)\tilde{I}_2 = \frac{i}{\omega C}\tilde{I}_1 \ .$$

Its substitution in (1) gives

$$\tilde{I}_1 = \frac{(R_2 - \frac{i}{\omega C})}{R_1 R_2 - \frac{i}{\omega C}(R_1 + R_2)}\tilde{V} \ .$$

The voltage drop through resistance R_1 is

$$\tilde{V}_1 = \tilde{I}_1 R_1 = \frac{R_1(R_2 - \frac{i}{\omega C})}{R_1 R_2 - \frac{i}{\omega C}(R_1 + R_2)}\tilde{V} \ ,$$

so the real voltage drop through R_1 is

$$V_1 = \sqrt{\frac{1 + (\omega R_2 C)^2}{(\omega R_1 R_2 C)^2 + (R_1 + R_2)^2}} \ R_1 V_0 \cos(\omega t + \varphi) \ ,$$

where

$$\varphi = \arctan\left[\frac{\omega C R_2^2}{R_1 R_2\ ^2\omega^2 C^2 + (R_1 + R_2)}\right] \ .$$

(b) When $V(t) = A\delta(t)$, we use the relation

$$\delta(t) = \frac{1}{2\pi}\int_{-\infty}^{\infty} e^{i\omega t} d\omega$$

and write the voltage drop through R_1 as

$$V_1 = \frac{A}{2\pi} \int_{-\infty}^{\infty} \frac{R_1(R_2 - \frac{i}{\omega C})}{R_1 R_2 - \frac{i}{\omega C}(R_1 + R_2)} e^{i\omega t} d\omega$$

$$= \frac{A}{2\pi} \int_{-\infty}^{\infty} \frac{(\omega - \frac{i}{R_2 C})}{(\omega - \omega_1)} e^{i\omega t} d\omega ,$$

where $\omega_1 = i\frac{R_1 + R_2}{CR_1 R_2}$. The integrand has a singular point at $\omega = \omega_1$. Using the residue theorem we find the solution $V_1 \propto \exp(i\omega_1 t) = \exp(-\frac{R_1 + R_2}{CR_1 R_2}t)$. Hence V_1 is zero for $t < 0$ and

$$V_1 \propto \exp\left(-\frac{R_1 + R_2}{CR_1 R_2}t\right)$$

for $t > 0$.

3085

A semi-infinite electrical network is formed from condensers C and inductances L, as shown in Fig. 3.84. The network starts from the left at the terminals A and B; it continues infinitely to the right. An alternating voltage $V_0 \cos \omega t$ is applied across the terminals A and B and this causes a current to flow through the network. Compute the power P, averaged over a cycle, that is fed thereby into the circuit. The answer will be quantitatively different in the regimes $\omega > \omega_0$, $\omega < \omega_0$, where ω_0 is a certain critical frequency formed out of C and L.

(*CUSPEA*)

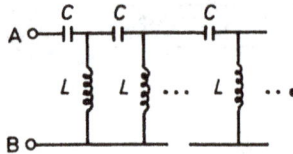

Fig. 3.84

Solution:

As the applied voltage is sinusoidal, the complex voltage and current are respectively

$$\tilde{V} = V_0 e^{i\omega t} , \qquad \tilde{I} = I_0 e^{i\omega t} .$$

The average power in a period is

$$\overline{P} = \frac{1}{2} \operatorname{Re}(\tilde{V}^* \tilde{I}) = \frac{1}{2} \operatorname{Re}\left(\frac{\tilde{V}^* \tilde{V}}{Z}\right) = \frac{V_0^2}{2} \operatorname{Re}\left(\frac{1}{Z}\right),$$

where the star $*$ denotes the complex conjugate and Z is the impedance of the circuit, $Z = \frac{\tilde{V}}{\tilde{I}}$. Let $Z_1 = \frac{1}{i\omega C}$, $Z_2 = i\omega L$, and assume any mutual inductance to be negligible. If L is the total impedance of the network, consider the equivalent circuit shown in Fig. 3.85 whose total impedance is still Z. Thus

$$Z = Z_1 + \frac{1}{\frac{1}{Z_2} + \frac{1}{Z}} = Z_1 + \frac{Z Z_2}{Z + Z_2},$$

or

$$Z^2 - Z_1 Z - Z_1 Z_2 = 0.$$

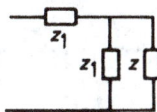

Fig. 3.85

As $Z > 0$, this equation has only one solution

$$Z = \frac{Z_1}{2} + \frac{\sqrt{Z_1^2 + 4 Z_1 Z_2}}{2}.$$

With $\frac{1}{2\sqrt{LC}} = \omega_0$ the solution becomes

$$Z = \frac{1}{2i\omega C}\left(1 + \sqrt{1 - \frac{\omega^2}{\omega_0^2}}\right).$$

For $\omega < \omega_0$, $\sqrt{1 - \frac{\omega^2}{\omega_0^2}}$ is a real number so that $\operatorname{Re}\left(\frac{1}{Z}\right) = 0$, i.e., $\overline{P} = 0$.
For $\omega > \omega_0$, $\operatorname{Re}\left(\frac{1}{Z}\right) = \frac{1}{2\omega L}\sqrt{\frac{\omega^2}{\omega_0^2} - 1}$ and

$$\overline{P} = \frac{V_0^2}{4\omega L}\sqrt{\frac{\omega^2}{\omega_0^2} - 1}.$$

3086

In the circuit shown in Fig. 3.86, L_1, L_2, and M are the self-inductances and mutual inductance of the windings of a transformer, R_1 and R_2 are the winding resistances, S is a switch and R is a resistive load in the secondary circuit. The input voltage is $V = V_0 \sin \omega t$.

(a) Calculate the amplitude of the current in the primary winding when the switch S is open.

(b) Calculate the amplitude of the steady-state current through R when S is closed.

(c) For an ideal transformer $R_1 = R_2 = 0$, and M, L_1, L_2 are simply related to N_1, N_2, the numbers of turns in the primary and secondary windings of the transformer. Putting these relations into (b), show that the results of (b) reduces to that expected from the turns ratio N_2/N_1 of the transformer.

(*CUSPEA*)

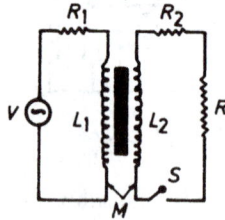

Fig. 3.86

Solution:

(a) When S is opened, we have

$$I_2 = 0, \quad I_1 = \frac{V_0}{\sqrt{R_1^2 + \omega^2 L_1^2}} .$$

(b) With S closed we have the circuit equations

$$V = I_1 R_1 + L_1 \frac{\partial I_1}{\partial t} + M \frac{\partial I_2}{\partial t}$$

$$0 = I_2 (R_2 + R) + L_2 \frac{\partial I_2}{\partial t} + M \frac{\partial I_1}{\partial t} .$$

As $V = V_0 \sin \omega t$, let

$$V = V_0 e^{-i\omega t}, \quad I_1 = I_{10} e^{-i\omega t}, \quad I_2 = I_{20} e^{-i\omega t} .$$

The circuit equations become

$$V = I_1(R_1 - i\omega L_1) - i\omega M I_2 ,$$
$$0 = -i\omega M I_1 + I_2[(R_2 + R) - i\omega L_2] .$$

Defining

$$\Delta = \begin{vmatrix} R_1 - i\omega L_1 & -i\omega M \\ -i\omega M & (R_2 + R) - i\omega L_2 \end{vmatrix}$$
$$= R_1(R_2 + R) + \omega^2(M^2 - L_1 L_2) - i\omega[L_1(R_2 + R) + L_2 R_1] ,$$

we have

$$I_2 = \frac{1}{\Delta} \begin{vmatrix} R_1 - i\omega L_1 & V \\ -i\omega M & 0 \end{vmatrix}$$
$$= \frac{i\omega M V}{\Delta} ,$$

and

$$I_{20} = \frac{\omega M V_0}{\sqrt{[\omega L_1(R + R_2) + \omega L_2 R_1]^2 + [\omega^2(M^2 - L_1 L_2) + R_1(R_2 + R)]^2}} .$$

(c) If the transformer is ideal, $R_1 = R_2 = 0$, $M^2 = L_1 L_2$, and we have

$$I_2 = \frac{\omega M V_0}{\omega L_1 R} = \frac{M V_0}{L_1 R} .$$

Then as $M \sim N_2 N_1$, $L \sim N_1^2$, we obtain

$$I_2 = \frac{N_2}{N_1} \frac{V_0}{R} .$$

This is just what is expected, namely the ideal transformer changes voltage V_0 into $\frac{N_2}{N_1} V_0$.

3087

Consider the circuit shown in Fig. 3.87.

(a) Find the impedance to a voltage V of frequency ω applied to the terminals.

(b) If one varies the frequency but not the amplitude of V, what is the maximum current that can flow? The minimum current? At what frequency will the minimum current be observed.

(*UC, Berkeley*)

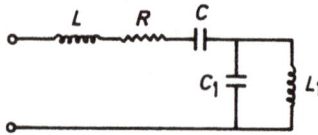

Fig. 3.87

Solution:

(a) The impedance is given by

$$Z(\omega) = j\omega L + R + \frac{1}{j\omega C} + \frac{\frac{1}{j\omega C_1} \cdot j\omega L_1}{\frac{1}{j\omega C_1} + j\omega L_1}$$

$$= R + j\omega L + \frac{1}{j\omega C} + \frac{j\omega L_1}{1 - \omega^2 L_1 C_1}$$

$$= R + j\left(\omega L - \frac{1}{\omega C} + \frac{\omega L_1}{1 - \omega^2 L_1 C_1}\right) .$$

(b) The complex current is

$$I = \frac{V}{Z} = \frac{V}{R + j(\omega L - \frac{1}{\omega C} + \frac{\omega L_1}{1 - \omega^2 L_1 C_1})} .$$

So its amplitude is

$$I_0 = \frac{V_0}{[R^2 + (\omega L - \frac{1}{\omega C} + \frac{\omega L_1}{1 - \omega^2 L_1 C_1})^2]^{1/2}} ,$$

where V_0 is the amplitude of the input voltage. Inspection shows that

$$(I_0)_{\max} = \frac{V_0}{R} , \qquad (I_0)_{\min} = 0 .$$

When I_0 is minimum, i.e., $I_0 = 0$,

$$\omega L - \frac{1}{\omega C} + \frac{\omega L_1}{1 - \omega^2 L_1 C_1} = \infty .$$

The solutions of this equation are $\omega = 0$, $\omega = \infty$, and $\omega = \frac{1}{\sqrt{L_1 C_1}}$. Discarding the first two solutions, we have $\omega = \frac{1}{\sqrt{L_1 C_1}}$ for the observation of the minimum current.

3088

In Fig. 3.88 a single-wire transmission (telegraph) line carries a current of angular frequency ω. The earth, assumed to be a perfect conductor, serves as the return wire. If the wire has resistance per unit length r, self-inductance per unit length l, and capacitance to ground per unit length C, find the voltage and current as functions of the length of the line.

(UC, Berkeley)

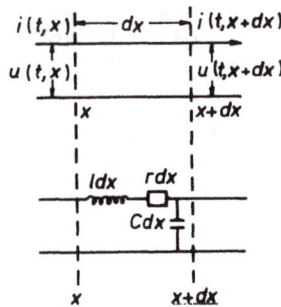

Fig. 3.88

Solution:

Take the origin at the starting point of the wire and its direction as the x direction and suppose the voltage amplitude at the starting point is V_0. Consider a segment x to $x + dx$. By Kirchhoff's law we have

$$u(t, x) = u(t, x + dx) + l dx \frac{\partial i(t, x)}{\partial t} + r i(t, x) dx ,$$

$$i(t, x) = i(t, x + dx) + C dx \frac{\partial u(t, x)}{\partial t} ,$$

i.e.

$$-\frac{\partial u}{\partial x} = l \frac{\partial i}{\partial t} + r i , \qquad -\frac{\partial i}{\partial x} = C \frac{\partial u}{\partial t} .$$

Assuming solution of the form $e^{-j(\omega t - Kx)}$, then

$$\frac{\partial}{\partial t} \sim -j\omega , \qquad \frac{\partial}{\partial x} \sim jK ,$$

and the above equations become

$$i(r - j\omega l) + jKu = 0 ,$$
$$i(jK) - j\omega Cu = 0 .$$

The condition that this system of equations has non-zero solutions is

$$\begin{vmatrix} r - j\omega l & jK \\ jK & -j\omega C \end{vmatrix} = -j\omega C(r - j\omega l) + K^2 = 0 ,$$

giving

$$K = \sqrt{\omega^2 lC + j\omega Cr} .$$

Let $K = \alpha + j\beta$, then

$$\alpha^2 - \beta^2 = \omega^2 lC ,$$
$$2\alpha\beta = \omega Cr ,$$

and we have

$$u = V_0 e^{-\beta x} e^{j(\alpha x - \omega t)}$$
$$i = \frac{\omega C}{K} u = \frac{\omega C V_0}{\sqrt{\alpha^2 + \beta^2}} e^{-\beta x} e^{j(\alpha x - \omega t + \varphi)} ,$$

where we have made use of the fact that $u = V_0$ when $x = t = 0$, and φ is given by

$$\tan\varphi = \frac{\beta}{\alpha} .$$

The expressions can be simplified if

$$r \ll \omega l ,$$

for we then have

$$K = \omega\sqrt{lC}\left(1 + j\frac{r}{\omega l}\right)^{\frac{1}{2}} \approx \omega\sqrt{lC} + j\frac{r}{2}\sqrt{\frac{C}{l}} .$$

Accordingly,

$$u = V_0 \exp[j\omega(\sqrt{lC}x - t)] \exp\left(\frac{-r}{2}\sqrt{\frac{C}{l}}x\right),$$

$$i = \frac{\omega C}{K}u \approx \sqrt{\frac{C}{l}}V_0 \exp\left[-j\omega(\sqrt{lC}x - t)\right] \exp\left(-\frac{r}{2}\sqrt{\frac{C}{l}}x\right).$$

3089

Consider two parallel perfect conductors of arbitrary but constant cross section (Fig. 3.89). A current flows down one conductor and returns via the other. Show that the product of the inductance per unit length, L, and the capacitance per unit length, C, is (in CGS units)

$$LC = \frac{\mu\varepsilon}{c^2},$$

where μ and ε are the permeability and dielectric constant of the medium surrounding the conductors and c is the velocity of light in vacuum.

(*Columbia*)

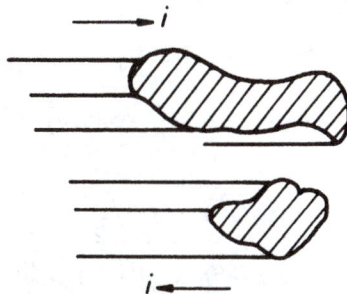

Fig. 3.89

Solution:

The conductors form a transmission line, which is equivalent to the circuit shown in Fig. 3.90.

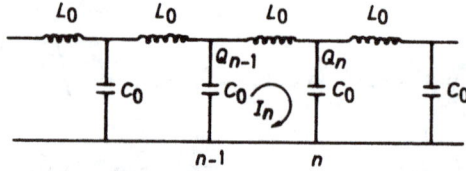

Fig. 3.90

Consider the n-th segment of the circuit. The following equations apply:

$$-L_0 \frac{dI_n}{dt} = \frac{Q_n}{C_0} - \frac{Q_{n-1}}{C_0} \; ,$$

$$I_{n-1} + \frac{dQ_{n-1}}{dt} = I_n \; ,$$

$$I_n + \frac{dQ_n}{dt} = I_{n+1} \; ,$$

from which we obtain

$$-L_0 \frac{d^2 I_n}{dt^2} = \frac{1}{C_0} \frac{d}{dt}(Q_n - Q_{n-1}) = \frac{1}{C_0}(I_{n+1} - I_n) - \frac{1}{C_0}(I_n - I_{n-1}) \; ,$$

or

$$L_0 C_0 \frac{d^2 I_n}{dt^2} = 2I_n - I_{n+1} - I_{n-1} \; .$$

Let $I_n = A_0 \cos(Kna - \omega t)$, where $K = \frac{\omega \sqrt{\mu \varepsilon}}{c}$, then the above gives

$$L_0 C_0 \omega^2 + 2 = -2 \cos(Ka) \; ,$$

or

$$L_0 C_0 \omega^2 = 4 \sin^2 \frac{Ka}{2} \; .$$

In the low frequency limit of $a \to 0$, $\sin(Ka/2) \sim \frac{Ka}{2}$ and we have

$$L_0 C_0 \omega^2 = K^2 a^2 \; .$$

As $\frac{\omega^2}{K^2} = \frac{c^2}{\mu \varepsilon}$, $\frac{L_0 C_0}{a^2} = \frac{\mu \varepsilon}{c^2}$. In this equation L_0/a and C_0/a denote the inductance and capacitance per unit length, respectively, of the transmission line. Replacing these by L and C, we obtain

$$LC = \mu \varepsilon / c^2 \; .$$

3090

Two circuits each contains a circular solenoid of length l, radius ρ ($\rho \ll l$), with N total turns. The solenoids are on the same axis, at distance d apart ($d \gg l$). The resistance of each circuit is R. Inductive efffects other than those associated with the solenoids are negligible.

(a) Calculate the self and mutual inductances of the circuits. Specify the appropriate units.

(b) Use L and M for the values found in (a). Calculate the magnitude and phase of the current which flows in the second circuit if an alternating emf of amplitude V, angular frequency ω is applied to the first. Assume ω is not too large.

(c) What is the order of magnitude to which ω can be increased before your calculation in (b) becomes invalid?

(UC, Berkeley)

Solution:

(a) As $l \gg \rho$, the magnetic induction inside the solenoid has

$$B = \mu_0 \frac{NI}{l}$$

and is along the axis of the solenoid. The magnetic flux linkage for the solenoid is

$$\Psi = NBS = N \cdot \mu_0 \frac{NI}{l} \cdot \pi \rho^2 = \frac{\mu_0 N^2 I \pi \rho^2}{l} \ ,$$

so the self-inductance is

$$L = \frac{\Psi}{I} = \frac{\pi \mu_0 N^2 \rho^2}{l} \ .$$

As $d \gg l$, the magnetic field produced by one solenoid at the location of the other can be approximated by that of a magnetic dipole. As the two solenoids are coaxial, this field may be expressed as $B_M = \frac{\mu_0}{4\pi} \cdot \frac{2m}{d^3}$ with $m = NI\pi\rho^2$, i.e.,

$$B_M = \frac{\mu_0 NI \rho^2}{2d^3} \ .$$

Hence

$$\Psi_M = NB_M S = N\frac{\mu_0 NI \rho^2}{2d^3} \pi \rho^2 = \frac{\mu_0 N^2 \rho^2 \pi \rho^2 I}{2d^3} \ ,$$

giving the mutual inductance as

$$M = \frac{\Psi_M}{I} = \frac{\pi \mu_0 N^2 \rho^4}{2d^3} \ .$$

The units of L and M are $H = A \cdot s/V$.

(b) Let the emf in the first circuit be $\varepsilon = V \cos \omega t = \mathrm{Re}\,(V e^{j\omega t})$. Then we have for the two circuits

$$L \frac{dI_1}{dt} + M \frac{dI_2}{dt} + I_1 R = V e^{j\omega t} \ ,$$

$$L \frac{dI_2}{dt} + M \frac{dI_1}{dt} + I_2 R = 0 \ .$$

As $I_1, I_2 \sim e^{j\omega t}$, we have $\frac{d}{dt} \rightarrow j\omega$ and the above equations become

$$j\omega L I_1 + j\omega M I_2 + I_1 R = V e^{j\omega t} \ , \tag{1}$$
$$j\omega L I_2 + j\omega M I_1 + I_2 R = 0 \ . \tag{2}$$

$(1) \pm (2)$ give

$$j\omega L(I_1 + I_2) + j\omega M(I_1 + I_2) + R(I_1 + I_2) = V e^{j\omega t} \ ,$$
$$j\omega L(I_1 - I_2) - j\omega M(I_1 - I_2) + R(I_1 - I_2) = V e^{j\omega t} \ .$$

Hence

$$I_1 + I_2 = \frac{V e^{j\omega t}}{j\omega(L + M) + R} \ , \qquad I_1 - I_2 = \frac{V e^{j\omega t}}{j\omega(L - M) + R} \ ,$$

whence

$$\begin{aligned}
I_2 &= \frac{1}{2} \left[\frac{V}{j\omega(L + M) + R} - \frac{V}{j\omega(L - M) + R} \right] e^{j\omega t} \\
&= \frac{-j\omega M V e^{j\omega t}}{[j\omega(L + M) + R][j\omega(L - M) + R]} \\
&= \frac{-j\omega M V e^{j\omega t}}{R^2 - \omega^2(L^2 - M^2) + 2j\omega LR} \ .
\end{aligned}$$

Writing $\mathrm{Re}\, I_2 = I_{20} \cos(\omega t + \varphi_0)$, we have

$$I_{20} = \frac{\omega M V}{\sqrt{[R^2 - \omega^2(L^2 - M^2)]^2 + 4\omega^2 L^2 R^2}} \ ,$$

$$\varphi_0 = \pi - \arctan \frac{2\omega L R}{R^2 - \omega^2(L^2 - M^2)} \ .$$

Using the given data and noting that $L \gg M$, we get

$$I_{20} \approx \frac{\omega M V}{R^2 + 2\omega^2 L^2} = \frac{\mu_0 \pi N^2 \rho^4 \omega V l^2}{2d^3 [R^2 l^2 + 2\omega^2 \mu_0^2 \pi^2 N^4 \rho^4]} \,,$$

$$\varphi_0 \approx \pi - \arctan \frac{2\omega L R}{R^2 - \omega^2 L^2}$$

$$= \pi - \arctan \frac{2\mu_0 \pi \omega R N^2 \rho^2 l}{R^2 l^2 - \omega^2 \mu_0^2 \pi^2 N^4 \rho^4} \,.$$

(c) The calculation in (b) is valid only under quasistationary conditions. This requires

$$d \ll \lambda = \frac{2\pi c}{\omega} \,,$$

or

$$\omega \ll \frac{2\pi d}{c} \,.$$

PART 4

ELECTROMAGNETIC WAVES

1. PLANE ELECTROMAGNETIC WAVES (4001–4009)

4001

The electric field of an electromagnetic wave in vacuum is given by

$$E_x = 0 \, ,$$

$$E_y = 30 \cos \left(2\pi \times 10^8 t - \frac{2\pi}{3} x \right) \, ,$$

$$E_z = 0 \, ,$$

where E is in volts/meter, t in seconds, and x in meters. Determine

(a) the frequency f,

(b) the wavelength λ,

(c) the direction of propagation of the wave,

(d) the direction of the magnetic field.

(*Wisconsin*)

Solution:

$$k = \frac{2\pi}{3} \text{ m}^{-1}, \quad \omega = 2\pi \times 10^8 \text{ s}^{-1} \, .$$

(a) $f = \dfrac{\omega}{2\pi} = 10^8$ Hz .

(b) $\lambda = \dfrac{2\pi}{k} = 3$ m .

(c) The wave is propagating along the positive x direction.

(d) As $\mathbf{E}$, $\mathbf{B}$, and $\mathbf{k}$ form a right-hand set, $\mathbf{B}$ is parallel to $\mathbf{k} \times \mathbf{E}$. As $\mathbf{k}$ and $\mathbf{E}$ are respectively in the x and y directions the magnetic field is in the z direction.

4002

The velocity of light c, and ε_0 and μ_0 are related by

(a) $c = \sqrt{\dfrac{\varepsilon_0}{\mu_0}}$; (b) $c = \sqrt{\dfrac{\mu_0}{\varepsilon_0}}$; (c) $c = \sqrt{\dfrac{1}{\varepsilon_0 \mu_0}}$.

(*CCT*)

Solution:

The answer is (c).

4003

Consider electromagnetic waves in free space of the form

$$E(x, y, z, t) = E_0(x, y)e^{ikz-i\omega t} ,$$

$$B(x, y, z, t) = B_0(x, y)e^{ikz-i\omega t} ,$$

where E_0 and B_0 are in the xy plane.

(a) Find the relation between k and ω, as well as the relation between $E_0(x, y)$ and $B_0(x, y)$. Show that $E_0(x, y)$ and $B_0(x, y)$ satisfy the equations for electrostatics and magnetostatics in free space.

(b) What are the boundary conditions for E and B on the surface of a perfect conductor?

(c) Consider a wave of the above type propagating along the transmission line shown in Fig. 4.1. Assume the central cylinder and the outer sheath are perfect conductors. Sketch the electromagnetic field pattern for a particular cross section. Indicate the signs of the charges and the directions of the currents in the conductors.

(d) Derive expressions for E and B in terms of the charge per unit length λ and the current i in the central conductor.

(SUNY, Buffalo)

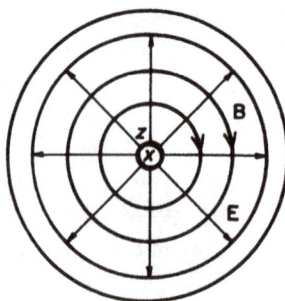

Fig. 4.1

Solution:

(a)

$$\nabla \times \mathbf{E} \equiv \begin{vmatrix} \mathbf{e}_x & \mathbf{e}_y & \mathbf{e}_z \\ \frac{\partial}{\partial x} & \frac{\partial}{\partial y} & \frac{\partial}{\partial z} \\ E_x & E_y & E_z \end{vmatrix} = \begin{vmatrix} \mathbf{e}_x & \mathbf{e}_y & \mathbf{e}_z \\ \frac{\partial}{\partial x} & \frac{\partial}{\partial y} & ik \\ E_{0x} & E_{0y} & 0 \end{vmatrix} e^{i(kz-\omega t)}$$

$$= \left[-ik\, E_{0y}\mathbf{e}_x + ik\, E_{0x}\mathbf{e}_y + \left(\frac{\partial E_{0y}}{\partial x} - \frac{\partial E_{0x}}{\partial y} \right) \mathbf{e}_z \right] e^{i(kz-\omega t)}$$

$$\stackrel{.}{=} \left[ik\mathbf{e}_z \times \mathbf{E}_0 + \nabla \times \mathbf{E}_0 \right] e^{i(kz-\omega t)} .$$

A similar expression is obtained for $\nabla \times \mathbf{B}$. Hence Maxwell's equations

$$\nabla \times \mathbf{E} = -\frac{\partial \mathbf{B}}{\partial t}, \qquad \nabla \times \mathbf{B} = \frac{1}{c^2}\frac{\partial \mathbf{E}}{\partial t}$$

can be written respectively as

$$ik\mathbf{e}_z \times \mathbf{E}_0(x,y) = i\omega\mathbf{B}_0(x,y) - \nabla \times \mathbf{E}_0 ,$$

$$ik\mathbf{e}_z \times \mathbf{B}_0(x,y) = -i\frac{\omega}{c^2}\,\mathbf{E}_0(x,y) - \nabla \times \mathbf{B}_0 .$$

Noting that $\nabla \times \mathbf{E}_0$ and $\nabla \times \mathbf{B}_0$ have only z-components while $\mathbf{e}_z \times \mathbf{E}_0$ and $\mathbf{e}_z \times \mathbf{B}_0$ are in the xy plane, we require

$$\nabla \times \mathbf{E}_0 = 0, \qquad \nabla \times \mathbf{B}_0 = 0, \tag{1}$$

so that

$$\mathbf{e}_z \times \mathbf{E}_0(x,y) = \frac{\omega}{k}\,\mathbf{B}_0(x,y), \tag{2}$$

$$\mathbf{e}_z \times \mathbf{B}_0(x,y) = -\frac{\omega}{kc^2}\,\mathbf{E}_0(x,y). \tag{3}$$

Taking the vector product of $\mathbf{e}_z$ and (2), we obtain

$$\mathbf{E}_0 = -\frac{\omega}{k}\mathbf{e}_z \times \mathbf{B}_0 .$$

Its substitution in (3) gives

$$\frac{\omega^2}{k^2 c^2} = 1,$$

or

$$k = \frac{\omega}{c} .$$

Equations (2) and (3) relate $\mathbf{E}_0$ and $\mathbf{B}_0$ and show that $\mathbf{E}_0$, $\mathbf{B}_0$, and $\mathbf{e}_z$ are mutually perpendicular forming a right-hand set. Furthermore, their amplitudes are related by

$$|E_0(x,y)| = \frac{\omega}{k}|B_0(x,y)| = c\,|B_0(x,y)|\,.$$

Maxwell's equations $\nabla \cdot \mathbf{E} = 0$, $\nabla \cdot \mathbf{B} = 0$ give

$$\nabla \cdot \mathbf{E}_0 = 0\,, \qquad \nabla \cdot \mathbf{B}_0 = 0\,. \tag{4}$$

Equations (1) and (4) show that $\mathbf{E}_0(x,y)$ and $\mathbf{B}_0(x,y)$ satisfy the equations for electrostatics and magnetostatics in free space.

(b) The boundary conditions for the surface of a perfect conductor are

$$\mathbf{n} \times \mathbf{E} = 0\,, \qquad \mathbf{n} \cdot \mathbf{D} = 0\,,$$

$$\mathbf{n} \times \mathbf{H} = \mathbf{I}_l\,, \qquad \mathbf{n} \cdot \mathbf{B} = 0\,,$$

where $\mathbf{n}$ is the outward normal unit vector at the conductor surface and $\mathbf{I}_l$ is the linear current density (current per unit width) on the conductor surface.

(c) For a particular cross section at $z = z_0$ and at a particular instant $t = t_0$, the electric field is $\mathbf{E}_0(x,y)\exp[i(kz_0 - \omega t_0)]$. Since $\mathbf{E}_0(x,y)$ satisfies the electrostatic equations the electric field is the same as that between oppositely charged coaxial cylindrical surfaces. Thus the lines of $\mathbf{E}_0(x,y)$ are radial. The magnetic induction satisfies (2), i.e.

$$\mathbf{B}_0(x,y) = \frac{1}{c}\mathbf{e}_z \times \mathbf{E}_0(x,y)\,,$$

so that magnetic lines of force will form concentric circles around the cylindrical axis. Suppose at (z_0, t_0) the central cylinder carries positive charge and the outer sheath carries negative charge then $\mathbf{E}$ and $\mathbf{B}$ have directions as shown in Fig. 4.1. The linear current density on the surface of the central conductor is given by $\mathbf{I}_l = \mathbf{n} \times \mathbf{H}$. As $\mathbf{n}$ is radially outwards, the current in the central cylinder is along the $+z$ direction while that in the outer sheath is along the $-z$ direction.

Using Maxwell's integral equations (Gauss' flux theorem and Ampère's circuital law) we have

$$\mathbf{E} = \frac{\lambda}{2\pi\varepsilon_0 r}\mathbf{e}_r\,, \qquad \mathbf{B} = \frac{\mu_0 I}{2\pi r}\mathbf{e}_\theta\,,$$

which give the charge per unit length λ and current I carried by the central conductor. The relation between **E** and **B** gives $I = c\lambda$.

4004

Consider a possible solution to Maxwell's equations given by

$$\mathbf{A}(x,t) = \mathbf{A}_0 e^{i(\mathbf{K}\cdot\mathbf{x}-\omega t)}, \qquad \phi(\mathbf{x},t) = 0,$$

where **A** is the vector potential and ϕ is the scalar potential. Further suppose $\mathbf{A}_0$, **K** and ω are constants in space-time. Give, and interpret, the constraints on $\mathbf{A}_0$, **K** and ω imposed by each of the Maxwell's equations given below.

(a) $\nabla \cdot \mathbf{B} = 0$;   (b) $\nabla \times \mathbf{E} + \dfrac{1}{c}\dfrac{\partial \mathbf{B}}{\partial t} = 0$;

(c) $\nabla \cdot \mathbf{E} = 0$;   (d) $\nabla \times \mathbf{B} - \dfrac{1}{c}\dfrac{\partial \mathbf{E}}{\partial t} = 0$.

(*Columbia*)

Solution:

The Maxwell's equations given in this problem are in Gaussian units, which will also be used below. As $\mathbf{A} = \mathbf{A}_0 \exp[i(K_x x + K_y y + K_z z - \omega t)]$, we have $\frac{\partial}{\partial x} = i(K_y z - K_z y)$, $\frac{\partial}{\partial y} = i(K_z x - K_x z)$, $\frac{\partial}{\partial z} = i(K_x y - K_y x)$, or $\nabla \times = i\mathbf{K} \times$. Hence the electromagnetic field can be represented by

$$\mathbf{B} = \nabla \times \mathbf{A} = i\mathbf{K} \times \mathbf{A}_0 e^{i(\mathbf{K}\cdot\mathbf{x}-\omega t)},$$

$$\mathbf{E} = -\nabla\phi - \frac{1}{c}\frac{\partial \mathbf{A}}{\partial t} = -\frac{1}{c}\frac{\partial \mathbf{A}}{\partial t} = i\frac{\omega}{c}\mathbf{A}_0 e^{i(\mathbf{K}\cdot\mathbf{x}-\omega t)}.$$

(a) Since $\nabla \cdot \mathbf{B} = -\mathbf{K}\cdot(\mathbf{K} \times \mathbf{A}_0)e^{i(\mathbf{K}\cdot\mathbf{x}-\omega t)} \equiv 0$ identically, no constraint is imposed by $\nabla \cdot \mathbf{B} = 0$.

(b) As $\nabla \times \mathbf{E} + \frac{1}{c}\frac{\partial \mathbf{B}}{\partial t} = i\mathbf{K} \times \mathbf{E} - \frac{i\omega}{c}\mathbf{B} = -\frac{\omega}{c}\mathbf{K} \times \mathbf{A} + \frac{\omega}{c}\mathbf{K} \times \mathbf{A} \equiv 0$, no constraint is imposed by the equation.

(c) As $\nabla \cdot \mathbf{E} = \frac{i\omega}{c}\nabla \cdot \mathbf{A} = \frac{i\omega}{c}\mathbf{K} \cdot \mathbf{A} = 0$, we require $\mathbf{K} \cdot \mathbf{A} = 0$.

(d) As $\nabla \times \mathbf{B} - \frac{1}{c}\frac{\partial \mathbf{E}}{\partial t} = -\mathbf{K} \times (\mathbf{K} \times \mathbf{A}) + \frac{\omega^2}{c^2}\mathbf{A} = -(\mathbf{K} \cdot \mathbf{A})\mathbf{K} + K^2\mathbf{A} - \frac{\omega^2}{c^2}\mathbf{A} = (K^2 - \frac{\omega^2}{c^2})\mathbf{A} = 0$ we require $K^2 = \frac{\omega^2}{c^2}$ or $K = \pm\frac{\omega}{c}$. Therefore the constraints imposed by Maxwell's equations are

$$K = \omega/c, \quad \text{and} \quad \mathbf{K} \cdot \mathbf{A} = 0 .$$

The second constraint means that $\mathbf{K}$ is perpendicular to $\mathbf{A}$. Hence $\mathbf{K}$ is perpendicular to $\mathbf{E}$. As $\mathbf{K}$ is also perpendicular to $\mathbf{B}$, this constraint shows that the solution is a transverse wave. The first constraint gives $\frac{|\mathbf{B}|}{|\mathbf{E}|} = \frac{K}{\frac{\omega}{c}} = 1$, showing that the wave is a plane electromagnetic wave. The $\pm$ signs correspond to $\pm\mathbf{x}$ directions of propagation.

4005

Consider a plane wave with vector potential $A_\mu(x) = a_\mu e^{i(\mathbf{K}\cdot\mathbf{x}-\omega t)}$, where a_μ is a constant four-vector. Further suppose that $\mathbf{K} = K\mathbf{e}_z$ and choose a (non-orthogonal) set of basis vectors for a_μ:

$$\varepsilon^{(1)\mu} = (0, 1, 0, 0),$$

$$\varepsilon^{(2)\mu} = (0, 0, 1, 0),$$

$$\varepsilon^{(L)\mu} = \frac{1}{K}\left(\frac{\omega}{c}, 0, 0, K\right) = \frac{1}{K}K^\mu,$$

$$\varepsilon^{(B)\mu} = \frac{1}{K}\left(K, 0, 0, -\frac{\omega}{c}\right),$$

where $\varepsilon^\mu = (\varepsilon^0, \boldsymbol{\varepsilon})$. Write

$$a_\mu = a_1\varepsilon^{(1)\mu} + a_2\varepsilon^{(2)\mu} + a_L\varepsilon^{(L)\mu} + a_B\varepsilon^{(B)\mu} .$$

What constraints, if any, does one get on a_1, a_2, a_L, a_B from

(a) $\nabla \cdot \mathbf{B} = 0$,

(b) $\nabla \times \mathbf{E} + \frac{1}{c}\frac{\partial \mathbf{B}}{\partial t} = 0$,

(c) $\nabla \times \mathbf{B} - \frac{1}{c}\frac{\partial \mathbf{E}}{\partial t} = 0$,

(d) $\nabla \cdot \mathbf{E} = 0$?

(e) Which of the parameters a_1, a_2, a_L, a_B are gauge dependent?

(f) Give the average energy density in terms of a_1, a_2, a_L, a_B after imposing (a)–(d).

(Columbia)

Solution:

We are given the four-vectors

$$K^\mu = (\omega/c, 0, 0, K), \qquad A_\mu = (\varphi, A_x, A_y, A_z) .$$

With $\mathbf{K} = K\mathbf{e}_z$, $\mathbf{K} \cdot \mathbf{x} = Kz$. For plane waves we also have $K = \frac{\omega}{c}$. Then for $\mu = 1$, we have

$$\varphi = \left[a_L \frac{\omega}{Kc} + a_B \left(\frac{K}{K}\right)\right] e^{i(Kz-\omega t)} = (a_L + a_B) e^{i(Kz-\omega t)} .$$

Similarly for $\mu = 2, 3, 4$, we have

$$\mathbf{A} = [a_1 \mathbf{e}_x + a_2 \mathbf{e}_y + (a_L - a_B)\mathbf{e}_z] e^{i(Kz-\omega t)} .$$

Hence

$$\mathbf{B} = \nabla \times \mathbf{A} = iK\mathbf{e}_z \times \mathbf{A} = iK(-a_2\mathbf{e}_x + a_1\mathbf{e}_y) e^{i(Kz-\omega t)} ,$$

$$\begin{aligned}
\mathbf{E} &= -\nabla\varphi - \frac{1}{c}\frac{\partial \mathbf{A}}{\partial t} = -iK(a_L + a_B)\mathbf{e}_z e^{i(Kz-\omega t)} + iK\mathbf{A} \\
&= iK(a_1\mathbf{e}_x + a_2\mathbf{e}_y - 2a_B\mathbf{e}_z) e^{i(Kz-\omega t)} .
\end{aligned}$$

(a) As $\nabla \cdot \mathbf{B} = \nabla \cdot (\nabla \times \mathbf{A}) \equiv 0$ identically, it imposes no constraint.

(b) As $\nabla \times \mathbf{E} + \frac{1}{c}\frac{\partial \mathbf{B}}{\partial t} = -\nabla \times \nabla\varphi - \frac{1}{c}\frac{\partial}{\partial t}(\nabla \times \mathbf{A}) + \frac{1}{c}\frac{\partial}{\partial t}(\nabla \times \mathbf{A}) \equiv 0$ identically, it does not lead to any constraint.

(c) As

$$\nabla \times \mathbf{B} = iK\mathbf{e}_z \times \mathbf{B} = K^2(a_1\mathbf{e}_x + a_2\mathbf{e}_y) e^{i(Kz-\omega t)} ,$$

$$\frac{1}{c}\frac{\partial \mathbf{E}}{\partial t} = -iK\mathbf{E} = K^2(a_1\mathbf{e}_x + a_2\mathbf{e}_y - 2a_B\mathbf{e}_z) e^{i(Kz-\omega t)} ,$$

$$\nabla \times \mathbf{B} - \frac{1}{c}\frac{\partial \mathbf{E}}{\partial t} = 0$$

demands that $a_B = 0$.

(d) As

$$\nabla \cdot \mathbf{E} = \frac{\partial E_z}{\partial z} = 2K^2 a_B e^{i(Kz-\omega t)} ,$$

$\nabla \cdot \mathbf{E} = 0$ also requires $a_B = 0$.

(e) Since a_1 and a_2 are not involved in the gauge equation $\nabla \cdot \mathbf{A} = 0$ for the Coulomb gauge and $\nabla \cdot \mathbf{A} + \frac{1}{c} \frac{\partial \varphi}{\partial t} = 0$ for the Lorentz gauge, a_1, a_2 are gauge-independent.

(f) The average energy density is

$$U = \frac{1}{16\pi} (|E|^2 + |B|^2) = \frac{K^2}{8\pi} (a_1^2 + a_2^2) .$$

4006

A plane wave of angular frequency ω and wave number $|\mathbf{K}|$ propagates in a neutral, homogeneous, anisotropic, non-conducting medium with $\mu = 1$.

(a) Show that $\mathbf{H}$ is orthogonal to $\mathbf{E}$, $\mathbf{D}$ and $\mathbf{K}$, and also that $\mathbf{D}$ and $\mathbf{H}$ are transverse but $\mathbf{E}$ is not.

(b) Let $D_k = \sum_{l=1}^{3} \varepsilon_{kl} E_l$, where ε_{kl} is a real symmetric tensor. Choose the principal axes of ε_{kl} as a coordinate system ($D_k = \varepsilon_k E_k$; $k = 1, 2, 3$). Define $\mathbf{K} = K\hat{S}$, where the components of the unit vector $\hat{S}$ along the principal axes are S_1, S_2, and S_3. If $V = \omega/K$ and $V_j = c/\sqrt{\varepsilon_j}$, show that the components of $\mathbf{E}$ satisfy

$$S_j \sum_{i=1}^{3} S_i E_i + \left(\frac{V^2}{V_j^2} - 1 \right) E_j = 0 .$$

Write down the equation for the phase velocity V in terms of $\hat{S}$ and V_j. Show that this equation has two finite roots for V^2, corresponding to two distinct modes of propagation in the direction $\hat{S}$.

(*Wisconsin*)

Solution:

Use the Gaussian system of units.

(a) Maxwell's equations for the given medium are

$$\nabla \times \mathbf{E} = -\frac{1}{c} \frac{\partial \mathbf{B}}{\partial t} ,$$

$$\nabla \times \mathbf{B} = \frac{1}{c} \frac{\partial \mathbf{D}}{\partial t} ,$$

$$\nabla \cdot \mathbf{B} = 0\,,$$

$$\nabla \cdot \mathbf{D} = 0\,.$$

The plane wave can be represented by $\sim e^{i(\mathbf{K}\cdot\mathbf{x}-\omega t)}$, so that $\nabla\times \equiv i\mathbf{K}\times$, $\nabla\cdot \equiv i\mathbf{K}\cdot$, $\frac{\partial}{\partial t} = -i\omega$ and the above equations reduce to

$$\mathbf{K} \times \mathbf{E} = \frac{\omega}{c}\mathbf{H}\,,$$

$$\mathbf{K} \times \mathbf{H} = -\frac{\omega}{c}\mathbf{D}\,,$$

$$\mathbf{K} \cdot \mathbf{B} = \mathbf{K} \cdot \mathbf{D} = 0\,,$$

as $\mu = 1$. From these we have

$$\mathbf{D} \cdot \mathbf{H} = -\frac{c}{\omega}(\mathbf{K} \times \mathbf{H}) \cdot \mathbf{H} \equiv 0\,, \qquad \mathbf{K} \cdot \mathbf{H} = \frac{c}{\omega}\mathbf{K} \cdot \mathbf{K} \times \mathbf{E} \equiv 0\,.$$

Hence $\mathbf{K}$, $\mathbf{D}$, and $\mathbf{H}$ are mutually perpendicular, i.e., $\mathbf{D}$ and $\mathbf{H}$ are transverse to $\mathbf{K}$. However, as $\mathbf{K} \times (\mathbf{K} \times \mathbf{E}) = \frac{\omega}{c}\mathbf{K} \times \mathbf{H}$,

$$\mathbf{K} \cdot \mathbf{E} = \frac{1}{K}\left(\frac{\omega}{c}\mathbf{K} \times \mathbf{H} + K^2\mathbf{E}\right)$$

$$= \frac{1}{K}\left[-\left(\frac{\omega}{c}\right)^2\mathbf{D} + K^2\mathbf{E}\right] \neq 0$$

unless $K^2 = (\frac{\omega}{c})^2$, $\mathbf{E}$ need not be transverse to $\mathbf{K}$.

(b) From the above we have

$$\mathbf{K}(\mathbf{K} \cdot \mathbf{E}) - K^2\mathbf{E} = -\frac{\omega^2}{c^2}\mathbf{D}\,.$$

As

$$\mathbf{K} = K(S_1, S_2, S_3)\,,$$
$$\mathbf{D} = (\epsilon_1 E_1, \epsilon_2 E_2, \epsilon_3 E_3)\,,$$
$$\mathbf{E} = (E_1, E_2, E_3)\,,$$

The j-th component of the equation is

$$S_j K^2 \sum_i S_i E_i - K^2 E_j + \frac{\omega^2}{c^2}\epsilon_j E_j = 0\,.$$

Putting $\omega^2/K^2 = V^2$ and $c^2/\epsilon_j = V_j^2$, it becomes

$$S_j \sum_{i=1}^{3} S_i E_i + \left(\frac{V^2}{V_j^2} - 1\right) E_j = 0.$$

For $j = 1, 2, 3$, we have

$$\left(S^2 + \frac{V^2}{V_1^2} - 1\right) E_1 + S_1 S_2 E_2 + S_1 S_3 E_3 = 0,$$

$$S_1 S_2 E_1 + \left(S_2^2 - \frac{V^2}{V_2^2} - 1\right) E_2 + S_1 S_3 E_3 = 0,$$

$$S_3 S_1 E_1 + S_3 S_2 E_2 + \left(S_3^2 + \frac{V^2}{V_3^2} - 1\right) E_3 = 0.$$

The sufficient and necessary condition for a non-zero solution is that the determinant

$$\begin{vmatrix} S_1^2 + \frac{V^2}{V_1^2} - 1 & S_1 S_2 & S_1 S_3 \\ S_2 S_1 & S_2^2 + \frac{V^2}{V_2^2} - 1 & S_2 S_3 \\ S_3 S_1 & S_3 S_2 & S_3^2 + \frac{V^2}{V_3^2} - 1 \end{vmatrix} = 0.$$

This gives

$$V^2\left[\frac{V^4}{V_1^2 V_2^2 V_3^2} + (S_1^2 - 1)\frac{V^2}{V_2^2 V_3^2} + (S_2^2 - 1)\cdot \frac{V^2}{V_1^2 V_3^2}\right.$$
$$\left. + (S_3^2 - 1)\cdot\frac{V^2}{V_1^2 V_2^2} + \left(\frac{S_1^2}{V_1^2} + \frac{S_2^2}{V_2^2} + \frac{S_3^2}{V_3^2}\right)\right] = 0,$$

which can be solved to find two finite roots for V^2 if $V^2 \neq 0$.

From $V^2 = \omega^2/K^2$ we can find two values of K^2 corresponding to the two roots of V^2. This shows that there are two distinct modes of propagation.

4007

Four identical coherent monochromatic wave sources A, B, C, D, as shown in Fig. 4.2 produce waves of the same wavelength λ. Two receivers R_1 and R_2 are at great (but equal) distances from B.

(a) Which receiver picks up the greater signal?

(b) Which receiver, if any, picks up the greater signal if source B is turned off?

(c) if source D is turned off?

(d) Which receiver can tell which source, B or D, has been turned off?

(*Wisconsin*)

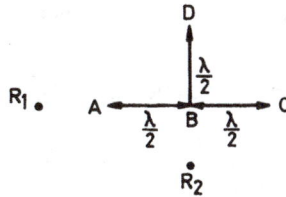

Fig. 4.2

Solution:

(a) Let r be the distance of R_1 and R_2 from B. We are given $r \gg \lambda$. Suppose the amplitude of the electric vector of the electromagnetic waves emitted by each source is E_0. The total amplitudes of the electric field at the receivers are

$$R_1: \quad E_{10} = E_0 \exp\left[iK\left(r - \frac{\lambda}{2}\right)\right] + E_0 e^{iKr} + E_0 \exp\left[iK\left(r + \frac{\lambda}{2}\right)\right]$$
$$+ E_0 \exp\left[iK\sqrt{r^2 + \frac{\lambda^2}{4}}\right],$$

$$R_2: \quad E_{20} = E_0 e^{iKr} + E_0 \exp\left[iK\left(r + \frac{\lambda}{2}\right)\right] + 2E_0 \exp\left[iK\sqrt{r^2 + \frac{\lambda^2}{4}}\right].$$

As $K\lambda = \frac{2\pi\nu\lambda}{c} = 2\pi$, $\exp[\pm i\frac{K\lambda}{2}] = e^{\pm i\pi} = -1$. With $r \gg \lambda$, $\sqrt{r^2 + \frac{\lambda^2}{4}} \approx r$. Thus

$$E_0 \exp\left[iK\sqrt{r^2 + \frac{\lambda^2}{4}}\right] \approx E_0 e^{iKr},$$

and we have

$$E_{10} \approx 0, \qquad E_{20} \approx 2E_0 e^{iKr}.$$

The intensity of a signal $I \propto |E|^2$, so the intensities received by R_1 and R_2 are respectively

$$I_0 = 0, \qquad I_2 \sim 4E_0^2 .$$

Hence R_2 picks up greater signal.

(b) If source B is turned off, then

$$E_{10} \approx -E_0 e^{iKr} , \qquad E_{20} \approx E_0 e^{iKr} .$$

Thus $I_1 = I_2 \sim E_{10}^2$, that is, the two receivers pick up signals of the same intensity.

(c) If source D is turned off, one has

$$E_{10} \approx -E_0 e^{iKr} , \qquad E_{20} \approx 3E_0 e^{iKr} ,$$

and

$$I_1 \sim E_{10}^2 , \qquad I_2 \sim 9E_0^2 .$$

Hence R_2 picks up greater signal.

(d) From the above, we can see that I_1 remains the same whether the sources B and D are on or off. Hence R_1 cannot determine the on–off state of B and D. On the other hand, the intensity of I_2 differs for the three cases above so the strength of the signal received by R_2 can determine the on–off state of the sources B and D.

4008

(a) Write down Maxwell's equations assuming that no dielectrics or magnetic materials are present. State your system of units. In all of the following you must justify your answer.

(b) If the signs of all the source charges are reversed, what happens to the electric and magnetic fields **E** and **B**?

(c) If the system is space inverted, i.e., $\mathbf{x} \to \mathbf{x}' = -\mathbf{x}$, what happens to the charge density and current density, ρ and $\mathbf{j}$, and to **E** and **B**?

(d) If the system is time reversed, i.e., $t \to t' = -t$, what happens to $\rho, \mathbf{j}, \mathbf{E}$ and **B**?

(SUNY, Buffalo)

Solution:

Use the MKSA system of units.

(a) In the absence of dielectric or magnetic materials Maxwell's equations are

$$\begin{cases} \nabla \cdot \mathbf{E} = \frac{\rho}{\varepsilon_0}, & \nabla \times \mathbf{E} = -\frac{\partial \mathbf{B}}{\partial t}, \\ \nabla \cdot \mathbf{B} = 0, & \nabla \times \mathbf{B} = \mu_0 \mathbf{j} + \frac{1}{c^2} \frac{\partial \mathbf{E}}{\partial t}. \end{cases}$$

(b) Under charge conjugation $e \to -e$, we have $\nabla \to \nabla' = \nabla$, $\frac{\partial}{\partial t} \to \frac{\partial}{\partial t'} = \frac{\partial}{\partial t}$, $\rho \to \rho' = -\rho$, $\mathbf{j} \to \mathbf{j}' = -\mathbf{j}$. Under this transformation Maxwell's equations remain the same:

$$\begin{cases} \nabla' \cdot \mathbf{E}' = \frac{\rho'}{\varepsilon_0}, & \nabla' \times \mathbf{E}' = -\frac{\partial \mathbf{B}'}{\partial t'}, \\ \nabla' \cdot \mathbf{B}' = 0, & \nabla' \times \mathbf{B}' = \mu_0 \mathbf{j}' + \frac{1}{c^2} \frac{\partial \mathbf{E}'}{\partial t'}. \end{cases}$$

A comparison of the first equations in (a) and (b), we see that, as $\rho' = -\rho$,

$$\mathbf{E}'(\mathbf{r}, t) = -\mathbf{E}(\mathbf{r}, t).$$

Substituting this in the fourth equation in (a), we see that

$$\nabla' \times \mathbf{B}' = \nabla \times \mathbf{B}' = -\mu_0 \mathbf{j} - \frac{1}{c^2} \frac{\partial \mathbf{E}}{\partial t}.$$

Hence

$$\mathbf{B}'(\mathbf{r}, t) = -\mathbf{B}(\mathbf{r}, t).$$

(c) Under space inversion

$$\mathbf{r} \to \mathbf{r}' = -\mathbf{r}, \qquad \nabla \to \nabla' = -\nabla,$$
$$\frac{\partial}{\partial t} \to \frac{\partial}{\partial t'} = \frac{\partial}{\partial t}, \qquad e \to e' = e.$$

Then

$$\rho(\mathbf{r}, t) \to \rho'(\mathbf{r}, t) = \rho, \qquad \mathbf{j} \to \mathbf{j}' = \rho' \mathbf{u}' = -\rho \mathbf{u} = -\mathbf{j},$$

$\mathbf{u}$ being the velocity of the charges in an elementary volume.

As Maxwell's equations remain the same under this transformation we have

$$\mathbf{E}'(\mathbf{r}, t) = -\mathbf{E}(\mathbf{r}, t), \qquad \mathbf{B}'(\mathbf{r}, t) = \mathbf{B}(\mathbf{r}, t).$$

(d) Under time reversal,

$$\frac{\partial}{\partial t} \rightarrow \frac{\partial}{\partial t'} = -\frac{\partial}{\partial t}, \qquad \nabla \rightarrow \nabla' = \nabla, \qquad e \rightarrow e' = e.$$

Then $\rho' = \rho$, $\mathbf{j}' = -\rho\mathbf{u} = -\mathbf{j}$, and we have from the covariance of Maxwell's equations that

$$\mathbf{E}'(\mathbf{r},\, t) = \mathbf{E}(\mathbf{r},\, t), \qquad \mathbf{B}'(\mathbf{r},\, t) = -\mathbf{B}(\mathbf{r},\, t).$$

4009

Let $\mathbf{A}_\omega$, ϕ_ω, $\mathbf{J}_\omega$ and ρ_ω be the temporal Fourier transforms of the vector potential, scalar potential, current density and charge density respectively. Show that

$$\phi_\omega(\mathbf{r}) = \frac{1}{4\pi\varepsilon_0} \int \rho_\omega(\mathbf{r}') \frac{e^{iK|\mathbf{r}-\mathbf{r}'|}}{|\mathbf{r}-\mathbf{r}'|}\, d^3r',$$

$$\mathbf{A}_\omega(\mathbf{r}) = \frac{\mu_0}{4\pi} \int \mathbf{J}_\omega(\mathbf{r}') \frac{e^{iK|\mathbf{r}-\mathbf{r}'|}}{|\mathbf{r}-\mathbf{r}'|}\, d^3r'. \qquad \left(K = \frac{|\omega|}{c}\right)$$

How is the law of charge-current conservation expressed in terms of ρ_ω and $\mathbf{j}_\omega$? In the far zone ($r \rightarrow \infty$) find the expressions for the magnetic and electric fields $\mathbf{B}_\omega(\mathbf{r})$ and $\mathbf{E}_\omega(\mathbf{r})$. Find these fields for a current density $\mathbf{J}_\omega(\mathbf{r}) = \mathbf{r}f(r)$.

(*Wisconsin*)

Solution:

If we express the current density $\mathbf{J}(\mathbf{r},\, t)$ as the Fourier integral

$$\mathbf{J}(\mathbf{r},\, t) = \int_{-\infty}^{\infty} \mathbf{J}_\omega(\mathbf{r})e^{-i\omega t}\, d\omega,$$

the retarded vector potential can be rewritten:

$$\mathbf{A}(\mathbf{r},\, t) = \frac{\mu_0}{4\pi} \int \frac{\mathbf{J}(\mathbf{r}',\, t - \frac{|\mathbf{r}-\mathbf{r}'|}{c})}{|\mathbf{r}-\mathbf{r}'|}\, d^3r'$$

$$= \frac{\mu_0}{4\pi} \int \frac{1}{|\mathbf{r}-\mathbf{r}'|}\, d^3r' \int_{-\infty}^{\infty} \mathbf{J}_\omega(\mathbf{r}') \exp\left[-i\omega\left(t - \frac{|\mathbf{r}-\mathbf{r}'|}{c}\right)\right] d\omega$$

$$= \frac{\mu_0}{4\pi} \int_{-\infty}^{\infty} e^{-i\omega t}\, d\omega \int \frac{\mathbf{J}_\omega(\mathbf{r}')e^{iK|\mathbf{r}-\mathbf{r}'|}}{|\mathbf{r}-\mathbf{r}'|}\, d^3r',$$

where the volume integral is over all space. Thus the Fourier transform of the vector potential is

$$\mathbf{A}_\omega(\mathbf{r}) = \frac{\mu_0}{4\pi} \int \frac{\mathbf{J}_\omega(\mathbf{r}') \exp\left[iK|\mathbf{r}-\mathbf{r}'|\right]}{|\mathbf{r}-\mathbf{r}'|} d^3r' \,.$$

Similarly the Fourier transform of the scalar potential $\phi(\mathbf{r}, t)$ is

$$\phi_\omega(\mathbf{r}) = \frac{1}{4\pi\varepsilon_0} \int \rho_\omega(\mathbf{r}') \frac{\exp[iK|\mathbf{r}-\mathbf{r}'|]}{|\mathbf{r}-\mathbf{r}'|} d^3r' \,.$$

The continuity equation that expresses charge-current conservation, $\frac{\partial \rho}{\partial t} + \nabla \cdot \mathbf{J} = 0$, is written in terms of Fourier integrals as

$$\frac{\partial}{\partial t} \int_{-\infty}^{\infty} \rho_\omega(\mathbf{r}) e^{-i\omega t} d\omega + \nabla \cdot \int_{-\infty}^{\infty} \mathbf{J}_\omega(\mathbf{r}) e^{-i\omega t} d\omega = 0 \,,$$

or

$$\int_{-\infty}^{\infty} [-i\omega\rho_\omega(\mathbf{r}) + \nabla \cdot \mathbf{J}_\omega(\mathbf{r})] e^{-i\omega t} d\omega = 0 \,.$$

Hence

$$\nabla \cdot \mathbf{J}_\omega - i\omega\rho_\omega = 0 \,.$$

In the far zone $r \to \infty$, we make the approximation

$$|\mathbf{r}-\mathbf{r}'| \approx r - \frac{\mathbf{r} \cdot \mathbf{r}'}{r} \,.$$

Then

$$\mathbf{J}_\omega \frac{e^{iK|\mathbf{r}-\mathbf{r}'|}}{|\mathbf{r}-\mathbf{r}'|} \approx \mathbf{J}_\omega \frac{e^{iKr}}{r} \left(1 - iK\frac{\mathbf{r} \cdot \mathbf{r}'}{r} - \cdots\right)\left(1 + \frac{\mathbf{r} \cdot \mathbf{r}'}{r^2} + \cdots\right)$$

$$\approx \mathbf{J}_\omega \frac{e^{iKr}}{r} \left(1 - iK\frac{\mathbf{r} \cdot \mathbf{r}'}{r}\right) \,.$$

Consider

$$\nabla' \cdot (x'\mathbf{J}_\omega) = J_{\omega x'} + x'\nabla \cdot \mathbf{J}_\omega \,.$$

As

$$\int \nabla' \cdot (x'\mathbf{J}_\omega) d^3r' = \oint x'\mathbf{J}_\omega \cdot d\mathbf{S}' = 0$$

for a finite current distribution,

$$\int \mathbf{J}_\omega d^3 r' = -\int \mathbf{r}' \nabla' \cdot \mathbf{J}_\omega d^3 r'$$

$$= -i\omega \int \mathbf{r}' \rho_\omega(\mathbf{r}') d^3 r' . \qquad (1)$$

Also

$$\mathbf{J}_\omega(\mathbf{r} \cdot \mathbf{r}') = \frac{1}{2}[\mathbf{J}_\omega(\mathbf{r} \cdot \mathbf{r}') - (\mathbf{r} \cdot \mathbf{J}_\omega)\mathbf{r}'] + \frac{1}{2}[\mathbf{J}_\omega(\mathbf{r} \cdot \mathbf{r}') + (\mathbf{r} \cdot \mathbf{J}_\omega)\mathbf{r}']$$

$$= \frac{1}{2}(\mathbf{r}' \times \mathbf{J}_\omega) \times \mathbf{r} + \frac{1}{2}[\mathbf{J}_\omega(\mathbf{r} \cdot \mathbf{r}') + (\mathbf{r} \cdot \mathbf{J}_\omega)\mathbf{r}'] .$$

The second term on the right-hand side would give rise to an electric quadrupole field. It is neglected as we are interested only in the lowest multipole field. Hence

$$\mathbf{A}_\omega(\mathbf{r}) \xrightarrow[r \to \infty]{} \frac{\mu_0}{4\pi}\left(-i\omega \mathbf{p}_\omega \frac{e^{iKr}}{r} - iK\mathbf{m}_\omega \times \frac{\mathbf{r}}{r} e^{iKr} \right) , \qquad (2)$$

where

$$\mathbf{p}_\omega = \int \mathbf{r}' \rho_\omega(\mathbf{r}') d^3 r' , \qquad \mathbf{m}_\omega = \frac{1}{2}\int \mathbf{r}' \times \mathbf{J}_\omega(\mathbf{r}') d^3 r'$$

are the electric and magnetic dipole moments of the sources. To find $\mathbf{E}_\omega(\mathbf{r})$ in the far zone, we use Maxwell's equation

$$\nabla \times \mathbf{B} = \mu_0 \epsilon_0 \frac{\partial \mathbf{E}}{\partial t} + \mu_0 \mathbf{J} = \frac{1}{c^2} \frac{\partial \mathbf{E}}{\partial t}$$

assuming the source to be finite. In terms of Fourier transforms, the above becomes

$$\nabla \times \int \mathbf{B}_\omega(\mathbf{r}) e^{-i\omega t} d\omega = \frac{1}{c^2} \frac{\partial}{\partial t} \int \mathbf{E}_\omega(\mathbf{r}) e^{-i\omega t} d\omega ,$$

or

$$\int \nabla \times \mathbf{B}_\omega(\mathbf{r}) e^{-i\omega t} d\omega = -\int \frac{i\omega}{c^2} \mathbf{E}_\omega(\mathbf{r}) e^{-i\omega t} d\omega ,$$

giving

$$\mathbf{E}_\omega(\mathbf{r}) = \frac{ic^2}{\omega} \nabla \times \mathbf{B}_\omega .$$

Similarly, from $\mathbf{B} = \nabla \times \mathbf{A}$ we have

$$\mathbf{B}_\omega(\mathbf{r}) = \nabla \times \mathbf{A}_\omega(\mathbf{r}) .$$

For a current density $\mathbf{J}_\omega(r) = \mathbf{r}f(r)$, we have

$$\mathbf{m}_\omega = \frac{1}{2} \int \mathbf{r}' \times \mathbf{r}'f(r')d^3r' = 0 .$$

Also,

$$\mathbf{p}_\omega = \int \mathbf{r}'\rho_\omega(\mathbf{r}')d^3r' = \frac{i}{\omega} \int \mathbf{J}_\omega(\mathbf{r}')d^3r' = \frac{i}{\omega} \int \mathbf{r}'f(r')d^3r' ,$$

using (1) and assuming the current distribution to be finite.

Hence, using (2),

$$\mathbf{A}_\omega(\mathbf{r}) \approx \frac{\mu_0 e^{iKr}}{4\pi r} \int \mathbf{r}'f(r')d^3r' .$$

Then

$$\mathbf{B}_\omega(\mathbf{r}) = \nabla \times \mathbf{A}_\omega(\mathbf{r}) \approx \frac{i\mu_0 K e^{iKr}}{4\pi r} \hat{\mathbf{r}} \times \int \mathbf{r}'f(r')d^3r' ,$$

$$\mathbf{E}_\omega(\mathbf{r}) = \frac{ic^2}{\omega}\nabla \times \mathbf{B}_\omega(\mathbf{r})$$

$$\approx -ic\frac{\mu_0 K e^{iKr}}{4\pi r} \hat{\mathbf{r}} \times \left[\hat{\mathbf{r}} \times \int \mathbf{r}'f(r')d^3r'\right] ,$$

where $\hat{\mathbf{r}} = \frac{\mathbf{r}}{r}$, terms of higher orders in $\frac{1}{r}$ having been neglected.

2. REFLECTION AND REFRACTION OF ELECTROMAGNETIC WAVES ON INTERFACE BETWEEN TWO MEDIA (4010–4024)

4010

(a) Write down Maxwell's equations in a non-conducting medium with constant permeability and susceptibility ($\rho = \mathbf{j} = 0$). Show that $\mathbf{E}$ and $\mathbf{B}$ each satisfies the wave equation, and find an expression for the wave velocity. Write down the plane wave solutions for $\mathbf{E}$ and $\mathbf{B}$ and show how $\mathbf{E}$ and $\mathbf{B}$ are related.

(b) Discuss the reflection and refraction of electromagnetic waves at a plane interface between the dielectrics and derive the relationships between the angles of incidence, reflection and refraction.

(SUNY, Buffalo)

Solution:

(a) Maxwell's equations in a source-free, homogeneous non-conducting medium are

$$
\begin{cases}
\nabla \times \mathbf{E} = -\frac{\partial \mathbf{B}}{\partial t}, & (1) \\
\nabla \times \mathbf{H} = \frac{\partial \mathbf{D}}{\partial t}, & (2) \\
\nabla \cdot \mathbf{D} = 0, & (3) \\
\nabla \cdot \mathbf{B} = 0, & (4)
\end{cases}
$$

where $\mathbf{D} = \varepsilon \mathbf{E}$, $\mathbf{B} = \mu \mathbf{H}$, ε, μ being constants. As

$$
\nabla \times (\nabla \times \mathbf{E}) = \nabla(\nabla \cdot \mathbf{E}) - \nabla^2 \mathbf{E} = -\nabla^2 \mathbf{E}
$$

and Eq. (2) can be written as

$$
\nabla \times \left(\frac{\partial \mathbf{B}}{\partial t}\right) = \mu\varepsilon \frac{\partial^2 \mathbf{E}}{\partial t^2},
$$

Eq. (1) gives

$$
\nabla^2 \mathbf{E} - \mu\varepsilon \frac{\partial^2 \mathbf{E}}{\partial t^2} = 0.
$$

Similarly, one finds

$$
\nabla^2 \mathbf{B} - \mu\varepsilon \frac{\partial^2 \mathbf{B}}{\partial t^2} = 0.
$$

Thus each of the field vectors $\mathbf{E}$ and $\mathbf{B}$ satisfies the wave equation. A comparison with the standard wave equation $\nabla^2 \mathbf{E} - \frac{1}{v^2} \frac{\partial^2 \mathbf{E}}{\partial t^2} = 0$ shows that the wave velocity is

$$
v = \frac{1}{\sqrt{\varepsilon\mu}}.
$$

Solutions corresponding to plane electromagnetic waves of angular frequency ω are

$$
\mathbf{E}(\mathbf{r}, t) = \mathbf{E}_0 e^{i(\mathbf{k}\cdot\mathbf{r} - \omega t)},
$$

$$
\mathbf{B}(\mathbf{r}, t) = \mathbf{B}_0 e^{i(\mathbf{k}\cdot\mathbf{r} - \omega t)},
$$

where the wave vector $\mathbf{k}$ and the amplitudes $\mathbf{E}_0$ and $\mathbf{B}_0$ form an orthogonal right-hand set. Furthermore $v = \frac{\omega}{k}$, and $\mathbf{E}$, $\mathbf{B}$ are related by

$$
\mathbf{B} = \sqrt{\mu\varepsilon}\, \frac{\mathbf{k}}{k} \times \mathbf{E}.
$$

(b) The boundary condition that the tangential component of **E**, $\mathbf{n} \times \mathbf{E} = \mathbf{E}_t$, **n** being a unit normal to the interface, is continuous across the interface requires that in general there will be a reflected and a refracted wave in addition to the incident wave at the interface. Furthermore experiments show that if the incident electromagnetic wave is a plane wave, $\mathbf{E}(\mathbf{r}, t) = \mathbf{E}_0 e^{i(\mathbf{k}\cdot\mathbf{r}-\omega t)}$, the reflected and refracted waves are also plane waves, which are represented respectively by

$$\mathbf{E}' = \mathbf{E}_0 e^{i(\mathbf{k}'\cdot\mathbf{x}-\omega' t)},$$

$$\mathbf{E}'' = \mathbf{E}_0'' e^{i(\mathbf{k}''\cdot\mathbf{x}-\omega'' t)}.$$

The boundary condition at the interface then gives

$$\mathbf{n} \times \left[\mathbf{E}_0 e^{i(\mathbf{k}\cdot\mathbf{r}-\omega t)} + \mathbf{E}_0' e^{i(\mathbf{k}'\cdot\mathbf{r}-\omega' t)} \right] = \mathbf{n} \times \mathbf{E}_0'' e^{i(\mathbf{k}''\cdot\mathbf{r}-\omega'' t)}.$$

This means that all the exponents in the equation must be the same, i.e.,

$$\omega = \omega' = \omega'', \qquad \mathbf{k} \cdot \mathbf{r} = \mathbf{k}' \cdot \mathbf{r} = \mathbf{k}'' \cdot \mathbf{r}.$$

Thus frequency is not changed by reflection and refraction.

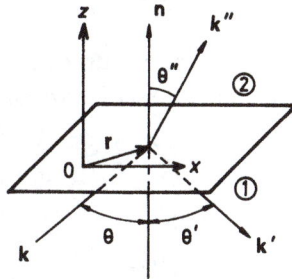

Fig. 4.3

Choose the origin on the interface so that the position vector **r** of the point where the incident wave strikes the interface is perpendicular to **k**. We then have

$$\mathbf{k} \cdot \mathbf{r} = \mathbf{k}' \cdot \mathbf{r} = \mathbf{k}'' \cdot \mathbf{r} = \mathbf{n} \cdot \mathbf{r} = 0. \tag{5}$$

This means that **k**, **k**', **k**'' and **n** are coplanar. Hence reflection and refraction occur in the vertical plane containing the incident wave, called the plane of incidence. Now choose a coordinate system with the origin at an arbitrary point O on the interface, the x-axis parallel to the incidence plane, i.e. the plane of $\mathbf{k}$ and $\mathbf{n}$, the z-axis parallel to the normal $\mathbf{n}$, and let θ, θ' and θ'' respectively be the angles of incidence, reflection and refraction, measured from the normal, as shown in Fig. 4.3. Then

$$\mathbf{r} = (x, y, 0),$$

$$\mathbf{k} = \omega\sqrt{\mu_1\varepsilon_1}\,(\sin\theta,\, 0,\, \cos\theta).$$

$$\mathbf{k}' = \omega\sqrt{\mu_1\varepsilon_1}\,(\sin\theta',\, 0,\, -\cos\theta').$$

$$\mathbf{k}'' = \omega\sqrt{\mu_2\varepsilon_2}\,(\sin\theta'',\, 0,\, \cos\theta'').$$

Equation (5) gives for arbitrary x and ω

$$\sqrt{\mu_1\varepsilon_1}\,\sin\theta = \sqrt{\mu_1\varepsilon_1}\,\sin\theta' = \sqrt{\mu_2\varepsilon_2}\,\sin\theta''.$$

Hence

$$\theta = \theta',$$

i.e., the angle of incidence is equal to that of reflection. This is called the law of reflection. We also have

$$\frac{\sin\theta}{\sin\theta''} = \frac{\sqrt{\mu_2\varepsilon_2}}{\sqrt{\mu_1\varepsilon_1}} = \frac{n_2}{n_1} = n_{21},$$

where $n \propto \sqrt{\varepsilon\mu}$ is called the index of refraction of a medium and n_{21} is the index of refraction of medium 2 relative to medium 1. This relation is known as the law of refraction.

4011

A plane electromagnetic wave of intensity I falls upon a glass plate with index of refraction n. The wave vector is at right angles to the surface (normal incidence).

(a) Show that the coefficient of reflection (of the intensity) at normal incidence is given by $R = \frac{(n-1)^2}{(n+1)^2}$ for a single interface.

(b) Neglecting any interference effects calculate the radiation pressure acting on the plate in terms of I and n.

(Chicago)

Solution:

The directions of the wave vectors of the incident, reflected and refracted waves are shown in Fig. 4.4. For normal incidence, $\theta = \theta' = \theta'' = 0$. Let the incident, reflected and refracted electromagnetic waves be represented respectively by

$$E = E_0 e^{i(\mathbf{k}\cdot\mathbf{x}-\omega t)},$$

$$E' = E_0' e^{i(\mathbf{k'}\cdot\mathbf{x}-\omega t)}, \qquad E'' = E_0'' e^{i(\mathbf{k''}\cdot\mathbf{x}-\omega t)}.$$

As the permeability of glass is very nearly equal to that of vacuum, i.e., $\mu = \mu_0$, the index of refraction of glass is $n = \sqrt{\varepsilon/\varepsilon_0}$, ε being its permittivity.

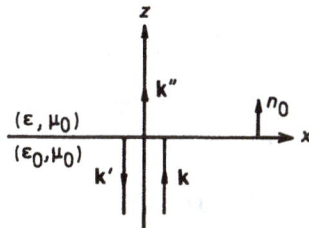

Fig. 4.4

(a) A plane electromagnetic wave can be decomposed into two polarized components with mutually perpendicular planes of polarization. In the interface we take an arbitrary direction as the x direction, and the direction perpendicular to it as the y direction, and decompose the incident wave into two polarized components with **E** parallel to these two directions. We also decompose the reflected and refracted waves in a similar manner. As **E**, **H** and **k** form a right-hand set, we have for the two polarizations:

x-polarization	y-polarization
$E_x,\ H_y$	$E_y,\ -H_x$
$E_x',\ -H_y'$	$E_y',\ H_x'$
$E_x'',\ H_y''$	$E_y'',\ -H_x''$

The boundary condition that E_t and H_t are continuous across the interface gives for the x-polarization

$$E_x + E_x' = E_x'', \tag{1}$$

$$H_y - H_y' = H_y'' . \tag{2}$$

For a plane wave we have $\sqrt{\mu}\,|\mathbf{H}| = \sqrt{\varepsilon}\,|\mathbf{E}|$. With $\mu \approx \mu_0$ and $\sqrt{\frac{\varepsilon}{\varepsilon_0}} = n$, (2) becomes

$$E_x - E_x' = n E_x'' . \tag{3}$$

(1) and (3) give

$$E_x' = \left(\frac{1-n}{1+n}\right) E_x .$$

Since for normal incidence, the plane of incidence is arbitrary, the same result holds for y-polarization. Hence for normal incidence, we have

$$E' = \left(\frac{1-n}{1+n}\right) E .$$

The intensity of a wave is given by the magnitude of the Poynting vector $\mathbf{N}$ over one period. We have

$$\mathbf{N} = \mathrm{Re}\,\mathbf{E} \times \mathrm{Re}\,\mathbf{H} = \frac{1}{4}\left(\mathbf{E} \times \mathbf{H} + \mathbf{E}^* \times \mathbf{H}^* + \mathbf{E} \times \mathbf{H}^* + \mathbf{E}^* \times \mathbf{H}\right).$$

As the first two terms in the last expression contain the time factor $e^{\pm 2i\omega t}$, they vanish on taking average over one period. Hence the intensity is

$$I = \langle N \rangle = \frac{1}{2}\,\mathrm{Re}\,|\mathbf{E} \times \mathbf{H}^*| = \frac{1}{2} E H^* = \frac{1}{2}\sqrt{\frac{\varepsilon}{\mu}}\,E_0^2 ,$$

$\mathbf{E}_0$ being the amplitude of the $\mathbf{E}$ field.

Therefore the coefficient of reflection is

$$R = \frac{E_0'^2}{E_0^2} = \left(\frac{1-n}{1+n}\right)^2 .$$

(b) The average momentum density of a wave is given by $G = \frac{\langle N \rangle}{v^2} = \frac{I}{v^2}$. So the average momentum impinging normally on a unit area per unit time is Gv. The radiation pressure exerted on the glass plate is therefore

$$P = Gc - (-G'c) - G''v$$
$$= Gc\left(1 + \frac{G'}{G} - \frac{G''}{G}\frac{v}{c}\right)$$
$$= \frac{I}{c}\left(1 + \frac{I'}{I} - \frac{c}{v}\frac{I''}{I}\right).$$

(1) + (3) gives

$$\frac{E_0''}{E_0} = \frac{2}{1+n} ,$$

or

$$\frac{I''}{I} = \sqrt{\frac{\varepsilon}{\varepsilon_0}} \frac{E_0''^2}{E_0^2} \frac{4n}{(1+n)^2} .$$

With $\frac{I'}{I} = \left(\frac{1-n}{1+n}\right)^2$ also, we have

$$P = \frac{I}{c}\left[1 + \left(\frac{1-n}{1+n}\right)^2 - \frac{4n^2}{(1+n)^2}\right]$$

$$= 2\frac{I}{c}\left(\frac{1-n}{1+n}\right) .$$

4012

(a) On the basis of Maxwell's equations, and taking into account the appropriate boundary conditions for an air-dielectric interface, show that the reflecting power of glass of index of refraction n for electromagnetic waves at normal incidence is $R = \frac{(n-1)^2}{(n+1)^2}$.

(b) Also show that there is no reflected wave if the incident light is polarized as shown in Fig. 4.5 (i.e., with the electric vector in the plane of incidence) and if $\tan\theta_1 = n$, where θ_1 is the angle of incidence. You can regard it as well-known that Fresnel's law holds.

(UC, Berkeley)

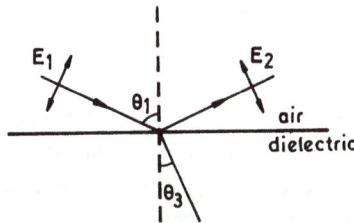

Fig. 4.5

Solution:

(a) Same as for (a) of Problem **4011**.

(b) For waves with the electric vector in the plane of incidence, the following Fresnel's formula applies,

$$\frac{E_2}{E_1} = \frac{\tan(\theta_3 - \theta_1)}{\tan(\theta_3 + \theta_1)} \ .$$

When $\theta_3 + \theta_1 = \frac{\pi}{2}$, $E_2 = 0$, i.e., the reflected wave vanishes. Snell's law gives

$$\sin\theta_1 = n \sin\theta_3 = n \cos\theta_1 \ ,$$

or

$$\tan\theta_1 = n \ .$$

Hence no reflection occurs if the incidence angle is $\theta_1 = \arctan n$.

4013

Calculate the reflection coefficient for an electromagnetic wave which is incident normally on an interface between vacuum and an insulator. (Let the permeability of the insulator be 1 and the dielectric constant be ε. Have the wave incident from the vacuum side.)

(Wisconsin)

Solution:

Referring to Problem **4011**, the reflection coefficient is

$$R = \left(\frac{1 - \sqrt{\varepsilon}}{1 + \sqrt{\varepsilon}}\right)^2 \ .$$

4014

A plane polarized electromagnetic wave travelling in a dielectric medium of refractive index n is reflected at normal incidence from the surface of a conductor. Find the phase change undergone by its electric vector if the refractive index of the conductor is $n_2 = n(1 + i\rho)$.

(SUNY, Buffalo)

Solution:

For normal incidence the plane of incidence is arbitrary. So we can take the electric vector as in the plane of the diagram in Fig. 4.6.

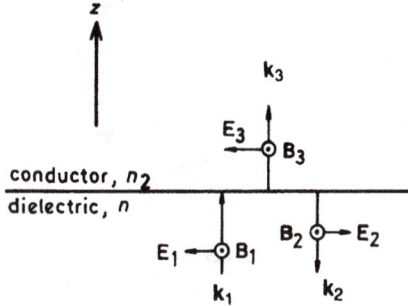

Fig. 4.6

The incident wave is represented by

$$\mathbf{E}_1 = \mathbf{E}_{10} e^{i(k_1 z - \omega t)} \,,$$

$$\mathbf{B}_1 = \mathbf{B}_{10} e^{i(k_1 z - \omega t)} \,,$$

$$B_{10} = \frac{n}{c} E_{10} \,, \qquad k_1 = \frac{\omega}{c} n \,;$$

the reflected wave by

$$\mathbf{E}_2 = \mathbf{E}_{20} e^{-i(k_2 z + \omega t)} \,,$$

$$\mathbf{B}_2 = \mathbf{B}_{20} e^{-i(k_2 z + \omega t)} \,,$$

$$B_{20} = \frac{n E_{20}}{c} \,, \qquad k_2 = \frac{\omega}{c} n \,;$$

and the transmitted wave by

$$\mathbf{E}_3 = \mathbf{E}_{30} e^{i(k_3 z - \omega t)} \,,$$

$$\mathbf{B}_3 = \mathbf{B}_{30} e^{i(k_3 z - \omega t)} \,,$$

$$B_{30} = \frac{n_2 E_{30}}{c} \,, \qquad k_3 = \frac{\omega}{c} n_2 \,.$$

The boundary condition at the interface is that E_t and H_t are continuous. Thus

$$E_{10} - E_{20} = E_{30} \,, \tag{1}$$

$$B_{10} + B_{20} = \frac{\mu}{\mu_2} B_{30} \approx B_{30} , \tag{2}$$

assuming the media to be non-ferromagnetic so that $\mu \approx \mu_2 \approx \mu_0$. Equation (2) can be written as

$$E_{10} + E_{20} = \frac{n_2}{n} E_{30} . \tag{3}$$

(1) and (3) give

$$E_{20} = \frac{n_2 - n}{n_2 + n} E_{10} = \frac{in\rho}{2n + in\rho} E_{10} = \frac{\rho}{\sqrt{\rho^2 + 4}} e^{i\varphi} E_{10} ,$$

with

$$\tan \varphi = \frac{2}{\rho} .$$

The phase shift of the electric vector of the reflected wave with respect to that of the incident wave is therefore

$$\varphi = \arctan \left(\frac{2}{\rho} \right) .$$

4015

In a region of empty space, the magnetic field (in Gaussian units) is described by

$$\mathbf{B} = B_0 e^{ax} \hat{\mathbf{e}}_z \sin w ,$$

where $w = ky - \omega t$.

(a) Calculate **E**.

(b) Find the speed of propagation v of this field.

(c) Is it possible to generate such a field? If so, how?

(*SUNY, Buffalo*)

Solution:

Express **B** as $\mathrm{Im}\,(B_0 e^{ax} e^{iw})\hat{\mathbf{e}}_z$.

(a) Using Maxwell's equation

$$\nabla \times \mathbf{B} = \frac{1}{c} \frac{\partial \mathbf{E}}{\partial t}$$

and the definition $k = \frac{\omega}{c}$ for empty space, we obtain

$$\mathbf{E} = \frac{ic}{\omega} \nabla \times \mathbf{B} = \frac{i}{k} \begin{vmatrix} \mathbf{e}_x & \mathbf{e}_y & \mathbf{e}_z \\ \frac{\partial}{\partial x} & \frac{\partial}{\partial y} & 0 \\ 0 & 0 & B_z \end{vmatrix},$$

where $\frac{\partial}{\partial z} = 0$ as $\mathbf{B}$ does not depend on z.

Hence

$$E_x = \mathrm{Im}\left(\frac{i}{k} B_0 e^{ax} ik e^{iw}\right) = -B_0 e^{ax} \sin w,$$

$$E_y = \mathrm{Im}\left(-\frac{i}{k} B_0 a e^{ax} e^{iw}\right) = -\frac{ac}{\omega} B_0 e^{ax} \cos w,$$

$$E_z = 0.$$

(b) If the wave form remains unchanged during propagation, we have

$$dw = k\,dy - w\,dt = 0,$$

or $\frac{dy}{dt} = \frac{\omega}{k} = c$. Hence the wave propagates along the y direction with a speed $v = c$.

(c) Such an electromagnetic wave can be generated by means of total reflection. Consider the plane interface between a dielectric of refractive index $n(> 1)$ and empty space. Let this be the yz plane and take the $+x$ direction as away from the dielectric. A plane wave polarized with $\mathbf{B}$ in the z direction travels in the dielectric and strikes the interface at incidence angle θ. The incident and refracted waves may be represented by

$$B_z = B_0 \exp[i(xk\cos\theta + yk\sin\theta - \omega t)],$$

$$B_z'' = B_0'' \exp[i(xk''\cos\theta'' + yk''\sin\theta'' - \omega t)],$$

where $k = \frac{\omega}{v} = \frac{\omega}{c} n$, $k'' = \frac{\omega}{c}$.

At the interface, $x = 0$ and y is arbitrary. The boundary condition that H_t is continuous requires that

$$k\sin\theta = k''\sin\theta'',$$

or

$$\sin\theta = \frac{1}{n}\sin\theta'' \le \frac{1}{n}.$$

As $n > 1$, if $\theta > \arcsin\left(\frac{1}{n}\right)$ total reflection occurs.

Under total reflection,

$$\sin\theta'' = n\sin\theta\,,$$

$$\cos\theta'' = \pm i\sqrt{n^2\sin^2\theta - 1}\,.$$

Then

$$B_z'' = B_0''\exp\left(\mp\frac{\omega}{c}\sqrt{n^2\sin^2\theta - 1}\,x\right)\exp\left[i\left(\frac{\omega}{c}n\sin\theta y - \omega t\right)\right].$$

As x increases with increasing penetration into the empty space, $-$ sign is to be used. This field has exactly the given form.

4016

A harmonic plane wave of frequency ν is incident normally on an interface between two dielectric media of indices of refraction n_1 and n_2. A fraction ρ of the energy is reflected and forms a standing wave when combined with the incoming wave. Recall that on reflection the electric field changes phase by π for $n_2 > n_1$.

(a) Find an expression for the total electric field as a function of the distance d from the interface. Determine the positions of the maxima and minima of $\langle E^2\rangle$.

(b) From the behavior of the electric field, determine the phase change on reflection of the magnetic field. Find $B(z,t)$ and $\langle B^2\rangle$.

(c) When O. Wiener did such an experiment using a photographic plate in 1890, a band of minimum darkening of the plate was found for $d = 0$. Was the darkening caused by the electric or the magnetic field?

(*Columbia*)

Solution:

(a) With the coordinates shown in Fig. 4.7 and writing z for d, the electric field of the incident wave is $E_0\cos(kz - \omega t)$. Because the electric field changes phase by π on reflection from the interface, the amplitude E_0' of the reflected wave is opposite in direction to E_0. A fraction ρ of the energy is reflected. As energy is proportional to E_0^2, we have

$$E_0'^2 = \rho E_0^2\,.$$

Thus the electric field of the reflected wave is

$$\mathbf{E}' = -\sqrt{\rho}\,\mathbf{E}_0 \cos(-kz - \omega t)\,.$$

Hence the total electric field in the first medium is

$$\mathbf{E} = \mathbf{E}_0 \cos(kz - \omega t) - \sqrt{\rho}\,\mathbf{E}_0 \cos(kz + \omega t)\,,$$

giving

$$E^2 = E_0^2 \cos^2(kz - \omega t) + \rho E_0^2 \cos^2(kz + \omega t) - \sqrt{\rho}\,E_0^2[\cos(2kz) + \cos(2\omega t)]\,.$$

Taking average over a period $T = \frac{2\pi}{\omega}$ we have

$$\langle E^2 \rangle = \frac{1}{T}\int_0^T E^2\,dt = \frac{(1 + \rho)E_0^2}{2} - \sqrt{\rho}\,E_0^2 \cos(2kz)\,.$$

When $kz = m\pi$, or $z = \frac{mc}{2\nu n_1}$, where m is an integer $0, 1, 2, \ldots$, $\langle E^2 \rangle$ will be minimum with the value

$$\langle E^2 \rangle_{\min} = \frac{(1 - \sqrt{\rho})^2}{2} E_0^2\,.$$

When $kz = \frac{(2m+1)\pi}{2}$, or $z = \frac{(2m+1)c}{4\nu n_1}$, where $m = 0, 1, 2, \ldots$, $\langle E^2 \rangle$ is maximum with the value

$$\langle E^2 \rangle_{\max} = \frac{(1 + \sqrt{\rho})^2}{2} E_0^2\,.$$

(b) As $\mathbf{E}$, $\mathbf{B}$, and $\mathbf{k}$ form an orthogonal right-hand set, we see from Fig. 4.7 that the amplitude B_0 of the magnetic field of the reflected wave is in the same direction as that of the incident wave B_0, hence no phase change occurs. The amplitudes of the magnetic fields are

$$B_0 = n_1 E_0\,, \qquad B_0' = n_1 E_0' = \sqrt{\rho}\,n_1 E_0 = \sqrt{\rho}\,B_0\,.$$

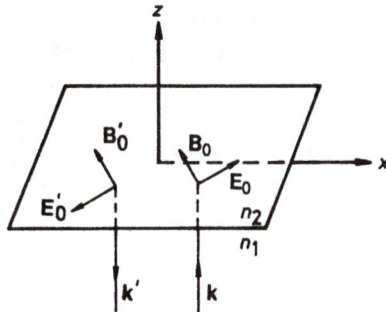

Fig. 4.7

The total magnetic field in the first medium is

$$\mathbf{B}(z,t) = \mathbf{B}_0 \cos(kz - \omega t) + \sqrt{\rho}\, \mathbf{B}_0 \cos(kz + \omega t),$$

giving

$$B^2 = n_1^2 E_0^2 \cos^2(kz - \omega t) + \rho n_1^2 E_0^2 \cos^2(kz + \omega t)$$
$$+ \sqrt{\rho}\, n_1^2 E_0^2 \left[(\cos(2kz) + \cos(2\omega t) \right],$$

with the average value

$$\langle B^2 \rangle = \frac{(1+\rho)n_1^2 E_0^2}{2} + \sqrt{\rho}\, n_1^2 \cos(2kz).$$

Hence $\langle B^2 \rangle$ will be maximum for $kz = m\pi$ and minimum for $kz = \frac{2m+1}{2}\pi$.

(c) The above shows that $\langle B^2 \rangle$ is maximum for $z = 0$ and $\langle E^2 \rangle$ is minimum for $z = 0$. Hence the darkening of the photographic plate, which is minimum at $z = 0$, is caused by the electric field.

4017

Beams of electromagnetic radiation, e.g. radar beams, light beams, eventually spread because of diffraction. Recall that a beam which propagates through a circular aperture of diamater D spreads with a diffraction angle $\theta_d = \frac{1.22}{D}\lambda_n$. In many dielectric media the index of refraction increases in large electric fields and can be well represented by $n = n_0 + n_2 E^2$.

Show that in such a nonlinear medium the diffraction of the beam can be counterbalanced by total internal reflection of the radiation to form a self-trapped beam. Calculate the threshold power for the existence of a self-trapped beam.

(*Princeton*)

Solution:

Consider a cylindrical surface of diameter D in the dielectric medium. Suppose that the electric field inside the cylinder is E and that outside is zero. As the index of refraction of the medium is $n = n_0 + n_2 E^2$, the index outside is n_0.

Consider a beam of radiation propagating along the axis of the cylinder. A ray making an angle θ with the axis will be totally reflected at the cylindrical surface if

$$n \sin\left(\frac{\pi}{2} - \theta\right) \geq n_0 ,$$

i.e.,

$$n \geq \frac{n_0}{\cos\theta} .$$

The diffraction spread $\theta_d = 1.22\lambda_n/D$ will be counterbalanced by the total internal reflection if $n = n_0 + n_2 E^2 \geq \frac{n_0}{\cos\theta_d}$. Hence we require an electric intensity greater than a critical value

$$E_c = \sqrt{\frac{n_0}{n_2} [(\cos\theta_d)^{-1} - 1]} .$$

Assume the radiation to be plane electromagnetic waves we have

$$\sqrt{\varepsilon}\, E = \sqrt{\mu}\, H .$$

Waves with the critical electric intensity have average Poynting vector

$$\langle N \rangle = \frac{1}{2} E H^* = \frac{1}{2}\sqrt{\frac{\varepsilon}{\mu}}\, E_c^2 .$$

Hence the threshold radiation power is

$$\langle P \rangle = \langle N \rangle \frac{\pi D^2}{4} = \frac{\pi D^2}{8} \sqrt{\frac{\varepsilon}{\mu}} \frac{n_0}{n_2} \left(\frac{1}{\cos\theta_d} - 1\right) .$$

As

$$n = \frac{n_0}{\cos\theta_d} \approx \sqrt{\frac{\varepsilon}{\varepsilon_0}} , \qquad \mu \approx \mu_0 ,$$

$$\langle P \rangle = \frac{\pi c \varepsilon_0 D^2}{8} \frac{n_0^2}{n_2} \cdot \frac{1 - \cos\theta_d}{\cos^2\theta_d} .$$

With $\theta_d = 1.22\lambda_n/D \ll 1$, we have

$$\langle P \rangle = \frac{\pi c \varepsilon_0 D^2}{16} \frac{n_0^2}{n_2} \theta_d^2 = \frac{\pi c \varepsilon_0}{16} \frac{n_0^2}{n_2} (1.22\lambda_n)^2 .$$

Since $n_0 \lambda_n = \lambda$ is the wavelength in vacuum, we obtain

$$\langle P \rangle = \frac{\pi c \varepsilon_0}{16 n_2} (1.22\lambda)^2 .$$

4018

Consider the propagation of a plane electromagnetic wave through a medium whose index of refraction depends on the state of circular polarization.

(a) Write down expressions for the right and left circularly polarized plane waves.

(b) Assume that the index of refraction in the medium is of the form

$$n_\pm = n \pm \beta ,$$

where n and β are real and the plus and minus signs refer to right and left circularly polarized plane waves respectively. Show that a linearly polarized plane wave incident on such a medium has its plane of polarization rotated as it travels through the medium. Find the angle through which the plane of polarizations is rotated as a function of the distance z into the medium.

(c) Consider a tenuous electronic plasma of uniform density n_0 with a strong static uniform magnetic induction $\mathbf{B}_0$ and transverse waves propagating parallel to the direction of $\mathbf{B}_0$. Assume that the amplitude of the electronic motion is small, that collisions are negligible, and that the amplitude of the $\mathbf{B}$ field of the waves is small compared with B_0. Find the indices of refraction for circularly polarized waves, and show that for high frequencies the indices can be written in the form assumed in part (b). Specify what you mean by high frequencies.

(SUNY, Buffalo)

Solution:

Take the z-axis along the direction of propagation of the wave.

(a) The electric vector of right circularly polarized light can be represented by the real part of

$$\mathbf{E}_R(z, t) = (E_0 \mathbf{e}_x + E_0 e^{-i\frac{\pi}{2}} \mathbf{e}_y) e^{-i\omega t + ik_+ z} ,$$

and that of the left circularly polarized light by the real part of

$$\mathbf{E}_L(z,t) = (E_0\mathbf{e}_x + E_0 e^{i\frac{\pi}{2}}\mathbf{e}_y)\,e^{-i\omega t + ik_- z},$$

where

$$k_+ = \frac{\omega}{c}n_+, \qquad k_- = \frac{\omega}{c}n_-.$$

(b) A linearly polarized light can be decomposed into right and left circularly polarized waves:

$$\mathbf{E}(z,t) = \mathbf{E}_R(z,t) + \mathbf{E}_L(z,t)$$
$$= (E_0\mathbf{e}_x + E_0 e^{-i\frac{\pi}{2}}\mathbf{e}_y)e^{-i\omega t + ik_+ z} + (E_0\mathbf{e}_x + E_0 e^{i\frac{\pi}{2}}\mathbf{e}_y)e^{-i\omega t + ik_- z}.$$

At the point of incidence $z = 0$, $\mathbf{E}(0,t) = 2E_0 e^{-i\omega t}\mathbf{e}_x$, which represents a wave with $\mathbf{E}$ polarized in the $\mathbf{e}_x$ direction. At a distance z into the medium, we have

$$\mathbf{E}(z,t) = E_0[(e^{i(k_+ - k_-)z} + 1)\mathbf{e}_x + (e^{i(k_+ - k_-)z - i\frac{\pi}{2}} + e^{i\frac{\pi}{2}})\mathbf{e}_y]\,e^{-i\omega t + ik_- z},$$

and thus

$$\frac{E_y}{E_x} = \frac{\cos\left[(k_+ - k_-)z - \frac{\pi}{2}\right]}{1 + \cos[(k_+ - k_-)z]} = \frac{\sin[(k_+ - k_-)z]}{1 + \cos[(k_+ - k_-)z]} = \tan\left[\frac{(k_+ - k_-)z}{2}\right].$$

Hence on traversing a distance z in the medium the plane of the electric vector has rotated by an angle

$$\varphi = \frac{k_+ - k_-}{2}z = \frac{1}{2}\cdot\frac{\omega}{c}(n_+ - n_-)z.$$

As $n_+ > n_-$ (assuming $\beta > 0$), $\varphi > 0$. That is, the rotation of the plane of polarization is anti-clockwise looking against the direction of propagation.

(c) The Lorentz force on an electron in the electromagnetic field of a plane electromagnetic wave is $-e(\mathbf{E} + \mathbf{v}\times\mathbf{B})$, where $\mathbf{v}$ is the velocity of the electron. As $\sqrt{\varepsilon_0}\,|\mathbf{E}| = \sqrt{\mu_0}\,|\mathbf{H}|$, or $|\mathbf{B}| = |\mathbf{E}|/c$, we have

$$\frac{|\mathbf{v}\times\mathbf{B}|}{|\mathbf{E}|} \approx \frac{v}{c} \ll 1.$$

Hence the magnetic force exerted by the wave on the electron may be neglected. The equation of the motion of an electron in $\mathbf{B}_0$ and the electromagnetic field of the wave, neglecting collisions, is

$$m\ddot{\mathbf{r}} = -e\mathbf{E} - e\mathbf{v}\times\mathbf{B}_0,$$

where $\mathbf{E}$ is the sum of $\mathbf{E_R}$ and $\mathbf{E_L}$ in (a). Consider an electron at an arbitrary point z. Then the solution of the equation of motion has the form

$$\mathbf{r} = \mathbf{r}_0 e^{-i\omega t}.$$

Substitution gives

$$-m\omega^2 \mathbf{r} = -e\mathbf{E} - e(-i\omega)\mathbf{r} \times \mathbf{B}_0.$$

The electron, oscillating in the field of the wave, acts as an oscillating dipole, the dipole moment per unit volume being $\mathbf{P} = -n_0 e \mathbf{r}$. The above equation then gives

$$m\omega^2 \mathbf{P} = -n_0 e^2 \mathbf{E} - i\omega e \mathbf{P} \times \mathbf{B}_0,$$

or, using $\mathbf{P} = \chi\varepsilon_0 \mathbf{E}$,

$$m\omega^2 \chi\varepsilon_0 \mathbf{E} + n_0 e^2 \mathbf{E} = -i\omega e \chi\varepsilon_0 \mathbf{E} \times \mathbf{B}_0.$$

Defining

$$\omega_P^2 = \frac{n_0 e^2}{m\varepsilon_0}, \qquad \omega_B = \frac{n_0 e}{\varepsilon_0 B_0},$$

and with $\mathbf{B}_0 = B_0 \mathbf{e}_z$, the above becomes

$$\chi \frac{\omega^2}{\omega_P^2} \mathbf{E} + \mathbf{E} = -i\chi \frac{\omega}{\omega_B} \mathbf{E} \times \mathbf{e}_z,$$

or

$$\left(1 + \chi \frac{\omega^2}{\omega_P^2}\right) E_x + i\chi \frac{\omega}{\omega_B} E_y = 0, \qquad (1)$$

$$\left(1 + \chi \frac{\omega^2}{\omega_P^2}\right) E_y - i\chi \frac{\omega}{\omega_B} E_x = 0. \qquad (2)$$

$(1)\pm i\times(2)$ gives

$$\left(1 + \chi \frac{\omega^2}{\omega_P^2}\right)(E_x \pm iE_y) \pm \frac{\chi\omega}{\omega_B}(E_x \pm iE_y) = 0.$$

Note that $E_x - iE_y = 0$ and $E_x + iE_y = 0$ represent the right and left circularly polarized waves respectively. Hence for the right circularly polarized component, whose polarizability is denoted by χ_+, $E_x + iE_y \neq 0$ so that

$$\left(1 + \chi_+ \frac{\omega^2}{\omega_P^2}\right) + \chi_+ \frac{\omega}{\omega_B} = 0,$$

or

$$\chi_+ = -\frac{1}{\frac{\omega^2}{\omega_P^2} + \frac{\omega}{\omega_B}}.$$

Similarly for the left polarization we have

$$\chi_- = -\frac{1}{\frac{\omega^2}{\omega_P^2} - \frac{\omega}{\omega_B}}.$$

The permittivity of a medium is given by $\varepsilon = (1 + \chi)\varepsilon_0$ so that the refractive index is

$$n = \sqrt{\frac{\varepsilon}{\varepsilon_0}} = \sqrt{1 + \chi}.$$

Hence for the two polarizations we have the refractive indices

$$n_{\pm} = \sqrt{1 - \frac{1}{\frac{\omega^2}{\omega_P^2} \pm \frac{\omega}{\omega_B}}} = \sqrt{1 - \frac{\omega_P^2}{\omega^2 \pm \frac{\omega_P^2 \omega}{\omega_B}}} = \sqrt{1 - \frac{\omega_P^2}{\omega(\omega \pm \omega_B')}},$$

where

$$\omega_B' = \frac{\omega_P^2}{\omega_B} = \frac{n_0 e^2}{m\varepsilon_0} \cdot \frac{\varepsilon_0 B_0}{n_0 e} = \frac{B_0 e}{m}.$$

For frequencies sufficiently high so that $\omega \gg \omega_P$, $\omega \gg \omega_B'$, we obtain approximately

$$n_{\pm} = \left[1 - \frac{\omega_P^2}{\omega^2} \left(1 \pm \frac{\omega_B'}{\omega} \right)^{-1} \right]^{\frac{1}{2}}$$

$$\approx 1 - \frac{1}{2} \frac{\omega_P^2}{\omega^2} \left(1 \mp \frac{\omega_B'}{\omega} \right)$$

$$= n \pm \beta$$

with $n \approx 1 - \frac{\omega_P^2}{2\omega^2}$, $\beta \approx \frac{\omega_P^2}{2\omega^2} \frac{\omega_B'}{\omega}$.

4019

Linearly polarized light of the form $E_x(z,t) = E_0 e^{i(kz - \omega t)}$ is incident normally onto a material which has index of refraction n_R for right-hand circularly polarized light and n_L for left-hand circularly polarized light.

Using Maxwell's equations calculate the intensity and polarization of the reflected light.

(*Wisconsin*)

Solution:

Using Maxwell's equations $\oint \mathbf{E} \cdot d\mathbf{r} = -\int \frac{\partial \mathbf{B}}{\partial t} \cdot d\mathbf{S}$ and $\oint \mathbf{H} \cdot d\mathbf{r} = \int \left(\frac{\partial \mathbf{D}}{\partial t} + \mathbf{J} \right) \cdot d\mathbf{S}$, we find that at the boundary of two dielectric media the tangential components of $\mathbf{E}$ and $\mathbf{H}$ are each continuous. Then as $\mathbf{E}$, $\mathbf{H}$ and the direction of propagation of a plane electromagnetic wave form an orthogonal right-hand set, we have for normal incidence

$$E + E'' = E' , \qquad H - H'' = H' ,$$

where the prime and double prime indicate the reflected and refracted components respectively. Also the following relation holds for plane waves,

$$H = \sqrt{\frac{\varepsilon}{\mu}} E = \sqrt{\frac{\varepsilon}{\varepsilon_0}} \sqrt{\frac{\varepsilon_0}{\mu}} E \approx n \sqrt{\frac{\varepsilon_0}{\mu_0}} E .$$

Hence the H equation can be written as

$$E - E'' = nE' ,$$

taking the first medium as air $(n = 1)$.

Eliminating E', we get

$$E'' = \frac{1 - n}{1 + n} E .$$

For normal incidence, the plane of incidence is arbitrary and this relation holds irrespective of the polarization state. Hence

$$E''_L = \frac{1 - n_L}{1 + n_L} E_L , \qquad E''_R = \frac{1 - n_R}{1 + n_R} E_R .$$

The incident light can be decomposed into left-hand and right-hand circularly polarized components:

$$E = \begin{pmatrix} E_0 \\ 0 \end{pmatrix} = E_0 \begin{pmatrix} 1 \\ 0 \end{pmatrix} = \frac{1}{2} E_0 \begin{pmatrix} 1 \\ i \end{pmatrix} + \frac{1}{2} E_0 \begin{pmatrix} 1 \\ -i \end{pmatrix} ,$$

where $\begin{pmatrix} 1 \\ i \end{pmatrix}$ represents the left-hand circularly polarized light and $\begin{pmatrix} 1 \\ -i \end{pmatrix}$ the right-hand one. Hence the reflected amplitude is

$$E'' = \frac{1}{2}E_0\frac{1-n_R}{1+n_R}\begin{pmatrix} 1 \\ i \end{pmatrix} + \frac{1}{2}E_0\frac{1-n_L}{1+n_L}\begin{pmatrix} 1 \\ -i \end{pmatrix}$$

$$= \frac{1}{2}E_0\begin{pmatrix} \frac{1-n_R}{1+n_R} + \frac{1-n_L}{1+n_L} \\ i\left(\frac{1-n_R}{1+n_R} - \frac{1-n_L}{1+n_L}\right) \end{pmatrix} .$$

This shows that the reflected light is elliptically polarized and the ratio of intensities is

$$\frac{I''}{I} = \frac{1}{4}\left[\left(\frac{1-n_R}{1+n_R} + \frac{1-n_L}{1+n_L}\right)^2 + \left(\frac{1-n_R}{1+n_R} - \frac{1-n_L}{1+n_L}\right)^2\right]$$

$$= \frac{1}{4}\left[2\left(\frac{1-n_R}{1+n_R}\right)^2 + 2\left(\frac{1-n_L}{1+n_L}\right)^2\right] = \frac{1}{2}\left[\left(\frac{1-n_R}{1+n_R}\right)^2 + \left(\frac{1-n_L}{1+n_L}\right)^2\right] .$$

4020

A dextrose solution is optically active and is characterized by a polarization vector (electric dipole moment per unit volume): $\mathbf{P} = \gamma\nabla \times \mathbf{E}$, where γ is a real constant which depends on the dextrose concentration. The solution is non-conducting ($j_{\text{free}} = 0$) and non-magnetic (magnetization vector $\mathbf{M} = 0$). Consider a plane electromagnetic wave of (real) angular frequency ω propagating in such a solution. For definiteness, assume that the wave propagates in the $+z$ direction. (Also assume that $\frac{\gamma\omega}{c} \ll 1$ so that square roots can be approximated by $\sqrt{1 + A} \approx 1 + \frac{1}{2}A$.)

(a) Find the two possible indices of refraction for such a wave. For each possible index, find the corresponding electric field.

(b) Suppose linearly polarized light is incident on the dextrose solution. After traveling a distance L through the solution, the light is still linearly polarized but the direction of polarization has been rotated by an angle ϕ (Faraday rotation). Find ϕ in terms of L, γ, and ω.

(Columbia)

Solution:

(a) $\mathbf{D}$, $\mathbf{E}$, $\mathbf{P}$, $\mathbf{B}$, $\mathbf{H}$, $\mathbf{M}$ are related by

$$\mathbf{D} = \varepsilon_0 \mathbf{E} + \mathbf{P}, \qquad \mathbf{B} = \mu_0 (\mathbf{H} + \mathbf{M}).$$

With $\mathbf{P} = \gamma \nabla \times \mathbf{E}$, $\mathbf{M} = 0$, we have

$$\mathbf{D} = \varepsilon_0 \mathbf{E} + \gamma \nabla \times \mathbf{E}, \qquad \mathbf{B} = \mu_0 \mathbf{H}.$$

For a source-free medium, two of Maxwell's equations are

$$\nabla \times \mathbf{H} = \dot{\mathbf{D}}, \qquad \nabla \cdot \mathbf{E} = 0.$$

The first equation gives

$$\nabla \times \mathbf{B} = \mu_0 \dot{\mathbf{D}} = \frac{1}{c^2} \dot{\mathbf{E}} + \gamma \mu_0 \nabla \times \dot{\mathbf{E}},$$

while the second gives

$$\nabla \times (\nabla \times \mathbf{E}) = -\nabla^2 \mathbf{E}.$$

Then from Maxwell's equation

$$\nabla \times \mathbf{E} = -\dot{\mathbf{B}},$$

we have

$$\nabla \times (\nabla \times \mathbf{E}) = -\nabla \times \dot{\mathbf{B}},$$

or

$$-\nabla^2 \mathbf{E} = -\frac{1}{c^2} \ddot{\mathbf{E}} - \gamma \mu_0 \nabla \times \ddot{\mathbf{E}}. \tag{1}$$

For a plane electromagnetic wave

$$\mathbf{E} = \mathbf{E}_0 e^{i(kz - \omega t)} = E_x \mathbf{e}_x + E_y \mathbf{e}_y,$$

the actions of the operators ∇ and $\frac{\partial}{\partial t}$ result in (see Problem **4004**)

$$\nabla \to ik\mathbf{e}_z, \qquad \frac{\partial}{\partial t} \to -i\omega.$$

Equation (1) then becomes

$$k^2 \mathbf{E} = \frac{\omega^2}{c^2} \mathbf{E} + i\gamma\mu_0\omega^2 k \mathbf{e}_z \times \mathbf{E},$$

which has component equations

$$\left(k^2 - \frac{\omega^2}{c^2}\right) E_x + i\gamma\mu_0\omega^2 k E_y = 0,$$

$$i\gamma\mu_0\omega^2 k E_x - \left(k^2 - \frac{\omega^2}{c^2}\right) E_y = 0.$$

These simultaneous equations have nonzero solutions if and only if

$$\begin{vmatrix} k^2 - \frac{\omega^2}{c^2} & i\gamma\mu_0\omega^2 k \\ i\gamma\mu_0\omega^2 k & -(k^2 - \frac{\omega^2}{c^2}) \end{vmatrix} = -\left(k^2 - \frac{\omega^2}{c^2}\right) + \gamma^2\mu_0^2\omega^4 k^2 = 0,$$

i.e.,

$$k^2 - \frac{\omega^2}{c^2} = \pm\gamma\mu_0\omega^2 k.$$

The top and bottom signs give

$$\left(k_\pm^2 - \frac{\omega^2}{c^2}\right)(E_x \pm iE_y) = 0.$$

Hence the wave is equivalent to two circularly polarized waves. For the right-hand circular polarization, $E_x + iE_y \neq 0$ and we have

$$k_+^2 = \frac{\omega^2}{c^2} + \gamma\mu_0\omega^2 k_+.$$

For the left-hand circular polarization, $E_x - iE_y \neq 0$ and we have

$$k_-^2 = \frac{\omega^2}{c^2} - \gamma\mu_0\omega^2 k_-.$$

Solving the equation for $k_\pm$

$$k_\pm^2 \mp \gamma\mu_0\omega^2 k_\pm - \frac{\omega^2}{c^2} = 0,$$

we have

$$k_{\pm} = \frac{1}{2}\left[\pm\gamma\mu_0\omega^2 \pm \sqrt{\gamma^2\mu_0^2\omega^4 + \frac{4\omega^2}{c^2}}\right].$$

As $k_{\pm}$ has to be positive we choose the positive sign in front of the square root. Hence

$$k_{\pm} = \frac{\omega}{c}\left[1 + \left(\frac{\gamma\mu_0\omega c}{2}\right)^2\right]^{\frac{1}{2}} \pm \frac{\gamma\mu_0\omega^2}{2}.$$

To convert to Gaussian units, we have to replace $\frac{1}{4\pi\epsilon_0} = \frac{\mu_0 c^2}{4\pi}$ by 1. Thus $\frac{\gamma\mu_0\omega c}{2}$ is to be replaced by $\frac{2\pi\gamma\omega}{c}$, which is assumed to be $\ll 1$. Therefore $k_{\pm} \approx \frac{\omega}{c} \pm \frac{\gamma\mu_0\omega^2}{2}$, and $n_{\pm} = \frac{c}{\omega}k_{\pm} \approx 1 \pm \frac{\gamma\mu_0\omega c}{2}$.

(b) If the traversing light is linearly polarized, the different refractive indices of the circularly polarized components mean that the components will rotate by different angles. Recombining them, the plane of polarization is seen to rotate as the medium is traversed. The angle rotated after traversing a distance L is (cf. Problem 4018)

$$\phi = \frac{1}{2}(\phi_1 + \phi_2) = \frac{1}{2}(k_+ - k_-)L \approx \frac{1}{2}\gamma\mu_0\omega^2 L.$$

4021

Some isotropic dielectrics become birefringent (doubly refracting) when they are placed in a static external magnetic field. Such magnetically-biased materials are said to be gyrotropic and are characterized by a permittivity ϵ and a constant "gyration vector" **g**. In general, **g** is proportional to the static magnetic field which is applied to the dielectric. Consider a monochromatic plane wave

$$\begin{bmatrix} \mathbf{E}(\mathbf{x}, t) \\ \mathbf{B}(\mathbf{x}, t) \end{bmatrix} = \begin{bmatrix} \mathbf{E}_0 \\ \mathbf{B}_0 \end{bmatrix} e^{i(K\hat{n}\cdot\mathbf{x} - \omega t)}$$

traveling through a gyrotropic material. ω is the given angular frequency of the wave, and $\hat{n}$ is the given direction of propagation. $\mathbf{E}_0$, $\mathbf{B}_0$, and K are constants to be determined. For a non-conducting ($\sigma = 0$) and non-permeable ($\mu = 1$) gyrotropic material, the electric displacement **D** and the electric field **E** are related by

$$\mathbf{D} = \epsilon\mathbf{E} + i(\mathbf{E} \times \mathbf{g}),$$

where the permittivity ε is a positive real number and where the "gyration vector" **g** is a constant real vector. Consider plane waves which propagate in the direction of **g**, with **g** pointing along the z-axis:

$$\mathbf{g} = g\mathbf{e}_z \quad \text{and} \quad \hat{\mathbf{n}} = \mathbf{e}_z .$$

(a) Starting from Maxwell's equations, find the possible values for the index of refraction $N \equiv Kc/\omega$. Express your answers in terms of the constants ε and g.

(b) For each possible value of N, find the corresponding polarization $\mathbf{E}_0$.

(*Columbia*)

Solution:

In Gaussian units Maxwell's equations for a source-free medium are

$$\nabla \cdot \mathbf{D} = 0, \qquad \nabla \cdot \mathbf{B} = 0,$$

$$\nabla \times \mathbf{E} = -\frac{1}{c}\frac{\partial \mathbf{B}}{\partial t}, \qquad \nabla \times \mathbf{B} = \frac{1}{c}\frac{\partial \mathbf{D}}{\partial t}.$$

where we have used $\mu = 1$ and $\mathbf{B} = \mathbf{H}$.

As the wave vector is $\mathbf{K} = K\mathbf{e}_z$ and the electromagnetic wave is represented by $\mathbf{E} = \mathbf{E}_0 e^{i(Kz - \omega t)}$, the above equations become (see Problem 4004)

$$\mathbf{K} \cdot \mathbf{D} = 0, \qquad \mathbf{K} \cdot \mathbf{B} = 0,$$

$$\mathbf{K} \times \mathbf{E} = \frac{\omega}{c}\mathbf{B}, \qquad \mathbf{K} \times \mathbf{B} = -\frac{\omega}{c}\mathbf{D}.$$

Thus

$$\mathbf{K} \times (\mathbf{K} \times \mathbf{E}) = \mathbf{K}(\mathbf{K} \cdot \mathbf{E}) - K^2\mathbf{E} = \frac{\omega}{c}\mathbf{K} \times \mathbf{B} = -\frac{\omega^2}{c^2}\mathbf{D},$$

or

$$K^2\mathbf{E} - \frac{\omega^2}{c^2}\mathbf{D} - \mathbf{K}(\mathbf{K} \cdot \mathbf{E}) = 0 .$$

Making use of $\mathbf{D} = \varepsilon\mathbf{E} + i(\mathbf{E} \times \mathbf{g})$, we have

$$\left(K^2 - \frac{\omega^2\varepsilon}{c^2}\right)\mathbf{E} - \mathbf{K}(\mathbf{K} \cdot \mathbf{E}) - i\frac{\omega^2}{c^2}(\mathbf{E} \times \mathbf{g}) = 0 ,$$

or, with $N = \frac{Kc}{\omega}$,

$$(N^2 - \varepsilon)\mathbf{E} - \frac{c^2}{\omega^2}\mathbf{K}(\mathbf{K} \cdot \mathbf{E}) - i(\mathbf{E} \times \mathbf{g}) = 0.$$

As $\mathbf{K} = K\mathbf{e}_z$, $\mathbf{g} = g\mathbf{e}_z$, $N = \frac{Kc}{\omega}$ the component equations are

$$(N^2 - \varepsilon)E_x - igE_y = 0, \tag{1}$$

$$igE_x + (N^2 - \varepsilon)E_y = 0, \tag{2}$$

$$\varepsilon E_z = 0. \tag{3}$$

Eq. (3) shows that $E_z = 0$. Hence the wave is transverse. For non-zero solutions of (1) and (2), we require

$$\det \begin{pmatrix} N^2 - \varepsilon & -ig \\ ig & N^2 - \varepsilon \end{pmatrix} = 0,$$

giving

$$(N^2 - \varepsilon)^2 = g^2,$$

i.e.,

$$N = \sqrt{\varepsilon \pm g}.$$

Thus the index of refraction has two values,

$$N_1 = \sqrt{\varepsilon + g}, \qquad N_2 = \sqrt{\varepsilon - g}.$$

Substituting in (1) we obtain

$$\text{for } N_1: \quad g(E_x - iE_y) = 0,$$

$$\text{for } N_2: \quad g(E_x + iE_y) = 0.$$

Since $g \neq 0$, N_1 is the refractive index of the right circularly polarized components and N_2 is that of the left circularly polarized component. $\mathbf{E}_0$ for the two components are respectively

$$E_{0x} = iE_{0y}, \qquad E_{0z} = 0,$$

$$E_{0x} = -iE_{0y}, \qquad E_{0z} = 0.$$

4022

A plane electromagnetic wave of angular frequency ω is incident normally on a slab of non-absorbing material. The surface lies in the xy plane. The material is anisotropic with

$$\varepsilon_{xx} = n_x^2 \varepsilon_0 \quad \varepsilon_{yy} = n_y^2 \varepsilon_0, \quad \varepsilon_{zz} = n_z^2 \varepsilon_0,$$

$$\varepsilon_{xy} = \varepsilon_{yz} = \varepsilon_{zx} = 0, \qquad n_x \neq n_y.$$

(a) If the incident plane wave is linearly polarized with its electric field at $45°$ to the x and y axes, what will be the state of polarization of the reflected wave for an infinitely thick slab?

(b) For a slab of thickness d, derive an equation for the relative amplitude and phase of the transmitted electric field vectors for polarization in the x and y directions.

(UC, Berkeley)

Solution:

Consider a plane electromagnetic wave incident from an anisotropic medium 1 into another anisotropic medium 2, and choose coordinate axes so that the incidence takes place in the xz plane, the interface being the $x0y$ plane, as shown in Fig. 4.8. The incident, reflected, and transmitted waves are represented as follows:

$$\text{incident wave: } e^{i(\mathbf{K} \cdot \mathbf{r} - \omega t)},$$
$$\text{reflected wave: } e^{i(\mathbf{K}' \cdot \mathbf{r} - \omega' t)},$$
$$\text{transmitted wave: } e^{i(\mathbf{K}'' \cdot \mathbf{r} - \omega'' t)}.$$

The boundary condition on the interface that the tangential components of **E** and **H** are continuous requires that

$$K_x = K_x' = K_x'',$$
$$K_z = K_z' = K_z''$$
$$\omega = \omega' = \omega''.$$

From these follow the laws of reflection and refraction:

$$K(\theta) \sin \theta = K'(\theta') \sin \theta',$$
$$K(\theta) \sin \theta = K''(\theta'') \sin \theta''.$$

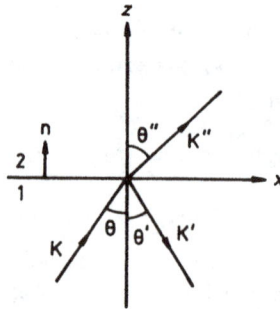

Fig. 4.8

(a) As the medium 1 given is air or vacuum, we have $K = K' = \frac{\omega}{c}$. For normal incidence

$$\theta = \theta' = \theta'' = 0,$$

so that

$$\mathbf{K} = K\mathbf{n}, \quad \mathbf{K}' = -K\mathbf{n}, \quad \mathbf{K}'' = K''\mathbf{n}. \quad (\mathbf{n} = \mathbf{e}_z)$$

From Maxwell's equation $\nabla \times \mathbf{H} = \dot{\mathbf{D}}$, we have (Problem **4004**)

$$\mathbf{K}'' \times \mathbf{H}'' = -\omega \mathbf{D}''. \tag{1}$$

As $\mathbf{K}''$ is parallel to $\mathbf{e}_z$, $\mathbf{D}''$ and $\mathbf{H}''$ are in the xy plane. Take the axes along the principal axes of the dielectric, then

$$D_i'' = \varepsilon_{ii} E_i'', \quad (i = x, y, z)$$

and

$$\mathbf{E}'' = E_x'' \mathbf{e}_x + E_y'' \mathbf{e}_y, \quad D_z'' = E_z'' = 0.$$

If the incident wave is linearly polarized with its electric field at 45° to the x and y axes, we have

$$\mathbf{E} = E_x \mathbf{e}_x + E_y \mathbf{e}_y,$$

with $E_x^2 + E_y^2 = E^2$, $E_x = E_y = \frac{E}{\sqrt{2}}$.

Let the reflected wave be $\mathbf{E}' = E_x' \mathbf{e}_x + E_y' \mathbf{e}_y$. The continuity of the tangential component of the electric field across the interface gives

$$E_x + E_x' = E_x'', \tag{2}$$

$$E_y + E_y' = E_y'' . \tag{3}$$

Equation (1) also holds for the incident and reflected wave. As medium 1 is isotropic with permittivity ε_0, we have

$$\mathbf{n} \times \mathbf{H} = -\varepsilon_0 \frac{\omega}{K} \mathbf{E} = -\varepsilon_0 \frac{\omega}{K} (E_x \mathbf{e}_x + E_y \mathbf{e}_y),$$

$$\mathbf{n} \times \mathbf{H}' = \varepsilon_0 \frac{\omega}{K} \mathbf{E}' = \varepsilon_0 \frac{\omega}{K} (E_x' \mathbf{e}_x + E_y' \mathbf{e}_y),$$

as well as

$$\mathbf{n} \times \mathbf{H}'' = -\frac{\omega}{K''} \mathbf{D}'' = -\frac{\omega}{K''} (\varepsilon_{xx} E_x'' \mathbf{e}_x + \varepsilon_{yy} E_y'' \mathbf{e}_y) .$$

The continuity of the tangential component of $\mathbf{H}$ across the interface, $\mathbf{n} \times (\mathbf{H} + \mathbf{H}') = \mathbf{n} \times \mathbf{H}''$, then gives

$$\frac{\varepsilon_0}{K} (E_x - E_x') = \frac{\varepsilon_{xx}}{K''} E_x'' ,$$

$$\frac{\varepsilon_0}{K} (E_y - E_y') = \frac{\varepsilon_{yy}}{K''} E_y'' .$$

Using $\varepsilon_{xx} = n_x^2 \varepsilon_0$, $\varepsilon_{yy} = n_y^2 \varepsilon_0$, $\varepsilon_{zz} = n_z^2 \varepsilon_0$ and $K'' = \frac{\omega}{v_z''} = \frac{\omega}{c} n_z = n_z K$, these equations become

$$E_x - E_x' = \frac{n_x^2}{n_z} E_x'' , \tag{4}$$

$$E_y - E_y' = \frac{n_y^2}{n_z} E_y'' . \tag{5}$$

Combining equations (2) to (5), we have

$$E_x' = \left(\frac{n_z - n_x^2}{n_z + n_x^2} \right) E_x , \qquad E_x'' = \frac{2n_z}{n_x^2 + n_z} E_x ,$$

$$E_y' = \left(\frac{n_z - n_y^2}{n_z + n_y^2} \right) E_y , \qquad E_y'' = \frac{2n_z}{n_y^2 + n_z} E_y .$$

As $E_x^2 + E_y^2 = E^2$ we have

$$\left(\frac{E_x'}{a} \right) + \left(\frac{E_y'}{b} \right)^2 = 1 ,$$

where

$$a = \left(\frac{n_z - n_x^2}{n_z + n_x^2}\right) E,$$

$$b = \left(\frac{n_z - n_y^2}{n_z + n_y^2}\right) E,$$

showing that the reflected wave is elliptically polarized with **E** parallel to the xy plane.

(b) For a slab of thickness d, the transmitted wave **E''** above becomes the wave incident on the plane $z = d$. Denote the three waves at the boundary by subscript 1, as in Fig. 4.9. We then have for the incident wave

$$\mathbf{K}_1 = K_1 \mathbf{n}, \quad K_1 = n_z K, \quad \mathbf{E}_1 = E_{1x}\mathbf{e}_x + E_{1y}\mathbf{e}_y,$$

$$E_{1x} = \frac{2n_z}{n_x^2 + n_z} E_x,$$

$$E_{1y} = \frac{2n_z}{n_y^2 + n_z} E_y,$$

$$\mathbf{n} \times \mathbf{H}_1 = -\frac{\omega}{K_1}\left(\varepsilon_{xx} E_{1x}\mathbf{e}_x + \varepsilon_{yy} E_{1y}\mathbf{e}_y\right);$$

for the refracted wave

$$\mathbf{K}_1' = -K_1\mathbf{n},$$

$$\mathbf{E}_1' = E_{1x}'\mathbf{e}_x + E_{1y}'\mathbf{e}_y,$$

$$\mathbf{n} \times \mathbf{H}_1' = \frac{\omega}{K_1}\left(\varepsilon_{xx} E_{1x}'\mathbf{e}_x + \varepsilon_{yy} E_{1y}'\mathbf{e}_y\right);$$

for the transmitted wave

$$\mathbf{K}_1'' = K\mathbf{n},$$

$$\mathbf{E}_1'' = E_{1x}''\mathbf{e}_x + E_{1y}''\mathbf{e}_y,$$

$$\mathbf{n} \times \mathbf{H}_1'' = -\varepsilon_0\frac{\omega}{K}\left(E_{1x}''\mathbf{e}_x + E_{1y}''\mathbf{e}_y\right).$$

The boundary conditions for the interface $z = d$ give

$$E_{1x}e^{iK_1 d} + E_{1x}'e^{-iK_1 d} = E_{1x}''e^{iKd}, \tag{6}$$

$$E_{1y}e^{iK_1 d} + E_{1y}'e^{-iK_1 d} = E_{1y}''e^{iKd}, \tag{7}$$

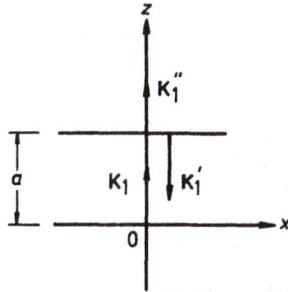

Fig. 4.9

$$\frac{\omega}{K_1}\varepsilon_{xx}\left(E_{1x}e^{iK_1 d} - E'_{1x}e^{-iK_1 d}\right) = \varepsilon_0 \frac{\omega}{K}E''_{1x}e^{iKd}\,,$$

$$\frac{\omega}{K_1}\varepsilon_{yy}\left(E_{1y}e^{iK_1 d} - E'_{1y}e^{-iK_1 d}\right) = \varepsilon_0 \frac{\omega}{K}E''_{1y}e^{iKd}\,.$$

The last two equations can be rewritten as

$$E_{1x}e^{iK_1 d} - E'_{1x}e^{-iK_1 d} = \frac{n_z}{n_x^2}E''_{1x}e^{iKd} \tag{8}$$

$$E_{1y}e^{iK_1 d} - E'_{1y}e^{-iK_1 d} = \frac{n_z}{n_y^2}E''_{1y}e^{iKd}\,. \tag{9}$$

The simultaneous equations (6), (7), (8), (9) give the amplitude and phase of the three electromagnetic waves at the second interface. In fact, the reflected wave K'_1 again becomes the incident wave on the plane $z = 0$ and reflection and transmission will again occur, and so on. Thus multi-reflection will occur between the upper and lower surfaces of the slab, with some energy transmitted out of the slab at each reflection.

4023

Consider an electromagnetic wave of angular frequency ω in a medium containing free electrons of density n_e.

(a) Find the current density induced by E (neglect interaction between electrons).

(b) From Maxwell's equations write the differential equations for the spatial dependence of a wave of frequency ω in such a medium.

(c) Find from these equations the necessary and sufficient condition that the electromagnetic waves propagate in this medium indefinitely.

(*Columbia*)

Solution:

(a) The equation of motion of an electron in the field of an electromagnetic wave is

$$m_e \frac{dv}{dt} = -eE,$$

where we have neglected the action of the magnetic field, which is of magnitude vE/c, as $v \ll c$. For a wave of angular frequency ω, $\frac{\partial}{\partial t} \rightarrow -i\omega$ and the above gives

$$v = -i\frac{e}{m_e\omega}E.$$

Thus the current density is

$$j = -n_e ev = i\frac{n_e e^2 E}{m_e\omega}.$$

(b) Maxwell's equations are

$$\nabla \cdot D = \rho, \tag{1}$$

$$\nabla \times E = -\frac{\partial B}{\partial t}, \tag{2}$$

$$\nabla \cdot B = 0, \tag{3}$$

$$\nabla \times B = \mu_0 j + \mu_0\varepsilon_0\frac{\partial E}{\partial t}. \tag{4}$$

Equations (2) and (4) give

$$\nabla \times (\nabla \times E) = \nabla(\nabla \cdot E) - \nabla^2 E = -\frac{\partial}{\partial t}(\nabla \times B) = -\frac{\partial}{\partial t}\left(\mu_0 j + \frac{1}{c^2}\frac{\partial E}{\partial t}\right),$$

as $c = (\mu_0\varepsilon_0)^{-\frac{1}{2}}$.

We can take the medium to be charge free apart from the free electrons. Thus (1) gives $\nabla \cdot E = \frac{\rho}{\varepsilon_0} = 0$. We can also write

$$\mu_0\frac{\partial j}{\partial t} = -\frac{\mu_0}{i\omega}\frac{\partial^2 j}{\partial t^2} = -\frac{\mu_0 n_e e^2}{m_e\omega^2}\frac{\partial^2 E}{\partial t^2}.$$

Hence

$$\nabla^2 \mathbf{E} - \frac{1}{c^2}\left(1 - \frac{\omega_P^2}{\omega^2}\right)\frac{\partial^2 \mathbf{E}}{\partial t^2} = 0$$

with

$$\omega_P^2 = \frac{n_e e^2}{m_e \varepsilon_0}\,.$$

Similarly, we obtain

$$\nabla^2 \mathbf{B} - \frac{1}{c^2}\left(1 - \frac{\omega_P^2}{\omega^2}\right)\frac{\partial^2 \mathbf{B}}{\partial t^2} = 0\,.$$

The wave equations can be written in the form

$$\nabla^2 \mathbf{E}_0 + \frac{\omega^2}{c^2}\left(1 - \frac{\omega_P^2}{\omega^2}\right)\mathbf{E}_0 = 0$$

by putting $\mathbf{E}(\mathbf{r}, t) = \mathbf{E}_0(\mathbf{r})\exp(-i\omega t)$, giving the spatial dependence.

(c) The solution of the last equation is of the form $\mathbf{E}_0(\mathbf{r}) \sim \exp(i\mathbf{K}\cdot\mathbf{r})$, giving

$$K^2 c^2 = \omega^2 - \omega_P^2\,.$$

The necessary and sufficient condition that the electromagnetic waves propagate in this medium indefinitely is that K is real, i.e. $\omega^2 > \omega_P^2$, or

$$n_e < \frac{\varepsilon_0 m_e \omega^2}{e^2}\,.$$

4024

An electromagnetic wave with electric field given by

$$E_y = E_0 e^{i(Kz - \omega t)}\,, \qquad E_x = E_z = 0\,,$$

propagates in a uniform medium consisting of n free electrons per unit volume. All other charges in the medium are fixed and do not affect the wave.

(a) Write down Maxwell's equations for the fields in the medium.

(b) Show that they can be satisfied by the wave provided $\omega^2 > \frac{ne^2}{m\varepsilon_0}$.

(c) Find the magnetic field and the wavelength of the electromagnetic wave for a given (allowable) ω. Neglect the magnetic force on the electrons.

(*Wisconsin*)

Solution:

(a) (b) Refer to Problem **4023** for the solution.

(c) Using Maxwell's equation

$$\nabla \times \mathbf{E} = -\frac{\partial \mathbf{B}}{\partial t},$$

as

$$\nabla \times \mathbf{E} = iK\mathbf{e}_z \times E_y\mathbf{e}_y = -iKE_z\mathbf{e}_x,$$

$$\frac{\partial \mathbf{B}}{\partial t} = -i\omega\mathbf{B},$$

we have

$$\mathbf{B} = -\frac{K}{\omega}E_0 e^{i(Kz-\omega t)}\mathbf{e}_x.$$

Note that we have used $\nabla \times = -i\mathbf{K} \times$ from Problem **4004**.

3. PROPAGATION OF ELECTROMAGNETIC WAVES IN A MEDIUM (4025–4045)

4025

What is the attenuation distance for a plane wave propagating in a good conductor? Express your answer in terms of the conductivity σ, permeability μ, and frequency ω.

(*Coulumbia*)

Solution:

For a ohmic conducting medium of permittivity ε, permeability μ and conductivity σ, the general wave equation to be used is

$$\nabla^2\mathbf{E} - \mu\varepsilon\ddot{\mathbf{E}} - \mu\sigma\dot{\mathbf{E}} = 0.$$

For plane electromagnetic waves of angular frequency ω, $\mathbf{E}(\mathbf{r},t) = \mathbf{E}_0(\mathbf{r})e^{-i\omega t}$, the above becomes

$$\nabla^2\mathbf{E}_0 + \mu\varepsilon\omega^2\left(1 + \frac{i\sigma}{\omega\varepsilon}\right)\mathbf{E}_0 = 0.$$

Comparing this with the wave equation for a dielectric, we see that for the conductor we have to replace

$$\mu\varepsilon \rightarrow \mu\varepsilon\left(1 + \frac{i\sigma}{\omega\varepsilon}\right),$$

if we wish to use the results for a dielectric.

Consider the plane wave as incident on the conductor along the inward normal, whose direction is taken to be the z-axis. Then in the conductor the electromagnetic wave can be represented as

$$\mathbf{E} = \mathbf{E}_0 e^{i(kz-\omega t)}.$$

The wave vector has magnitude

$$k = \frac{\omega}{v} = \omega\sqrt{\mu\varepsilon}\left(1 + \frac{i\sigma}{\omega\varepsilon}\right)^{\frac{1}{2}}.$$

Let $k = \beta + i\alpha$. We have

$$\beta^2 - \alpha^2 = \omega^2\mu\varepsilon, \qquad \alpha\beta = \frac{1}{2}\omega\mu\sigma.$$

For a good conductor, i.e. for $\frac{\sigma}{\varepsilon\omega} \gg 1$, we have the solution

$$\alpha = \beta = \pm\sqrt{\frac{\omega\varepsilon\sigma}{2}}.$$

In the conductor we then have

$$\mathbf{E} = \mathbf{E}_0 e^{-\alpha z} e^{i(\beta z-\omega t)}.$$

By the definition of the wave vector, β has to take the positive sign. As the wave cannot amplify in the conductor, α has also to take the positive sign. The attenuation length δ is the distance the wave travels for its amplitude to reduce to e^{-1} of its initial value. Thus

$$\delta = \frac{1}{\alpha} = \sqrt{\frac{2}{\omega\mu\sigma}}.$$

4026

Given a plane polarized electric wave

$$\mathbf{E} = \mathbf{E}_0 \exp\left\{ i\omega\left[t - \frac{n}{c}(\mathbf{K} \cdot \mathbf{r}) \right] \right\},$$

derive from Maxwell's equations the relations between $\mathbf{E}$, $\mathbf{K}$ and the $\mathbf{H}$ field. Obtain an expression for the index of refraction n in terms of ω, ε, μ, σ (the conductivity).

(*Wisconsin*)

Solution:

Maxwell's equations for a charge-free ohmic conducting medium are

$$\begin{cases} \nabla \times \mathbf{E} = -\frac{\partial \mathbf{B}}{\partial t}, & (1) \\ \nabla \times \mathbf{H} = \mathbf{J} + \frac{\partial \mathbf{D}}{\partial t}, & (2) \\ \nabla \cdot \mathbf{D} = 0, & (3) \\ \nabla \cdot \mathbf{B} = 0, & (4) \end{cases}$$

with

$$\mathbf{D} = \varepsilon\mathbf{E}, \qquad \mathbf{B} = \mu\mathbf{H}, \qquad \mathbf{J} = \sigma\mathbf{E}.$$

For the given type of wave we have $\frac{\partial}{\partial t} \to i\omega$, $\nabla \to -i\frac{n\omega}{c}\mathbf{K}$ (cf. Problem 4004). Equations (3) and (4) then give

$$\mathbf{E} \cdot \mathbf{K} = \mathbf{B} \cdot \mathbf{K} = 0,$$

and (1) gives

$$i\frac{n\omega}{c}\mathbf{K} \times \mathbf{E} = i\omega\mu\mathbf{H},$$

or

$$\mathbf{H} = \frac{n}{\mu c}\mathbf{K} \times \mathbf{E}.$$

Taking curl of both sides of (1) and using (2) and (3) we have

$$\nabla^2\mathbf{E} = \mu\sigma\frac{\partial \mathbf{E}}{\partial t} + \mu\varepsilon\frac{\partial^2 \mathbf{E}}{\partial t^2},$$

or

$$\nabla^2\mathbf{E} - \left(\mu\varepsilon - i\frac{\mu\sigma}{\omega} \right)\frac{\partial^2 \mathbf{E}}{\partial t^2} = 0,$$

which is the equation for a wave propagating with phase velocity v given by

$$v^2 = \left(\mu\varepsilon - i\frac{\mu\sigma}{\omega}\right)^{-1} = \frac{1}{\mu\varepsilon}\left(1 - \frac{i\sigma}{\omega\varepsilon}\right)^{-1}.$$

Hence the index of refraction of the medium is

$$n = \frac{c}{v} = \left[\frac{\mu\varepsilon}{\mu_0\varepsilon_0}\left(1 - \frac{i\sigma}{\omega\varepsilon}\right)\right]^{\frac{1}{2}}.$$

Writing $n = \sqrt{\frac{\mu\varepsilon}{\mu_0\varepsilon_0}}(\beta - i\alpha)$, we have

$$\beta^2 - \alpha^2 = 1, \qquad \alpha\beta = \frac{\sigma}{2\omega\varepsilon}.$$

Solving for α and β, we find

$$n = \sqrt{\frac{\mu\varepsilon}{2\mu_0\varepsilon_0}}\left[\sqrt{\left(1 + \frac{\sigma^2}{\varepsilon^2\omega^2}\right)^{1/2} + 1} - i\sqrt{\left(1 + \frac{\sigma^2}{\varepsilon^2\omega^2}\right)^{1/2} - 1}\right].$$

4027

A plane polarized electromagnetic wave $E = E_{y0}e^{i(Kz-\omega t)}$ is incident normally on a semi-infinite material with permeability μ, dielectric constants ε, and conductivity σ.

(a) From Maxwell's equations derive an expression for the electric field in the material when σ is large and real as for a metal at low frequencies.

(b) Do the same for a dilute plasma where the conductivity is limited by the inertia rather than the scattering of the electrons and the conductivity is

$$\sigma = i\frac{ne^2}{m\omega}.$$

(c) From these solutions comment on the optical properties of metals in the ultraviolet.

(Wisconsin)

Solution:

Assume the medium to be ohmic and charge-free, then $\mathbf{j} = \sigma\mathbf{E}$, $\rho = 0$ and Maxwell's equations are

$$\nabla \times \mathbf{E} = -\frac{\partial \mathbf{B}}{\partial t}, \qquad \nabla \times \mathbf{H} = \frac{\partial \mathbf{D}}{\partial t} + \mathbf{j},$$

$$\nabla \cdot \mathbf{D} = \nabla \cdot \mathbf{B} = 0.$$

Assume also that the medium is linear, isotropic and homogeneous so that

$$\mathbf{D} = \varepsilon\mathbf{E}, \qquad \mathbf{B} = \mu\mathbf{H}.$$

For a sinusoidal electromagnetic wave

$$\mathbf{E}(\mathbf{r}, t) = \mathbf{E}(\mathbf{r})e^{-i\omega t}, \qquad \mathbf{B}(\mathbf{r}, t) = \mathbf{B}(\mathbf{r})e^{-i\omega t}$$

in the conducting medium Maxwell's equations become

$$\nabla \times \mathbf{E}(\mathbf{r}) = i\omega\mu\mathbf{H}(\mathbf{r}),$$
$$\nabla \times \mathbf{H}(\mathbf{r}) = -i\omega\varepsilon\mathbf{E}(\mathbf{r}) + \sigma\mathbf{E}(\mathbf{r}),$$
$$\nabla \cdot \mathbf{E}(\mathbf{r}) = \nabla \cdot \mathbf{B}(\mathbf{r}) = 0.$$

Using these we obtain

$$\nabla \times (\nabla \times \mathbf{E}) = \nabla(\nabla \cdot \mathbf{E}) - \nabla^2\mathbf{E} = -\nabla^2\mathbf{E}$$
$$= i\omega\mu\nabla \times \mathbf{H} = (\omega^2\varepsilon\mu + i\omega\mu\sigma)\mathbf{E},$$

i.e.

$$\nabla^2\mathbf{E} + (\omega^2\varepsilon\mu + i\omega\mu\sigma)\mathbf{E} = 0. \tag{1}$$

Putting

$$K''^2 = \omega^2\mu\varepsilon'', \qquad \varepsilon'' = \varepsilon + i\frac{\sigma}{\omega},$$

we can write Eq. (1) as

$$\nabla^2\mathbf{E}(\mathbf{r}) + K''^2\mathbf{E}(\mathbf{r}) = 0.$$

This is Helmholtz' equation with the plane wave solution

$$\mathbf{E}(\mathbf{r}) = \mathbf{E}_0 e^{i\mathbf{K}''\cdot\mathbf{r}},$$

where the propagation vector $\mathbf{K}''$ has a complex magnitude

$$K'' = \omega\left[\mu\left(\varepsilon + \frac{i\sigma}{\omega}\right)\right]^{\frac{1}{2}} = \beta + i\alpha, \text{ say}.$$

β and α are given by the simultaneous equations

$$\begin{cases} \beta^2 - \alpha^2 = \omega^2\mu\varepsilon, \\ \beta\alpha = \frac{1}{2}\omega\mu\sigma, \end{cases}$$

which have solution

$$\beta = \omega\sqrt{\mu\varepsilon}\left[\frac{1}{2}\left(1 + \sqrt{1 + \frac{\sigma^2}{\varepsilon^2\omega^2}}\right)\right]^{1/2},$$

$$\alpha = \omega\sqrt{\mu\varepsilon}\left[\frac{1}{2}\left(-1 + \sqrt{1 + \frac{\sigma^2}{\varepsilon^2\omega^2}}\right)\right]^{1/2}.$$

The given incident wave is $E = E_{y0}e^{i(Kz-\omega t)}$, so that $\mathbf{K} = K\mathbf{e}_z$, $\mathbf{E}_0 = E_{y0}\mathbf{e}_y$, $\mathbf{H}_0 = H_{x0}\mathbf{e}_x$. Let the reflected and transmitted waves be

$$\mathbf{E}' = \mathbf{E}_0'e^{i(\mathbf{K}'\cdot\mathbf{r}-\omega t)}, \qquad \mathbf{H}' = \mathbf{H}_0'e^{i(\mathbf{K}'\cdot\mathbf{r}-\omega t)},$$

$$\mathbf{E}'' = \mathbf{E}_0''e^{i(\mathbf{K}''\cdot\mathbf{r}-\omega t)}, \qquad \mathbf{H}'' = \mathbf{H}_0''e^{i(\mathbf{K}''\cdot\mathbf{r}-\omega t)}.$$

As $\mathbf{E}$, $\mathbf{B}$, $\mathbf{K}$ form a right-hand set and the incident wave is polarized with $\mathbf{E}$ in the y direction, the vectors are as shown in Fig. 4.10. To satisfy the boundary condition that the tangential component of $\mathbf{E}$ is continuous at all points of the interface, we require that the exponents involved should be the same, which in general gives rise to the laws of reflection and refraction, and that the amplitudes should satisfy the following:

$$\begin{cases} E_{y0} + E_{y0}' = E_{y0}'', & (2) \\ H_{x0} - H_{x0}' = H_{x0}''. & (3) \end{cases}$$

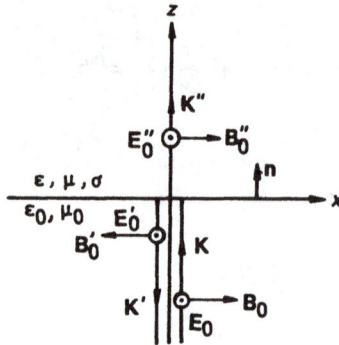

Fig. 4.10

As the waves are plane electromagnetic waves we have

$$H_{x0} = \sqrt{\frac{\varepsilon_0}{\mu_0}}\, E_{y0}\,, \qquad H'_{x0} = \sqrt{\frac{\varepsilon_0}{\mu_0}}\, E'_{y0}\,,$$

$$H''_{x0} = \frac{K''}{\omega\mu}\, E''_{y0}\,,$$

and (3) can be written as

$$\sqrt{\frac{\varepsilon_0}{\mu_0}}\,(E_{y0} - E'_{y0}) = \frac{K''}{\omega\mu}\, E''_{y0}\,. \tag{4}$$

Equations (2) and (4) give

$$E''_{y0} = \frac{2E_{y0}}{1 + \sqrt{\frac{\mu_0}{\varepsilon_0}}\,\frac{K''}{\omega\mu}}\,. \tag{5}$$

(a) If σ is large and real as for a metal at low frequencies, we have

$$\frac{\sigma}{\varepsilon\omega} \gg 1\,, \qquad \sqrt{1 + \frac{\sigma^2}{\varepsilon^2\omega^2}} \pm 1 \approx \frac{\sigma}{\varepsilon\omega}\,,$$

and thus

$$\beta = \alpha \approx \sqrt{\frac{\omega\mu\sigma}{2}}\,,$$

or

$$K'' = \sqrt{\frac{\omega \mu \sigma}{2}}(1 + i).$$

Hence

$$E''_{y0} = \frac{2E_{y0}}{\left(1 + \sqrt{\frac{\mu_0 \sigma}{2\epsilon_0 \omega \mu}}\right) + i\sqrt{\frac{\mu_0 \sigma}{2\epsilon_0 \omega \mu}}} \approx 2E_{y0}\sqrt{\frac{2\epsilon_0 \mu \omega}{\mu_0 \sigma}}(1 + i)^{-1}$$

$$= 2E_{y0}\sqrt{\frac{\epsilon_0 \mu \omega}{\mu_0 \sigma}} e^{-i\frac{\pi}{4}},$$

since, as $\mu \approx \mu_0$, $\epsilon_0 \sim \epsilon$, $\sqrt{\frac{\mu_0 \sigma}{2\epsilon_0 \mu \omega}} \sim \sqrt{\frac{\sigma}{\epsilon \omega}} \gg 1$. Equation (5) then gives

$$E''_{y0} \approx 2\sqrt{\frac{\epsilon_0 \mu \omega}{\mu_0 \sigma}} E_{y0} e^{-\alpha z} e^{i(\beta z - \omega t - \frac{\pi}{4})} \mathbf{e}_y$$

for the electric field in the conducting medium.

(b) For a dilute plasma $\mu \approx \mu_0$, $\epsilon \approx \epsilon_0$, with $\sigma = i\frac{ne^2}{m\omega}$, (1) becomes

$$\nabla^2 \mathbf{E} + \mu_0 \epsilon_0 (\omega^2 - \omega_P^2)\mathbf{E} = 0,$$

where $\omega_P^2 = \frac{ne^2}{m\epsilon_0}$ is the (angular) plasma frequency of the medium. Thus

$$K''^2 = \frac{1}{c^2}(\omega^2 - \omega_P^2).$$

If $\omega_P < \omega$, K'' is real and (5) becomes

$$E''_{y0} = \frac{2E_{y0}}{1 + \frac{cK''}{\omega}} = \frac{2E_{y0}}{1 + (1 - \omega_P^2/\omega^2)^{1/2}},$$

giving

$$\mathbf{E}''_y = E''_{y0} e^{i(K'' z - \omega t)} \mathbf{e}_y.$$

If $\omega_P^2 > \omega^2$, K'' is imaginary and

$$E''_{y0} = \frac{2E_{y0}}{1 + i(\omega_P^2/\omega^2 - 1)^{1/2}},$$

$$\mathbf{E}''_y = E''_{y0} e^{-|K''|z} e^{-i\omega t}.$$

(c) The typical electron number density of metals is $n \approx 10^{22}/\text{cm}^3$. The corresponding plasma frequency is

$$\omega_P = \left(\frac{ne^2}{m\varepsilon_0}\right)^{1/2} = \left(\frac{10^{22} \times 10^6 \times (1.6 \times 10^{-19})^2}{9.1 \times 10^{-31} \times 8.85 \times 10^{-12}}\right)^{1/2}$$
$$\approx 0.56 \times 10^{16} \text{ s}^{-1}.$$

For ultraviolet light, the angular frequency is $\omega > 10^{16}$ s^{-1}. So the condition $\omega_P < \omega$ is satisfied and ultraviolet light can generally propagate in metals.

4028

A plane electromagnetic wave of frequency ω and wave number K propagates in the $+z$ direction. For $z < 0$, the medium is air with $\varepsilon = \varepsilon_0$ and conductivity $\sigma = 0$. For $z > 0$, the medium is a lossy dielectric with $\varepsilon > \varepsilon_0$ and $\sigma > 0$. Assume that $\mu = \mu_0$ in both media.

(a) Find the dispersion relation (i.e., the relationship between ω and K) in the lossy medium.

(b) Find the limiting values of K for a very good conductor and a very poor conductor.

(c) Find the e^{-1} penetration depth δ for plane wave power in the lossy medium.

(d) Find the power transmission coefficient T for transmission from $z < 0$ to $z > 0$, assuming $\sigma \ll \varepsilon\omega$ in the lossy medium.

(e) Most microwave ovens operate at 2.45 GHz. At this frequency, beef may be described approximately by $\varepsilon/\varepsilon_0 = 49$ and $\sigma = 2$ mho m^{-1}. Evaluate T and δ for these quantities, using approximations where needed. Does your answer for δ indicate an advantage of microwave heating over infrared heating (broiling)?

(*MIT*)

Solution:

(a) In a lossy medium, the wave number K' is complex, $\mathbf{K}' = (\beta + i\alpha)\mathbf{e}_z$. From Problem 4025, we see that K' is related to ω by $K'^2 = \omega^2\mu(\varepsilon + i\frac{\sigma}{\omega})$. Thus

$$\beta^2 - \alpha^2 = \omega^2\mu_0\varepsilon,$$

$$\alpha\beta = \frac{1}{2}\omega\mu_0\sigma \, .$$

Solving the simultaneous equations we have

$$\beta = \omega\sqrt{\mu_0\varepsilon}\left[\frac{1}{2}\left(1 + \sqrt{1 + \frac{\sigma^2}{\varepsilon^2\omega^2}}\right)\right]^{1/2} \, ,$$

$$\alpha = \omega\sqrt{\mu_0\varepsilon}\left[\frac{1}{2}\left(-1 + \sqrt{1 + \frac{\sigma^2}{\varepsilon^2\omega^2}}\right)\right]^{1/2} \, .$$

As refractive index is defined as $n = \frac{cK}{\omega} = \frac{c}{\omega}(\beta + i\alpha)$, these equations give the dispersion relation for the medium.

(b) For a very good conductor, $\frac{\sigma}{\varepsilon\omega} \gg 1$, and we have

$$\beta = \alpha \approx \sqrt{\frac{\omega\mu_0\sigma}{2}} \, .$$

For a very poor conductor, $\frac{\sigma}{\varepsilon\omega} \ll 1$, and we have

$$\beta \approx \omega\sqrt{\mu_0\varepsilon} \, , \qquad \alpha \approx \frac{\sigma}{2}\sqrt{\frac{\mu_0}{\varepsilon}} \, .$$

(c) The transmitted wave can be represented as $E_2 = E_{20}e^{-\alpha z}e^{i(\beta z - \omega t)}$. Thus the e^{-1} penetration depth is

$$\delta = \frac{1}{\alpha} = \frac{1}{\omega\sqrt{\mu_0\varepsilon}}\left[\frac{1}{2}\left(-1 + \sqrt{1 + \frac{\sigma^2}{\varepsilon^2\omega^2}}\right)\right]^{-1/2} \, .$$

For a very good conductor: $\delta \approx \sqrt{\frac{2}{\omega\mu_0\sigma}}$.

For a very poor conductor: $\delta \approx \frac{2}{\sigma}\sqrt{\frac{\varepsilon}{\mu_0}}$.

(d) The solution of Problem **4011** gives

$$\frac{E_{20}}{E_{10}} = \frac{2}{1 + n'} \, ,$$

where n' is the index of refraction of the lossy medium. Here n' is complex

$$n' = \frac{c}{\omega}(\beta + i\alpha) \, .$$

For $\sigma \ll \varepsilon \omega$,

$$n' = \frac{1}{\omega\sqrt{\mu_0\varepsilon_0}} \left(\omega\sqrt{\mu_0\varepsilon} + \frac{i\sigma}{2}\sqrt{\frac{\mu_0}{\varepsilon}} \right)$$

$$= \sqrt{\frac{\varepsilon}{\varepsilon_0}} \left(1 + \frac{i}{2}\frac{\sigma}{\varepsilon\omega} \right)$$

$$= n\left(1 + \frac{i}{2}\frac{\sigma}{\varepsilon\omega} \right),$$

where

$$n = \sqrt{\frac{\varepsilon}{\varepsilon_0}}.$$

The average energy incident on or leaving unit area of the interface in unit time is the magnitude of the average Poynting vector $\overline{S}$ (Problem **4011**):

For incident wave: $\overline{S}_1 = \frac{1}{2}\sqrt{\frac{\varepsilon_0}{\mu_0}}|E_{10}|^2.$

For transmitted wave: $\overline{S}_2 = \frac{1}{2}\sqrt{\frac{\varepsilon}{\mu_0}}|E_{20}|^2.$

The power transmision coefficient is therefore

$$T = \frac{\overline{S}_2}{\overline{S}_1} = \sqrt{\frac{\varepsilon}{\varepsilon_0}}\left|\frac{E_{20}}{E_{10}}\right|^2 = \frac{4n}{|1+n'|^2}$$

$$= \frac{4n}{(1+n)^2 + n^2\sigma^2/4\varepsilon^2\omega^2}.$$

(e) To cook beef in the microwave oven given, we have

$$\varepsilon\omega = 49 \times \frac{10^{-9}}{36\pi} \times 2\pi \times 2.45 \times 10^9 \text{ mho/m} \approx 7 \text{ mho/m} > \sigma.$$

If beef can be treated as a poor conductor, the penetration depth and the power transmission coefficient are respectively

$$\delta = \frac{2}{\sigma}\sqrt{\frac{\varepsilon}{\mu_0}} = \frac{2 \times 7}{2}\sqrt{\frac{8.85 \times 10^{-12}}{12.6 \times 10^{-7}}} \approx 1.85 \text{ cm},$$

$$T = \frac{4n}{(1+n)^2 + n^2\sigma^2/4\varepsilon^2\omega} \approx \frac{4 \times 7}{8^2 + 7^2 \times 2^2/4 \times 7^2} \approx 0.43.$$

$$\left(n = \sqrt{\frac{\varepsilon}{\varepsilon_0}} = 7 \right)$$

The wavelength of infrared rays is approximately 10^{-3} cm, so its frequency is $\sim 3 \times 10^{13}$ Hz. For beef in an infrared oven, $\varepsilon \omega \approx \frac{7 \times 3 \times 10^{13}}{2.45 \times 10^9} \approx 10^5$ mho/m $\gg \sigma$, so it is still a poor conductor. Thus the penetration depth and power transmission coefficient of infrared rays in beef will be similar to those for the microwaves. Hence for cooking beef, the effects of the two types of wave are about the same as far as energy penetration and absorption are concerned. No advantage of microwave heating over infrared heating is indicated.

4029

(a) X-rays which strike a metal surface at an angle of incidence to the normal greater than a critical angle θ_0 are totally reflected. Assuming that a metal contains n free electrons per unit volume, calculate θ_0 as a function of the angular frequency ω of the X-rays.

(b) If ω and θ are such that total reflection does not occur, calculate what fraction of the incident wave is reflected. You may assume, for simplicity, that the polarization vector of the X-rays is perpendicular to the plane of incidence.

(*Princeton*)

Solution:

(a) The equation of the motion of an electron in the field of the X-rays is

$$m\ddot{\mathbf{x}} = -e\mathbf{E} = -e\mathbf{E}_0 e^{-i\omega t} \, .$$

Its solution has the form $\mathbf{x} = \mathbf{x}_0 e^{-i\omega t}$. Substitution gives

$$m\omega^2 \mathbf{x} = e\mathbf{E} \, .$$

Each electron acts as a Hertzian dipole, so the polarization vector of the metal is

$$\mathbf{P} = -ne\mathbf{x} = \chi \varepsilon_0 \mathbf{E} \, ,$$

giving the polarizability as

$$\chi = -\frac{ne^2}{m\varepsilon_0 \omega^2} \, .$$

Let $\omega_P^2 = \frac{ne^2}{m\epsilon_0}$, then the index of refraction of the metal is

$$n = \sqrt{1+\chi} = \left(1 - \frac{\omega_P^2}{\omega^2}\right)^{1/2},$$

and the critical angle is

$$\theta_0 = \arcsin\left(1 - \frac{\omega_P^2}{\omega^2}\right)^{1/2}.$$

(b) As the X-rays are assumed to be polarized with **E** perpendicular to the plane of incidence, **E** is tangential to the metal surface. Letting a prime and a double-prime indicate the reflected and refracted rays respectively, we have

$$E + E' = E'', \tag{1}$$

$$H \cos\theta - H' \cos\theta' = H'' \cos\theta''. \tag{2}$$

Note that **E**, **H** and the direction of propagation form a right-hand set. As

$$\theta = \theta', \qquad \sqrt{\epsilon}\, E = \sqrt{\mu}\, H, \qquad \mu \approx \mu_0,$$

(2) can be written as

$$E - E' = \sqrt{\frac{\epsilon}{\epsilon_0}}\, \frac{\cos\theta''}{\cos\theta}\, E'' = \frac{n \cos\theta''}{\cos\theta}\, E''. \tag{3}$$

(1) and (3) together give

$$\frac{E'}{E} = \frac{\cos\theta - n\cos\theta''}{\cos\theta + n\cos\theta''}.$$

As $\theta = \theta'$ and the intensity is $\frac{1}{2}\sqrt{\frac{\epsilon}{\mu}}\, E_0^2$, the reflection coefficient is

$$R = \left(\frac{E'}{E}\right)^2 = \left(\frac{\cos\theta - n\cos\theta''}{\cos\theta + n\cos\theta}\right)^2.$$

4030

Consider a space which is partially filled with a material which has continuous but coordinate-dependent susceptibility χ and conductivity σ given by (χ_∞, λ, σ_∞ are positive constants):

$$\chi(z) = \begin{cases} 0, & -\infty < z \leq 0, \\ \chi_\infty(1 - e^{-\lambda z}), & 0 < z < \infty; \end{cases}$$

$$\sigma(z) = \begin{cases} 0, & -\infty < z \leq 0, \\ \sigma_\infty(1 - e^{-\lambda z}), & 0 < z < \infty. \end{cases}$$

The space is infinite in the x, y directions. Also $\mu = 1$ in all space. An s-polarized plane wave (i.e., **E** is perpendicular to the plane of incidence) traveling from minus to plus infinity is incident on the surface at $z = 0$ with an angle of incidence θ (angle between the normal and k_0), ($k_0 c = \omega$):

$$\mathbf{E}_y^I(\mathbf{r}, t) = A \exp[i(x k_0 \sin\theta + z k_0 \cos\theta - \omega t)]\mathbf{e}_y .$$

The reflected wave is given by

$$\mathbf{E}_y^R(\mathbf{r}, t) = R \exp[i(x k_0 \sin\theta - z k_0 \cos\theta - \omega t)]\mathbf{e}_y ,$$

and the transmitted wave by

$$\mathbf{E}_y^T(\mathbf{r}, t) = E(z) \exp[i(x k' \sin\gamma - \omega t)]\mathbf{e}_y .$$

A and R are the incident and reflected amplitudes. $E(z)$ is a function which you are to determine. γ is the angle between the normal and k'.

(a) Find expressions for the incident, reflected and transmitted magnetic fields in terms of the above parameters.

(b) Match the boundary conditions at $z = 0$ for the components of the fields. (Hint: Remember Snell's law!)

(c) Use Maxwell's equations and the relationships

$$\mathbf{D}(\mathbf{r}, t) = \varepsilon(\mathbf{r})\mathbf{E}(\mathbf{r}, t), \qquad \varepsilon(\mathbf{r}) = 1 + 4\pi\chi(\mathbf{r}) + \frac{4\pi i}{\omega}\sigma(\mathbf{r})$$

to find the wave equation for $\mathbf{E}_y^T(\mathbf{r}, t)$.

(SUNY, Buffalo)

Solution:

(a) The incident and reflected wave vectors are respectively

$$\mathbf{k}^I = (k_0 \sin \theta, \, 0, \, k_0 \cos \theta), \qquad \mathbf{k}^R = (k_0 \sin \theta, \, 0, \, -k_0 \cos \theta).$$

For sinusoidal plane electromagnetic waves, we have (Problem **4004**) $\nabla \rightarrow i\mathbf{k}$, $\frac{\partial}{\partial t} \rightarrow -i\omega$. Maxwell's equation $\nabla \times \mathbf{E} = -\frac{1}{c} \frac{\partial \mathbf{B}}{\partial t}$ then gives

$$i\mathbf{k} \times \mathbf{E} = -\frac{1}{c}(-i\omega)\mathbf{B},$$

or

$$\mathbf{B} = \frac{c}{\omega}\mathbf{k} \times \mathbf{E}.$$

Thus

$$\mathbf{B}^I = \frac{c}{\omega}\mathbf{k}^I \times \mathbf{E} = (-\mathbf{e}_x \cos \theta + \mathbf{e}_z \sin \theta)E_y(\mathbf{r}, t)$$
$$= (-\mathbf{e}_x \cos \theta + \mathbf{e}_z \sin \theta)A \exp[i(xk_0 \sin \theta + zk_0 \cos \theta - \omega t)],$$

$$\mathbf{B}^R = (\mathbf{e}_x \cos \theta + \mathbf{e}_z \sin \theta)R \exp\left[i(xk_0 \sin \theta - zk_0 \cos \theta - \omega t)\right].$$

The magnetic field of the transmitted wave is

$$\mathbf{B}^T = \frac{c}{i\omega}\nabla \times \mathbf{E}^T$$
$$= \frac{c}{i\omega}\left[\mathbf{e}_z E(z)(ik' \sin \gamma) - \mathbf{e}_x \frac{\partial E(z)}{\partial z}\right]\exp[i(k'x \sin \gamma - \omega t)].$$

(b) E_t and H_t are continuous across the boundary, i.e., for $z = 0$

$$E_y^I(\mathbf{r}, t) + E_y^R(\mathbf{r}, t) = E_y^T(\mathbf{r}, t),$$

$$[\mathbf{B}^I(\mathbf{r}, t) + \mathbf{B}^R(\mathbf{r}, t)] \cdot \mathbf{e}_x = \mathbf{B}^T(\mathbf{r}, t) \cdot \mathbf{e}_x.$$

B_n is also continuous across the boundary:

$$[\mathbf{B}^I(\mathbf{r}, t) + \mathbf{B}^R(\mathbf{r}, t)] \cdot \mathbf{e}_z = \mathbf{B}^T(\mathbf{r}, t) \cdot \mathbf{e}_z.$$

Also, Snell's law applies with $z = 0$:

$$k_0 \sin \theta = k' \sin \gamma.$$

Combining the above we obtain

$$A + R = E(0),$$

$$k_0(R - A)\cos\theta = i\left(\frac{\partial E(z)}{\partial z}\right)_{z=0}.$$

(c) Combining Maxwell's equations

$$\nabla \times \mathbf{E} = -\frac{1}{c}\frac{\partial \mathbf{B}}{\partial t}, \qquad \nabla \times \mathbf{H} = \frac{1}{c}\frac{\partial \mathbf{D}}{\partial t} + \frac{4\pi}{c}\mathbf{J},$$

where

$$\mathbf{D} = \varepsilon(\mathbf{r})\mathbf{E}, \quad \mathbf{B} = \mu\mathbf{H} = \mathbf{H}, \quad \mathbf{J} = \sigma\mathbf{E},$$

we have

$$\nabla \times (\nabla \times \mathbf{E}) = -\frac{1}{c}\frac{\partial}{\partial t}(\nabla \times \mathbf{H}) = \frac{\omega^2}{c^2}\left(\varepsilon + i\frac{4\pi\sigma}{\omega}\right)\mathbf{E}.$$

As

$$\nabla \times (\nabla \times \mathbf{E}) = \nabla(\nabla \cdot \mathbf{E}) - \nabla^2\mathbf{E} = -\nabla^2\mathbf{E},$$

for a charge-free medium, the above becomes

$$\nabla^2\mathbf{E} + \frac{\omega^2}{c^2}\left(\varepsilon + i\frac{4\pi\sigma}{\omega}\right)\mathbf{E} = 0.$$

This is the wave equation for a charge-free conducting medium. Apply this to the transmitted wave. As

$$\nabla^2\mathbf{E}^{\mathrm{T}} = \left(\frac{\partial^2}{\partial x^2} + \frac{\partial^2}{\partial y^2} + \frac{\partial^2}{\partial z^2}\right)E(z)\exp[i(xk'\sin\gamma - \omega t)]\mathbf{e}_y$$

$$= \left[-E(z)k'^2\sin^2\gamma + \frac{\partial^2 E(z)}{\partial z^2}\right]\exp[i(xk'\sin\gamma - \omega t)]\mathbf{e}_y,$$

$$\varepsilon = 1 + 4\pi\chi$$

by the definition of electric susceptibility,

$$\chi = \chi_\infty(1 - e^{\lambda z}), \qquad \sigma = \sigma_\infty(1 - e^{-\lambda z}),$$

we have the equation for $E(z)$:

$$\frac{\partial^2 E(z)}{\partial z^2} + \frac{\omega^2}{c^2}\left[1 + 4\pi\left(\chi_\infty + \frac{i\sigma_\infty}{\omega}\right)(1 - e^{-\lambda z})\right]E(z) - k'^2\sin^2\gamma\, E(z) = 0.$$

4031

A plane polarized electromagnetic wave is incident on a perfect conductor at an angle θ. The electric field is given by

$$\mathbf{E} = \mathbf{E}_0 \,\text{Re}\, \exp\left[i(\mathbf{k} \cdot \mathbf{r} - \omega t)\right].$$

$\mathbf{E}$ is in the plane of incidence as shown in Fig. 4.11. Starting with the boundary conditions imposed on an electromagnetic field by a conductor, derive the following properties of the reflected wave: direction of propagation, amplitude, polarization and phase.

$$(MIT)$$

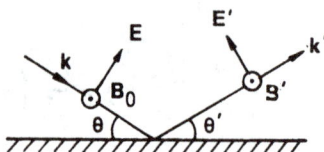

Fig. 4.11

Solution:

In a perfect conductor, $\mathbf{E}'' = \mathbf{B}'' = 0$. Since the normal component of $\mathbf{B}$ is continuous across the interface, the magnetic vector $\mathbf{B}'$ of the reflected wave has only tangential component, as shown in Fig. 4.11. As for a plane electromagnetic wave, $\mathbf{E}$, $\mathbf{B}$ and $\mathbf{k}$ form an orthogonal right-hand set, $\mathbf{E}'$ and $\mathbf{k}'$ must then be in a plane containing $\mathbf{k}$ and perpendicular to the boundary (the plane of incidence). Also, because of the continuity of the tangential component of $\mathbf{E}$ across the interface, the electric vector $\mathbf{E}'$ of the reflected wave must have the direction shown in Fig. 4.11 and we thus have

$$E \sin \theta - E' \sin \theta' = 0.$$

In addition, for the boundary conditions to be satisfied, the exponents in the expressions for $\mathbf{E}$ and $\mathbf{E}'$ must be equal at the boundary. This requires that

$$\mathbf{k} \cdot \mathbf{r} = \mathbf{k}' \cdot \mathbf{r},$$

or

$$k \cos \theta = k' \cos \theta',$$

taking $\mathbf{r}$ in the interface and in the plane incidence.

As $k = k' = \frac{\omega}{c}$, $\cos\theta = \cos\theta'$, or $\theta = \theta'$, from which follows $E = E'$. Therefore, the direction of propagation of the reflected wave, given by the vector $\mathbf{k}'$, makes the same angle with the surface of the conductor as that of the incident wave, given by $\mathbf{k}$; both are in the plane of incidence. The magnitude E' of the electric field of the reflected wave is the same as that of the incident wave, and the reflected wave remains linearly polarized. However, as $\mathbf{E}_t = -\mathbf{E}'_t$, a phase change of π occurs on reflection.

4032

(a) Consider a long straight cylindrical wire of electrical conductivity σ and radius a carrying a uniform axial current of density J. Calculate the magnitude and direction of the Poynting vector at the surface of the wire.

(b) Consider a thick conducting slab (conductivity σ) exposed to a plane EM wave with peak amplitudes E_0, B_0. Calculate the Poynting vector within the slab, averaged in time over a wave period. Consider σ large, i.e. $\sigma \gg \omega\varepsilon_0$.

(c) In part (b), if σ is infinite, what is the value of the average Poynting vector everywhere in space?

(*Wisconsin*)

Solution:

(a) Use cylindrical coordinates (r, θ, z) with the z-axis along the axis of the wire and let the current flow along the $+z$ direction. Assume the conductor to be ohmic, i.e., $\mathbf{J} = \sigma\mathbf{E}$. Then $\mathbf{E} = \frac{\mathbf{J}}{\sigma} = \frac{J}{\sigma}\mathbf{e}_z$ inside the wire. Due to the continuity of the tangential component of $\mathbf{E}$ across the interface, we also have $\mathbf{E} = \frac{J}{\sigma}\mathbf{e}_z$ just outside the surface of the wire. Using Ampère's circuital law $\oint \mathbf{B} \cdot d\mathbf{l} = \mu_0 I$ we find the magnetic field near the surface of the wire as

$$\mathbf{B} = \frac{\mu_0 I}{2\pi a}\mathbf{e}_\theta = \frac{\mu_0 J\pi a^2}{2\pi a}\mathbf{e}_\theta = \frac{\mu_0 J a}{2}\mathbf{e}_\theta .$$

Hence the Poynting vector at the surface of the wire is

$$\mathbf{S} = \frac{1}{\mu_0}\mathbf{E} \times \mathbf{B} = \frac{J}{\sigma}\mathbf{e}_z \times \frac{J a}{2}\mathbf{e}_\theta = -\frac{J^2 a}{2\sigma}\mathbf{e}_r .$$

(b) For simplicity suppose that the normal to the surface of the slab is parallel to the direction of wave propagation, i.e., along the $+z$ direction.

Then the wave vector in the conductor is

$$\mathbf{K} = \boldsymbol{\beta} + i\boldsymbol{\alpha} = (\beta + i\alpha)\mathbf{e}_z \, .$$

As σ is large, we have (Problem **4027**)

$$\alpha = \beta \approx \sqrt{\frac{\omega\mu_0\sigma}{2}} \, ,$$

taking $\mu \approx \mu_0$ (nonferromagnetic).

The electric field inside the conductor is

$$\mathbf{E}(\mathbf{r}, t) = \mathbf{E}_0(\mathbf{r}) e^{-\alpha z} e^{i(\beta z - \omega t)}$$

and the magnetic field is

$$\mathbf{H} = \frac{1}{\omega\mu_0} \mathbf{K} \times \mathbf{E} = \frac{1}{\omega\mu_0} (\beta + i\alpha)\mathbf{e}_z \times \mathbf{E}$$

$$\approx \sqrt{\frac{\sigma}{\omega\mu_0}} e^{i\frac{\pi}{4}} \mathbf{e}_z \times \mathbf{E} \, ,$$

so the Poynting vector is

$$\mathbf{S} = \mathbf{E} \times \mathbf{H} = \sqrt{\frac{\sigma}{\omega\mu_0}} e^{i\frac{\pi}{4}} E^2 \mathbf{e}_z \, ,$$

as $\mathbf{E} \cdot \mathbf{K} = \mathbf{E} \cdot \mathbf{e}_z = 0$ for a plane wave.

Averaging over one period, we obtain (Problem **4042**)

$$\overline{\mathbf{S}} = \frac{1}{2} \mathrm{Re}\,(\mathbf{E}^* \times \mathbf{H}) = \frac{1}{2} \sqrt{\frac{\sigma}{\omega\mu_0}} \cos\left(\frac{\pi}{4}\right) E_0^2 e^{-2\alpha z} \mathbf{e}_z$$

$$= \frac{\sqrt{2}}{4} \sqrt{\frac{\sigma}{\omega\mu_0}} E_0^2 e^{-2\alpha z} \mathbf{e}_z \, .$$

(c) As $\sigma \to \infty$, $\alpha \to \infty$ and $\sqrt{\sigma}\, e^{-2\alpha z} \to 0$. That is, $\overline{\mathbf{S}}$ inside the conducting slab becomes zero. In this case, the wave will be totally reflected at the surface of the slab. Moreover, outside the slab the incident and reflected waves will combine to form stationary waves. Hence $\overline{\mathbf{S}} = 0$ everywhere.

4033

A slowly varying magnetic field, $B = B_0 \cos \omega t$, in the y direction induces eddy currents in a slab of material occupying the half plane $z > 0$. The slab has permeability μ and conductivity σ. Starting from the Maxwell equations, determine the attenuation of the eddy currents with depth into the slab and the phase relation between the currents and the inducing field.

(*UC, Berkeley*)

Solution:

From Maxwell's equations for a conductor of constants μ, ε, σ

$$\nabla \cdot \mathbf{E} = 0, \qquad \nabla \times \mathbf{E} = -\frac{\partial \mathbf{B}}{\partial t},$$

$$\nabla \cdot \mathbf{B} = 0, \qquad \nabla \times \mathbf{B} = \mu \sigma \mathbf{E} + \frac{\varepsilon \partial \mathbf{E}}{\partial t}.$$

We find

$$\nabla \times (\nabla \times \mathbf{B}) = -\nabla^2 \mathbf{B} = -\mu \sigma \frac{\partial \mathbf{B}}{\partial t} - \mu \varepsilon \frac{\partial^2 \mathbf{B}}{\partial t^2}.$$

With the given geometry and magnetic field, we expect

$$\mathbf{B}' = B_0' \exp[i(kz - \omega t)] \mathbf{e}_y$$

in the conducting material and the above equation to reduce to

$$\frac{\partial^2 \mathbf{B}}{\partial z^2} - \mu \sigma \frac{\partial \mathbf{B}}{\partial t} - \mu \varepsilon \frac{\partial^2 \mathbf{B}}{\partial t^2} = 0,$$

and further to

$$-k^2 + i\mu\sigma\omega + \mu\varepsilon\omega^2 = 0.$$

Hence

$$k = \omega \sqrt{\mu\varepsilon} \left(1 + \frac{i\sigma}{\varepsilon\omega}\right)^{\frac{1}{2}} \equiv \alpha + i\beta.$$

Since the given frequency is low we can take $\varepsilon\omega \ll \sigma$. Accordingly we have

$$\alpha + i\beta \approx \sqrt{i\mu\sigma\omega} = \sqrt{\frac{\mu\sigma\omega}{2}}(1 + i),$$

or

$$\alpha \approx \beta \approx \sqrt{\frac{\mu\sigma\omega}{2}}.$$

Therefore in the conducting material we have

$$\mathbf{B}' = B_0' e^{-\beta z} e^{i(\alpha z - \omega t)} \mathbf{e}_y .$$

Thus the magnetic field will attenuate with increasing depth with attenuation coefficient β. The last Maxwell's equation above gives

$$\nabla \times \mathbf{B}' = \mu \sigma \mathbf{E}' - i \mu \epsilon \omega \mathbf{E}' \approx \mu \sigma \mathbf{E}'$$

as $\sigma \gg \epsilon \omega$. Thus

$$\begin{aligned}
\mathbf{E}' &\approx \frac{1}{\mu \sigma} \nabla \times \mathbf{B}' = -\frac{1}{\mu \sigma} \frac{\partial B_y'}{\partial z} \mathbf{e}_x \\
&= \frac{-ik}{\mu \sigma} B_y' \mathbf{e}_x = \sqrt{\frac{\omega}{\mu \sigma}} e^{i \frac{\pi}{4}} e^{-i \frac{\pi}{2}} B_y' \mathbf{e}_x \\
&= \sqrt{\frac{\omega}{\mu \sigma}} B_0' e^{-\beta z} e^{i(\alpha z - \omega t - \frac{\pi}{4})} \mathbf{e}_x .
\end{aligned}$$

Hence the induced current density is

$$\mathbf{J} = \sigma \mathbf{E}' = \sqrt{\frac{\sigma \omega}{\mu}} B_0' e^{-\beta z} e^{i(\alpha z - \omega t - \frac{\pi}{4})} \mathbf{e}_x .$$

Thus there is a phase difference of $\frac{\pi}{4}$ between the current and the inducing magnetic field.

4034

Given a hollow copper box of dimensions shown in Fig. 4.12.

(a) How many electromagnetic modes of wavelength λ are there in the range $(4/\sqrt{5}) \leq \lambda \leq (8/\sqrt{13})$ cm?

(b) Find the wavelengths.

(c) Identify the modes by sketches of the $\mathbf{E}$ field.

(d) Approximately how many modes are there in the range $(0.01) \leq \lambda \leq (0.011)$ cm?

(*UC, Berkeley*)

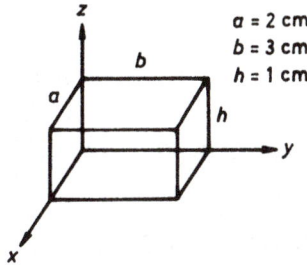

Fig. 4.12

Solution:

(a) For this cavity resonator the wavelength of the stationary wave mode (m, n, p) is given by

$$\lambda_{m,n,p} = \frac{2}{\sqrt{(\frac{m}{a})^2 + (\frac{n}{b})^2 + (\frac{p}{h})^2}}$$

$$= \frac{2}{\sqrt{\frac{m^2}{4} + \frac{n^2}{9} + p^2}} \text{ cm}.$$

For $\frac{4}{\sqrt{5}} \le \lambda \le \frac{8}{\sqrt{13}}$, $\frac{13}{16} \le \frac{m^2}{4} + \frac{n^2}{9} + p^2 \le \frac{5}{4}$.
As the integers m, n, p must be either 0 or positive with $mn + np + pm \ne 0$, we have

$$m = 1, \ n = 3; \qquad m = 2, \ n = 1; \quad \text{for } p = 0;$$
$$m = 1, \ n = 0; \qquad m = 0, \ n = 1; \quad \text{for } p = 1.$$

However, each set of m, n, p corresponds to a TE and a TM mode. Hence in the wavelength range $\frac{4}{\sqrt{5}} \le \lambda \le \frac{8}{\sqrt{13}}$ cm there are eight resonant modes: 2 for each $(1, 3, 0)$, $(2, 1, 0)$, $(1, 0, 1)$ and $(0, 1, 1)$.

(b) The wavelengths of the four double modes are respectively $\frac{4}{\sqrt{5}}$, $\frac{6}{\sqrt{10}}$, $\frac{4}{\sqrt{5}}$, $\frac{6}{\sqrt{10}}$ cm. However there are only two different resonant wavelengths.

(c) The **E** field in the cavity has components

$$E_x = A_1 \cos(k_x x) \sin(k_y y) \sin(k_z z),$$
$$E_y = A_2 \sin(k_x x) \cos(k_y y) \sin(k_z z),$$
$$E_z = A_3 \sin(k_x x) \sin(k_y y) \cos(k_z z),$$

with

$$k_x = \frac{m\pi}{a}, \quad k_y = \frac{n\pi}{b}, \quad k_z = \frac{p\pi}{c}, \quad k_x A_1 + k_y A_2 + k_z A_3 = 0.$$

The four electric modes have **E** fields as follows:

mode $(1, 3, 0)$: $E_x = 0, \ E_y = 0, \ E_z = A_3 \sin\left(\frac{\pi}{2}x\right) \sin(\pi y)$;

mode $(2, 1, 0)$: $E_x = 0, \ E_y = 0, \ E_z = A_3 \sin(\pi x) \sin\left(\frac{\pi}{3}y\right)$;

mode $(1, 0, 1)$: $E_x = 0, \ E_y = A_2 \sin\left(\frac{\pi}{2}x\right) \sin(\pi z), \ E_z = 0$;

mode $(0, 1, 1)$: $E_x = A_1 \sin\left(\frac{\pi}{3}y\right) \sin(\pi z), \ E_y = 0, \ E_z = 0.$

(d) If

$$0.01 \text{ cm} \ \leq \ \frac{2}{\sqrt{\frac{m^2}{4} + \frac{n^2}{9} + p^2}} \ \leq 0.011 \text{ cm},$$

we have

$$181.8^2 \leq \frac{m^2}{4} + \frac{n^2}{9} + p^2 \leq 200^2 .$$

This corresponds to an ellipsoid shell in the *mnp*-space where each unit cell with positive m, n, p represents two electromagnetic modes, one TE and one TM, with frequency less than or equal to $2(\frac{m^2}{4} + \frac{n^2}{9} + p^2)^{-\frac{1}{2}}$, of volume

$$
\begin{aligned}
\Delta V &= V_2 - V_1 \\
&= \frac{4}{3}\pi(2 \times 200 \times 3 \times 200 \times 200 - 2 \times 181.8 \times 3 \times 181.8 \times 181.8) \\
&= \frac{4}{3}\pi \times 2 \times 3 \times (200^3 - 181.8^3) \approx 5 \times 10^7 .
\end{aligned}
$$

Hence in the given range of wavelengths there are $2 \cdot \frac{1}{8} \cdot \Delta V = 1.25 \times 10^7$ modes, where the factor $\frac{1}{8}$ is for the requirement that m, n, p should all be non-negative.

4035

Estimate the number of distinct standing light which can exist between frequencies 1.0×10^{15} Hz and 1.2×10^{15} Hz in a cavity of volume 1 cm^3.

(UC, Berkeley)

Solution:

Consider a cubical cavity resonator of sides of length a. The resonant frequency f is given by

$$4\pi^2 f^2 = \frac{\pi^2}{\mu \varepsilon a^2} (m^2 + n^2 + p^2),$$

where m, n, p are positive integers.

Each set of positive integers m, n, p with

$$m^2 + n^2 + p^2 \leq r^2 = \frac{4a^2 f^2}{v^2}$$

corresponds to a frequency $\leq f(r)$, where $v = \frac{1}{\sqrt{\mu \varepsilon}}$.

For wavelengths short compared with a, we can consider an m, n, p-space where each unit cell represents a set of m, n, p. Then the number of modes N with frequencies $\leq f(r)$ is equal to the volume of $\frac{1}{8}$ of a sphere of radius r in this space:

$$N = \frac{1}{8} \cdot \frac{4}{3} \pi r^3 = \frac{4 \pi f^3 V}{3 v^3},$$

where $V = a^3$ is the volume of the cavity.

However, each set of m, n, p actually corresponds to two modes of the same frequency, one electric and one magnetic. Thus

$$N = \frac{8 \pi f^3 V}{3 v^3}.$$

Under the condition of short wavelengths, this formula can be applied to a cavity of any shape.

For this problem, we have $V = 1$ cm^3 and shall assume the dielectric of the cavity to be air. Then

$$N = \frac{8 \pi f^3}{3 c^3}.$$

Hence the number of modes between the two given frequencies is

$$\Delta N = \frac{8\pi}{3 \times (3 \times 10^{10})^3} [(1.2 \times 10^{15})^3 - (1.0 \times 10^{15})^3]$$

$$= 2.26 \times 10^{14}.$$

4036

Consider a rectangular waveguide, infinitely long in the x-directions, with a width (y-direction) 2 cm and a height (z-direction) 1 cm. The walls are perfect conductor, as in Fig. 4.13.

(a) What are the boundary conditions on the components of **B** and **E** at the walls?

(b) Write the wave equation which describes the **E** and **B** fields of the lowest mode. (Hint: The lowest mode has the electric field in the z-direction only.)

(c) For the lowest mode that can propagate, find the phase velocity and the group velocity.

(d) The possible modes of propagation separate naturally into two classes. What are these two classes and how do they differ physically?

(*Princeton*)

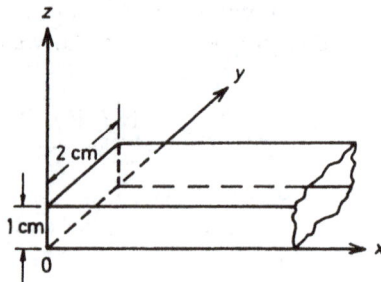

Fig. 4.13

Solution:

(a) The boundary conditions are that the tangential component of **E** and the normal component of **B** are zero on the surface of a perfect conductor. In this case

$$B_y = 0, \quad E_x = E_z = 0, \text{ for } y = 0, 2 \text{ cm};$$
$$B_z = 0, \quad E_x = E_y = 0, \text{ for } z = 0, 1 \text{ cm}.$$

It follows from $\nabla \cdot \mathbf{E} = 0$ that $\frac{\partial E_y}{\partial y} = 0$ for $y = 0, 2$ cm and $\frac{\partial E_z}{\partial z} = 0$ for $z = 0, 1$ cm also.

(b) For sinusoidal waves of angular frequency ω, the wave equation reduces to Helmholtz's equation

$$\nabla^2 \mathbf{E} + k^2 \mathbf{E} = 0$$

with

$$k^2 = \frac{\omega^2}{c^2},$$

and Maxwell's equation

$$\nabla \times \mathbf{E} = -\frac{\partial \mathbf{B}}{\partial t}$$

reduces to

$$\mathbf{B} = -\frac{i}{\omega} \nabla \times \mathbf{E}.$$

For the lowest mode, $E_x = E_y = 0$, $E = E_z$. Thus it is a TE wave, given by the wave equation $\nabla^2 E_z + k^2 E_z = 0$. The magnetic vector is then given by

$$B_x = \frac{-i}{\omega} \frac{\partial E_z}{\partial y}, \qquad B_y = \frac{i}{\omega} \frac{\partial E_z}{\partial x}, \qquad B_z = 0.$$

(c) For the lowest mode, the wave can be represented by

$$E_z = Y(y) Z(z) e^{i(k'x - \omega t)}.$$

Helmholtz's equation can then be separated into

$$\frac{d^2 Y}{dy^2} + k_1^2 Y = 0, \qquad \frac{d^2 Z}{dz^2} + k_2^2 Z = 0,$$

with $k_1^2 + k_2^2 = k^2 - k'^2$. The solutions are

$$Y = A_1 \cos(k_1 y) + A_2 \sin(k_1 y),$$
$$Z = B_1 \cos(k_2 z) + B_2 \sin(k_2 z).$$

The boundary conditions that

$$E_z = 0 \text{ for } y = 0, 2,$$

$$\frac{\partial E_z}{\partial z} = 0 \text{ for } z = 0, 1$$

give $A_1 = B_2 = 0$, $k_1 = \frac{m}{2}\pi$, $k_2 = n\pi$, m, n being 0 or positive integers. Hence

$$k'^2 = \frac{\omega^2}{c^2} - \left[\left(\frac{m}{2}\right)^2 + n^2\right]\pi^2,$$

$$E_z = C\sin\left(\frac{m}{2}\pi y\right)\cos(n\pi z)e^{i(k'x - \omega t)}.$$

Let the phase velocity in the waveguide be v. Then $k' = \frac{\omega}{v}$, or

$$\omega = v\left\{\frac{\omega^2}{c^2} - \left[\left(\frac{m}{2}\right)^2 + n^2\right]\pi^2\right\}^{\frac{1}{2}}.$$

n can be allowed to have zero value without E_z vanishing identically. Hence the lowest mode is TE_{10}, whose phase velocity is

$$v = \frac{\omega}{\sqrt{\left(\frac{\omega^2}{c^2} - \frac{\pi^2}{4}\right)}} > c.$$

The group velocity is

$$v_g = \frac{d\omega}{dk'} = \left(\frac{dk'}{d\omega}\right)^{-1} = \frac{c^2}{\omega}k' = \frac{c^2}{\omega^2}\sqrt{\frac{\omega^2}{c^2} - \frac{\pi^2}{4}} = \frac{c^2}{v}.$$

(d) Electromagnetic waves propagating in a waveguide can be classified into two groups. One with the electric field purely transverse but the magnetic field having a longitudinal component (TE or M mode), the other with the magnetic field purely transverse but the electric field having a longitudinal component (TM or E mode). For the type of guiding system under consideration, it is not possible to propagate waves that are transverse in both electric and magnetic fields (TEM mode).

4037

As in Fig. 4.14 an electromagnetic wave is propagating in the TE mode in the rectangular waveguide. The walls of the waveguide are conducting and the inside is vacuum.

(a) What is the cutoff frequency in this mode?

(b) If the inside is filled with a material with dielectric constant ε, how does the cutoff frequency change?

(Columbia)

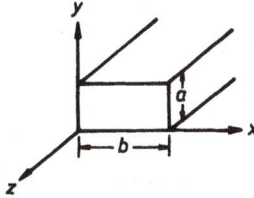

Fig. 4.14

Solution:

In TE modes $E_z = 0$, $H_z \neq 0$, using the coordinate system shown in Fig. 4.14. The transverse, i.e. x and y, component waves in the waveguide are standing waves, while the z component is a traveling wave. Let m and n denote the numbers of half-waves in the x and y directions respectively. The wave numbers of the standing waves are then

$$k_x = \frac{m\pi}{b}, \qquad k_y = \frac{n\pi}{a},$$

while the wave number of the traveling wave is

$$k_z^2 = k^2 - (k_x^2 + k_y^2),$$

where $k^2 = \mu\varepsilon\omega^2$.

(a) If the inside of the waveguide is vacuum, we have

$$k^2 = \mu_0\varepsilon_0\omega^2,$$

or

$$k_z^2 = \mu_0\varepsilon_0\omega^2 - \left[\left(\frac{m\pi}{b}\right)^2 + \left(\frac{n\pi}{a}\right)^2\right].$$

If $k_z^2 < 0$, k_z is purely imaginary and the traveling wave $\sim e^{ik_z z}$ becomes exponentially attenuating, i.e., no propagation. Hence the cutoff frequency is given by

$$\omega_{mn} = \frac{\pi}{\sqrt{\varepsilon_0\mu}}\sqrt{\left(\frac{m}{b}\right)^2 + \left(\frac{n}{a}\right)^2}.$$

(b) If the inside of the waveguide is filled with a dielectric, we can still use the results for vacuum with the substitution $\varepsilon_0 \to \varepsilon$, $\mu_0 \to \mu$. Since $\mu \sim \mu_0$ generally, the cutoff frequency is now given by

$$\omega_{mn} = \frac{\pi}{\sqrt{\varepsilon\mu_0}} \sqrt{\left(\frac{m}{b}\right)^2 + \left(\frac{n}{a}\right)^2}.$$

4038

(a) Give the wave equation and the boundary conditions satisfied by an electromagnetic wave propagating in the z direction in a waveguide with sides a and b. Assume that the waveguide is perfectly conducting with $\varepsilon = \mu = 1$ inside.

(b) Determine the lowest angular frequency ω at which a transverse electric (TE) wave polarized in the x (vertical) direction can propagate in this waveguide.

(Wisconsin)

Solution:

(a) Refer to Problem **4036**.

(b) From Problem **4037** we see that the TE_{10} mode has the lowest frequency for $a > b$ and that its cutoff angular frequency is $\omega_{10} = \frac{c\pi}{a}$.

4039

(a) Write out Maxwell's equations for a non-conducting medium with permeability μ and dielectric constant ε, and derive a wave equation for the propagation of electromagnetic waves in this medium. Give the plane wave solutions for **E** and **B**.

(b) Determine the electric and magnetic fields for the lowest TE mode of a square waveguide (side l) filled with the foregoing medium. State the boundary conditions which you use.

(c) For what range of the frequencies ω is the mode in (b) the only TE mode which can be excited? What happens to the other modes?

(Wisconsin)

Solution:

(a) Refer to Problem **4010**.

(b) Use the coordinate system shown in Fig. 4.15. The boundary conditions are given by the continuity of the tangential component of **E** across an interface and $\nabla \cdot \mathbf{E} = 0$ as

$$E_y = E_z = 0, \qquad \frac{\partial E_x}{\partial x} = 0 \text{ for } x = 0, l,$$

$$E_x = E_z = 0, \qquad \frac{\partial E_y}{\partial y} = 0 \text{ for } y = 0, l.$$

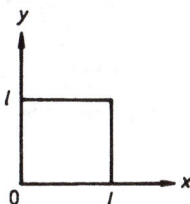

Fig. 4.15

The electromagnetic wave propagating inside the waveguide is a traveling wave along the z-direction, and can be represented as

$$E(x, y, z, t) = E(x, y)e^{i(k_z z - \omega t)}.$$

The wave equation then reduces to

$$\left(\frac{\partial^2}{\partial x^2} + \frac{\partial^2}{\partial y^2} \right) E(x, y) + (k^2 - k_z^2)E(x, y) = 0,$$

where $k^2 = \mu \varepsilon \omega^2$.

Let $u(x, y)$ be a component (x, or y) of $E(x, y)$. Taking $u(x, y) = X(x)Y(y)$, we have

$$\frac{d^2 X}{dx^2} + k_x^2 X = 0,$$

$$\frac{d^2 Y}{dy^2} + k_y^2 Y = 0,$$

with

$$k_x^2 + k_y^2 + k_z^2 = k^2.$$

Hence

$$u(x,y) = [C_1 \cos(k_x x) + D_1 \sin(k_x x)][C_2 \cos(k_y y) + D_2(\sin k_y y)].$$

The boundary conditions require that

$$E_x = A_1 \cos(k_x x) \sin(k_y y) e^{i(k_z z - \omega t)},$$
$$E_y = A_2 \sin(k_x x) \cos(k_y y) e^{i(k_z z - \omega t)},$$
$$E_z = A_3 \sin(k_x x) \sin(k_y y) e^{i(k_z z - \omega t)},$$

$$k_x = \frac{m\pi}{l}, \qquad k_y = \frac{n\pi}{l}, \qquad m, n = 0, 1, 2, \ldots.$$

We thus have $k_z = \left[\mu\varepsilon\omega^2 - (m^2 + n^2)\frac{\pi^2}{l^2}\right]^{\frac{1}{2}}$. For propagation we require k_z to be real. Hence the lowest TE modes are those for which $m, n = 0, 1$ or $1, 0$, i.e., the TE_{01} or TE_{10} mode.

For the TE_{10} mode the electric field is

$$E_x = 0, \qquad E_z = 0, \qquad E_y = A_2 \sin\left(\frac{\pi x}{l}\right) e^{i(k_z z - \omega t)}.$$

The magnetic field is obtained using $\mathbf{H} = -\frac{ic}{\omega\mu} \nabla \times \mathbf{E}$ to be

$$H_y = 0,$$
$$H_x = -\frac{ck_z}{\omega\mu} A_2 \sin\left(\frac{\pi x}{l}\right) e^{i(k_z z - \omega t)},$$
$$H_z = -\frac{ic\pi}{\omega\mu l} A_2 \cos\left(\frac{\pi x}{l}\right) e^{i(k_z z - \omega t)}.$$

Similar results can be obtained for the TE_{01} mode.

(c) The cutoff (angular) frequency of the TE_{10} or TE_{01} mode is

$$\omega_1 = \frac{\pi}{\sqrt{\mu\varepsilon}} \sqrt{\left(\frac{1}{l}\right)^2 + \left(\frac{0}{l}\right)^2} = \frac{\pi}{\sqrt{\mu\varepsilon}\, l},$$

and the cutoff frequency of the TE_{11} mode is

$$\omega_2 = \frac{\pi}{\sqrt{\mu\varepsilon}} \sqrt{\left(\frac{1}{l}\right)^2 + \left(\frac{1}{l}\right)^2} = \frac{\sqrt{2}\pi}{\sqrt{\mu\varepsilon}\, l}.$$

Hence if the TE_{10} and TE_{01} modes are to be the only propagating waves in the waveguide then we require that

$$\frac{\pi}{\sqrt{\mu\varepsilon}\,l} \le \omega < \frac{\sqrt{2}\pi}{\sqrt{\mu\varepsilon}\,l}\,.$$

For the other modes, k_z will become imaginary, $k_z = ik_z'$, and the propagating factor will become $e^{-k_z'z}$. Such waves will attenuate rapidly and cannot be propagated in the waveguide.

4040

A waveguide is constructed so that the cross section of the guide forms a triangle with sides of length a, a, and $\sqrt{2}\,a$ (see Fig. 4.16). The walls are perfect conductors and $\varepsilon = \varepsilon_0$, $\mu = \mu_0$ inside the guide. Determine the allowed modes for TE, TM and TEM electromagnetic waves propagating in the guide. For allowed modes find $\mathbf{E}(x,y,z,t)$, $\mathbf{B}(x,y,z,t)$ and the cutoff frequencies. If some modes are not allowed, explain why not.

(*Princeton*)

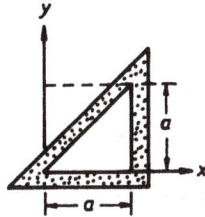

Fig. 4.16

Solution:

We first consider a square waveguide whose cross section has sides of length a. The electric vector of the electromagnetic wave propagating along $+z$ direction is given by

$$E_x = A_1 \cos(k_1 x)\sin(k_2 y)e^{i(k_3 z - \omega t)}\,,$$
$$E_y = A_2 \sin(k_1 x)\cos(k_2 y)e^{i(k_3 z - \omega t)}\,,$$
$$E_z = A_3 \sin(k_1 x)\sin(k_2 y)e^{i(k_3 z - \omega t)}\,,$$

with

$$k_1^2 + k_2^2 + k_3^2 = k^2 = \mu_0 \varepsilon_0 \omega^2 = \frac{\omega^2}{c^2},$$

$$k_1 A_1 + k_2 A_2 - i k_3 A_3 = 0,$$

$$k_1 = \frac{m\pi}{a},$$

$$k_2 = \frac{n\pi}{a}.$$

The boundary conditions being satisfied are

$$E_x = E_z = 0 \text{ for } y = 0 \text{ and } E_y = E_z = 0 \text{ for } x = a.$$

For the waveguide with triangular cross section, we have to choose from the above those that satisfy additional boundary conditions on the $y = x$ plane: $E_z = 0$, $E_x \cos \frac{\pi}{4} + E_y \sin \frac{\pi}{4} = 0$ for $y = x$. The former condition gives $A_3 = 0$, while the latter gives $A_1 = A_2$ and $\tan(k_1 x) = -\tan(k_2 x)$, or $A_1 = -A_2$ and $\tan(k_1 x) = \tan(k_2 x)$, i.e., either $k_1 = -k_2$, $A_1 = A_2$, or $k_1 = k_2$, $A_1 = -A_2$. Thus for the waveguide under consideration we have

$$E_x = -A \cos(k_1 x) \sin(k_1 y) e^{i(k_3 z - \omega t)},$$

$$E_y = A \sin(k_1 x) \cos(k_1 y) e^{i(k_3 z - \omega t)},$$

$$E_z = 0,$$

with

$$k_1 = \frac{n\pi}{a}, \qquad k_3 = \sqrt{\frac{\omega^2}{c^2} - 2\frac{n^2 \pi^2}{a^2}}.$$

The associated magnetic field can be found using $\nabla \times \mathbf{E} = -\frac{\partial \mathbf{B}}{\partial t}$, or $\mathbf{k} \times \mathbf{E} = \omega \mathbf{B}$ as

$$B_x = -\frac{k_3}{\omega} E_y = -\frac{k_3}{\omega} A \sin(k_1 x) \cos(k_1 y) e^{i(k_3 z - \omega t)},$$

$$B_y = \frac{k_3}{\omega} E_x = -\frac{k_3}{\omega} A \cos(k_1 x) \sin(k_1 y) e^{i(k_3 z - \omega t)},$$

$$B_z = \frac{1}{\omega}(k_1 E_y - k_2 E_x) = \frac{k_1}{\omega} A[\sin(k_1 x) \cos(k_1 y)$$

$$+ \cos(k_1 x) \sin(k_1 y)] e^{i(k_3 z - \omega t)}$$

$$= \frac{k_1}{\omega} A \sin[k_1 (x + y)] e^{i(k_3 z - \omega t)}.$$

Thus the allowed modes are $TE_{n,-n}$ or $TE_{n,n}$, but not TM. The cutoff frequencies are

$$\omega_n = \sqrt{2}\,\frac{n\pi c}{a}\,.$$

4041

As in Fig. 4.17, two coaxial cylindrical conductors with r_1 and r_2 form a waveguide. The region between the conductors is vacuum for $z < 0$ and is filled with a dielectric medium with dielectric constant $\varepsilon \neq 1$ for $z > 0$.

(a) Describe the TEM mode for $z < 0$ and $z > 0$.

(b) If an electromagnetic wave in such a mode is incident from the left on the interface, calculate the transmitted and reflected waves.

(c) What fraction of the incident energy is transmitted? What fraction is reflected?

(*Columbia*)

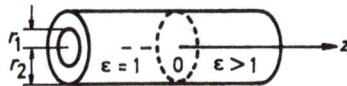

Fig. 4.17

Solution:

Interpret ε as the relative dielectric constant (permittivity $= \varepsilon\varepsilon_0$) and use SI units.

(a) Consider first the region $z > 0$. Assume $\mu = \mu_0$. For sinusoidal waves $\frac{\partial}{\partial t} \rightarrow -i\omega$, and the wave equation becomes

$$\left(\nabla^2 + \varepsilon\frac{\omega^2}{c^2}\right)\left\{\begin{matrix} \mathbf{E}' \\ \mathbf{B}' \end{matrix}\right\} = 0\,,$$

where ε is the relative dielectric constant of the medium, i.e. permittivity $= \varepsilon\varepsilon_0$. Because of cylindrical symmetry, special solutions of the above equation are

$$\mathbf{E}'(\mathbf{r},t) = \mathbf{E}'(x,y)e^{i(k'z-\omega t)}\,,$$

$$\mathbf{B}'(\mathbf{r}, t) = \mathbf{B}'(x, y)e^{i(k'z - \omega t)},$$

with

$$k'^2 = \varepsilon \frac{\omega^2}{c^2}.$$

Let

$$\nabla^2 = \nabla_t^2 + \frac{\partial^2}{\partial z^2},$$

∇_t^2 being the transverse part of the Laplace operator ∇^2. Decompose the electromagnetic field into transverse and longitudinal components:

$$\mathbf{E}' = \mathbf{E}_t' + E_z \mathbf{e}_z, \qquad \mathbf{B}' = \mathbf{B}_t' + B_z \mathbf{e}_z.$$

For TEM waves $B_z' = E_z' = 0$. Then Maxwell's equation for a charge-free medium $\nabla \cdot \mathbf{E}' = 0$ reduces to

$$\nabla_t \cdot \mathbf{E}_t' = 0.$$

Also from Maxwell's equation $\nabla \times \mathbf{E}' = -\frac{\partial \mathbf{B}'}{\partial t} = i\omega \mathbf{B}'$ we have

$$\nabla_t \times \mathbf{E}_t' = 0.$$

These equations allow us to introduce a scalar function ϕ such that

$$\mathbf{E}_t' = -\nabla \phi, \qquad \nabla^2 \phi = 0.$$

Furthermore, symmetry requires that ϕ is a function of r only and the last equation reduces to

$$\frac{1}{r} \frac{\partial}{\partial r} \left(r \frac{\partial \phi}{\partial r} \right) = 0,$$

whose solution is

$$\phi = C \ln r + C'$$

C, C' being constants.
Then the electric field is

$$\mathbf{E}_t'(\mathbf{r}, t) = \frac{C}{r} e^{i(k'z - \omega t)} \mathbf{e}_r,$$

and the associated magnetic field is given by $\nabla \times \mathbf{E}' = -\frac{\partial \mathbf{B}}{\partial t}$ with $\nabla \rightarrow ik'$, $\frac{\partial}{\partial t} \rightarrow -i\omega$ as

$$\mathbf{B}_t'(\mathbf{r}, t) = \frac{\sqrt{\varepsilon}}{c} \mathbf{e}_z \times \mathbf{E}_t' = \frac{C\sqrt{\varepsilon}}{rc} e^{i(k'z - \omega t)} \mathbf{e}_\theta.$$

Therefore, in the $z > 0$ region which is filled with a medium of relative dielectric constant ε, the TEM waves can be represented as

$$\mathbf{E}'(x,t) = \frac{C}{r} e^{i(k'z - \omega t)} \mathbf{e}_r ,$$

$$\mathbf{B}'(x,t) = \frac{C\sqrt{\varepsilon}}{rc} e^{i(k'z - \omega t)} \mathbf{e}_\theta .$$

Similarly, for the $z < 0$ region, $\varepsilon = 1$ and the TEM waves are given by

$$\mathbf{E}(x,t) = \frac{A}{r} e^{i(kz - \omega t)} \mathbf{e}_r ,$$

$$\mathbf{B}(x,t) = \frac{A}{rc} e^{i(kz - \omega t)} \mathbf{e}_\theta ,$$

where A and C are constants, and $k = \frac{\omega}{c}$.

(b) Consider a TEM wave incident normally on the interface $z = 0$ from the vacuum side. Assuming that the transmitted and reflected waves both remain in the TEM mode, the incident and transmitted waves are given by **E, B** and **E', B'** respectively. Let the reflected wave be represented as

$$\mathbf{E}''(\mathbf{r},t) = \frac{D}{r} e^{-i(kz + \omega t)} \mathbf{e}_r ,$$

$$\mathbf{B}''(\mathbf{r},t) = -\frac{D}{rc} e^{-i(kz + \omega t)} \mathbf{e}_\theta .$$

Note that the negative sign for **B''** is introduced so that **E''**, **B''** and $\mathbf{k}'' = -\mathbf{k}$ form a right-hand set.

The boundary conditions that E_t and H_t are continuous across the interface give

$$(E_r + E_r'' - E_r')|_{z=0} = 0 ,$$

$$(B_\theta + B_\theta'' - B_\theta')|_{z=0} = 0 ,$$

and hence

$$C = \frac{2A}{1 + \sqrt{\varepsilon}} , \qquad D = \frac{1 - \sqrt{\varepsilon}}{1 + \sqrt{\varepsilon}} A .$$

(c) The coefficients of reflection and transmission are therefore respectively

$$R = \frac{|E'' H''^*|}{|EH^*|} = \frac{|E'' B''^*|}{|EB^*|} = \left(\frac{D}{A}\right)^2 = \left(\frac{1 - \sqrt{\varepsilon}}{1 + \sqrt{\varepsilon}}\right)^2 ,$$

$$T = \frac{|E' H'^*|}{|EH^*|} = \frac{|E' B'^*|}{|EB^*|} = \left(\frac{C}{A}\right)^2 \sqrt{\varepsilon} = \frac{4\sqrt{\varepsilon}}{(1 + \sqrt{\varepsilon})^2} .$$

As the incident, reflected and transmitted waves are all in the same direction, R and T respectively give the fractions of the incident energy that are reflected and transmitted. Note that $R + T = 1$ as required by energy conservation.

4042

A waveguide is made of two perfectly conducting coaxial cylinders with the radiation propagating in the space between them. Show that it is possible to have a mode in which both the electric and magnetic fields are perpendicular to the axis of the cylinders. Is there a cutoff frequency for this mode? Calculate the velocity of propagation of this mode and the time-averaged power flow along the line.

(Columbia)

Solution:

Take a coordinate system with the z-axis along the axis of the cylinder and for simplicity take the region between the cylinders as free space. As was shown in the solution of Problem **4041**, it is possible to obtain solutions of the wave equation which have $E_z = B_z = 0$ without the other components being identically zero. Hence it is possible to have TEM waves propagating in the space between the cylinders. Furthermore, the TEM waves can be represented as

$$\mathbf{E} = \frac{A}{r} e^{i(kz - \omega t)} \mathbf{e}_r , \qquad \mathbf{B} = \frac{A}{rc} e^{i(kz - \omega t)} \mathbf{e}_\theta ,$$

where A is a constant and k is a real number equal to $\frac{\omega}{c}$. Thus there is no cutoff frequency for the TEM waves and the phase velocity of the waves is c.

The Poynting vector averaged over one period is

$$\langle \mathbf{N} \rangle = \langle \operatorname{Re} \mathbf{E} \times \operatorname{Re} \mathbf{H} \rangle = \frac{1}{4} \langle (\mathbf{E} + \mathbf{E}^*) \times (\mathbf{H} + \mathbf{H}^*) \rangle$$

$$= \frac{1}{4} \left(\langle \mathbf{E} \times \mathbf{H} \rangle + \langle \mathbf{E}^* \times \mathbf{H}^* \rangle + \langle \mathbf{E} \times \mathbf{H}^* \rangle + \langle \mathbf{E}^* \times \mathbf{H} \rangle \right)$$

$$= \frac{1}{2} \mathbf{E} \times \mathbf{H}^* = \frac{1}{2\mu_0} \mathbf{E} \times \mathbf{B}^* ,$$

where we have used the fact that $\mathbf{E} \times \mathbf{H}$ and $\mathbf{E}^* \times \mathbf{H}^*$ vanish on averaging over one cycle. Thus

$$\langle \mathbf{N} \rangle = \frac{1}{2} \sqrt{\frac{\varepsilon_0}{\mu_0}} \frac{A^2}{r^2} \mathbf{e}_z .$$

The average power flow is then

$$\int_a^b \langle N \rangle 2\pi r \, dr = \sqrt{\frac{\varepsilon_0}{\mu_0}} \pi A^2 \ln \left(\frac{b}{a} \right) ,$$

where a and b ($b > a$) are the radii of the two cylinders.

4043

A transmission line consists of two parallel conductors of arbitrary but constant cross-sections. Current flows down one conductor and returns by way of the other. The conductors are immersed in an insulating medium of dielectric constant ε and permeability μ, as shown in Fig. 4.18.

(a) Derive wave equations for the $\mathbf{E}$ and $\mathbf{B}$ fields in the medium for waves propagating in the z direction.

(b) Obtain the speed of propagation of the waves.

(c) Under what conditions can one define a voltage between the two conductors? (Note: to define a voltage all the points on a given plane $z = $ constant on one conductor must be on an equipotential. Those on the other conductor may be on another equipotential.)

(*Princeton*)

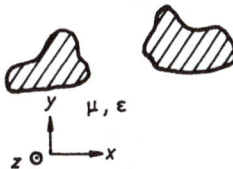

Fig. 4.18

Solution:

(a) From Maxwell's equations for a source-free medium,

$$\nabla \times \mathbf{E} = -\frac{\partial \mathbf{B}}{\partial t} , \qquad \nabla \times \mathbf{H} = \frac{\partial \mathbf{D}}{\partial t} ,$$

$$\nabla \cdot \mathbf{D} = 0, \qquad \nabla \cdot \mathbf{B} = 0,$$

we obtain

$$\nabla \times (\nabla \times \mathbf{E}) = -\frac{\partial}{\partial t} \nabla \times \mathbf{B} = -\mu\varepsilon \frac{\partial^2 \mathbf{E}}{\partial t^2}.$$

As

$$\nabla \times (\nabla \times \mathbf{E}) = \nabla(\nabla \cdot \mathbf{E}) - \nabla^2 \mathbf{E} = -\nabla^2 \mathbf{E},$$

we have

$$\nabla^2 \mathbf{E} - \mu\varepsilon \frac{\partial^2 \mathbf{E}}{\partial t^2} = 0.$$

The same wave equation applies to **B**. For a transmission line the waves can be taken to be purely transverse (TEM). We can write

$$\mathbf{E}(\mathbf{r}, t) = \mathbf{E}_0(x, y)e^{i(kz - \omega t)}.$$

The wave equation then becomes

$$\left(\frac{\partial^2}{\partial x^2} + \frac{\partial^2}{\partial y^2} \right) \mathbf{E}_0 + (\mu\varepsilon\omega^2 - k^2)\mathbf{E}_0 = 0.$$

(b) The phase velocity v of the waves is obtained from the wave equation:

$$\frac{1}{v^2} = \mu\varepsilon,$$

or

$$v = \frac{1}{\sqrt{\varepsilon\mu}}.$$

(c) The required condition is $\lambda \gg l$, l being the dimension of the transverse cross-section of the conductors.

4044

The spectral lines from an atom in a magnetic field are split. In the direction of the field the higher frequency light is:

(a) unpolarized,   (b) linearly polarized,   (c) circularly polarized.

$$(CCT)$$

Solution:

The answer is (c).

4045

To go through the ionosphere an electromagnetic wave should have a frequency of at least

10, 10^4, 10^7, 10^9 Hz.

Solution:

To go through the ionosphere, the angular frquency ω of a wave should be greater than the plasma frequency $\omega_p = \sqrt{\frac{Ne^2}{\epsilon_0 m}}$. The maximum electron density of a typical layer is $N \sim 10^{13}$ m^{-3}. For an electron, $\frac{e^2}{\epsilon_0 m} = 3 \times 10^3$ m^{3}s^{-2}. Hence

$$\omega_p \approx \sqrt{3 \times 10^{16}} \approx 1.7 \times 10^8 \text{ s}^{-1}.$$

Thus the answer is 10^7 Hz.

4. ELECTROMAGNETIC RADIATION AND RADIATING SYSTEMS (4046–4067)

4046

A measuring device is disturbed by the following influences. How would you separately protect the device from each one?

(a) High frequency electric fields.

(b) Low frequency electric fields.

(c) High frequecy magnetic fields.

(d) Low frequency magnetic fields.

(e) D.C. magnetic fields.

Solution:

(a), (c) High frequency electric and magnetic fields usually come together in the form of electromagnetic radiation. To protect a measuring device from it, the former is enclosed in a grounded shell made of a good conductor.

(b) The same protection as in (a) can be used. The thickness of the conductor should be at least a few times the depth of penetration.

(d), (e) Enclose the device in a shell made of μ-metal (a Ni–Fe alloy containing Mo, Cu, Si) or, even better, of a superconductor.

4047

(a) What is the rate of energy radiation per unit area from each side of a thin uniform alternating current sheet?

(b) Show what effective radiation resistance in ohms is acting on a square area of this current sheet.

(c) Find the force per unit area on each side of the current sheet (due to the radiation) for a surface current density of 1000 amperes per unit length.

(*Wisconsin*)

Solution:

(a) Take the y-axis along the current and the z-axis perpendicular to the current sheet as shown in Fig. 4.19. Let the current per unit width be $\alpha = \alpha e^{-i\omega t} \mathbf{e}_y$. Consider a unit square area with sides parallel to the x and y axes. At large distances from the current sheet, the current in the area may be considered as a Hertzian dipole of dipole moment $\mathbf{p}$ given by

$$\dot{\mathbf{p}} = \alpha e^{-i\omega t} \mathbf{e}_y .$$

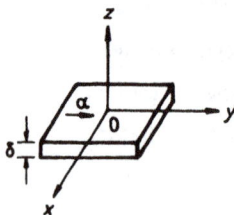

Fig. 4.19

Hence the power radiated, averaged over one period, from unit area of the sheet is

$$P = \frac{1}{4\pi\varepsilon_0} \frac{\omega^2 |\dot{\mathbf{p}}|^2}{3c^3} = \frac{\alpha^2 \omega^2}{12\pi\varepsilon_0 c^3} .$$

As the thickness δ is very small, the radiation is emitted mainly from the top and bottom surfaces of the area so that the power radiated per unit area from each side of the thin sheet is

$$\frac{P}{2} = \frac{\alpha^2 \omega^2}{24\pi\varepsilon_0 c^3} .$$

(b) The average power is related to the amplitude of the ac, I, by

$$P = \frac{1}{2} I^2 R ,$$

where R is the resistance. Hence the effective radiation resistance per unit area

$$R = \frac{2P}{\alpha^2} = \frac{\omega^2}{6\pi\varepsilon_0 c^3} .$$

(c) Electromagnetic radiation of energy density U carries a momentum $\frac{U}{c}$. Hence the loss of momentum per unit time per unit area of one surface of the sheet is $\frac{P}{2c}$. Momentum conservation requires a pressure exerting on the sheet of the same amount:

$$F = \frac{P}{2c} = \frac{\alpha^2 \omega^2}{24\pi\varepsilon_0 c^4} .$$

Taking the frequency of the alternating current as $f = 50$ Hz and with $\alpha = 1000$ A, $\varepsilon_0 = 8.85 \times 10^{-12}$ F/m, we have

$$F \approx 1.83 \times 10^{-14} \text{ N} .$$

4048

Radio station WGBH–FM radiates a power of 100 kW at about 90 MHz from its antenna on Great Blue Hill, approximately 20 km from M.I.T. Obtain a rough estimate of the strength of its electric field at M.I.T. in volts per centimeter.

(*MIT*)

Solution:

The intensity of electromagnetic radiation is given by $\langle N \rangle$, N being the magnitude of the Poynting vector. For plane electromagnetic waves, this becomes

$$I = \frac{1}{2}\epsilon_0 E_0^2 c \ .$$

The total power radiated is then $P = 4\pi R^2 I = 2\pi\epsilon_0 c R^2 E_0^2$ where R is the distance from the antenna. Hence the amplitude of the electric field at M.I.T. is

$$E_0 = \left(\frac{P}{2\pi\epsilon_0 c R^2}\right)^{\frac{1}{2}} = \left(\frac{10^5}{2\pi \times 8.85 \times 10^{-12} \times 3 \times 10^8 \times (2 \times 10^4)^2}\right)^{\frac{1}{2}} \text{ V/m}$$

$$= 1.2 \times 10^{-3} \text{ V/m} \ .$$

4049

An oscillating electric dipole $\mathbf{P}(t)$ develops radiation fields

$$\mathbf{B}(\mathbf{r},t) = -\frac{\mu_0}{4\pi rc}\mathbf{e}_r \times \frac{\partial^2}{\partial t^2}\mathbf{P}\left(t - \frac{r}{c}\right),$$

$$\mathbf{E}(\mathbf{r},t) = -c\mathbf{e}_r \times \mathbf{B}(\mathbf{r},t) \ .$$

(a) A charge q at the origin is driven by a linearly polarized electromagnetic wave of angular frequency ω and electric field amplitude $\mathbf{E}_0$. Obtain in vector form the radiated eletromagnetic fields.

(b) Sketch the directions of $\mathbf{E}$ and $\mathbf{B}$ at a field position $\mathbf{r}$. Describe the state of polarization of the radiated fields.

(c) Find the angular dependence of the radiation intensity in terms of the spherical angles θ and ϕ, where the z-axis is the direction of propagation of the incident wave and the x axis is the direction of polarization of the incident wave.

(UC, Berkeley)

Solution:

(a) For an oscillating charge of low speed we can neglect the influence of the magnetic field of the incident radiation. Then the equation of the motion of the charge q, of mass m, in the field of the incident wave is

$$m\ddot{x} = qE_0 e^{-i\omega t} \ .$$

The charge will oscillate with the same frequency: $x = x_0 e^{-i\omega t}$. Hence the displacement of the charge is

$$x = -\frac{qE_0}{m\omega^2} e^{-i\omega t} .$$

This gives rise to an electric dipole of moment

$$P(t) = qx = -\frac{q^2 E_0}{m\omega^2} e^{-i\omega t} .$$

As

$$\frac{\partial^2}{\partial t^2} P\left(t - \frac{r}{c}\right) = \left.\frac{\partial^2 P(t')}{\partial t'^2}\right|_{t'=t-\frac{r}{c}} = \frac{q^2 E_0}{m} e^{i(kr-\omega t)} ,$$

where $k = \frac{\omega}{c}$, we have

$$\mathbf{B}(\mathbf{r}, t) = -\frac{\mu_0 q^2}{4\pi mrc} e^{i(kr-\omega t)} \mathbf{e}_r \times \mathbf{E}_0 ,$$

$$\mathbf{E}(r, t) = \frac{\mu_0 q^2}{4\pi mr} e^{i(kr-\omega t)} \mathbf{e}_r \times (\mathbf{e}_r \times \mathbf{E}_0)$$

$$= \frac{\mu_0 q^2}{4\pi mr} e^{i(kr-\omega t)} [(\mathbf{E}_0 \cdot \mathbf{e}_r)\mathbf{e}_r - \mathbf{E}_0] .$$

(b) The directions of $\mathbf{E}$ and $\mathbf{B}$ are as shown in Fig. 4.20, i.e., $\mathbf{E}$ is in the plane of $\mathbf{P}$ and $\mathbf{r}$, and $\mathbf{B}$ is perpendicular to it. Thus the radiation emitted is linearly polarized.

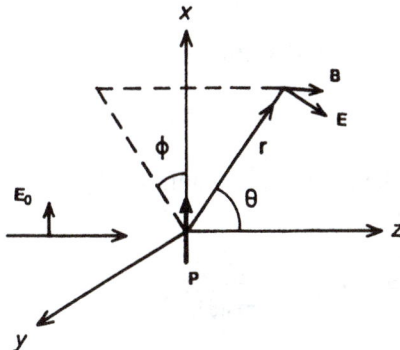

Fig. 4.20

(c) As $e_x = (\cos\phi\cos\theta,\ \cos\phi\sin\theta,\ -\sin\phi)$ in spherical coordinates,

$$e_r \times e_x = \cos\theta\cos\phi e_\phi - \sin\phi e_\theta .$$

The average Poynting vector is

$$\langle N\rangle = \frac{1}{2}\,\mathrm{Re}\,(E^* \times H) = \frac{1}{2\mu_0}\,\mathrm{Re}\,[-c(e_r \times B^*) \times B] .$$

As $e_r \cdot B = 0$, $e_r \times E_0 = E_0(e_r \times e_x)$, the average radiation intensity is

$$I = \langle N\rangle = \frac{c}{2\mu_0}\,|B|^2 = \frac{\mu_0 q^4 E_0^2}{32\pi^2 c m^2 r^2}\,(\cos^2\theta\cos^2\phi + \sin^2\phi) .$$

4050

A massive atom with an atomic polarizability $\alpha(\omega)$ is subjected to an electromagnetic field (the atom being located at the origin)

$$E = E_0 e^{i(kx-\omega t)} e_z$$

Find the asymptotic electric and magnetic fields radiated by the atom and calculate the energy radiated per unit solid angle. State any approximations used in this calculation, and state when (and why) they will break down as ω is increased.

(*Wisconsin*)

Solution:

The atom acts as a Hertzian dipole at the origin with dipole moment

$$P = \alpha E = \alpha E_0 e^{-i\omega t} e_z .$$

At a large distance r the asymptotic (radiation) electric and magnetic fields radiated by the atom are

$$B(r,t) = -\frac{\alpha E_0 \omega^2}{4\pi\varepsilon_0 c^3 r}\,\sin\theta e^{-i\omega t} e_\phi ,$$

$$E(r,t) = -\frac{\alpha E_0 \omega^2}{4\pi\varepsilon_0 c^2 r}\,\sin\theta e^{-i\omega t} e_\theta .$$

The energy radiated per unit solid angle is (Problem **4049**)

$$\frac{dW}{d\Omega} = \frac{\langle N \rangle}{r^{-2}} = \frac{c}{2\mu_0 r^{-2}} |\mathbf{B}|^2 = \frac{\alpha^2 E_0^2 \omega^4}{32\pi^2 \varepsilon_0 c^3} \sin^2 \theta .$$

The approximation used is $r \gg \lambda \gg l$, where l is the linear dimension of the atom and $\lambda = 2\pi c/\omega$. As ω is increased, λ will decrease and eventually become smaller than l, thus invalidating the approximation.

4051

A radially pulsating charged sphere

(a) emits electromagnetic radiation

(b) creates a static magnetic field

(c) can set a nearby electrified particle into motion.

(*CCT*)

Solution:

The answer is (a).

4052

A charge radiates whenever

(a) it is moving in whatever manner

(b) it is being accelerated

(c) it is bound in an atom.

(*CCT*)

Solution:

The answer is (b).

4053

Radiation emitted by an antenna has angular distribution characteristic of dipole radiation when

(a) the wavelength is long compared with the antenna

(b) the wavelength is short compared with the antenna

(c) the antenna has the appropriate shape.

(CCT)

Solution:

The answer is (a).

4054

The frequency of a television transmitter is 100 kHz, 1 MHz, 10 MHz, 100 MHz.

$(Columbia)$

Solution:

The answer is 100 MHz.

4055

A small circuit loop of wire of radius a carries a current $i = i_0 \cos \omega t$ (see Fig. 4.21). The loop is located in the xy plane.

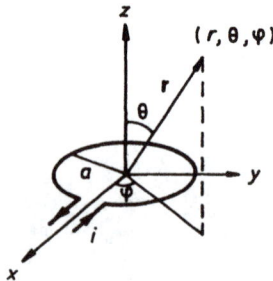

Fig. 4.21

(a) Calculate the first non-zero multipole moment of the system.

(b) Give the form of the vector potential for this system for $r \to \infty$, calculate the asymptotic electric and magnetic fields, and determine the angular distribution of the outgoing radiation.

(c) Describe the main features of the radiation pattern.

(d) Calculate the average power radiated.

(*Wisconsin*)

Solution:

(a) The first non-zero multipole moment of the small circuit loop is its magnetic dipole moment

$$\mathbf{m} = \pi a^2 i_0 \cos(\omega t)\mathbf{e}_z = \pi a^2 i_0 \, \text{Re}\,(e^{-i\omega t})\mathbf{e}_z \,.$$

(b) Use spherical coordinates with the origin at the center of the loop. The vector potential at a point $\mathbf{r} = (r, \theta, \phi)$ for $r \to \infty$ is

$$\mathbf{A}(\mathbf{r}, t) = \frac{ik\mu_0 e^{ikr}}{4\pi r} \, \mathbf{e}_r \times \mathbf{m} \,.$$

where $k = \frac{\omega}{c}$. As $\mathbf{e}_z = (\cos\theta, -\sin\theta, 0)$ we have

$$\mathbf{A}(\mathbf{r}, t) = -i\frac{\mu_0 \omega i_0 a^2 \sin\theta}{4cr} e^{i(kr-\omega t)}\mathbf{e}_\varphi \,,$$

whence the radiation field vectors are

$$\mathbf{B} = \nabla \times \mathbf{A} = ik\mathbf{e}_r \times \mathbf{A} = \frac{\mu_0 \omega^2 i_0 a^2 \sin\theta}{4c^2 r} e^{i(kr-\omega t)}\mathbf{e}_\theta \,,$$

$$\mathbf{E} = c\mathbf{B} \times \mathbf{e}_r = \frac{\mu_0 \omega^2 i_0 a^2 \sin\theta}{4cr} e^{i(kr-\omega t)}\mathbf{e}_\varphi \,.$$

The average Poynting vector at r is (Problem 4042)

$$\langle \mathbf{N} \rangle = \frac{1}{2} \, \text{Re}\,(\mathbf{E}^* \times \mathbf{H}) = \frac{c}{2\mu_0} \, \text{Re}\,\{(\mathbf{B} \times \mathbf{e}_r) \times \mathbf{B}\}$$

$$= \frac{c}{2\mu_0} |\mathbf{B}|^2 \mathbf{e}_r = \frac{\mu_0 \omega^4 a^4 i_0^2}{32c^3 r^2} \sin^2\theta \, \mathbf{e}_r \,.$$

The average power radiated per unit solid angle is then

$$\frac{dP}{d\Omega} = \frac{\langle \mathbf{N} \rangle}{r^{-2}} = \frac{\mu_0 \omega^4 a^4 i_0^2}{32c^2} \sin^2\theta \,.$$

(c) The radiated energy is distributed according to $\sin^2\theta$. In the plane $\theta = 90°$ the radiation is most strong, and there is no radiation along the axis of the loop ($\theta = 0°$ or $180°$), as illustrated in Fig. 4.22 where the length of a vector at θ is proportional to the radiation per unit solid angle per unit time in that direction. The actual angular distribution is given by the surface obtained by rotating the curve about the z-axis.

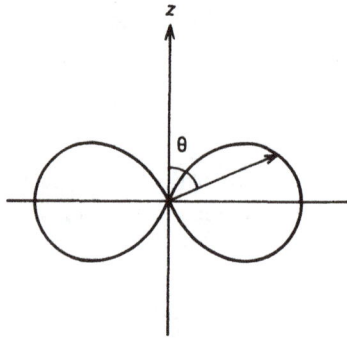

Fig. 4.22

(d) The average radiated power is

$$P = \int \frac{dP}{d\Omega} d\Omega = \frac{\mu_0 \omega^4 a^4 i_0^2}{32 c^3} \cdot 2\pi \int_0^\pi \sin^3 \theta d\theta = \frac{\pi \mu_0 \omega^4 a^4 i_0^2}{12 c^3}.$$

4056

As in Fig. 4.23, a current-fed antenna is operated in the $\lambda/4$ mode ($a = \lambda/4$). Find the pattern (angular distribution) of the radiated power.

(*Chicago*)

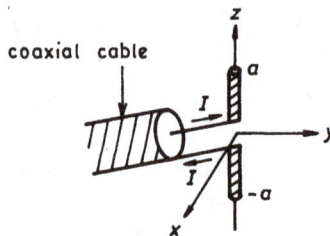

Fig. 4.23

Solution:

As $l \sim \lambda$ the antenna cannot be treated as a dipole. In the $\lambda/4$ mode, $a = \frac{\lambda}{4}$ and the current is in the form of a stationary wave with nodes at the ends of the antenna, i.e.,

$$I(z, t') = I_0 \cos\left(\frac{\pi}{2} \frac{z}{a}\right) e^{-i\omega t'}.$$

The vector potential at a point **r** is given by

$$\mathbf{A}(\mathbf{r}, t) = \frac{\mu_0}{4\pi} \int \frac{\mathbf{J} dV'}{r} = \frac{\mu_0}{4\pi} \int \frac{I\left(z', t - \frac{r}{c}\right)}{r} dz' \mathbf{e}_z.$$

At a large distance **r**,

$$r \approx r_0 - z' \cos\theta,$$

where r_0 is the distance from the centre of the antenna. Then

$$e^{-i\omega(t - \frac{r}{c})} = e^{i(kr - \omega t)} \approx e^{i(kr_0 - \omega t)} e^{-ikz' \cos\theta},$$

where $k = \frac{\omega}{c}$, and

$$\frac{1}{r} \approx \frac{1}{r_0}\left(1 - \frac{z'}{r_0}\cos\theta\right)^{-1} \approx \frac{1}{r_0}\left(1 + \frac{z'}{r_0}\cos\theta\right) \approx \frac{1}{r_0},$$

neglecting terms of order $\frac{z'}{r_0^2}$. Hence

$$A(\mathbf{r}, t) \approx \frac{\mu_0}{4\pi} \frac{I_0 e^{i(kr_0 - \omega t)}}{r_0} \int_{-a}^{a} \cos\left(\frac{\pi}{2}\frac{z'}{a}\right) e^{-ikz'\cos\theta} dz'.$$

Using

$$\int e^{ax}\cos(bx)dx = \frac{e^{ax}}{a^2 + b^2}\left[a\cos(bx) + b\sin(bx)\right],$$

we have

$$A(\mathbf{r}, t) = \frac{\mu_0}{4\pi} \frac{I_0 e^{i(kr_0 - \omega t)}}{r_0} \frac{\left(\frac{\pi}{2a}\right)}{\left(\frac{\pi}{2a}\right)^2 - (k\cos\theta)^2} \cdot 2\cos(ka\cos\theta).$$

As $ka = \frac{\omega}{c}\frac{\pi}{4} = \frac{\pi}{2}$,

$$\mathbf{A}(\mathbf{r}, t) = \frac{\mu_0}{2\pi} \frac{I_0 e^{i(kr_0 - \omega t)}}{r_0} \frac{c}{\omega} \frac{\cos(\frac{\pi}{2}\cos\theta)}{\sin^2\theta} \mathbf{e}_z.$$

In spherical coordinates we have

$$\mathbf{e}_z = (\cos\theta, -\sin\theta, 0),$$

so that

$$\mathbf{A} = A\mathbf{e}_z = A\cos\theta\mathbf{e}_r - A\sin\theta\mathbf{e}_\theta = A_r\mathbf{e}_r + A_\theta\mathbf{e}_\theta,$$

$$\nabla\times\mathbf{A} = \frac{1}{r_0}\left[\frac{\partial}{\partial r_0}(r_0 A_\theta) - \frac{\partial A_r}{\partial\theta}\right]\mathbf{e}_\varphi \approx \frac{1}{r_0}\frac{\partial}{\partial r_0}(r_0 A_\theta)\mathbf{e}_\varphi,$$

neglecting the second term which varies as r_0^{-2} as we are interested only in the radiation field which varies as r_0^{-1}. Thus

$$B = B_\varphi = i\frac{\mu_0}{2\pi}\cdot\frac{\cos(\frac{\pi}{2}\cos\theta)}{\sin\theta}\frac{I_0 e^{i(kr_0-\omega t)}}{r_0}.$$

The intensity averaged over one cycle is then (Problem **4049**)

$$\langle N\rangle = \frac{c}{2\mu_0}|\mathbf{B}|^2 = \frac{\mu_0 c}{8\pi^2}\cdot\frac{\cos^2(\frac{\pi}{2}\cos\theta)}{\sin^2\theta}\cdot\frac{1}{r_0^2}.$$

Hence the radiated power per unit solid angle is

$$\frac{dP}{d\Omega} = \frac{\langle N\rangle}{r_0^{-2}},$$

which has an angular distribution given by

$$\frac{\cos^2(\frac{\pi}{2}\cos\theta)}{\sin^2\theta}.$$

4057

(a) What is the average power radiated by an electric current element of magnitude Il, where the length l of the element is very short compared with the wavelength of the radiation and I is varying as $\cos(\omega t)$?

(b) In Fig. 4.24 if we now identify the xy plane with the surface of the earth (regarded as a perfect conductor at λ), what is the average power radiated?

(c) What is the optimal height for maximum radiated power, and the corresponding gain in power radiated due to the ground plane?

(*Princeton*)

Solution:

(a) The system can be considered as a Hertzian dipole of moment $\mathbf{p} = \mathbf{p}_0 e^{-i\omega t'}$ such that $\dot{\mathbf{p}} = -i\omega \mathbf{p} = I_0 e^{-i\omega t'} \mathbf{l}$. The average radiated power is

$$\overline{P} = \frac{1}{4\pi\varepsilon_0} \frac{\omega^2}{3c^3} |\dot{\mathbf{p}}|^2 = \frac{\omega^4 I_0^2 l^2}{12\pi\varepsilon_0 c^3} \; .$$

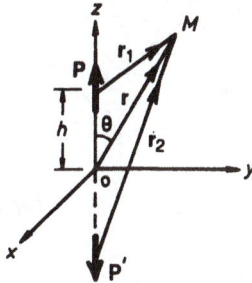

Fig. 4.24

(b) If the earth is regarded as a perfect conductor, the induced charges on the surface of the earth are such that their effect can be replaced by that of an image dipole $\mathbf{p}'$ as shown in Fig. 4.24 provided ω is not too large, where $\mathbf{p}' = \mathbf{p}$. The electromagnetic field at a large distance $\mathbf{r}$ is a coherent superposition of the fields of these two dipoles i.e.,

$$\mathbf{E}_{\text{total}} = \mathbf{E} + \mathbf{E}', \qquad \mathbf{B}_{\text{total}} = \mathbf{B} + \mathbf{B}'.$$

The average Poynting vector at $\mathbf{r}$, a distant point M, is (Problem **4042**)

$$\overline{\mathbf{S}}_{\text{total}} = \frac{1}{\mu_0} \langle \mathbf{E}_{\text{total}} \times \mathbf{B}_{\text{total}} \rangle = \frac{1}{2\mu_0} \text{Re} \left(\mathbf{E}_{\text{total}}^* \times \mathbf{B}_{\text{total}} \right),$$

or

$$\overline{\mathbf{S}}_{\text{total}} = \overline{\mathbf{S}} + \overline{\mathbf{S}}' + \frac{1}{2\mu_0} \text{Re} \left[\mathbf{E}^* \times \mathbf{B}' + \mathbf{E}'^* \times \mathbf{B} \right],$$

where $\mathbf{S}$, $\mathbf{S}'$ are the Poynting vectors at the distant point M due to $\mathbf{p}$ and $\mathbf{p}'$ respectively. The radiation field vectors at $\mathbf{r}$ due to a dipole $\mathbf{p}$ at the origin are

$$\mathbf{E} = \frac{(\mathbf{k} \times \mathbf{p}) \times \mathbf{k}}{4\pi\varepsilon_0 r}, \qquad \mathbf{B} = \frac{\mu_0}{4\pi} \frac{\omega \mathbf{k} \times \mathbf{p}}{r},$$

where **k** has magnitude $\frac{\omega}{c}$ and the direction of **r**, and

$$\mathbf{p} = \mathbf{p}_0 e^{-i\omega t'} = \mathbf{p}_0 e^{-i\omega(t-\frac{r}{c})} = \mathbf{p}_0 e^{i(kr-\omega t)}.$$

As $|\mathbf{r}| \gg h$, we can make the approximation $|\mathbf{r}_1| = |\mathbf{r}_2| \approx |\mathbf{r}|$ and write

$$\mathbf{E} = -\frac{1}{4\pi\varepsilon_0} \frac{\omega^2}{c^2} \frac{\sin\theta}{r} p_0 e^{i(kr_1-\omega t)} \mathbf{e}_\theta,$$

$$\mathbf{B} = -\frac{\mu_0}{4\pi} \frac{\omega^2}{c} \frac{\sin\theta}{r} p_0 e^{i(kr_1-\omega t)} \mathbf{e}_\varphi,$$

$$\mathbf{E}' = -\frac{1}{4\pi\varepsilon_0} \frac{\omega^2}{c^2} \frac{\sin\theta}{r} p_0 e^{i(kr_2-\omega t)} \mathbf{e}_\theta,$$

$$\mathbf{B}' = -\frac{\mu_0}{4\pi} \frac{\omega^2}{c} \frac{\sin\theta}{r} p_0 e^{i(kr_2-\omega t)} \mathbf{e}_\varphi.$$

Using these we have

$$\overline{\mathbf{S}}_{\text{total}} = \frac{\omega^4 p_0^2 \sin^2\theta}{32\pi^2\varepsilon_0 c^3 r^2} \{2 + 2\cos[k(r_1 - r_2)]\}\mathbf{e}_r.$$

Under the same approximation, $r_2 - r_1 \approx 2h\cos\theta$. To calculate the radiated power $\overline{P}$ we integrate over the half space above the ground:

$$\overline{P} = \int_0^{\frac{\pi}{2}} \overline{\mathbf{S}}_{\text{total}} \cdot 2\pi r \sin\theta \cdot r d\theta$$

$$= \frac{p_0^2 \omega^4}{8\pi\varepsilon_0 c^3} \int_0^{\frac{\pi}{2}} \sin^3\theta[1 + \cos(2kh\cos\theta)]d\theta$$

$$= \frac{p_0^2 \omega^4}{8\pi\varepsilon_0 c^3} \left[\frac{2}{3} + \int_0^{\frac{\pi}{2}} \sin^3\theta \cos(2kh\cos\theta)d\theta\right].$$

Putting $\beta = 2kh$, $x = \cos\theta$ in the second term we get

$$\int_0^{\frac{\pi}{2}} \sin^3\theta \cos(2kh\cos\theta)d\theta = \int_0^1 (1 - x^2)\cos(\beta x)dx$$

$$= \frac{2}{\beta^2}\left(\frac{\sin\beta}{\beta} - \cos\beta\right).$$

Hence the average power radiated by the system is

$$\overline{P} = \frac{I_0^2 l^2 \omega^2}{8\pi\varepsilon_0 c^3}\left\{\frac{2}{3} + \frac{c^2}{2h^2\omega^2}\left[\frac{c\sin(\frac{2h\omega}{c})}{2h\omega} - \cos\left(\frac{2h\omega}{c}\right)\right]\right\}.$$

(c) The optimal height h for maximum radiated power is given by $\frac{d\overline{P}}{dh} = 0$, or by

$$\frac{d}{d\beta}\left[\frac{1}{\beta^2}\left(\frac{\sin\beta}{\beta} - \cos\beta\right)\right] = 0,$$

giving

$$-3\frac{\sin\beta}{\beta} + 3\cos\beta + \beta\sin\beta = 0,$$

or

$$\tan\beta = \frac{3\beta}{3 - \beta^2}. \tag{1}$$

This equation can be solved numerically to find β, and hence the optimal height $h = \frac{\beta c}{2\omega}$. At optimal height we have

$$\sin\beta = \frac{3\beta}{\sqrt{\beta^4 + 3\beta^2 + 9}}, \qquad \cos\beta = \frac{3 - \beta^2}{\sqrt{\beta^4 + 3\beta^2 + 9}},$$

so that the maximum radiated power is

$$\overline{P}_{\text{max}} = \frac{I_0^2 l^2 \omega^2}{4\pi\varepsilon_0 c^3}\left[\frac{1}{3} + (\beta^4 + 3\beta^2 + 9)^{-\frac{1}{2}}\right]$$

with β given by Eq. (1).

For megahertz waves, $\lambda \sim \frac{3\times10^8}{10^6} = 300$ m. So we can usually assume $h \ll \lambda$, or $\beta = \frac{2h\omega}{c} = \frac{4\pi h}{\lambda} \ll 1$. For such waves, (1) is identically satisfied and the average power radiated is

$$\overline{P} = \frac{I_0^2 l^2 \omega^2}{6\pi\varepsilon_0 c^3}.$$

4058

A thin linear antenna of length d is excited in such a way that the sinusoidal current makes a full wavelength of oscillation as shown in Fig. 4.25 (frequency $\omega = 2\pi c/d$).

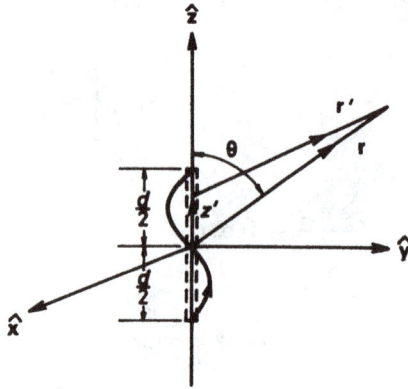

$$J = I_0\,\delta(x)\delta(y)\sin(\tfrac{2\pi z}{d})\hat{z}\,e^{i\omega t'}$$

Fig. 4.25

(a) Calculate exactly the power radiated per unit solid angle and plot the angular distribution as a function of θ.

(b) Compare your result of (a) with those obtained from a multipole expansion.

(MIT)

Solution:

The retarded vector potential $A(r,t)$ is given by

$$A = \frac{\mu_0}{4\pi}\int\frac{J(r',t-\frac{r'}{c})}{r'}dV' = \frac{\mu_0}{4\pi}\int\frac{I(r',t-\frac{r'}{c})}{r'}dl$$

$$= \frac{\mu_0}{4\pi}I_0 e^{i\omega t}e_z\int_{-\lambda/2}^{\lambda/2}\frac{\sin(kz')}{r'}e^{-ikr'}dz'\,,$$

where $r' = \sqrt{r^2 - 2rz'\cos\theta + z'^2}$, $k = \frac{2\pi}{d} = \frac{2\pi}{\lambda} = \frac{\omega}{c}$.

As we are only interested in the radiation field, which varies as $\frac{1}{r}$, we shall ignore terms of orders higher than $\frac{1}{r}$. Accordingly, we use the approximations

$$r' = \sqrt{(r - z'\cos\theta)^2 + z'^2\sin^2\theta} \approx r - z'\cos\theta\,,$$

$$\frac{1}{r'} \approx \frac{1}{r}$$

and write

$$\mathbf{A} = \frac{\mu_0 I_0 e^{i\omega t}}{4\pi} \mathbf{e}_z \int_{-\lambda/2}^{\lambda/2} \sin(kz') e^{-ik(r - z'\cos\theta)} dz'$$

$$= \frac{\mu_0}{4\pi} \frac{I_0 e^{i(\omega t - kr)}}{r} \mathbf{e}_z \int_{-\lambda/2}^{\lambda/2} \sin(kz') e^{ikz'\cos\theta} dz' .$$

(a) Integration yields (cf. Problem 4056)

$$\mathbf{A} = i \frac{\mu_0}{2\pi} \frac{I_0}{kr} \frac{\sin(\pi\cos\theta)}{\sin^2\theta} e^{i(\omega t - kr)} \mathbf{e}_z .$$

Defining $\mathbf{k} = k\mathbf{e}_r$ and with $\mathbf{e}_z = (\cos\theta, -\sin\theta, 0)$ in spherical coordinates, we have

$$\mathbf{B} = \nabla \times \mathbf{A} = -i\mathbf{k} \times \mathbf{A} = -\frac{\mu_0}{2\pi} \frac{I_0}{r} \frac{\sin(\pi\cos\theta)}{\sin\theta} e^{i(\omega t - kr)} \mathbf{e}_\varphi .$$

From Maxwell's equation $\nabla \times \mathbf{H} = \dot{\mathbf{D}}$, or

$$\nabla \times \mathbf{B} = -i\mathbf{k} \times \mathbf{B} = i\mu_0\varepsilon_0\omega\mathbf{E}$$

$$= i\frac{k}{c}\mathbf{E} ,$$

we find

$$\mathbf{E} = -c\frac{\mathbf{k}}{k} \times \mathbf{B} = c\mathbf{B} \times \mathbf{e}_r .$$

Hence the average Poynting vector is (Problem 4042)

$$\langle \mathbf{S} \rangle = \frac{1}{2\mu_0} R_e(\mathbf{E}^* \times \mathbf{B}) = \frac{c}{2\mu_0} |\mathbf{B}|^2 \mathbf{e}_r$$

$$= \frac{\mu_0}{8\pi^2} \frac{cI_0^2}{r^2} \left[\frac{\sin(\pi\cos\theta)}{\sin\theta} \right]^2 \mathbf{e}_r$$

and the power radiated per unit solid angle is

$$\frac{dP}{d\Omega} = \frac{\langle S \rangle}{r^{-2}} = \frac{I_0^2}{8\pi^2} \sqrt{\frac{\mu_0}{\varepsilon_0}} \left[\frac{\sin(\pi\cos\theta)}{\sin\theta} \right]^2 .$$

In the formula the factor $\sqrt{\mu_0/\varepsilon_0}$ is the characteristic impedance of electromagnetic waves in vacuum. The curve of $\frac{dP}{d\Omega}$ versus θ is sketched in Fig. 4.26.

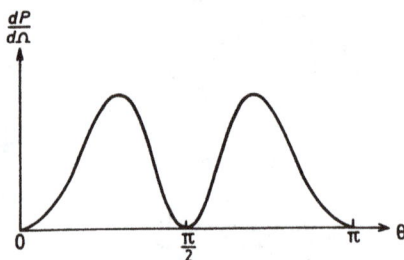

Fig. 4.26

(b) If we had used multipole expansion, we would have

$$\mathbf{A} = \frac{\mu_0}{4\pi} \frac{I_0}{r} e^{i(\omega t - kr)} \mathbf{e}_z \int_{-\frac{\lambda}{2}}^{\frac{\lambda}{2}} \sin(kz') \left[1 + \frac{ikz' \cos\theta}{1!} + \frac{(ikz' \cos\theta)^2}{2!} + \cdots \right] dz'$$

$$\approx \frac{i\mu_0}{4\pi} \frac{I_0}{r} e^{i(\omega t - kr)} \lambda \cos\theta \, \mathbf{e}_z \,,$$

neglecting terms of order $(\frac{z'}{\lambda})^2$ and higher in the expansion of $\exp(ikz' \cos\theta)$.

Then

$$\mathbf{B} = -i\mathbf{k} \times \mathbf{A} = \frac{\mu_0}{2} \frac{I_0}{r} e^{i(\omega t - kr)} \cos\theta \, \mathbf{e}_\varphi \,,$$

giving

$$\frac{dP}{d\Omega} = \frac{\mu_0}{8} c I_0^2 \cos^2\theta = \frac{I_0^2}{8} \sqrt{\frac{\mu_0}{\varepsilon_0}} \cos^2\theta \,.$$

The $\frac{dP}{d\Omega}$ vs. θ curve is shown in Fig. 4.27.

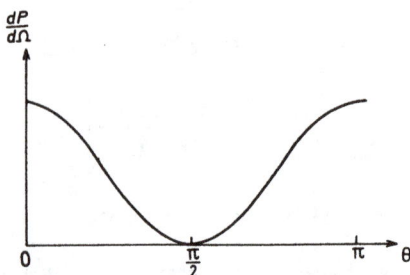

Fig. 4.27

Comparing the two figures we see that the multipole expansion method gives good approximation only in the neighborhood of $\frac{\pi}{2}$.

4059

Consider the situation shown in Fig. 4.28 where a perfectly conducting thin wire connects two small metallic balls. Suppose the charge density is given by

$$\rho(\mathbf{x}, t) = [\delta(z - a) - \delta(z + a)]\delta(x)\delta(y)Q\cos(\omega_0 t).$$

The current flows between the metallic balls through the thin wire. a, Q and ω_0 are constants.

(a) Calculate $\frac{d\overline{P}}{d\Omega}$, the average power emitted per unit solid angle in the dipole approximation.

(b) When is the dipole approximation valid?

(c) Calculate $\frac{d\overline{P}}{d\Omega}$ exactly.

(*Columbia*)

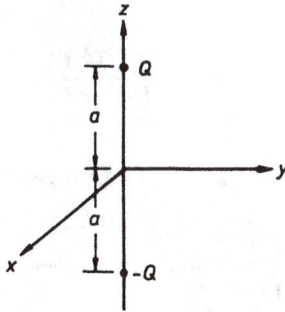

Fig. 4.28

Solution:

(a) The moment of the dipole is $\mathbf{p} = 2Qa\cos(\omega_0 t)\mathbf{e}_z$, or the real part of $2Qae^{-i\omega_0 t}\mathbf{e}_z$. The average power per unit solid angle at $\mathbf{R}$, as shown in Fig. 4.29, is

$$\frac{d\overline{P}}{d\Omega} = \frac{\langle S \rangle}{R^{-2}} = \frac{|\ddot{\mathbf{p}}|^2}{32\pi^2\varepsilon_0 c^3}\sin^2\theta = \frac{Q^2 a^2 \omega_0^4 \sin^2\theta}{8\pi^2\varepsilon_0 c^3}.$$

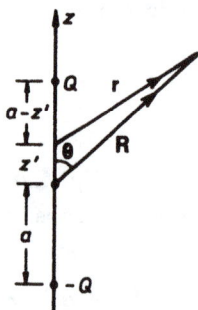

Fig. 4.29

(b) The dipole approximation is valid if $R \gg \lambda \gg a$.

(c) A current flows through the wire connecting the two small metallic balls of density

$$\mathbf{j}(\mathbf{x},t) = \mathbf{e}_z \frac{d}{dt} \int \rho \, dz = -i\omega_0 Q \delta(x)\delta(y)e^{-i\omega_0 t}\mathbf{e}_z, \qquad |z| \le a,$$

and produces a vector potential

$$\mathbf{A}(\mathbf{x},t) = \frac{\mu_0}{4\pi} \int_V \frac{\mathbf{j}(\mathbf{x}',t')}{r} dV',$$

where $t' = t - \frac{r}{c}$, $\mathbf{r} = \mathbf{x} - \mathbf{x}'$, and V is the region occupied by the current distribution. Thus

$$\mathbf{A}(\mathbf{x},t) = \frac{-\mu_0}{4\pi} \int_V \frac{i\omega_0 Q \delta(x')\delta(y')e^{-i\omega_0(t-\frac{r}{c})}}{r} dx' dy' dz' \mathbf{e}_z$$

$$= -i\frac{\mu_0 \omega_0 Q e^{-i\omega_0 t}}{4\pi} \int_{-a}^{a} \frac{e^{ik_0 r}}{r} dz' \mathbf{e}_z,$$

where $k_0 = \frac{\omega_0}{c}$. Let $R = |\mathbf{x}|$, then $r^2 = R^2 - 2Rz'\cos\theta + z'^2$, as shown in Fig. 4.29. Hence

$$\mathbf{A}(\mathbf{x},t) = -i\frac{\mu_0 \omega_0 Q e^{-i\omega_0 t}}{4\pi} \int_{-a}^{a} \frac{e^{ik_0\sqrt{R^2+z'^2-2Rz'\cos\theta}}}{\sqrt{R^2 + z'^2 - 2Rz'\cos\theta}} dz' \mathbf{e}_z.$$

This is the exact solution. To find the integral analytically, we assume $R \gg a$ and use the approximation $\frac{1}{r} \approx \frac{1}{R}$, $\sqrt{R^2 + z'^2 - 2Rz' \cos\theta} \approx R - z' \cos\theta$. Then

$$
\begin{aligned}
\mathbf{A}(\mathbf{x}, t) &= \frac{-i\mu_0\omega_0 Q e^{i(k_0 R - \omega_0 t)}}{4\pi R} \int_{-a}^{a} e^{-ik_0 z' \cos\theta} dz' \mathbf{e}_z \\
&= -\frac{iQ e^{i(k_0 R - \omega_0 t)}}{2\pi\varepsilon_0 cR} \frac{\sin(k_0 a \cos\theta)}{\cos\theta} \mathbf{e}_z .
\end{aligned}
$$

In spherical coordinates, $\mathbf{e}_z = \cos\theta \mathbf{e}_R - \sin\theta \mathbf{e}_\theta$. We can then write $\mathbf{A} = A_R \mathbf{e}_R + A_\theta \mathbf{e}_\theta$ with A_R, A_θ independent of the angle φ.

The magnetic field is given by

$$
\mathbf{B} = \nabla \times \mathbf{A} = \frac{1}{R}\left[\frac{\partial}{\partial R}(RA_\theta) - \frac{\partial}{\partial\theta} A_R \right] \mathbf{e}_\varphi .
$$

As we are only interested in the radiation field which varies as $\frac{1}{R}$, we can neglect the second differential on the right-hand side. Hence

$$
B_\phi \approx \frac{1}{R} \frac{\partial(RA_\theta)}{\partial R} = -\frac{k_0 Q e^{i(k_0 R - \omega t)}}{2\pi\varepsilon_0 cR} \cdot \frac{\sin(ka \cos\theta)\sin\theta}{\cos\theta} ,
$$

so that

$$
\overline{\mathbf{S}} = \frac{c}{2\mu_0} |\mathbf{B}|^2 \mathbf{e}_R = \frac{\omega_0^2 Q^2}{8\pi\varepsilon_0 cR^2} \frac{\sin^2(ka \cos\theta)\sin^2\theta}{\cos^2\theta} \mathbf{e}_R ,
$$

and finally

$$
\frac{d\overline{P}}{d\Omega} = \frac{\overline{S}}{R^{-2}} = \frac{\omega_0^2 Q^2}{8\pi^2\varepsilon_0 c} \frac{\sin^2\theta \sin^2(ka\cos\theta)}{\cos^2\theta} .
$$

If the condition $\lambda \gg a$ is also satisfied, then $\sin(ka\cos\theta) \approx ka\cos\theta$ and the above expression reduces to that for the dipole approximation.

4060

Two equal point charges $+q$ oscillate along the z-axis with their positions given by

$$
z_1 = z_0 \sin(\omega t), \quad z_2 = -z_0 \sin(\omega t), \quad x_i = y_i = 0, \quad (i = 1, 2).
$$

The radiation field is observed at a position $\mathbf{r}$ with respect to the origin (Fig. 4.30). Assume that $|\mathbf{r}| \gg \lambda \gg z_0$, where λ is the wavelength of the emitted radiation.

(a) Find the electric field $\mathbf{E}$ and magnetic field $\mathbf{B}$.

(b) Compute the power radiated per unit solid angle in the direction of $\mathbf{r}$.

(c) What is the total radiated power? How does the dependence on ω compare to that for dipole radiation?

$$(MIT)$$

Solution:

(a) As $|\mathbf{r}| \gg \lambda \gg z_0$, multipole expansion may be used to calculate the electromagnetic field. For the radiation field we need to consider only components which vary as $\frac{1}{r}$. The electric dipole moment of the system is

$$\mathbf{P} = (qz_1 + qz_2)\mathbf{e}_z = 0.$$

Hence the dipole field is zero.

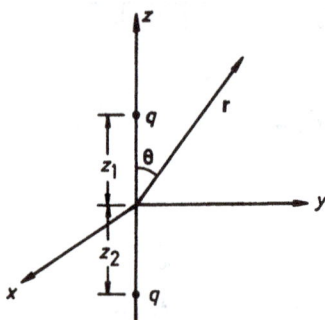

Fig. 4.30

The vector potential of the electric quadrupole radiation field is given by

$$\mathbf{A}(\mathbf{r}, t) = -\frac{\mu_0}{4\pi} \frac{\omega}{2r} e^{-i\omega t'} \int (\mathbf{k} \cdot \mathbf{r}')\mathbf{r}'\rho dV',$$

where

$$t' = t - \frac{r}{c}, \qquad \mathbf{k} = \frac{\omega}{c}\frac{\mathbf{r}}{r}.$$

Hence the magnetic induction is

$$\mathbf{B} = \nabla \times \mathbf{A} = ik \times \mathbf{A} = -i\frac{\mu_0}{4\pi}\frac{\omega k^2}{2r^3}e^{i(kr-\omega t)}\int \mathbf{r} \times \mathbf{r}'(\mathbf{r}\cdot\mathbf{r}')\rho dV'$$

$$= -i\frac{\mu_0}{4\pi}\frac{\omega^3}{2r^3c^2}e^{i(kr-\omega t)}\sum_n \mathbf{r} \times \mathbf{r}'(\mathbf{r}\cdot\mathbf{r}')q_n .$$

As

$$\mathbf{r} = r\mathbf{e}_r , \qquad \mathbf{r}' = \pm z_0(\mathbf{e}_r \cos\theta - \mathbf{e}_\theta \sin\theta)$$

in spherical coordinates, we have

$$\mathbf{B} = i\frac{\mu_0}{4\pi}\frac{\omega^3}{2r^3c^2}e^{i(kr-\omega t)}2r^2 z_0^2 q \sin\theta\cos\theta\mathbf{e}_\varphi$$

$$= i\frac{\mu_0}{4\pi}\frac{\omega^3 z_0^2 q}{rc^2}\sin\theta\cos\theta\, e^{i(kr-\omega t)}\mathbf{e}_\varphi .$$

Then using Maxwell's equations $\nabla \times \mathbf{H} = \dot{\mathbf{D}}$ or

$$\mathbf{E} = c\mathbf{B} \times \mathbf{e}_r ,$$

we find

$$\mathbf{E} = \frac{i\mu_0}{4\pi}\frac{\omega^3 z_0^2 q}{rc}\sin\theta\cos\theta e^{i(kr-\omega t)}\mathbf{e}_\theta .$$

Actually $\mathbf{E}$ and $\mathbf{B}$ are given by the real parts of the above expressions.

(b) The average Poynting vector is

$$\langle \mathbf{N} \rangle = \frac{1}{2\mu_0}\text{Re}(\mathbf{E} \times \mathbf{B}^*)$$

$$= \frac{\mu_0}{32\pi^2}\frac{\omega^6 z_0^4 q^2}{r^2 c^3}\sin^2\theta\cos^2\theta\mathbf{e}_r ,$$

so the average power radiated per unit solid angle is

$$\frac{dP}{d\Omega} = \frac{\langle \mathbf{N} \rangle}{r^{-2}} = \frac{\mu_0}{32\pi^2}\frac{\omega^6 z_0^4 q^2}{c^3}\sin^2\theta\cos^2\theta .$$

(c) The total radiated power is

$$P = \int \frac{dP}{d\Omega}d\Omega = \frac{\mu_0}{32\pi^2}\frac{\omega^6 z_0^4 q^2}{c^3}\int_0^\pi 2\pi\sin^3\theta\cos^2\theta d\theta$$

$$= \frac{\mu_0}{60\pi}\frac{\omega^6 z_0^4 q^2}{c^3} .$$

The total radiated power varies as ω^6 for electric quadrupole radiation, and as ω^4 for electric dipole radiation.

4061

Two point charges of charge e are located at the ends of a line of length $2l$ that rotates with a constant angular velocity $\omega/2$ about an axis perpendicular to the line and through its center as shown in Fig. 4.31.

(a) Find (1) the electric dipole moment, (2) the magnetic dipole moment, (3) the electric quadrupole moment.

(b) What type of radiation is emitted by this system? What is the frequency?

(c) Suppose the radiation is observed far from the charges at an angle θ relative to the axis of rotation. What is the polarization for $\theta = 0°$, $90°$, $0 < \theta < 90°$?

(*Princeton*)

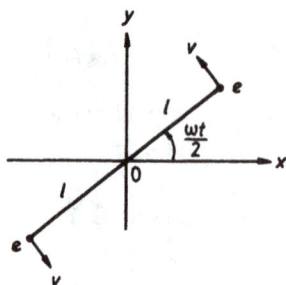

Fig. 4.31

Solution:

(a) (1) The electric dipole moment is

$$\mathbf{P} = e\mathbf{r}_1' + e\mathbf{r}_2' = 0.$$

(2) The magnetic dipole moment is

$$\mathbf{m} = IS\mathbf{e}_z = \frac{2e}{T} \cdot (\pi l^2)\mathbf{e}_z = \frac{1}{2}e\omega l^2\mathbf{e}_z,$$

which is constant.

(3) The position vectors of the two point charges are

$$\mathbf{r}_1' = -\mathbf{r}_2' = l\cos\left(\frac{\omega t}{2}\right)\mathbf{e}_x + l\sin\left(\frac{\omega t}{2}\right)\mathbf{e}_y .$$

The electric quadrupole moment tensor has components given by

$$Q_{ij} = \int (3x_i'x_j' - r'^2\delta_{ij})\rho dV' = \sum_n (3x_i'x_j' - r'^2 \cdot \delta_{ij})q_n ,$$

where $r'^2 = |\mathbf{r}_1'|^2 = |\mathbf{r}_2'|^2 = l^2$.
Thus the non-zero components are

$$Q_{11} = el^2[1 + 3\cos(\omega t')] ,$$
$$Q_{12} = Q_{21} = 3el^2\sin(\omega t') ,$$
$$Q_{22} = el^2[1 - 3\cos(\omega t')] .$$

(b) Because $\mathbf{P} = 0$ and $\mathbf{m}$ is a constant vector, they will not produce radiation. Thus the emitted radiation is that of an electric quadrupole with frequency ω.

(c) At a point $\mathbf{r}(r, \theta, \varphi)$ far away from the charges the magnetic induction of the radiation field is given by

$$\mathbf{B} = -i\frac{\mu_0}{4\pi}\frac{\omega k}{6r}\mathbf{k} \times \mathbf{Q} ,$$

where $k = \frac{\omega}{c}\mathbf{e}_r$, $\mathbf{Q}$ has components $Q_i = \frac{1}{r}\sum_j Q_{ij}x_j$. Writing Q_{ij} as the real parts of

$$Q_{11} = el^2(1 + 3e^{-i\omega t'}) ,$$
$$Q_{12} = Q_{21} = 3el^2 ie^{-i\omega t'} ,$$
$$Q_{22} = el^2(1 - 3e^{-i\omega t'}) ,$$

as $\mathbf{e}_r = r(\sin\theta\cos\varphi, \sin\theta\sin\varphi, \cos\theta)$, we have

$$Q_1 = el^2\sin\theta(\cos\varphi + 3e^{-i(\omega t' - \varphi)}) ,$$
$$Q_2 = el^2\sin\theta(\sin\varphi + 3ie^{-i(\omega t' - \varphi)}) ,$$

with $t' = t - \frac{r}{c}$, or $-\omega t' = kr - \omega t$. Note in calculating $\mathbf{B}$, we omit terms in Q_i which are constant in the retarded time t' as they do not contribute to emission of radiation.

(1) For $\theta = 0°$, $Q_1 = Q_2 = 0$ giving $B = 0$ so there is no radiation emitted at $\theta = 0°$.

(2) For $\theta = 90°$,

$$Q_1 \sim 3el^2 e^{-i(\omega t' - \varphi)},$$

$$Q_2 \sim 3el^2 i e^{-i(\omega t' - \varphi)},$$

$$e_r = (\cos \varphi, \sin \varphi, 0),$$

so that the radiation field is given by the real parts of the following:

$$\mathbf{B} = -i \frac{\mu_0}{4\pi} \frac{\omega^3}{6rc^2}(Q_2 \cos \varphi - Q_1 \sin \varphi)\mathbf{e}_z$$

$$= \frac{\mu_0}{8\pi} \frac{\omega^3 el^2}{rc^2} e^{-i(\omega t' - 2\varphi)}\mathbf{e}_z ,$$

and

$$\mathbf{E} = c\mathbf{B} \times \mathbf{e}_r = \frac{\mu_0}{8\pi} \frac{\omega^3 el^2}{rc} e^{-i(\omega t' - 2\varphi)}(-\sin \varphi \mathbf{e}_x + \cos \varphi \mathbf{e}_y).$$

As $E_x^2 + E_y^2 = $ constant, the radiation is circularly polarized.

(3) For $0° < \theta < 90°$,

$e_r \times Q = -\mathbf{e}_x Q_2 \cos \theta + \mathbf{e}_y Q_1 \cos \theta + \mathbf{e}_z(Q_2 \cos \varphi - Q_1 \sin \varphi) \sin \theta$, so that the radiation field is the real parts of the following:

$$B_x = -\frac{\mu_0}{8\pi} \frac{\omega^3 el^2}{rc^2} \sin \theta \cos \theta e^{-i(\omega t' - \varphi)},$$

$$B_y = -i\frac{\mu_0}{8\pi} \frac{\omega^3 el^2}{rc^2} \sin \theta \cos \theta e^{-i(\omega t' - \phi)},$$

$$B_z = \frac{\mu_0}{8\pi} \frac{\omega^3 el^2}{rc^2} \sin^2 \theta e^{-i(\omega t' - 2\varphi)}.$$

As all the three components of $\mathbf{E}$ and $\mathbf{B}$ are time-dependent, the radiation is not polarized.

4062

(a) Name the lowest electric multipole in the radiation field emitted by the following time-varying charge distributions.

(1) A uniform charged spherical shell whose radius varies as

$$R = R_0 + R_1 \cos(\omega t).$$

(2) Two identically charged particles moving about a common center with constant speed on the opposite sides of a circle.

(b) A loop with one positive and two negative charges as shown in Fig. 4.32 rotates with angular velocity ω about an axis through the center and perpendicular to the loop. What is the frequency of its electric quadrupole radiation?

(*MIT*)

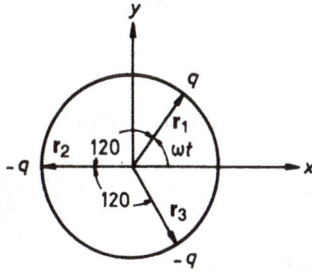

Fig. 4.32

Solution:

(a) (1) For a uniformly charged spherical shell, on account of the spherical symmetry,

$$\mathbf{P} = \mathbf{D} = 0.$$

Hence all the electric multipole moments are zero.

(2) Take coordinates as shown in Fig. 4.33 and let the line joining the charged particles be rotating about the z-axis with angular speed ω. The radius vectors of the two particles are then

$$\mathbf{r}'_1 = R\cos(\omega t)\mathbf{e}_x + R\sin(\omega t)\mathbf{e}_y,$$
$$\mathbf{r}'_2 = -[R\cos(\omega t)\mathbf{e}_x + R\sin(\omega t)\mathbf{e}_y],$$

where $R = |\mathbf{r}'_1| = |\mathbf{r}'_2|$.

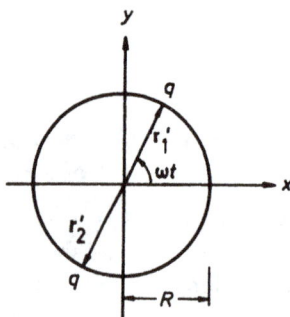

Fig. 4.33

The electric dipole moment of the system is

$$\mathbf{P} = q(\mathbf{r}_1' + \mathbf{r}_2') = 0.$$

The components of the electric quadrupole moment are given by

$$Q_{ij} = \int (3x_i'x_j' - r'^2\delta_{ij})\rho dV' = \sum_n (3x_i'x_j' - r'^2\delta_{ij})q_n,$$

where $r'^2 = R^2 = x_1'^2 + x_2'^2 + x_3'^2$. Thus

$$
\begin{aligned}
Q_{11} &= 2qR^2[2\cos^2(\omega t) - \sin^2(\omega t)], \\
Q_{22} &= 2qR^2[2\sin^2(\omega t) - \cos^2(\omega t)], \\
Q_{33} &= -2qR^2, \\
Q_{12} &= Q_{21} = 3qR^2\sin(2\omega t), \\
Q_{13} &= Q_{31} = Q_{23} = Q_{32} = 0.
\end{aligned}
$$

Hence the lowest electric multipole is a quadrupole.

(b) Take fixed coordinates as shown in Fig. 4.32. Then the position vectors of the three point charges are as follows:

$$
\begin{aligned}
q_1 = q: \quad & \mathbf{r}_1' = R\cos(\omega t)\mathbf{e}_x + R\sin(\omega t)\mathbf{e}_y, \\
q_2 = -q: \quad & \mathbf{r}_2' = R\cos\left(\omega t + \frac{2\pi}{3}\right)\mathbf{e}_x + R\sin\left(\omega t + \frac{2\pi}{3}\right)\mathbf{e}_y, \\
q_3 = -q: \quad & \mathbf{r}_3' = R\cos\left(\omega t + \frac{4\pi}{3}\right)\mathbf{e}_x + R\sin\left(\omega t + \frac{4\pi}{3}\right)\mathbf{e}_y.
\end{aligned}
$$

To determine the frequency of the quadrupole radiation, we only have to find a component of the quadrupole moment of the charge system, for example

$$Q_{12} = 3R^2 \left[q_1 \cos(\omega t) \sin(\omega t) + q_2 \cos\left(\omega t + \frac{2\pi}{3}\right) \sin\left(\omega t + \frac{2\pi}{3}\right) \right.$$

$$\left. + q_3 \cos\left(\omega t + \frac{4\pi}{3}\right) \sin\left(\omega t + \frac{4\pi}{3}\right) \right]$$

$$= \frac{3R^2 q}{2} \left[\sin(2\omega t) - \sin\left(2\omega t + \frac{4\pi}{3}\right) - \sin\left(2\omega t + \frac{8\pi}{3}\right) \right].$$

Thus the frequency of the quadrupole radiation is 2ω.

4063

An electric dipole oscillates with a frequency ω and amplitude P_0. It is placed at a distant $a/2$ from an infinite perfectly conducting plane and the dipole is parallel to the plane. Find the electromagnetic field and the time-averaged angular distribution of the emitted radiation for distances $r \gg \lambda$.

(Princeton)

Solution:

Use Cartesian coordinates as shown in Fig. 4.34. The action of the conducting plane on the $x > 0$ space is equivalent to that of an image dipole at $(-\frac{a}{2}, 0, 0)$ of moment

$$\mathbf{P}' = -\mathbf{P} = -P_0 e^{-i\omega t} \mathbf{e}_z .$$

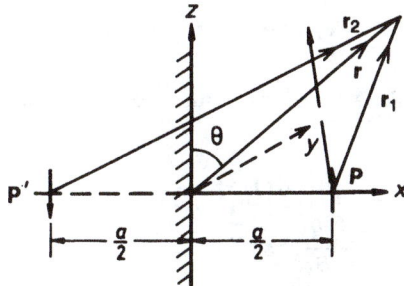

Fig. 4.34

The vector potential at a point $\mathbf{r}$ is

$$\mathbf{A} = \frac{\mu_0}{4\pi}\left(\frac{\dot{\mathbf{P}}}{r_1} + \frac{\dot{\mathbf{P}}}{r_2}\right)$$

$$= -i\frac{\mu_0}{4\pi}\omega P_0\left(\frac{e^{ikr_1}}{r_1} - \frac{e^{ikr_2}}{r_2}\right)e^{-i\omega t}\mathbf{e}_z .$$

As we are only interested in the radiation field which dominates at $r \gg a$, we use the approximation

$$r_1 \approx r - \frac{a}{2}\mathbf{e}_x \cdot \mathbf{e}_r , \qquad r_2 \approx r + \frac{a}{2}\mathbf{e}_x \cdot \mathbf{e}_r , \qquad \frac{1}{r_1} \approx \frac{1}{r_2} \approx \frac{1}{r} ,$$

$\mathbf{e}_r$, $\mathbf{e}_\theta$, $\mathbf{e}_\varphi$ being the unit vectors in spherical coordinates. As $\mathbf{e}_x = \mathbf{e}_r \sin\theta\cos\varphi + \mathbf{e}_\theta\cos\theta\cos\varphi - \mathbf{e}_\varphi\sin\varphi$, we have

$$r_1 \approx r - \frac{a}{2}\sin\theta\cos\varphi ,$$

$$r_2 \approx r + \frac{a}{2}\sin\theta\cos\varphi ,$$

and

$$\mathbf{A} \approx i\frac{\mu_0}{4\pi}\frac{\omega P_0}{r}\left(e^{i\frac{ka}{2}\sin\theta\cos\varphi} - e^{-i\frac{ka}{2}\sin\theta\sin\varphi}\right)e^{i(kr-\omega t)}\mathbf{e}_z$$

$$= -\frac{\mu_0}{2\pi}\frac{\omega P_0}{r}e^{i(kr-\omega t)}\sin\left(\frac{ka}{2}\sin\theta\cos\varphi\right)\mathbf{e}_z .$$

In spherical coordinates

$$\mathbf{e}_z = \mathbf{e}_r\cos\theta - \mathbf{e}_\theta\sin\theta .$$

To obtain $\mathbf{B} = \nabla \times \mathbf{A}$, we neglect terms of orders higher than $\frac{1}{r}$ and obtain

$$\mathbf{B}(\mathbf{r}, t) \approx \frac{\mathbf{e}_\varphi}{r}\frac{\partial}{\partial r}(rA_\theta)$$

$$= -\frac{\mathbf{e}_\varphi}{r}\frac{\partial}{\partial r}(r\sin\theta A)$$

$$= \frac{i\omega^2 P_0 e^{i(kr-\omega t)}}{2\pi\varepsilon_0 c^3 r}\sin\theta\sin\left(\frac{k}{2}a\sin\theta\cos\varphi\right)\mathbf{e}_\varphi .$$

The associated electric field intensity is

$$\mathbf{E}(\mathbf{r},t) = c\mathbf{B} \times \mathbf{e}_r$$
$$\approx \frac{i\omega^2 P_0 e^{i(kr-\omega t)}}{2\pi\varepsilon_0 c^2 r} \sin\theta \sin\left(\frac{k}{2}a\sin\theta\cos\varphi\right)\mathbf{e}_\theta .$$

The average Poynting vector is

$$\overline{\mathbf{S}} = \frac{\varepsilon_0 c}{2}|\mathbf{E}|^2\mathbf{e}_r = \frac{\omega^4 P_0^2 \sin^2\theta}{8\pi^2\varepsilon_0 c^3 r^2}\sin^2\left(\frac{k}{2}a\sin\theta\cos\varphi\right)\mathbf{e}_r .$$

The angular distribution of the radiation is therefore given by

$$\frac{d\overline{P}}{d\Omega} = \frac{\overline{S}}{r^{-2}} = \frac{\omega^4 P_0^2 \sin^2\theta}{8\pi^2\varepsilon_0 c^3}\sin^2\left(\frac{k}{2}a\sin\theta\cos\varphi\right) .$$

If $\lambda \gg a$, then $\sin(\frac{k}{2}a\sin\theta\cos\varphi) \approx \frac{k}{2}a\sin\theta\cos\varphi$ and we have the approximate expression

$$\frac{d\overline{P}}{d\Omega} \approx \frac{\omega^6 P_0^2 a^2 \sin^4\theta\cos^2\varphi}{32\pi^2\varepsilon_0 c^5} .$$

4064

A small electric dipole of dipole moment $\mathbf{P}$ and oscillating with frequency ν is placed at height $\lambda/2$ above an infinite perfectly conducting plane, as shown in Fig. 4.35, where λ is the wavelength corresponding to the frequency ν. The dipole points in the positive z-direction, which is normal to the plane, regarded as the xy-plane. The size of the dipole is assumed very small compared with λ. Find expressions for the electric and magnetic fields, and for the flux of energy at distances r very large compared with λ as a function of r and the unit vector $\mathbf{n}$ in the direction from the origin to the point of observation.

(*UC, Berkeley*)

Solution:

The effect of the conducting plane is equivalent to an image dipole of moment $\mathbf{P}' = -\mathbf{P} = -P_0 e^{-i\omega t}\mathbf{e}_z$, where $\omega = 2\pi\nu$, at $z = -\frac{\lambda}{2}$. Consider a point of observation M of position vector $\mathbf{r}$ (r, θ, φ) and let the distances from P and P' to point M be r_1 and r_2 respectively. For $r \gg \lambda$ we have

$$r_1 \approx r - \frac{\lambda}{2}\cos\theta ,$$

$$r_2 \approx r + \frac{\lambda}{2}\cos\theta .$$

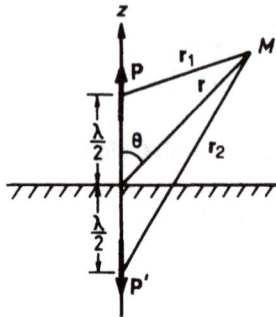

Fig. 4.35

Using the solution of Problem 4063 with $\varphi = 0$, $a = \lambda$ and noting that $k\lambda = 2\pi$, we have

$$\mathbf{B}(\mathbf{r}, t) \approx \frac{i\omega^2 P_0 e^{i(kr-\omega t)}}{2\pi\varepsilon_0 c^3 r} \sin\theta \sin(\pi\cos\theta)\mathbf{e}_\varphi ,$$

$$\mathbf{E}(\mathbf{r}, t) = c\mathbf{B} \times \mathbf{n}$$
$$\approx \frac{i\omega^2 P_0 e^{i(kr-\omega t)}}{2\pi\varepsilon_0 c^2 r} \sin\theta \sin(\pi\cos\theta)\mathbf{e}_\theta ,$$

and the average energy flux density (Problem 4011)

$$\overline{S} = \frac{c\varepsilon_0}{2}|\mathbf{E}|^2\mathbf{n} = \frac{\omega^4 P_0^2 \sin^2\theta}{8\varepsilon_0 c^3}\sin^2(\pi\cos\theta)\mathbf{n} ,$$

where $\omega = 2\pi\nu$.

4065

Two electric dipole oscillators vibrate with the same frequency ω, but their phases differ by $\frac{\pi}{2}$. The amplitudes of the dipole moments are both equal to $\mathbf{P}_0$, but the two vectors are at an angle ψ_0 to each other, (let $\mathbf{P}_1$ be along the x-axis and $\mathbf{P}_2$ in the xy plane) as in Fig. 4.36. For an oscillating dipole $\mathbf{P}$ at the origin, the $\mathbf{B}$-field in the radiation zone is given by

$$\mathbf{B} = k^2 \frac{1}{r} e^{ikr} \left(\frac{\mathbf{r}}{r} \times \mathbf{P} \right).$$

Find (a) the average angular distribution, and (b) the average total intensity of the emitted radiation in the radiation zone.

(SUNY, Buffalo)

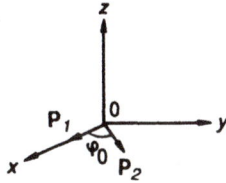

Fig. 4.36

Solution:

The electric dipole moments of the two oscillators are

$$\mathbf{P}_1 = P_0 e^{-i\omega t} \mathbf{e}_x$$
$$\mathbf{P}_2 = P_0 (\cos \psi_0 \mathbf{e}_x + \sin \psi_0 \mathbf{e}_y) e^{-i(\omega t - \frac{\pi}{2})}.$$

The dipole moment of the whole system is

$$\mathbf{P} = \mathbf{P}_1 + \mathbf{P}_2$$

and the magnetic field in Gaussian units is given by

$$\mathbf{B} = k^2 \frac{1}{r} e^{ikr} (\mathbf{e}_r \times \mathbf{P}).$$

As

$$\mathbf{e}_x = \mathbf{e}_r \sin \theta \cos \varphi + \mathbf{e}_\theta \cos \theta \cos \varphi - \mathbf{e}_\varphi \sin \varphi,$$
$$\mathbf{e}_y = \mathbf{e}_r \sin \theta \sin \varphi + \mathbf{e}_\theta \cos \theta \sin \varphi + \mathbf{e}_\varphi \cos \varphi,$$

$$e^{i\frac{\pi}{2}} = i,$$

we have

$$\mathbf{e}_r \times \mathbf{P}_1 = P_0 e^{-i\omega t}(\mathbf{e}_\varphi \cos\theta \cos\varphi + \mathbf{e}_\theta \sin\varphi),$$
$$\mathbf{e}_r \times \mathbf{P}_2 = i P_0 e^{-i\omega t}\left[\mathbf{e}_\varphi \cos(\varphi - \psi_0)\cos\theta + \mathbf{e}_\theta \sin(\varphi - \psi_0)\right],$$

so that

$$\mathbf{B} = \frac{k^2 P_0}{r}\left\{[\sin\varphi + i\sin(\varphi - \psi_0)]\mathbf{e}_\theta\right.$$
$$\left. + [\cos\varphi + i\cos(\varphi - \psi_0)]\cos\theta\mathbf{e}_\varphi\right\} e^{i(kr - \omega t)}.$$

(a) The average power per unit solid angle is then

$$\frac{d\overline{S}}{d\Omega} = \frac{c}{4\pi}\mathbf{B}\cdot\mathbf{B}^* r^2$$
$$= \frac{P_0^2 k^4 c}{8\pi}\left\{2 - \sin^2\theta[\cos^2\varphi + \cos^2(\varphi - \psi_0)]\right\}.$$

(b) The average total power of the emitted radiation is

$$\int_0^{2\pi} d\varphi \int_0^\pi \frac{d\overline{S}}{d\Omega}\sin\theta d\theta = \frac{2}{3}k^4 P_0^2 c.$$

4066

A system of N atoms with electric polarizability α is located along the x-axis as shown in Fig. 4.37. The separation between the atoms is a. The system is illuminated with plane polarized light traveling in $+x$ direction with the electric field along the z-axis, viz.

$$\mathbf{E} = (0, 0, E_0 e^{i(kx - \omega t)}).$$

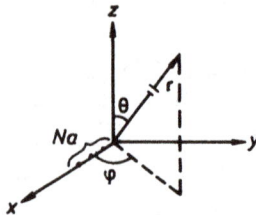

Fig. 4.37

(a) Calculate the angular distribution of the radiated power that would be measured by a detector located far from the atoms ($r \gg \lambda$ and $r \gg Na$). Express the result as a function of the polar and azimuth angles θ and ϕ shown in the figure.

(b) Calculate and sketch the θ dependence of the radiated power in the yz plane. Excluding the trivial case $E = 0$, find the conditions for no radiated power in the yz plane.

(c) Compute a general expression for the ϕ dependence of the radiated power in the xy plane and sketch the dependence for the case $ka \gg 1$.

(MIT)

Solution:

(a) The position of the m-th atom is

$$\mathbf{x}_m = (ma, 0, 0).$$

Under the illumination of the plane wave, its dipole moment is

$$\mathbf{P}_m = \alpha \mathbf{E}(\mathbf{x}_m, t) = \alpha E_0 e^{i(kma - \omega t)} \mathbf{e}_z.$$

The vector potential produced by the N atoms is

$$\mathbf{A} = \frac{\mu_0}{4\pi} \sum_{m=0}^{N-1} \frac{\dot{\mathbf{P}}_m}{r_m}$$

$$= -i \frac{\mu_0}{4\pi} \omega \alpha E_0 \mathbf{e}_z \sum_{m=0}^{N-1} \frac{1}{r_m} \exp\left\{ i \left[kma - \omega \left(t - \frac{r_m}{c} \right) \right] \right\}.$$

For $r \gg \lambda$, $r \gg Na$, we approximate

$$r_m \approx r - ma \sin\theta \cos\varphi,$$

$$\frac{1}{r_m} \approx \frac{1}{r}.$$

Then

$$\mathbf{A} = -i \frac{\mu_0}{4\pi} \frac{\omega \alpha E_0 e^{i(kr - \omega t)}}{r} \mathbf{e}_z \sum_{m=0}^{N-1} \exp[ikma(1 - \sin\theta \cos\varphi)].$$

To find the radiation field we need to retain in $\mathbf{B} = \nabla \times \mathbf{A}$ only terms $\sim \frac{1}{r}$. Hence, according to Problem 4063, we have

$$\mathbf{B}(\mathbf{r}, t) = -\frac{1}{r}\frac{\partial}{\partial r}(rA\sin\theta)\mathbf{e}_\varphi$$
$$= -\frac{\omega^2\alpha E_0\sin\theta\, e^{i(kr-\omega t)}}{4\pi\varepsilon_0 c^3 r}\sum_{m=0}^{N-1}e^{ikma(1-\sin\theta\cos\varphi)}\mathbf{e}_\varphi .$$

Using the identity

$$\left|\sum_{m=0}^{N-1}e^{imx}\right|^2 = \left|\frac{1-e^{iNx}}{1-e^{ix}}\right|^2 = \frac{\sin^2(\frac{Nx}{2})}{\sin^2\frac{x}{2}} ,$$

we find the average Poynting vector of the radiation as

$$\overline{\mathbf{S}} = \frac{\varepsilon_0 c^3}{2}|\mathbf{B}|^2\mathbf{e}_r$$
$$= \frac{\omega^4\alpha^2 E_0^2\sin^2\theta}{32\pi^2\varepsilon_0 c^3 r^2}\cdot\frac{\sin^2[\frac{1}{2}Nka(1-\sin\theta\cos\varphi)]}{\sin^2[\frac{1}{2}ka(1-\sin\theta\cos\varphi)]}\mathbf{e}_r .$$

The angular distribution is given by the average power radiated per unit solid angle

$$\frac{d\overline{P}}{d\Omega} = \frac{\omega^4\alpha^2 E_0^2\sin^2\theta}{32\pi^2\varepsilon_0 c^3}\frac{\sin^2[\frac{1}{2}Nka(1-\sin\theta\cos\varphi)]}{\sin^2[\frac{1}{2}ka(1-\sin\theta\cos\varphi)]} .$$

(b) In the yz plane $\varphi = 90°$, $\cos\varphi = 0$, the angular distribution of the radiation is given by

$$\frac{d\overline{P}}{d\Omega}\propto\sin^2\theta\,\frac{\sin^2[\frac{1}{2}Nka]}{\sin^2(\frac{1}{2}ka)}\sim\sin^2\theta ,$$

which is shown in Fig. 4.38.

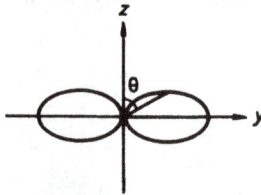

Fig. 4.38

For $\frac{d\overline{P}}{d\Omega} = 0$, we require $\sin(\frac{1}{2}Nka) = 0$, i.e. the condition for no radiation in the yz plane is

$$\frac{1}{2}Nka = n\pi \qquad n = 0, 1, 2, \ldots .$$

(c) In xy plane $\theta = 90°$, $\sin\theta = 1$, the angular distribution is given by

$$\frac{d\overline{P}}{d\Omega} \propto \frac{\sin^2[\frac{1}{2}Nka(1 - \cos\varphi)]}{\sin^2[\frac{1}{2}ka(1 - \cos\varphi)]} = \frac{\sin^2[Nka\sin^2\frac{\varphi}{2}]}{\sin^2[ka\sin^2\frac{\varphi}{2}]} .$$

As $\lim\limits_{k\to\infty} \frac{\sin^2(kx)}{(kx)^2} = \pi\delta(x)$, we have

$$\frac{\sin^2(Nkx)}{\sin^2(kx)} = \frac{N^2\sin^2(Nkx)/(Nkx)^2}{\sin^2(kx)/(kx)^2} \xrightarrow{k \to \infty} \frac{\pi N^2\delta(Nx)}{\pi\delta(x)} = N^2 .$$

Hence, for $ka \gg 1$ the angular distribution of the radiation in the xy plane is isotropic, i.e. the distribution is a circle as illustrated in Fig. 4.39.

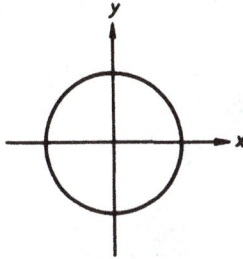

Fig. 4.39

4067

A complicated charge distribution rotates rigidly about a fixed axis with angular velocity ω_0. No point in the distribution is further than a distance d from the axis. The motion is non-relativistic, i.e. $\omega_0 d \ll c$ (see Fig. 4.40.)

(a) What frequencies of electromagnetic radiation may be seen by an observer at a distance $r \gg d$?

(b) Give an order of magnitude estimate of the relative amount of power radiated at each frequency (averaged over both time and angle of observation).

(*MIT*)

Fig. 4.40

Solution:

As can be seen from Fig. 4.40, d is the distance far away from the axis of rotation of the system with $v = \omega_0 d \ll c$, so the radiation of this system can be considered as a multipole radiation.

Let $\sum(x, y, z)$ be the observer's frame and $\sum'(x', y', z')$ a frame fixed on the system. The radius vector of a point in the distribution may be expressed as

$$\xi = x e_x + y e_x + z e_z = x' e'_x + y' e'_y + z' e'_z \tag{1}$$

in the $\sum$ and $\sum'$ frames. We take the axis of rotation as the common z-axis of the two frames and that at $t = 0$ the x'- and x-axes, the y'- and y-axes coincide. We then have the transformation equations

$$\begin{cases} x = x' \cos(\omega_0 t) - y' \sin(\omega_0 t), \\ y = x' \sin(\omega_0 t) + y' \cos(\omega_0 t), \\ z = z'. \end{cases} \tag{2}$$

We can now find the electric dipole moment $\mathbf{P}(t)$, electric quadrupole moment $\mathbf{D}(t)$ and magnetic dipole moment $\mathbf{m}(t)$ in the $\sum$ frame. The electric dipole moment is

$$\mathbf{P}(t) = \int \rho \boldsymbol{\xi} dV = \int \rho(x\mathbf{e}_x + y\mathbf{e}_y + z\mathbf{e}_z)dV$$

$$= \int \rho\{[x'\cos(\omega_0 t) - y'\sin(\omega_0 t)]\mathbf{e}_x + [x'\sin(\omega_0 t) + y'\cos(\omega_0 t)]\mathbf{e}_y + z'\mathbf{e}_z\}dV$$

$$= [P'_{x'}\cos(\omega_0 t) - P'_{y'}\sin(\omega_0 t)]\mathbf{e}_x + [P'_{x'}\sin(\omega_0 t) + P'_{y'}\cos(\omega_0 t)]\mathbf{e}_y + P'_{z'}\mathbf{e}_z , \tag{3}$$

where $P'_{x'}$, $P'_{y'}$ and $P'_{z'}$ are the x, y and z components of the electric dipole moment in the $\sum'$ frame in which the charge distribution is at rest. Eq. (3) shows that $\mathbf{P}(t)$ oscillates with the frequency ω_0. Hence the electric dipole radiation is a monochromatic radiation of angular frequency ω_0.

As the angular velocity ω_0 is a constant, the rotation of the charge system produces a stable current only. Hence the magnetic dipole moment of the system, $\mathbf{m}$, is a constant vector, independent of time. Therefore no magnetic dipole radiation, which is $\propto \ddot{\mathbf{m}}$, is emitted.

For the electric quadrupole moment $\mathbf{D}(t)$, the components are given by

$$D_{ij}(t) = \int [3x_i x_j - (x_1^2 + x_2^2 + x_3^2)\delta_{ij}]\rho dV .$$
$$(i, j = 1, 2, 3, \ x_1 = x, \ x_2 = y, \ x_3 = z)$$

For example,

$$D_{12} = D_{21} = 3\int \rho xy dV$$

$$= 3\int \rho[x'\cos(\omega_0 t) - y'\sin(\omega_0 t)][x'\sin(\omega_0 t) + y'\cos(\omega_0 t)]dV$$

$$= 3\int \rho\left[\frac{1}{2}(x'^2 - y'^2)\sin(2\omega_0 t) + x'y'\cos(2\omega_0 t)\right]dV .$$

In the $\sum'$ frame the quadrupole components are

$$D'_{12} \equiv D'_{x'y'} = 3\int \rho x'y' dV ,$$

$$D'_{11} - D'_{22} = D'_{x'x'} - D'_{y'y'}$$

$$= \int \rho(2x_1'^2 - x_2'^2 - x_3'^2 - 2x_2'^2 - x_1'^2 - x_3'^2)dV$$

$$= 3 \int \rho(x_1'^2 - x_2'^2)dV .$$

Note that under the rotation the charge element ρdV does not change. Thus

$$D_{12} = D_{21} = \frac{D'_{11} - D'_{22}}{2} \sin(2\omega_0 t) + D'_{12} \cos(2\omega_0 t) . \tag{4}$$

Similarly, we have

$$D_{13} = D_{31} = 3 \int \rho xz dV = 3 \int \rho[x' \cos(\omega_0 t) - y' \sin(\omega_0 t)] z' dV$$

$$= D'_{13} \cos(\omega_0 t) - D'_{23} \sin(\omega_0 t) . \tag{5}$$

And other components of $\mathbf{D}(t)$ can be similarly obtained. It is seen from Eq. (4) and Eq. (5) that the electric quadrupole radiation is a mixture of two monochromatic angular frequencies ω_0 and $2\omega_0$.

In short, Eqs. (3), (4) and (5) show that the frequencies of electromagnetic radiation of the system are ω_0 (electric dipole radiation, which is dominant) and $2\omega_0$ (electric quadrupole radiation).

The fields of the successive multipole radiations are reduced in magnitude by a factor $kd = \frac{\omega_0}{c} d$. So the electric quadrupole radiation is weaker than the electric dipole radiation by a factor $(\frac{\omega_0}{c} d)^2$.

PART 5

RELATIVITY, PARTICLE-FIELD INTERACTIONS

1. THE LORENTZ TRANSFORMATION (5001–5017)

5001

The radar speed trap operates on a frequency of 10^9 Hz. What is the beat frequency between the transmitted signal and one received after reflection from a car moving at 30 m/sec?

(Wisconsin)

Solution:

Suppose the car is moving towards the radar with velocity v. Let the radar frequency be ν_0 and the frequency of the signal as received by the car be ν_1. The situation is the same as if the car were stationary and the radar moved toward it with velocity v. Hence the relativistic Doppler effect gives

$$\nu_1 = \nu_0 \sqrt{\frac{1 + v/c}{1 - v/c}} \approx \nu_0 \left(1 + \frac{v}{c}\right) ,$$

correct to the first power of v/c. Now the car acts like a source of frequency ν_1, so the frequency of the reflected signal as received by the radar (also correct to the first power of v/c) is

$$\nu_2 = \nu_1 \sqrt{\frac{1 + v/c}{1 - v/c}} \approx \nu_1 \left(1 + \frac{v}{c}\right) \approx \nu_0 \left(1 + \frac{v}{c}\right)^2 \approx \nu_0 \left(1 + \frac{2v}{c}\right) .$$

Thus the beat frequency is

$$\nu_2 - \nu_0 = \nu_0 \cdot \frac{2v}{c} = 10^9 \times \frac{2 \times 30}{3 \times 10^8} = 200 \text{ Hz} .$$

The result is the same if we had assumed the car to be moving away from the stationary radar. For then we would have to replace v by $-v$ in the above and obtain $\nu_0 - \nu_2 \approx \nu_0 \frac{2v}{c}$.

5002

A plane monochromatic electromagnetic wave propagating in free space is incident normally on the plane of the surface of a medium of index of refraction n. Relative to stationary observer, the electric field of the incident wave is given by the real part of $E_x^0 e^{i(kz - \omega t)}$, where z is the coordinate along the normal to the surface. Obtain the frequency of the reflected wave in the case that the medium and its surface are moving with velocity v along the positive z direction, with respect to the observer.

(SUNY, Buffalo)

Solution:

Let the observer's frame and a frame fixed on the moving medium be Σ and Σ' respectively. Σ' moves with velocity v relative to Σ along the z direction. Let the propagation four-vectors of the incident and reflected waves in Σ and Σ' be respectively

$$k_i = (0, 0, k, \frac{\omega}{c}), \quad k_r = (0, 0, -k_2, \frac{\omega_2}{c}),$$

$$k_i' = (0, 0, k', \frac{\omega'}{c}), \quad k_r' = (0, 0, -k_2', \frac{\omega_2'}{c}),$$

where $k = \frac{\omega}{c}$, $k_2 = \frac{\omega_2}{c}$, $k' = \frac{\omega'}{c}$, $k_2' = \frac{\omega_2'}{c}$.

Lorentz transformation for a four-vector gives

$$\frac{\omega'}{c} = \gamma \left(\frac{\omega}{c} - \beta k \right) = \gamma \frac{\omega}{c} (1 - \beta),$$

$$\frac{\omega_2}{c} = \gamma \left[\frac{\omega_2'}{c} + \beta(-k_2') \right] = \gamma \frac{\omega_2'}{c} (1 - \beta),$$

with $\beta = \frac{v}{c}$, $\gamma = (1 - \beta^2)^{-\frac{1}{2}}$.

In Σ', no change of frequency occurs on reflection, i.e., $\omega_2' = \omega'$. Hence

$$\omega_2 = \gamma \omega' (1 - \beta) = \gamma^2 \omega (1 - \beta)^2 = \left(\frac{1 - \beta}{1 + \beta} \right) \omega,$$

being the angular frequency of the reflected wave as observed by the observer.

5003

In the inertial frame of the fixed stars, a spaceship travels along the x-axis, with $x(t)$ being its position at time t. Of course, the velocity v and acceleration a in this frame are $v = \frac{dx}{dt}$ and $a = \frac{d^2x}{dt^2}$. Suppose the motion to be such that the acceleration as determined by the space passengers is constant in time. What this means is the following. At any instant we transform to an inertial frame in which the spaceship is momentarily at rest. Let g be the acceleration of the spaceship in that frame at that instant. Now suppose that g, so defined instant by instant, is a constant.

You are given the constant g. In the fixed star frame the spaceship starts with initial velocity $v = 0$ when $x = 0$. What is the distance x traveled when it has achieved a velocity v?

Allow for relativistic kinematics, so that v is not necessarily small compared with the speed of light c.

(CUSPEA)

Solution:

Consider two inertial frames Σ and Σ' with Σ' moving with a constant velocity v along the x direction relative to Σ. Let the velocity and acceleration of an object moving in the x direction be $u, a = \frac{du}{dt}$, and $u', a' = \frac{du'}{dt'}$ in the two frames respectively. Lorentz transformation gives

$$x = \gamma(x' + \beta ct'), \quad ct = \gamma(ct' + \beta x') ,$$

where $\beta = \frac{v}{c}, \gamma = (1-\beta^2)^{-\frac{1}{2}}$. Then the velocity of the object is transformed according to

$$u = \frac{u' + v}{1 + \frac{vu'}{c^2}} . \tag{1}$$

Differentiating the above, we have

$$dt = \gamma \left(dt' + \frac{v}{c^2} dx' \right) = \gamma dt' \left(1 + \frac{v}{c^2} u' \right) ,$$

$$du = \frac{du'}{\gamma^2 \left(1 + \frac{vu'}{c^2} \right)^2} ,$$

whose ratio gives the transformation of acceleration:

$$a = \frac{a'}{\gamma^3 \left(1 + \frac{vu'}{c^2} \right)^3} . \tag{2}$$

Now assume that Σ is the inertial frame attached to the fixed stars and Σ' is the inertial frame in which the spaceship is momentarily at rest. Then in Σ'

$$u' = 0, \quad a' = g ,$$

and Eqs. (1) and (2) give

$$u = v, \quad a = \frac{g}{\gamma^3}$$

with $\gamma = \left(1 - \frac{u^2}{c^2} \right)^{-\frac{1}{2}}$. As the velocity of the spaceship is increased from 0 to v in Σ, the distance traveled is

$$x = \int u dt = \int_0^v \frac{u du}{a} = \frac{1}{g} \int_0^v \frac{u du}{(1 - \frac{u^2}{c^2})^{3/2}} = \frac{c^2}{g} \left\{ \frac{1}{\sqrt{1 - \frac{v^2}{c^2}}} - 1 \right\} .$$

5004

As observed in an inertial frame S, two spaceships are traveling in opposite directions along straight, parallel trajectories separated by a distance d as shown in Fig. 5.1. The speed of each ship is $c/2$, where c is the speed of light.

(a) At the instant (as viewed from S) when the ships are at the points of closest approach (indicated by the dotted line in Fig. 5.1) ship (1) ejects a small package which has speed $3c/4$ (also as viewed from S). From the point of view of an observer in ship (1), at what angle must the package be aimed in order to be received by ship (2)? Assume the observer in ship (1) has a coordinate system whose axes are parallel to those of S and, as shown in Fig. 5.1, the direction of motion is parallel to the y axis.

(b) What is the speed of the package as seen by the observer in ship (1)?

(CUSPEA)

Solution:

(a) In the inertial frame S, the y-component of the velocity of the package should be $c/2$ in order that the package will have the same y coordinate as ship (2) as the package passes through the distance $\Delta x = d$. The velocity of the package in S can be expressed in the form

$$\mathbf{u} = u_x \mathbf{e}_x + u_y \mathbf{e}_y$$

with $u_y = c/2$. As $u = |\mathbf{u}| = \frac{3}{4} c$,

$$u_x = \sqrt{u^2 - u_y^2} = \frac{\sqrt{5}}{4} c .$$

Let S' be the inertial frame fixed on ship (1). In S' the velocity of the package is

$$\mathbf{u}' = u_x' \mathbf{e}_x' + u_y' \mathbf{e}_y' .$$

S' moves with speed $c/2$ relative to S along the $-y$ direction, i.e. the velocity of S' relative to S is $\mathbf{v} = -c\mathbf{e}_y/2$. Velocity transformation then gives

$$u_y' = \frac{u_y - v}{1 - \frac{v u_y}{c^2}} = \frac{\frac{c}{2} + \frac{c}{2}}{1 + \frac{1}{4}} = \frac{4}{5} c ,$$

$$u_x' = \frac{u_x \sqrt{1 - \frac{v^2}{c^2}}}{1 - \frac{v u_y}{c^2}} = \frac{\frac{\sqrt{5}}{4} c \sqrt{1 - \frac{1}{4}}}{1 + \frac{1}{4}} = \sqrt{\frac{3}{20}} c .$$

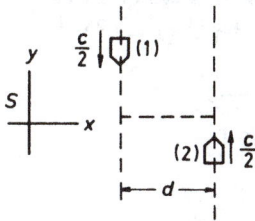

Fig. 5.1

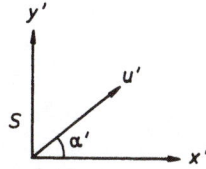

Fig. 5.2

Let α' be the angle between the velocity $\mathbf{u}'$ of the package in S' and the x' axis as shown in Fig. 5.2. Then

$$\tan \alpha' = \frac{u'_y}{u'_x} = \frac{8}{\sqrt{15}}, \quad \text{or} \quad \alpha' = \arctan\left(\frac{8}{\sqrt{15}}\right).$$

(b) In S' the speed of the package is

$$u' = |\mathbf{u}'| = \sqrt{u'^2_x + u'^2_y} = \frac{\sqrt{79}}{10}\, c\ .$$

5005

(a) Write down the equations of conservation of momentum and energy for the Compton effect (a photon striking a stationary electron).

(b) Find the scattered photon's energy for the case of $180°$ back scattering. (Assume the recoiling electron proceeds with approximately the speed of light.)

(*Wisconsin*)

Solution:

(a) Conservation of momentum is expressed by the equations

$$\frac{h\nu}{c} = \frac{h\nu'}{c} \cos \theta + \gamma m v \cos \varphi\ ,$$

$$\frac{h\nu'}{c} \sin \theta = \gamma m v \sin \varphi\ ,$$

where θ is the angle between the directions of motion of the incident and scattered photons, φ is that between the incident photon and the recoiling electron, as shown in Fig. 5.3, m is the rest mass of an electron, $\beta = \frac{v}{c}$, v being the speed of the recoil electron, and $\gamma = (1 - \beta^2)^{-\frac{1}{2}}$.

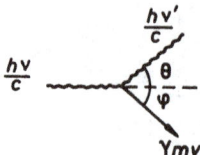

Fig. 5.3

Conservation of energy is expressed by the equation

$$h\nu + mc^2 = h\nu' + \gamma mc^2 .$$

(b) For back scattering, $\theta = 180°$, $\varphi = 0°$. The above equations reduce to

$$h\nu + h\nu' = \gamma\beta mc^2 , \tag{1}$$

$$h\nu - h\nu' = (\gamma - 1)mc^2 . \tag{2}$$

Squaring both sides of Eq. (1) we have

$$(h\nu + h\nu')^2 = \gamma^2\beta^2 m^2 c^4 = (\gamma^2 - 1)m^2 c^4 . \tag{3}$$

Combining Eqs. (2) and (3), we have

$$4h^2\nu\nu' = 2mc^2 h(\nu - \nu') ,$$

or

$$h\nu' = \frac{h\nu}{\frac{2h\nu}{mc^2} + 1} ,$$

which is the energy of the scattered photon.

5006

A charged particle is constrained to move with constant velocity v in the x-direction (with $y = y_0$, $z = 0$ fixed). It moves above an infinite perfectly conducting metal sheet that undulates with "wavelength" L along the x-direction. A distant observer is located in the $z = 0$ plane and detects the electromagnetic radiation emitted at angle θ (the angle between the velocity vector and a vector drawn from the charge to the observer) as shown in Fig. 5.4. What is the wavelength λ of the radiation detected by the observer?

(*Princeton*)

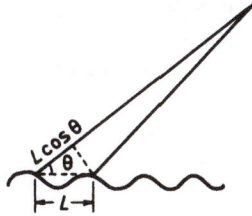

Fig. 5.4

Solution:

The induced charges in the metal sheet will move on the surface of the sheet along the general direction of motion of the charged particle. The acceleration of the induced charges moving on the undulating surface will lead to emission of bremsstrahlung (braking radiation). The radiation detected by a distant observer located along the θ direction is that resulting from the constructive interference in that direction. Hence, the wavelength of the radiation satisfies the condition

$$\frac{L}{v} - \frac{L \cos \theta}{c} = m \frac{\lambda}{c} \ ,$$

where m is an integer, or

$$\lambda_m = \frac{L}{m} \left(\frac{c}{v} - \cos \theta \right) \ .$$

For $m = 1$, $\lambda_1 = L\left(\frac{c}{v} - \cos \theta \right)$.

We can also approach the problem by regarding the effect of the metal sheet as that of an image charge, which together with the real charge forms an oscillating dipole of velocity $\mathbf{v} = v\mathbf{e}_x$ and frequency of vibration $f_0 = \frac{v}{L}$. From the formula of Doppler shift the frequency detected by the observer is

$$f(\theta) = \frac{1 + \beta \cos \theta}{\sqrt{1 - \beta^2}} f_0 \approx \left(1 + \frac{v}{c} \cos \theta \right) f \ ,$$

and the corresponding wavelength is

$$\lambda(\theta) = \frac{c}{f} = \frac{cL}{v} \left(1 - \frac{v}{c} \cos \theta \right) \ .$$

This result is the same as the foregoing λ_1.

5007

Two large parallel plates (non-conducting), separated by a distance d and oriented as shown in Fig. 5.5, move together along x-axis with velocity v, not necessarily small compared with c. The upper and lower plates have uniform surface charge densities $+\sigma$ and $-\sigma$ respectively in the rest frame of the plates. Find the magnitude and direction of the electric and magnetic fields between the plates (neglecting edge effects).

(*Columbia*)

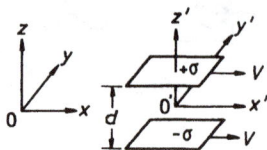

Fig. 5.5

Solution:

Let the electromagnetic fields be $\mathbf{E}'$, $\mathbf{B}'$ in frame S' $(0\ x'\ y'\ z')$ where the plates are at rest; and be $\mathbf{E}$, $\mathbf{B}$ in the laboratory frame S $(0\ x\ y\ z)$. The field vectors transform according to

$$E_x = E'_x, \quad B_x = B'_x ,$$

$$E_y = \gamma(E'_y + \beta c B'_z), \quad B_y = \gamma\left(B'_y - \frac{\beta}{c} E'_z\right) ,$$

$$E_z = \gamma(E'_z - \beta c B'_y), \quad B_z = \gamma\left(B'_z + \frac{\beta}{c} E'_y\right) ,$$

where $\beta = \frac{v}{c}$, $\gamma = (1 - \beta^2)^{-\frac{1}{2}}$.
In the rest frame S',

$$B'_x = B'_y = B'_z = 0 ,$$

$$E'_x = E'_y = 0, \quad E'_z = -\frac{\sigma}{\varepsilon_0} ,$$

so that

$$E_x = 0, \quad B_x = 0 ,$$

$$E_y = 0, \quad B_y = \frac{\gamma\beta}{\varepsilon_0 c}\sigma ,$$

$$E_z = -\frac{\gamma\sigma}{\varepsilon_0}, \quad B_z = 0 .$$

Hence in the laboratory frame, the electric intensity is in the $-z$ direction and has magnitude $\frac{\gamma\sigma}{\epsilon_0}$, while the magnetic induction is in the $+y$ direction and has magnitude $\frac{\gamma v}{\epsilon_0 c^2}\sigma$, where $\gamma = (1 - \frac{v^2}{c^2})^{-\frac{1}{2}}$.

5008

Show that $E^2 - B^2$ and $\mathbf{E} \cdot \mathbf{B}$ are invariant under a Lorentz transformation.

(*UC, Berkeley*)

Solution:

Decompose the electromagnetic field into longitudinal and transverse components with respect to the direction of the relative velocity between two inertial frames Σ and Σ'. In Σ, we have

$$\mathbf{E} = \mathbf{E}_\perp + \mathbf{E}_\parallel , \quad \mathbf{B} = \mathbf{B}_\perp + \mathbf{B}_\parallel .$$

In Σ', which moves with velocity v relative to Σ, we have (in Gaussian units)

$$\mathbf{E}'_\parallel = \mathbf{E}_\parallel , \quad \mathbf{B}'_\parallel = \mathbf{B}_\parallel ,$$

$$\mathbf{E}'_\perp = \gamma \left(\mathbf{E}_\perp + \frac{\mathbf{v}}{c} \times \mathbf{B}_\perp \right) ,$$

$$\mathbf{B}'_\perp = \gamma \left(\mathbf{B}_\perp - \frac{\mathbf{v}}{c} \times \mathbf{E}_\perp \right) ,$$

where

$$\gamma = \frac{1}{\sqrt{1 - v^2/c^2}} .$$

Thus

$$\begin{aligned} \mathbf{E}' \cdot \mathbf{B}' &= \mathbf{E}'_\parallel \cdot \mathbf{B}'_\parallel + \mathbf{E}'_\perp \cdot \mathbf{B}'_\perp \\ &= \mathbf{E}_\parallel \cdot \mathbf{B}_\parallel + \gamma^2 \left[\mathbf{E}_\perp \cdot \mathbf{B}_\perp - \frac{(\mathbf{v} \times \mathbf{B}_\perp) \cdot (\mathbf{v} \times \mathbf{E}_\perp)}{c^2} \right] . \end{aligned}$$

As $\mathbf{v}$ is perpendicular to both $\mathbf{B}_\perp$ and $\mathbf{E}_\perp$, we have

$$(\mathbf{v} \times \mathbf{B}_\perp) \cdot (\mathbf{v} \times \mathbf{E}_\perp) = v^2 \mathbf{E}_\perp \cdot \mathbf{B}_\perp ,$$

so that

$$\mathbf{E}' \cdot \mathbf{B}' = \mathbf{E}_\parallel \cdot \mathbf{B}_\parallel + \gamma^2 \left(1 - \frac{v^2}{c^2} \right) \mathbf{E}_\perp \cdot \mathbf{B}_\perp = \mathbf{E}_\parallel \cdot \mathbf{B}_\parallel + \mathbf{E}_\perp \cdot \mathbf{B}_\perp = \mathbf{E} \cdot \mathbf{B} \ .$$

From the expression for $\mathbf{E}'_\perp$, we have

$$\mathbf{E}'^2_\perp = \gamma^2 \left[\mathbf{E}^2_\perp + \frac{1}{c^2} (\mathbf{v} \times \mathbf{B}_\perp)^2 + 2\mathbf{E}_\perp \cdot \left(\frac{\mathbf{v}}{c} \times \mathbf{B}_\perp \right) \right]$$

$$= \gamma^2 \left(\mathbf{E}^2_\perp + \frac{v^2}{c^2} \mathbf{B}^2_\perp + \frac{2\mathbf{E}_\perp \cdot (\mathbf{v} \times \mathbf{B}_\perp)}{c} \right) ,$$

and

$$\mathbf{B}'^2_\perp = \gamma^2 \left(\mathbf{B}^2_\perp + \frac{v^2}{c^2} \mathbf{E}^2_\perp - \frac{2\mathbf{B}_\perp \cdot (\mathbf{v} \times \mathbf{E}_\perp)}{c} \right) .$$

Hence

$$\mathbf{E}'^2 - \mathbf{B}'^2 = (\mathbf{E}'_\parallel + \mathbf{E}'_\perp)^2 - (\mathbf{B}'_\parallel + \mathbf{B}'_\perp)^2$$

$$= \mathbf{E}'^2_\parallel - \mathbf{B}'^2_\parallel + \mathbf{E}'^2_\perp - \mathbf{B}'^2_\perp$$

$$= (\mathbf{E}^2_\parallel - \mathbf{B}^2_\parallel) + \gamma^2 \left(\mathbf{E}^2_\perp + \frac{v^2}{c^2} \mathbf{B}^2_\perp - \mathbf{B}^2_\perp - \frac{v^2}{c^2} \mathbf{E}^2_\perp \right)$$

$$+ \frac{2\gamma^2}{c} [\mathbf{E}_\perp \cdot \mathbf{v} \times \mathbf{B}_\perp + \mathbf{B}_\perp \cdot \mathbf{v} \times \mathbf{E}_\perp]$$

$$= (\mathbf{E}^2_\parallel - \mathbf{B}^2_\parallel) + (\mathbf{E}^2_\perp - \mathbf{B}^2_\perp)$$

$$= \mathbf{E}^2 - \mathbf{B}^2 ,$$

since $\mathbf{E}_\perp \cdot \mathbf{v} \times \mathbf{B}_\perp = \mathbf{E}_\perp \times \mathbf{v} \cdot \mathbf{B}_\perp = -\mathbf{B}_\perp \cdot \mathbf{v} \times \mathbf{E}_\perp$ for a box product.

Therefore $\mathbf{E}^2 - \mathbf{B}^2$ and $\mathbf{E} \cdot \mathbf{B}$ are invariant under a Lorentz transformation. Note that in SI units it is $\mathbf{E}^2 - c^2\mathbf{B}^2$ that is Lorentz invariant.

5009

(a) A classical electromagnetic wave satisfies the relations

$$\mathbf{E} \cdot \mathbf{B} = 0, \quad \mathbf{E}^2 = c^2\mathbf{B}^2$$

between the electric and magnetic fields. Show that these relations, if satisfied in any one Lorentz frame, are valid in all frames.

(b) If **K** is a unit three-vector in the direction of propagation of the wave, then according to classical electromagnetism, $\mathbf{K} \cdot \mathbf{E} = \mathbf{K} \cdot \mathbf{B} = 0$. Show that this statement is also invariant under Lorentz transformation by showing its equivalence to the manifestly Lorentz invariant statement $n^{\mu} F_{\mu\nu} = 0$, where n^{μ} is a four-vector oriented in the direction of propagation of the wave and $F_{\mu\nu}$ is the field strength tensor.

Parts (a) and (b) together show that what looks like a light wave in one frame looks like one in any frame.

(c) Consider an electromagnetic wave which in some frame has the form

$$E_x = cB_y = f(ct - z),,$$

where $\lim_{z \to \pm\infty} f(z) \to 0$. What would be the values of the fields in a different coordinate system moving with velocity v in the z direction relative to the frame in which the fields are as given above? Give an expression for the energy and momentum densities of the wave in the original frame and in the frame moving with velocity v, show that the total energy-momentum of the wave transforms as a four-vector under the transformation between the two frames. (Assume the extent of the wave in the x-y plane is large but finite, so that its total energy and momentum are finite.)

(*Princeton*)

Solution:

(a) It has been shown in Problem 5008 that $\mathbf{E} \cdot \mathbf{B}$, $\mathbf{E}^2 - c^2\mathbf{B}^2$ are Lorentz invariant. Hence in another Lorentz frame Σ' we have

$$\mathbf{E'} \cdot \mathbf{B'} = \mathbf{E} \cdot \mathbf{B} = 0, \quad \mathbf{E'}^2 - c^2\mathbf{B'}^2 = \mathbf{E}^2 - c^2\mathbf{B}^2 = 0 ,$$

i.e.

$$\mathbf{E'} \cdot \mathbf{B'} = 0, \quad \mathbf{E'}^2 = c^2\mathbf{B'}^2 .$$

(b) The electromagnetic field tensor can be represented by the matrix

$$F_{\mu\nu} = \begin{pmatrix} 0 & -cB_3 & cB_2 & -E_1 \\ cB_3 & 0 & -cB_1 & -E_2 \\ -cB_2 & cB_1 & 0 & -E_3 \\ E_1 & E_2 & E_3 & 0 \end{pmatrix} .$$

Using the electromagnetic wave propagation four-vector $K^{\mu} = (K_1, K_2, K_3, \frac{\omega}{c})$ where $K = \frac{\omega}{c}$, we can express $n^{\mu} \equiv \frac{1}{K} K^{\mu}$. $n^{\mu} F_{\mu\nu} = 0$ which is then equivalent to $K^{\mu} F_{\mu\nu} = 0$. For $\nu = 1$, we have

$$+K_2 cB_3 - K_3 cB_2 + \frac{\omega}{c} E_1 = 0 ,$$

or

$$E_1 = \frac{c}{K}(\mathbf{B} \times \mathbf{K})_1 .$$

Similar expressions are obtained for E_2 and E_3. Hence

$$\mathbf{E} = \frac{c}{K}(\mathbf{B} \times \mathbf{K}) . \tag{1}$$

For $\nu = 4$, we have

$$-K_1 E_1 - K_2 E_2 - K_3 E_3 = 0 ,$$

or

$$\mathbf{K} \cdot \mathbf{E} = 0 . \tag{2}$$

Since $n^\mu F_{\mu\nu} = 0$ is Lorentz covariant, it has the same form in all inertial frames. This means that Eqs. (1) and (2) are valid in all inertial frames. Now Eq. (1) gives

$$\mathbf{E}^2 = \frac{c^2}{K^2}(\mathbf{B} \times \mathbf{K}) \cdot (\mathbf{B} \times \mathbf{K}) = c^2 \mathbf{B}^2 - c^2(\mathbf{B} \cdot \mathbf{K})^2 . \tag{3}$$

From (a) we see that if $\mathbf{E}^2 = c^2 \mathbf{B}^2$ is valid in an inertial frame it is valid in all inertial frames. Since this relation is given for one inertial frame, Eq. (3) means that the relation $\mathbf{K} \cdot \mathbf{B} = 0$ is satisfied in all inertial frames.

(c) In frame Σ one has

$$E_x = f(ct - z), \quad E_y = E_z = 0 ,$$

$$B_x = 0, \quad B_y = \frac{1}{c}f(ct - z), \quad B_z = 0 .$$

Suppose a frame Σ' moves with velocity v relative to Σ frame along the z-axis. Then Lorentz transformation gives

$$E_z' = E_z = 0, \quad E_y' = \gamma(E_y + vB_x) = 0 ,$$
$$E_x' = \gamma(E_x - vB_y) = \gamma(1 - \beta)f(ct - z) ,$$

$$B_z' = B_z = 0, \quad B_x' = \gamma\left(B_x + \frac{\beta}{c}E_y\right) = 0 ,$$

$$B_y' = \gamma\left(B_y - \frac{\beta}{c}E_x\right) = \frac{\gamma}{c}(1 - \beta)f(ct - z) ,$$

where

$$\beta = \frac{v}{c}, \quad \gamma = (1-\beta^2)^{-\frac{1}{2}} .$$

The energy densities in frames Σ and Σ' are respectively given by

$$w = \frac{1}{2}\left(\varepsilon_0 E^2 + \frac{B^2}{\mu_0}\right) = \varepsilon_0 f^2(ct-z),$$

$$w' = \frac{1}{2}\left(\varepsilon_0 E'^2 + \frac{B'^2}{\mu_0}\right) = \varepsilon_0 \gamma^2 (1-\beta)^2 f^2(ct-z).$$

The inverse Lorentz transformation $z = \gamma(z'+\beta ct')$, $ct = \gamma(ct'+\beta z')$ gives

$$ct - z = \gamma(1-\beta)(ct'-z') . \tag{4}$$

Thus

$$w' = \varepsilon_0 \gamma^2 (1-\beta)^2 f^2[\gamma(1-\beta)(ct'-z')] .$$

The momentum density is

$$g = \frac{\mathbf{E}\times\mathbf{H}}{c^2} = \frac{\mathbf{E}\times\mathbf{B}}{c^2\mu_0} = \varepsilon_0 \mathbf{E}\times\mathbf{B} = \varepsilon_0 E_x B_y \mathbf{e}_z .$$

Hence the momentum density as seen in Σ and Σ' has components

$$g_x = g_y = 0, \quad g_z = \frac{\varepsilon_0}{c} f^2(ct-z) ,$$

$$g'_x = g'_y = 0, \quad g'_z = \frac{\varepsilon_0}{c} \gamma^2 (1-\beta)^2 f^2[\gamma(1-\beta)(ct'-z')] .$$

The total energy and the total momentum are

$$W = \int_V w dV = \varepsilon_0 \int_V f^2(ct-z)dV ,$$

$$G_x = G_y = 0, \quad G_z = \frac{\varepsilon_0}{c} \int_V f^2(ct-z)dV = \frac{W}{c}$$

in Σ and

$$W' = \int_{V'} w' dV'$$

$$= \varepsilon_0 \gamma^2 (1-\beta)^2 \int_{V'} f^2 \left[\gamma\left(1-\frac{v}{c}\right)(ct'-z')\right]dV' ,$$

$$G'_x = G'_y = 0 \, ,$$

$$G'_z = \frac{\varepsilon_0}{c} \gamma^2 (1 - \beta)^2 \int_{V'} f^2 [\gamma(1 - \beta)(ct' - z')] dV' = \frac{W'}{c}$$

in Σ'.

As the wave has finite extension, V and V' must contain the same finite number of waves in the direction of propagation, i.e. the z direction. As Eq. (4) requires

$$dV = \gamma(1 - \beta) dV' \, ,$$

$$W' = \gamma(1 - \beta) W \, .$$

Similarly

$$G'_z = \frac{W'}{c} = \gamma(1 - \beta) G_z \, .$$

Thus the transformation equations for total energy-momentum are

$$G'_x = G_x, \quad G'_y = G_y, \quad G'_z = \gamma \left(G_z - \beta \frac{W}{c} \right), \quad \frac{W'}{c} = \gamma \left(\frac{W}{c} - \beta G_z \right).$$

That is, $(\mathbf{G}, \frac{W}{c})$ transforms like a four-vector.

5010

An infinitely long perfectly conducting straight wire of radius r carries a constant current i and charge density zero as seen by a fixed observer A. The current is due to an electron stream of uniform density moving with high (relativistic) velocity U. A second observer B travels parallel to the wire with high (relativistic) velocity v. As seen by the observer B:

(a) What is the electromagnetic field?

(b) What is the charge density in the wire implied by this field?

(c) With what velocities do the electron and ion streams move?

(d) How do you account for the presence of a charge density seen by B but not by A?

(Princeton)

Solution:

(a) Let Σ and Σ' be the rest frames of the observers A and B respectively, the common x-axis being along the axis of the conducting wire, which is fixed in Σ, as shown in Fig. 5.6. In Σ, $\rho = 0$, $\mathbf{j} = \frac{i}{\pi r_0^2}\mathbf{e}_x$, so the electric and magnetic fields in Σ are respectively

$$\mathbf{E} = 0 ,$$

$$\mathbf{B}(r) = \begin{cases} \frac{\mu_0 i r}{2\pi r_0^2}\mathbf{e}_\varphi , & (r < r_0) \\ \frac{\mu_0 i}{2\pi r}\mathbf{e}_\varphi , & (r > r_0) \end{cases}$$

where $\mathbf{e}_x$, $\mathbf{e}_r$, and $\mathbf{e}_\varphi$ form an orthogonal system. Lorentz transformation gives the electromagnetic field as seen in Σ' as

$$E'_\parallel = E_\parallel = 0 , \quad B'_\parallel = B_\parallel = 0 ,$$

$$\mathbf{E}' = \mathbf{E}'_\perp = \gamma\left(\mathbf{E}_\perp + \mathbf{v} \times \mathbf{B}_\perp\right) = -\gamma v B\mathbf{e}_r = \begin{cases} -\frac{\mu_0 \gamma i v r}{2\pi r_0^2}\mathbf{e}_r , & (r < r_0) \\ -\frac{\mu_0 \gamma i v}{2\pi r}\mathbf{e}_r , & (r > r_0) \end{cases}$$

$$\mathbf{B}' = \mathbf{B}'_\perp = \gamma\left(\mathbf{B}_\perp - \frac{\mathbf{v} \times \mathbf{E}_\perp}{c^2}\right) = \gamma B\mathbf{e}_\varphi = \begin{cases} \frac{\mu_0 i \gamma r}{2\pi r_0^2}\mathbf{e}_\varphi , & (r < r_0) \\ \frac{\mu_0 i \gamma}{2\pi r}\mathbf{e}_\varphi , & (r > r_0) \end{cases}$$

where $\gamma = \frac{1}{\sqrt{1 - v^2/c^2}}$, and the lengths r and r_0 are not changed by the transformation.

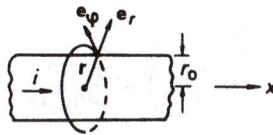

Fig. 5.6

(b) Let the charge density of the wire in Σ' be ρ', then the electric field produced by ρ' for $r < r_0$ is given by Gauss' law

$$2\pi r E'_r = \rho' \pi r^2 / \varepsilon_0$$

to be

$$\mathbf{E}' = \frac{\rho' r}{2\varepsilon_0}\mathbf{e}_r . \quad (r < r_0)$$

Comparing this with the expression for $\mathbf{E}'$ above we have

$$\rho' = -\frac{v i \gamma}{\pi r_0^2 c^2} ,$$

where we have used $\mu_0 \varepsilon_0 = \frac{1}{c^2}$.

(c) In Σ the velocity of the electron stream is $\mathbf{v_e} = -U\mathbf{e_x}$, while the ions are stationary, i.e. $v_i = 0$. Using the Lorentz transformation of velocity we have in Σ'

$$\mathbf{v'_e} = -\frac{v+U}{1+\frac{vU}{c^2}}\mathbf{e_x}, \quad \mathbf{v'_i} = -v\mathbf{e_x}. \tag{6}$$

(d) The charge density is zero in Σ. That is, the positive charges of the positive ions are neutralized by the negative charges of the electrons. Thus $\rho_e + \rho_i = 0$, where ρ_e and ρ_i are the charge densities of the electrons and ions. As

$$\rho_e = \frac{j}{-U} = -\frac{i}{\pi r_0^2 U}$$

we have

$$\rho_i = \frac{i}{\pi r_0^2 U}.$$

However, the positive ions are at rest in Σ and do not give rise to a current. Hence

$$\mathbf{j_e} = \mathbf{j} = \frac{i}{\pi r_0^2}\mathbf{e_x}, \quad \mathbf{j_i} = 0.$$

$(\frac{i}{c}, \rho)$ form a four-vector, so the charge densities of the electrons and ions in Σ' are respectively

$$\rho'_e = \gamma\left(\rho_e - \frac{v}{c^2}j_e\right) = -\frac{i\gamma}{\pi r_0^2 U} - \frac{vi\gamma}{\pi r_0^2 c^2},$$

$$\rho'_i = \gamma\rho_i = \frac{i\gamma}{\pi r_0^2 U}.$$

Obviously, $\rho'_e + \rho'_i \neq 0$, but the sum of ρ'_e and ρ'_i is just the charge density ρ' detected by B.

5011

(a) Derive the repulsive force on an electron at a distance $r < a$ from the axis of a cylindrical column of electrons of uniform charge density ρ_0 and radius a.

(b) An observer in the laboratory sees a beam of circular cross section and density ρ moving at velocity v. What force does he see on an electron of the beam at distance $r < a$ from the axis?

(c) If v is near the velocity of light, what is the force of part (b) as seen by an observer moving with the beam? Compare this force with the answer to part (b) and comment.

(d) If $n = 2 \times 10^{10}$ cm^{-3} and $v = 0.99c$ ($c = $ light velocity), what gradient of a transverse magnetic field would just hold this beam from spreading in one of its dimensions?

(*Wisconsin*)

Solution:

(a) Use cylindrical coordinates with the z-axis along the axis of the cylindrical column of electrons. By Gauss' flux theorem and the symmetry we obtain the electric field at a distance $r < a$ from the axis:

$$\mathbf{E}(r) = \frac{\rho_0 r}{2\varepsilon_0} \mathbf{e}_r . \quad (r < a)$$

Thus the force on an electron at that point is

$$\mathbf{F} = -e\mathbf{E} = -\frac{e\rho_0 r}{2\varepsilon_0} \mathbf{e}_r .$$

Note that this is a repulsive force as ρ_0 itself is negative.

(b) Let the rest frame of the column of electrons and the laboratory frame be Σ' and Σ respectively with Σ' moving with velocity v relative to Σ along the z-axis. By transforming the current-charge density four-vector we find $\rho = \gamma\rho_0$, where $\gamma = (1 - \frac{v^2}{c^2})^{-\frac{1}{2}}$. In Σ' the electric and magnetic fields are $\mathbf{E}' = \frac{\rho_0 r'}{2\varepsilon_0} \mathbf{e}_r$, $\mathbf{B}' = 0$. In Σ, one has

$$\mathbf{E}_\perp = \gamma(\mathbf{E}'_\perp - \mathbf{v} \times \mathbf{B}'_\perp) = \gamma\mathbf{E}' , \quad \mathbf{E}_\| = \mathbf{E}'_\| = 0 ,$$

$$\mathbf{B}_\perp = \gamma\left(\mathbf{B}'_\perp + \frac{\mathbf{v}}{c^2} \times \mathbf{E}'_\perp\right) = \gamma\frac{\mathbf{v}}{c^2} \times \mathbf{E}' , \quad \mathbf{B}_\| = \mathbf{B}'_\| = 0 .$$

Thus the force on an electron of the beam at $r < a$ is given by

$$\mathbf{F} = -e\mathbf{E} - e\mathbf{v} \times \mathbf{B} = -e\gamma\mathbf{E}' - e\mathbf{v} \times \left(\gamma\frac{\mathbf{v}}{c^2} \times \mathbf{E}'\right)$$

$$= -e\gamma\mathbf{E}' + e\gamma\frac{v^2}{c^2} \mathbf{E}' ,$$

as $\mathbf{v} = v\mathbf{e}_z$ is perpendicular to $\mathbf{E}'$.

As there is no transverse Lorentz contraction, $r = r'$. Hence

$$\mathbf{F} = -\frac{e\mathbf{E}'}{\gamma} = -\frac{e\rho r}{2\varepsilon_0\gamma^2}\,\mathbf{e}_r\,.$$

(c) In Σ' the force on the electron is

$$\mathbf{F}' = -e\mathbf{E}' = -\frac{e\rho_0 r}{2\varepsilon_0}\,\mathbf{e}_r = -\frac{e\rho r}{2\varepsilon_0\gamma}\,\mathbf{e}_r\,.$$

As $\gamma > 1$, $F' > F$. Actually, in the rest frame Σ' only the electric field exerts a force on the electron, while in the laboratory frame Σ, although the electric force is larger, there is also a magnetic force acting opposite in direction to the electric force. As a result the total force on the electron is smaller as seen in Σ.

(d) In Σ the force on the electron is

$$\mathbf{F} = -\frac{e\rho r}{2\varepsilon_0\gamma^2}\,\mathbf{e}_r\,.$$

The additional magnetic field $\mathbf{B}_0$ necessary to keep it stationary is given by

$$-e\mathbf{v} \times \mathbf{B}_0 + \mathbf{F} = 0\,,$$

i.e.,

$$e\mathbf{v} \times \mathbf{B}_0 + \frac{e\rho r}{2\varepsilon_0\gamma^2}\,\mathbf{e}_r = 0\,.$$

As $v = v\mathbf{e}_z$, the above requires

$$\mathbf{B}_0 = \frac{\rho r}{2\varepsilon_0\gamma^2 v}\,\mathbf{e}_\theta\,.$$

The gradient of the magnetic field is then

$$\frac{dB_0}{dr} = \frac{\rho}{2\varepsilon_0\gamma^2 v} = -\frac{ne}{2\varepsilon_0\gamma^2 v}\,.$$

With $n = 2 \times 10^{10} \times 10^6$ m^{-3}, $v = 0.99c$, $\varepsilon_0 = 8.84 \times 10^{-12}$ C/Vm, we obtain

$$\left|\frac{dB_0}{dr}\right| = \left|-\frac{2 \times 10^{16} \times 1.6 \times 10^{-19}}{2 \times 8.84 \times 10^{-12} \times \frac{1}{1-0.99^2} \times 0.99 \times 3 \times 10^8}\right|$$

$$= 0.0121 \text{ T/m} = 1.21 \text{ Gs/cm}\,.$$

5012

The uniformly distributed charge per unit length in an infinite ion beam of constant circular cross section is q. Calculate the force on a single beam ion that is located at radius r, assuming that the beam radius R is greater than r and that the ions all have the same velocity v.

(*UC, Berkeley*)

Solution:

Use cylindrical coordinates with the z-axis along the axis of the ion beam such that the flow of the ions is in the $+z$ direction. Let Σ' and Σ be the rest frame of the ions and the laboratory frame respectively, the former moving with velocity v relative to the latter in the $+z$ direction. The charge per unit length in Σ is q. In Σ' it is given by $q = \gamma(\frac{q' + \beta j'}{c}) = \gamma q'$, or $q' = q/\gamma$, where $\gamma = (1 - \beta^2)^{-\frac{1}{2}}$, $\beta = \frac{v}{c}$. In Σ' the electronic field is given by Gauss' law $2\pi E'_r = \frac{r^2}{R^2}\frac{q'}{\varepsilon_0}$ to be

$$\mathbf{E}' = \frac{rq'}{2\pi\varepsilon_0 R^2}\,\mathbf{e}_r\,. \quad (r < R)$$

As the ions are stationary,

$$\mathbf{B}' = 0\,.$$

Transforming to Σ we have $\mathbf{E}_\perp = \gamma(\mathbf{E}'_\perp - \mathbf{v}\times\mathbf{B}'_\perp) = \gamma\mathbf{E}'_\perp$, $\mathbf{E}_\parallel = \mathbf{E}'_\parallel = 0$, or

$$\mathbf{E} = \gamma\mathbf{E}' = \frac{r\gamma q'}{2\pi\varepsilon_0 R^2}\,\mathbf{e}_r = \frac{rq}{2\pi\varepsilon_0 R^2}\,\mathbf{e}_r\,,$$

and $\mathbf{B}_\perp = \gamma(\mathbf{B}'_\perp + \frac{\mathbf{v}\times\mathbf{E}'_\perp}{c^2}) = \gamma\frac{\mathbf{v}\times\mathbf{E}'_\perp}{c^2}$, $\mathbf{B}_\parallel = \mathbf{B}'_\parallel = 0$, or

$$\mathbf{B} = \gamma\frac{v}{c^2}\cdot\frac{rq'}{2\pi\varepsilon_0 R^2}\,\mathbf{e}_\theta = \frac{v}{c}\cdot\frac{rq}{2\pi\varepsilon_0 c R^2}\,\mathbf{e}_\theta\,.$$

Note that, as r is transverse to $\mathbf{v}$, $r' = r$. Hence the total force acting on an ion of charge Q at distance $r < R$ from the axis in the laboratory frame is

$$\begin{aligned}
\mathbf{F} &= Q\mathbf{E} + Q\mathbf{v}\times\mathbf{B} \\
&= \left(Q\cdot\frac{qr}{2\pi\varepsilon_0 R^2} - Q\frac{v^2}{c^2}\frac{qr}{2\pi\varepsilon_0 R^2}\right)\mathbf{e}_r \\
&= \frac{Qqr}{2\pi\varepsilon_0 R^2}\left(1 - \frac{v^2}{c^2}\right)\mathbf{e}_r = \frac{Qqr}{2\pi\varepsilon_0 R^2\gamma^2}\,\mathbf{e}_r\,.
\end{aligned}$$

If $v \ll c$, then $\mathbf{F} = \frac{Qqr}{2\pi\varepsilon_0 R^2} \mathbf{e}_r$, which is what one would obtain if both the charge and the ion beam were stationary.

5013

Given a uniform beam of charged particles q/l charges per unit length, moving with velocity v, uniformly distributed within a circular cylinder of radius R. What is the

(a) electric field $\mathbf{E}$

(b) magnetic field $\mathbf{B}$

(c) energy density

(d) momentum density

of the field throughout space?

(UC, Berkeley)

Solution:

(a), (b) Referring to Problems **5011** and **5012**, we have

$$\mathbf{E} = \begin{cases} \frac{qr}{2\pi\varepsilon_0 R^2 l} \mathbf{e}_r \,, & (r < R) \\ \frac{q}{2\pi\varepsilon_0 r l} \mathbf{e}_r \,. & (r > R) \end{cases}$$

$$\mathbf{B} = \begin{cases} \frac{vqr}{2\pi\varepsilon_0 c^2 R^2 l} \mathbf{e}_\theta \,, & (r < R) \\ \frac{vq}{2\pi\varepsilon_0 c^2 r l} \mathbf{e}_\theta \,. & (r > R) \end{cases}$$

(c) The energy density is

$$\begin{aligned} w &= \frac{1}{2} \left(\varepsilon_0 E^2 + \frac{1}{\mu_0} B^2 \right) \\ &= \begin{cases} \left(1 + \frac{v^2}{c^2}\right) \frac{q^2 r^2}{8\pi^2 \varepsilon_0 R^4 l^2} \,, & (r < R) \\ \left(1 + \frac{v^2}{c^2}\right) \frac{q^2}{8\pi^2 \varepsilon_0 r^2 l^2} \,. & (r > R) \end{cases} \end{aligned}$$

(d) The momentum density is

$$\begin{aligned} \mathbf{g} &= \varepsilon_0 \mathbf{E} \times \mathbf{B} \\ &= \begin{cases} \frac{vq^2 r^2}{4\pi^2 \varepsilon_0 c^2 R^4 l^2} \mathbf{e}_z \,, & (r < R) \\ \frac{vq^2}{4\pi^2 \varepsilon_0 c^2 r^2 l^2} \mathbf{e}_z \,. & (r > R) \end{cases} \end{aligned}$$

5014

Calculate the net radial force on an individual electron in an infinitely long cylindrical beam of relativistic electrons of constant density n moving with uniform velocity **v**. Consider both the electric and magnetic forces.

(*Wisconsin*)

Solution:

The charge density of the electron beam is $\rho = -en$. As shown in Problem **5011**, the net radial force on an individual electron is

$$\mathbf{F} = -\frac{e\rho r}{2\varepsilon_0 \gamma^2}\mathbf{e}_r = \frac{e^2 nr}{2\pi\varepsilon_0 \gamma^2}\,\mathbf{e}_r \ ,$$

where

$$\gamma = \frac{1}{\sqrt{1 - v^2/c^2}}\ .$$

5015

A perfectly conducting sphere of radius R moves with constant velocity $\mathbf{v} = v\mathbf{e}_x$ ($v \ll c$) through a uniform magnetic field $\mathbf{B} = B\mathbf{e}_y$. Find the surface charge density induced on the sphere to lowest order in v/c.

(*MIT*)

Solution:

Let Σ' and Σ be the rest frame of the conducting sphere and the laboratory frame respectively. In Σ we have $\mathbf{B} = B\mathbf{e}_y$, $\mathbf{E} = 0$. Transforming to Σ' we have

$$E'_\parallel = E_\parallel = 0 \ , \quad \mathbf{E}'_\perp = \gamma(\mathbf{E}_\perp + \mathbf{v} \times \mathbf{B}_\perp) = \gamma v B\mathbf{e}_z \ ,$$

$$B'_\parallel = B_\parallel = 0 \ , \quad \mathbf{B}'_\perp = \gamma\left(\mathbf{B}_\perp - \frac{\mathbf{v} \times \mathbf{E}_\perp}{c^2}\right) = \gamma B\mathbf{e}_y \ .$$

Hence

$$\mathbf{E}' = \gamma v B\mathbf{e}_z \ , \quad \mathbf{B}' = \gamma B\mathbf{e}_y \ .$$

In the lowest order approximation, $\gamma = (1 - \frac{v^2}{c^2})^{-1/2} \approx 1$, one has

$$\mathbf{E}' \approx v B\mathbf{e}_z \ , \quad \mathbf{B}' = B\mathbf{e}_y \ .$$

In Σ' the electric field external to the sphere $\mathbf{E}'$ is uniform so the potential outside the sphere is (see Problem 1065)

$$\varphi' = -E'r \, \cos \, \theta + \frac{E'R^3}{r^2} \, \cos \, \theta \, ,$$

with θ as shown in Fig. 5.7.

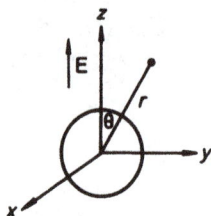

Fig. 5.7

The surface charge density on the conductor is given by the boundary condition for $\mathbf{D}$:

$$\sigma' = -\varepsilon_0 \left. \frac{\partial \varphi'}{\partial r} \right|_{r=R} = 3\varepsilon_0 v B \, \cos \, \theta \, .$$

On transforming back to Σ, as the relative velocity of Σ' is along the x direction, the angle θ remains unchanged. Hence the surface charge density induced on the sphere to lowest order in v/c is

$$\sigma = \gamma \sigma' \approx \sigma' = 3\varepsilon_0 v B \, \cos \, \theta \, .$$

5016

Let a particle of charge q and rest mass m be released with zero initial velocity in a region of space containing an electric field $\mathbf{E}$ in the y direction and a magnetic field $\mathbf{B}$ in the z direction.

(a) Describe the conditions necessary for the existence of a Lorentz frame in which (1) $\mathbf{E} = 0$ and (2) $\mathbf{B} = 0$.

(b) Describe the motion that would ensue in the original frame if case (a) (1) attains.

(c) Solve for the momentum as a function of time in the frame with $\mathbf{B} = 0$ for case (a)(2).

(*UC, Berkeley*)

Solution:

(a) Let Σ be the laboratory frame and Σ' be a frame moving with relative velocity v along the x direction. In Σ we have

$$\mathbf{E} = E\mathbf{e}_y , \quad \mathbf{B} = B\mathbf{e}_z .$$

Lorentz transformation gives the electromagnetic field in Σ' as

$$E'_x = E_x = 0, \quad E'_y = \gamma(E_y - vB_z) = \gamma(E - vB), \quad E'_z = \gamma(E_x + vB_y) = 0 ,$$

$$B'_x = B_x = 0, \quad B'_y = \gamma\left(B_y + \frac{v}{c^2} E_z\right) = 0,$$

$$B'_z = \gamma\left(B_z - \frac{v}{c^2} E_y\right) = \gamma\left(B - \frac{v}{c^2} E\right) .$$

(1) For $E' = 0$ in Σ' we require that

$$E - vB = 0 ,$$

or $v = \frac{E}{B}$. However, as $v \leq c$, for such a frame Σ' to exist we require that $E \leq cB$.

(2) For $\mathbf{B}' = 0$ in Σ', we require that $B - \frac{v}{c^2} E = 0$, or $v = \frac{c^2 B}{E}$. Then for such a frame Σ' to exist we require that

$$cB \leq E .$$

(b) If $\mathbf{E}' = 0$, the motion of the charge q in Σ' is described by

$$\frac{d}{dt'} \left(\frac{m\mathbf{u}'}{\sqrt{1 - u'^2/c^2}}\right) = q\mathbf{u}' \times \mathbf{B}' , \tag{1}$$

$$\frac{d}{dt'} \left(\frac{mc^2}{\sqrt{1 - u'^2/c^2}}\right) = q(\mathbf{u}' \times \mathbf{B}') \cdot \mathbf{u}' = 0 , \tag{2}$$

where $\mathbf{u}'$ is the velocity of the particle in Σ'. Equation (2) means that

$$\frac{mc^2}{\sqrt{1 - \frac{u'^2}{c^2}}} = \text{constant} .$$

Hence $u' = $ constant as well. This implies that the magnitude of the velocity of the particle does not change, while its direction changes. As the initial velocity of the particle in Σ is zero, the velocity of the particle in Σ' at the initial time $t' = 0$ is by the transformation equations

$$u'_x = \frac{u_x - v}{1 - \frac{vu_x}{c^2}} = -v, \quad u'_y = \frac{u_y}{\gamma(1 - \frac{vu_x}{c^2})} = 0, \quad u'_z = 0$$

to be

$$\mathbf{u}'_0 = -v\mathbf{e}_x = -\frac{E}{B}\mathbf{e}_x .$$

Thus the magnitude of the velocity of the particle will always be $u' = \frac{E}{B}$, and we have

$$\sqrt{1 - u'^2/c^2} = \frac{\sqrt{B^2c^2 - E^2}}{Bc} = \text{constant} .$$

Then Eq. (1) reduces to

$$\dot{\mathbf{u}}' = \frac{q}{m}\sqrt{1 - u'^2/c^2}\, \mathbf{u}' \times \mathbf{B}' . \tag{3}$$

From the transformation equations for $\mathbf{B}$ we have $\mathbf{B}' = \frac{1}{c}\sqrt{B^2c^2 - E^2}\,\mathbf{e}_z$. Hence Eq. (3) gives rise to

$$\dot{u}'_x = \omega u'_y , \tag{4}$$

$$\dot{u}'_y = -\omega u'_x , \tag{5}$$

$$\dot{u}'_z = 0 , \tag{6}$$

where

$$\omega = \frac{qB'\sqrt{1 - u'^2/c^2}}{m} = \frac{q(c^2B^2 - E^2)}{c^2 mB} .$$

Equation (6) shows that $u'_z = \text{constant}$. As $u'_z = 0$ at $t' = 0$, $u'_z = 0$ for all times.

(4) + (5)$\times i$ gives

$$\dot{u}'_x + i\dot{u}'_y = -i\omega(u'_x + iu'_y) ,$$

or

$$\dot{\xi} = -i\omega\xi ,$$

where

$$\xi = u'_x + iu'_y .$$

The solution is

$$\xi = -u_0' e^{-i\omega t'} ,$$

or

$$u_x' = u_0' \cos(\omega t'), \quad u_y' = -u_0' \sin(\omega t'),$$

where u_0' is a constant.
As $u_0' = \frac{E}{B}$ at $t' = 0$ we find that

$$u_x' = \frac{E}{B} \cos(\omega t'), \quad u_y' = -\frac{E}{B} \sin(\omega t').$$

These equations show that the particle will undergo circular motion in the xy plane with a radius

$$R = \frac{u'}{\omega} = \frac{E}{B\omega} = \frac{c^2 m E}{q(c^2 B^2 - E^2)} .$$

In Σ, because of Lorentz contraction in the x direction, the orbit is an ellipse with the minor axis along the x-axis.

(c) Consider a frame Σ' in which $\mathbf{B}' = 0$. Let $\mathbf{p}'$ be the momentum of the particle. The equation of motion is then

$$\frac{d\mathbf{p}'}{dt'} = q\mathbf{E}' .$$

The quantity $E^2 - c^2 B^2$ is Lorentz invariant as shown in Problem **5008**. Hence $\mathbf{E}' = \sqrt{E^2 - c^2 B^2} \, \mathbf{e}_y$, using also the result of (a). Then the equation of motion has component equations

$$\frac{dp_x'}{dt'} = \frac{dp_z'}{dt'} = 0 , \tag{7}$$

$$\frac{dp_y'}{dt'} = q \sqrt{E^2 - c^2 B^2} . \tag{8}$$

Equation (7) shows that both p_x' and p_z' are constant, being independent of time. The particle is initially at rest in Σ, so its initial velocity is opposite that of Σ', i.e., $\mathbf{u}_0' = -\frac{c^2 B}{E} \mathbf{e}_x$, as shown in (a)(2). Hence

$$p_x' = \frac{m u_0'}{\sqrt{1 - u'^2/c^2}} = \frac{c^2 m B}{\sqrt{E^2 - c^2 B^2}} ,$$

$$p_z' = 0 .$$

Equation (8) gives

$$p_y(t') = q\sqrt{E^2 - c^2 B^2}\, t',$$

where we have used the initial condition $u'_{0y} = 0$ at $t' = 0$.

5017

Consider an arbitrary plane electromagnetic wave propagating in vacuum in x-direction. Let $A(x - ct)$ be the vector potential of the wave; there are no sources, so adopt a gauge in which the scalar potential is identically zero. Assume that the wave does not extend throughout all spaces, in particular $A = 0$ for sufficiently large values of $x - ct$. The wave strikes a particle with charge e which is initially at rest and accelerates it to a velocity which may be relativistic.

(a) Show that $A_x = 0$.

(b) Show that $\mathbf{p}_\perp = -e\mathbf{A}$, where $\mathbf{p}_\perp$ is the particle momentum in the yz plane. (Note: Since this is a relativistic problem, do not solve it with non-relativistic mechanics.)

(*UC, Berkeley*)

Solution:

(a) As $\mathbf{A} = \mathbf{A}(x - ct)$,

$$\frac{\partial \mathbf{A}}{\partial x} = \frac{\partial \mathbf{A}}{\partial (x - ct)}, \qquad \frac{\partial \mathbf{A}}{\partial t} = -c\frac{\partial \mathbf{A}}{\partial (x - ct)}.$$

With the gauge condition $\varphi = 0$,

$$\mathbf{E} = -\nabla\varphi - \frac{\partial \mathbf{A}}{\partial t} = -\frac{\partial \mathbf{A}}{\partial t}.$$

As plane electromagnetic waves are transverse, $E_x = 0$. Thus

$$-\frac{\partial A_x}{\partial t} = \frac{c\partial A_x}{\partial (x - ct)} = 0,$$

showing that $A_x(x - ct) = $ constant.

Since the wave does not extend throughout all space, the vector potential vanishes for sufficiently large values of $x - ct$. Hence the above constant is zero, i.e., $A_x = 0$ at all points of space.

(b) Let **r** be the displacement of the charged particle at time t and write the vector potential as $\mathbf{A}(\mathbf{r}, t)$. We have

$$\frac{d\mathbf{A}}{dt} = \frac{\partial \mathbf{A}}{\partial t} + \sum_{j=1}^{3} \frac{\partial \mathbf{A}}{\partial x_j} \frac{\partial x_j}{\partial t} = \frac{\partial \mathbf{A}}{\partial t} + (\mathbf{v} \cdot \nabla)\mathbf{A} \,. \tag{1}$$

The equation of motion of a particle of charge e and momentum **p** in the electromagnetic field is

$$\frac{d\mathbf{p}}{dt} = e(\mathbf{E} + \mathbf{v} \times \mathbf{B}) \,. \tag{2}$$

Treating **r** and **v** as independent variables, we have

$$\mathbf{v} \times (\nabla \times \mathbf{A}) = \nabla(\mathbf{v} \cdot \mathbf{A}) - (\mathbf{v} \cdot \nabla)\mathbf{A} \,. \tag{3}$$

Equations (1)–(3) give

$$\frac{d\mathbf{p}}{dt} = e\nabla(\mathbf{v} \cdot \mathbf{A}) - e\frac{d\mathbf{A}}{dt} \,.$$

Consider the transverse component of the particle momentum, $\mathbf{p}_\perp = p_y \mathbf{e}_y + p_z \mathbf{e}_z$.

The vector potential $\mathbf{A}(x - ct)$ of the plane electromagnetic wave is independent of the coordinates y and z, and has no longitudinal component (see (a)). As **r** and **v** are to be treated as independent variables also, we have

$$\frac{\partial}{\partial y}(\mathbf{v} \cdot \mathbf{A}) = v_y \frac{\partial A_y}{\partial y} + v_z \frac{\partial A_z}{\partial z} = 0 \,,$$

and similarly $\frac{\partial}{\partial z}(\mathbf{v} \cdot \mathbf{A}) = 0$. Hence

$$\frac{d\mathbf{p}_\perp}{dt} = -e\frac{d\mathbf{A}_\perp}{dt} \,.$$

Integration gives

$$\mathbf{p}_\perp = -e\mathbf{A}_\perp + \mathbf{C} \,.$$

Since the initial velocity of the particle is zero, the constant **C** is zero. Furthermore with $A_x = 0$, $\mathbf{A}_\perp = \mathbf{A}$. So we can write the above as

$$\mathbf{p}_\perp = -v\mathbf{A} \,.$$

2. ELECTROMAGNETIC FIELD OF A CHARGED PARTICLE (5018–5025)

5018

Show that the electromagnetic field of a particle of charge q moving with constant velocity $\mathbf{v}$ is given by

$$E_x = \frac{q}{\chi}\gamma(x - vt), \quad B_x = 0,$$

$$E_y = \frac{q}{\chi}\gamma y, \qquad B_y = -\frac{q}{\chi}\beta\gamma z,$$

$$E_z = \frac{q}{\chi}\gamma z, \qquad B_z = \frac{q}{\chi}\beta\gamma y,$$

where

$$\beta = \frac{v}{c}, \quad \gamma = \left(1 - \frac{v^2}{c^2}\right)^{-\frac{1}{2}},$$

$$\chi = \{\gamma^2(x - vt)^2 + y^2 + z^2\}^{3/2},$$

and we have chosen the x-axis along $\mathbf{v}$ (note that we use units such that the proportionality constant in Coulomb's law is $K = 1$).

(SUNY, Buffalo)

Solution:

Let Σ be the observer's frame and Σ' the rest frame of the particle. In the units used we have in Σ'

$$\mathbf{E}'(\mathbf{x}') = \frac{q\mathbf{x}'}{r'^3}, \quad \mathbf{B}'(\mathbf{x}') = 0,$$

where

$$r' = |\mathbf{x}'|.$$

The Lorentz transformation for time-space between Σ and Σ' is given by

$$x' = \gamma(x - vt),$$
$$y' = y,$$
$$z' = z,$$

so that

$$r'^2 = x'^2 + y'^2 + z'^2 = \gamma^2(x - vt)^2 + y^2 + z^2.$$

The (inverse) Lorentz transformation for electromagnetic field gives

$$E_x = E'_x = \frac{qx'}{r'^3} = \frac{q}{\chi}\gamma(x - vt),$$

$$E_y = \gamma(E'_y + \beta B'_z) = \frac{qy}{\chi}\gamma,$$

$$E_z = \gamma(E'_z - \beta B'_y) = \frac{qz}{\chi}\gamma,$$

$$B_x = B'_x = 0,$$

$$B_y = \gamma(B'_y - \beta E'_z) = -\frac{q}{\chi}\beta\gamma z,$$

$$B_z = \gamma(B'_z + \beta E'_y) = \frac{q}{\chi}\beta\gamma y.$$

5019

(a) Consider two positrons in a beam at SLAC. The beam has energy of about 50 GeV ($\gamma \approx 10^5$). In the beam frame (rest frame) they are separated by a distance d, and positron e_2^+ is traveling directly ahead of e_1^+, as shown in Fig. 5.8. Write down expressions at e_1^+ giving the effect of e_2^+. Specifically, give the following vectors: $\mathbf{E}$, $\mathbf{B}$, the Lorentz force $\mathbf{F}$, and the acceleration $\mathbf{a}$. Do this in two reference frames:
1. the rest frame,   2. the laboratory frame.
The results will differ by various relativistic factors. Give intuitive explanations of these factors.

(b) The problem is the same as in part (a) except this time the two positrons are traveling side by side as sketched in Fig. 5.9.

(*UC, Berkeley*)

Solution:

(a) Let Σ' and Σ be the beam rest frame and the laboratory frame respectively. In Σ' the effects exerted by e_2^+ on e_1^+ are

$$\mathbf{E}' = -\frac{1}{4\pi\varepsilon_0}\frac{e}{d^2}\mathbf{e}_z,$$

$$\mathbf{B}' = 0,$$

$$\mathbf{F}' = e\mathbf{E}' = -\frac{1}{4\pi\varepsilon_0}\frac{e^2}{d^2}\mathbf{e}_z,$$

$$\mathbf{a}' = \frac{\mathbf{F}'}{m} = -\frac{1}{4\pi\varepsilon_0 m}\frac{e^2}{d^2}\mathbf{e}_z.$$

Thus in Σ', e_1^+ is a nonrelativistic particle that will undergo rectilinear accelerated motion under the action of the electrostatic field $\mathbf{E}'$ established by e_2^+.

The Lorentz transformation for electromagnetic field $E_\parallel = E'_\parallel$, $B_\parallel = B'_\parallel$, gives

$$\mathbf{E} = \mathbf{E}' = -\frac{1}{4\pi\varepsilon_0}\frac{e}{d^2}\,\mathbf{e}_z, \quad \mathbf{B} = 0.$$

Hence the force on e_1^+ is

$$\mathbf{F} = e\mathbf{E} = -\frac{1}{4\pi\varepsilon_0}\frac{e^2}{d^2}\,\mathbf{e}_z = \mathbf{F}'.$$

As $\gamma = 10^5$, e_1^+ is a relativistic particle in Σ and must satisfy the relativistic equations of motion

$$\frac{d}{dt}\left(\frac{m\mathbf{v}}{\sqrt{1-v^2/c^2}}\right) = \mathbf{F}, \quad \text{or} \quad mc\frac{d}{dt}(\gamma\beta) = F,$$

$$\frac{d}{dt}\left(\frac{mc^2}{\sqrt{1-v^2/c^2}}\right) = \mathbf{F}\cdot\mathbf{v}, \quad \text{or} \quad mc\frac{d\gamma}{dt} = F\beta$$

where $\beta = \frac{v}{c}$, $\gamma = (1-\beta^2)^{-\frac{1}{2}}$, since $\mathbf{F} = F\mathbf{e}_z$, $\mathbf{v} = v\mathbf{e}_z$. We then have

$$a = c\frac{d\beta}{dt} = \frac{c}{\gamma}\left[\frac{d}{dt}(\gamma\beta) - \beta\frac{d\gamma}{dt}\right] = \frac{1}{m\gamma}(F - F\beta^2) = \frac{F}{m\gamma^3},$$

or

$$\mathbf{a} = \frac{d\mathbf{v}}{dt} = \frac{\mathbf{F}}{m\gamma^3} = -\frac{e^2}{4\pi\varepsilon_0 r^3 m d^2}\,\mathbf{e}_z = \frac{\mathbf{a}}{\gamma^3}.$$

It follows that when the motion of the two positrons is as shown in Fig. 5.8, the electromagnetic field and the Lorentz force are the same in Σ and Σ'. However, due to the relativistic effect the acceleration of e_1^+ in the laboratory frame is only $\frac{1}{\gamma^3}$ times that in the rest frame. As $\frac{1}{\gamma^3} \approx 10^{-15}$, a is extremely small. In other words, the influence of the force exerted by a neighboring collinear charge on a charge moving with high speed will be small. The whole beam travels together in a state of high velocity and high energy.

Fig. 5.8

(b) In the case shown in Fig. 5.9, we have in the rest frame Σ' the various vectors at e_1^+:

$$\mathbf{E}' = -\frac{e}{4\pi\varepsilon_0 d^2}\,\mathbf{e}_z\,,\quad \mathbf{B}' = 0\,,$$

$$\mathbf{F}' = e\mathbf{E}' = -\frac{e^2}{4\pi\varepsilon_0 d^2}\,\mathbf{e}_z\,,$$

$$\mathbf{a}' = \frac{\mathbf{F}}{m} = -\frac{e^2}{4\pi\varepsilon_0 m d^2}\,\mathbf{e}_z\,.$$

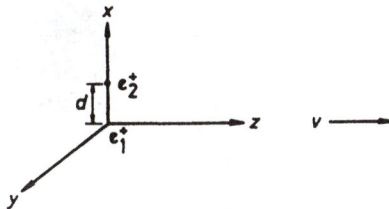

Fig. 5.9

In the laboratory frame Σ, as

$$\mathbf{E}_\perp = \gamma(\mathbf{E}'_\perp - \mathbf{v}\times\mathbf{B}'_\perp)\,,$$

$$\mathbf{B}_\perp = \gamma\left(\mathbf{B}'_\perp + \frac{\mathbf{v}\times\mathbf{E}'_\perp}{c^2}\right)\,,$$

we have

$$\mathbf{E} = \gamma\mathbf{E}' = -\frac{\gamma e}{4\pi\varepsilon_0 d^2}\,\mathbf{e}_x\,,$$

$$\mathbf{B} = \gamma\frac{v}{c^2}E'\mathbf{e}_y = -\frac{\gamma v e}{4\pi\varepsilon_0 c^2 d^2}\,\mathbf{e}_y\,,$$

$$\mathbf{F} = e(\mathbf{E} + \mathbf{v}\times\mathbf{B}) = -\frac{e^2}{4\pi\varepsilon_0 d^2\gamma}\,\mathbf{e}_x = \frac{\mathbf{F}'}{\gamma}\,.$$

In this case $\mathbf{F}\cdot\mathbf{v} = 0$, so that $\gamma = $ constant and

$$a = \frac{mc}{m\gamma}\frac{d}{dt}(\gamma\beta) = \frac{F}{m\gamma}\,,$$

or

$$\mathbf{a} = \frac{\mathbf{F}}{m\gamma} = -\frac{e^2}{4\pi\varepsilon_0 d^2 m\gamma^2}\,\mathbf{e}_x = \frac{\mathbf{a}'}{\gamma^2}\,.$$

These results show that when the two positrons are traveling side by side, all the vectors in Σ, as compared with the corresponding vectors in Σ', will involve the Lorentz factor γ which is a constant of the motion. In the laboratory frame, both the electric and magnetic fields exist, the former being increased by γ from that in the rest frame. As to the effects of $\mathbf{E}$ and $\mathbf{B}$ on e_1^+, they tend to cancel each other, which reduces the force on the acceleration of e_1^+ by factors $\frac{1}{\gamma}$ and $\frac{1}{\gamma^2}$ respectively, as compared with those in the rest frame.

5020

In Fig. 5.10 a point charge e moves with constant velocity v in the z direction so that at time t it is at the point Q with coordinates $x = 0$, $y = 0$, $z = vt$. Find at the time t and at the point P with coordinates $x = b$, $y = 0$, $z = 0$ (see Fig. 5.10)

(a) the scalar potential ϕ,

(b) the vector potential $\mathbf{A}$,

(c) the electric field in the x direction, E_x.

(*Wisconsin*)

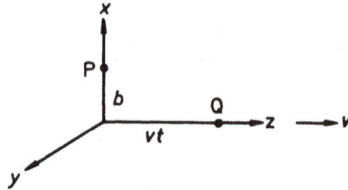

Fig. 5.10

Solution:

(a) The Liénard-Wiechert potentials at P due to the charge are given by

$$\phi = \frac{e}{4\pi\varepsilon_0 \left[r - \frac{\mathbf{v}}{c} \cdot \mathbf{r}\right]}, \quad \mathbf{A} = \frac{e\mathbf{v}}{4\pi\varepsilon_0 c^2 \left[r - \frac{\mathbf{v}}{c} \cdot \mathbf{r}\right]}$$

where $\mathbf{r}$ is the radius vector from the retarded position of the charge to the field point P, i.e.,

$$\mathbf{r} = b\mathbf{e}_x - v\left(t - \frac{r}{c}\right)\mathbf{e}_z = b\mathbf{e}_x - vt'\mathbf{e}_z ,$$

with

$$t' = t - \frac{r}{c}.$$

Thus

$$r^2 = \mathbf{r} \cdot \mathbf{r} = b^2 + v^2 \left(t - \frac{r}{c} \right)^2 = b^2 + v^2 \left(t^2 - \frac{2rt}{c} + \frac{r^2}{c^2} \right),$$

or

$$\left(1 - \frac{v^2}{c^2} \right) r^2 + 2 \frac{v^2 t}{c} r - b^2 - v^2 t^2 = 0.$$

This is the retardation condition, with the solutions

$$r = \frac{-\beta v t \pm \sqrt{(1 - \beta^2) b^2 + v^2 t^2}}{1 - \beta^2},$$

where $\beta = \frac{v}{c}$.
However the upper sign is to be taken since $r \geq 0$. As $\mathbf{v} = v \mathbf{e}_z$,

$$r - \frac{\mathbf{v} \cdot \mathbf{r}}{c} = r + \frac{v^2 t'}{c} = r + v\beta \left(t - \frac{r}{c} \right) = (1 - \beta^2) r + v \beta t$$
$$= \sqrt{(1 - \beta^2) b^2 + v^2 t^2}.$$

The scalar potential ϕ is then

$$\phi = \frac{e}{4\pi\varepsilon_0 \sqrt{(1 - \beta^2) b^2 + v^2 t^2}}.$$

(b) The vector potential $\mathbf{A}$ is

$$\mathbf{A} = \frac{ev}{4\pi\varepsilon_0 c^2 \sqrt{(1 - \beta^2) b^2 + v^2 t^2}} \mathbf{e}_z.$$

(c) The electric field at P is obtained by differentiating the Liénard-Wiechert potentials:

$$\mathbf{E}(t) = -\nabla\phi - \frac{\partial \mathbf{A}}{\partial t}.$$

For the spatial differentiation, b is to be first replaced by x. We then have

$$(\nabla\phi)_b = \left(\frac{\partial \phi}{\partial x} \right)_b \mathbf{e}_x = -\frac{e(1 - \beta^2) b}{4\pi\varepsilon_0 [(1 - \beta^2) b^2 + v^2 t^2]^{3/2}} \mathbf{e}_x.$$

As **A** is in the z-direction, it does not contribute to E_x. Hence

$$E_x = \frac{e(1 - \beta^2)b}{4\pi\varepsilon_0[(1 - \beta^2)b^2 + v^2t^2]^{3/2}}.$$

5021

For a particle of charge e moving non-relativistically, find

(a) The time-averaged power radiated per unit solid angle, $dP/d\Omega$, in terms of velocity βc, acceleration $\dot\beta c$, and the unit vector **n'** from the charge toward the observer;

(b) $dP/d\Omega$, if the particle moves as $z(t) = a\cos(\omega_0 t)$;

(c) $dP/d\Omega$ for circular motion of radius R in the xy plane with constant angular frequency ω_0.

(d) Sketch the angular distribution of the radiation in each case.

(e) Qualitatively, how is $dP/d\Omega$ changed if the motion is relativistic?

(*Princeton*)

Solution:

(a) For a non-relativistic particle of charge e the radiation field is given by

$$\mathbf{E} = \frac{e\mathbf{n}' \times (\mathbf{n}' \times \dot{\boldsymbol\beta}c)}{4\pi\varepsilon_0 c^2 r} = \frac{e}{4\pi\varepsilon_0 cr}\mathbf{n}' \times (\mathbf{n}' \times \dot{\boldsymbol\beta}),$$

$$\mathbf{B} = \frac{1}{c}\mathbf{n}' \times \mathbf{E},$$

where r is the distance of the observer from the charge. The Poynting vector at the observer is then

$$\mathbf{N} = \mathbf{E} \times \mathbf{H} = \frac{1}{\mu_0}\mathbf{E} \times \mathbf{B} = \frac{1}{\mu_0 c}E^2\mathbf{n}' = \frac{e^2}{16\pi^2\varepsilon_0 cr^2}|\mathbf{n}' \times (\mathbf{n}' \times \dot{\boldsymbol\beta})|^2\mathbf{n}'.$$

Let θ be the angle between **n'** and $\dot{\boldsymbol\beta}$, then

$$\frac{dP}{d\Omega} = \frac{\mathbf{N} \cdot \mathbf{n}'}{r^{-2}} = \frac{e^2}{16\pi^2\varepsilon_0 c}|\dot{\boldsymbol\beta}|^2 \sin^2\theta.$$

This result is not changed by time averging unless the motion of the charge is periodic.

(b) If $z = a\cos(\omega_0 t)$, then $\dot{\beta}c = \ddot{z} = -a\omega_0^2\cos(\omega_0 t)$ and as $\frac{1}{T}\int_0^T \cos^2(\omega_0 t)\,dt = \frac{1}{2}$, where T is the period,

$$\langle(\dot{\beta}c)^2\rangle = \frac{1}{2}Q^2\omega_0^4.$$

Hence

$$\frac{dP}{d\Omega} = \frac{e^2 a^2 \omega_0^4}{32\pi^2\varepsilon_0 c^3}\sin^2\theta.$$

(c) The circular motion of the particle in the xy plane may be considered as superposition of two mutually perpendicular harmonic oscillations:

$$\mathbf{R}(t') = R\,\cos\,(\omega_0 t')\,\mathbf{e}_x + R\,\sin\,(\omega_0 t')\,\mathbf{e}_y.$$

In spherical coordinates let the observer have radius vector $\mathbf{r}(r,\theta,\varphi)$ from the center of the circle, which is also the origin of the coordinate system. The angles between $\mathbf{r}$ or $\mathbf{n}'$ and $\boldsymbol{\beta}$ for the two oscillations are given by

$$\cos\theta_1 = \sin\theta\,\cos\,\varphi,$$

$$\cos\theta_2 = \sin\,\theta\,\cos\left(\frac{\pi}{2} - \varphi\right) = \sin\,\theta\,\sin\varphi.$$

Using the results of (b) we have

$$\begin{aligned}
\frac{dP}{d\Omega} &= \frac{e^2 R^2 \omega_0^4}{32\pi^2\varepsilon_0 c^3}\,(\sin^2\theta_1 + \sin^2\theta_2) \\
&= \frac{e^2 R^2 \omega_0^4}{32\pi^2\varepsilon_0 c^3}\,(1 + \cos^2\theta).
\end{aligned}$$

(d) For the cases (a) and (b), the curves $\rho = \frac{dP}{d\Omega}$ vs. θ are sketched in Figs. 5.11 and 5.12 respectively, where z is the direction of $\dot{\beta}$.

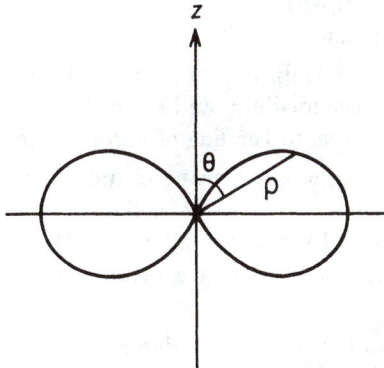

Fig. 5.11

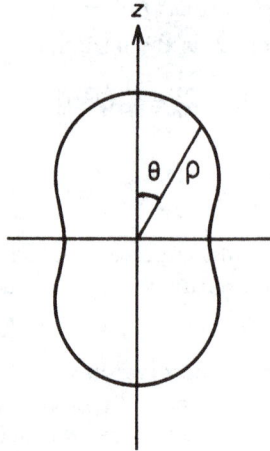

Fig. 5.12

(e) For $\beta \approx 0$, the direction of maximum intensity is along $\theta = \frac{\pi}{2}$. As $\beta \to 1$, the direction of maximum intensity tends more and more toward the direction $\theta = 0$, i.e., the direction of $\dot{\beta}$. In fact the radiation will be concentrated mainly in a cone with $\Delta\theta \sim \frac{1}{\gamma}$ about the direction of $\dot{\beta}$. However there is no radiation exactly along that direction.

5022

Čerenkov radiation is emitted by a high energy charged particle which moves through a medium with a velocity greater than the velocity of electromagnetic wave propagation in the medium.

(a) Derive the relationship between the particle velocity $v = \beta c$, the index of refraction n of the medium, and the angle θ at which the Čerenkov radiation is emitted relative to the line of flight of the particle.

(b) Hydrogen gas at one atmosphere and at 20°C has an index of refraction $n = 1 + 1.35 \times 10^{-4}$. What is the minimum kinetic energy in MeV which an electron (of rest mass 0.5 MeV/c^2) would need in order to emit Čerenkov radiation in traversing a medium of hydrogen gas at 20°C and one atmosphere?

(c) A Čerenkov radiation particle detector is made by fitting a long pipe of one atmosphere, 20°C hydrogen gas with an optical system capable of detecting the emitted light and of measuring the angle of emission θ to an accuracy of $\delta\theta = 10^{-3}$ radian. A beam of charged particles with momentum of 100 GeV/c are passed through the counter. Since the momentum is known, the measurement of the Čerenkov angle is, in effect, a measurement of the particle rest mass m_0. For particles with m_0 near 1 GeV/c^2, and to first order in small quantities, what is the fractional error (i.e., $\delta m_0/m_0$) in the determination of m_0 with the Čerenkov counter?

$$(CUSPEA)$$

Solution:

(a) As shown in Fig. 5.13, the radiation emitted by the charge at Q' at time t' arrives at P at time t when the charge is at Q. As the radiation propagates at the speed c/n and the particle has speed v where $v > c/n$, we have

$$Q'P = \frac{c}{n}(t - t'), \quad Q'Q = v(t - t'),$$

or

$$\frac{Q'P}{Q'Q} = \cos\theta = \frac{c}{vn} = \frac{1}{\beta n},$$

where $\beta = \frac{v}{c}$. At all the points intermediate between Q' and Q the radiation emitted will arrive at the line QP at time t. Hence QP forms the wavefront of all radiation emitted prior to t.

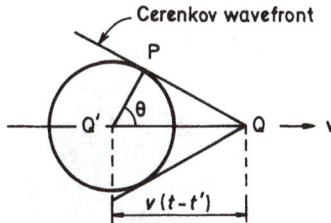

Fig. 5.13

(b) As $|\cos\theta| \leq 1$, we require $\beta \geq \frac{1}{n}$ for emission of Čerenkov radiation. Hence we require

$$\gamma = \frac{1}{\sqrt{1 - \beta^2}} \geq \left(1 - \frac{1}{n^2}\right)^{-\frac{1}{2}} = \frac{n}{\sqrt{n^2 - 1}}.$$

Thus the particle must have a kinetic energy greater than

$$T = (\gamma - 1)\, m_0 c^2$$

$$= \left[\frac{n}{\sqrt{(n+1)(n-1)}} - 1 \right] m_0 c^2$$

$$\approx \left(\frac{1}{\sqrt{2 \times 1.35 \times 10^{-4}}} - 1 \right) \times 0.5$$

$$\approx 29.93 \text{ MeV} .$$

(c) For a relativistic particle of momentum $P \gg m_0 c$,

$$\gamma = \frac{E}{m_0 c^2} = \frac{\sqrt{P^2 c^2 + m_0^2 c^4}}{m_0 c^2} \approx \frac{P}{m_0 c} .$$

With P fixed we have

$$d\gamma = -\frac{P}{c} \cdot \frac{dm_0}{m_0^2} .$$

Now $\beta = \frac{\gamma \beta}{\gamma} = \sqrt{\frac{\gamma^2 - 1}{\gamma^2}} = \sqrt{1 - \frac{1}{\gamma^2}} \approx 1 - \frac{1}{2\gamma^2}$ for $\gamma \gg 1$, so that

$$d\beta = \frac{d\gamma}{\gamma^3} .$$

For the Čerenkov radiation emitted by the particle, we have

$$\cos \theta = \frac{1}{\beta n} ,$$

or

$$d\beta = n\beta^2 \sin \theta d\theta .$$

Combining the above we have

$$\left| \frac{dm_0}{m_0} \right| = \frac{m_0 c}{P} d\gamma \approx \frac{d\gamma}{\gamma} = \gamma^2 \beta = n\beta^2 \gamma^2 \sin \theta d\theta = \beta \gamma^2 \tan \theta \, d\theta .$$

With $\gamma = \frac{Pc}{m_0 c^2} = \frac{100}{1} = 100$, $n = 1 + 1.35 \times 10^{-4}$, we have

$$\beta \approx 1 - \frac{1}{2 \times 10^4} = 1 - 5 \times 10^{-5} ,$$

$$\cos \theta = \frac{1}{\beta n} = (1 - 5 \times 10^{-5})^{-1} (1 + 1.35 \times 10^{-4})^{-1} \approx 1 - 8.5 \times 10^{-5} ,$$

$$\tan \theta = \sqrt{\frac{1}{\cos^2 \theta} - 1} = \sqrt{(1 - 8.5 \times 10^{-5})^{-2} - 1}$$

$$\approx \sqrt{1.7 \times 10^{-4}} \approx 1.3 \times 10^{-2},$$

and hence

$$\left| \frac{dm_0}{m_0} \right| = (1 - 5 \times 10^{-5}) \times 10^4 \times 1.3 \times 10^{-2} \times 10^{-3} = 0.13.$$

5023

A waveguide is formed by two infinite parallel perfectly conducting planes separated by a distance a. The gap between the planes is filled with a gas whose index of refraction is n. (This is taken to be frequency independent.)

(a) Consider the guided plane wave modes in which the field strengths are independent of the y variable. (The y axis is into the paper as shown in Fig. 5.14.) For a given wavelength λ find the allowed frequency ω. For each such mode find the phase velocity v_p and the group velocity v_g.

Fig. 5.14

(b) A uniform charged wire, which extends infinitely along the y direction (Fig. 5.15), moves in the midplane of the gap with velocity $v > c/n$. It emits Čerenkov radiation. At any fixed point in the gap this reveals itself as time-varying electric and magnetic fields. How does the magnitude of the electric field vary with time at a point in the midplane of the gap? Sketch the frequency spectrum and give the principal frequency.

Fig. 5.15

(c) Any electromagnetic disturbance (independent of y) must be expressible as a superposition of the waveguide modes considered in part (a). What is the mode corresponding to the principal frequency of the Čerenkov spectrum considered in part (b)?

(*Princeton*)

Solution:

(a) Take the midplane of the gap between the two planes as the xy plane. As the field strength does not depend on y and the wave is guided along the x direction, we can write

$$\mathbf{E} = \mathbf{E}(z)\, e^{i(k_x x - \omega t)}$$

with $k_x = \frac{2\pi}{\lambda}$. $\mathbf{E}$ satisfies the wave equation

$$\nabla^2 \mathbf{E} - \frac{n^2}{c^2}\frac{\partial^2 \mathbf{E}}{\partial t^2} = 0, \quad \text{or} \quad \frac{\partial^2 \mathbf{E}}{\partial z^2} + k'^2 \mathbf{E} = 0,$$

where

$$k'^2 = \frac{n^2}{c^2}\omega^2 - k_x^2,$$

subject to the boundary conditions

$$E_x = E_y = 0, \quad \text{for } z = 0,\, a.$$

$\mathbf{E}$ is also subject to the condition $\nabla \cdot \mathbf{E} = 0$, i.e., $\frac{\partial E_z}{\partial z} = -ik_x E_x$. This gives rise to another boundary condition that

$$\frac{\partial E_z}{\partial z} = 0 \quad \text{for} \quad z = 0,\, a.$$

Consider the equation for E_x:

$$\frac{\partial^2 E_x}{\partial z^2} + k'^2 E_x = 0.$$

The solution is

$$E_x = E_{x0}\left[\sin(k'z) + A\,\cos(k'z)\right] e^{i(k_x x - \omega t)}.$$

The boundary conditons give

$$A = 0, \quad k'a = m\pi. \quad (m = 0, 1, 2, 3 \ldots)$$

Hence

$$E_x = E_{x0} \sin\left(\frac{m\pi}{a} z\right) e^{i(k_x x - \omega t)}.$$

Similarly

$$E_y = E_{y0} \sin\left(\frac{m\pi}{a} z\right) e^{i(k_x x - \omega t)},$$

$$E_z = E_{z0} \cos\left(\frac{m\pi}{a} z\right) e^{i(k_x x - \omega t)}.$$

We also have

$$\frac{n^2 \omega^2}{c^2} = k_x^2 + k'^2 = \left(\frac{2\pi}{\lambda}\right)^2 + \left(\frac{m\pi}{a}\right)^2.$$

Thus for a given wavelength λ the allowed angular frequencies are the series of discrete values

$$\omega_m = \frac{c}{n} \sqrt{\left(\frac{2\pi}{\lambda}\right)^2 + \left(\frac{m\pi}{a}\right)^2}.$$

The phase velocity is then

$$v_p = \frac{\omega_m}{k_x} = \frac{c}{n}\left[1 + \left(\frac{m\lambda}{2a}\right)^2\right]^{1/2}$$

and the group velocity is

$$v_g = \frac{d\omega_m}{dk_x} = \frac{c}{n}\left[1 + \left(\frac{m\lambda}{2a}\right)^2\right]^{-1/2}.$$

(b) In vacuum the electric field at a field point at time t produced by a particle of charge q moving with uniform velocity $\mathbf{v}$ is

$$\mathbf{E} = \frac{q}{4\pi\varepsilon_0} \frac{\alpha \mathbf{R}}{s^3},$$

where $\mathbf{R}$ is the radius vector from the location of the charge at time t to the field point,

$$\alpha = 1 - \left(\frac{v}{c}\right)^2, \quad s = \left[\alpha R^2 + \frac{1}{c^2}(\mathbf{v} \cdot \mathbf{R})^2\right]^{\frac{1}{2}}.$$

If the charge moves in a medium of permittivity ε and refractive index n, the above expression is to be modified to

$$\mathbf{E} = \frac{q}{4\pi\varepsilon} \frac{\alpha \mathbf{R}}{s^3} ,$$

where

$$\alpha = 1 - \left(\frac{vn}{c}\right)^2, \quad s = \left[\alpha R^2 + \left(\frac{n}{c} \mathbf{v} \cdot \mathbf{R}\right)^2\right]^{\frac{1}{2}}.$$

Let φ be the angle between $\mathbf{v}$ and $\mathbf{R}$, then

$$s = \left[1 - \left(\frac{vn}{c}\right)^2 \sin^2\varphi\right]^{\frac{1}{2}} R .$$

If $v > \frac{c}{n}$, s will become imaginary except for the region of space with $\sin\varphi \leq \frac{c}{vn}$. As the particle speed is greater than the speed of propagation of electromagnetic waves in the medium, the field point must be to the rear of the particle at time t (Problem 5022). Thus the field will exist only within a rear cone of half angle $\varphi = \arcsin\left(\frac{c}{vn}\right)$ with the vertex at the location of the particle at time t. On the surface of this cone $\mathbf{E} \to \infty$. This surface is the surface of the Čerenkov shock wave and contains the Čerenkov radiation field. The infinitely long charged wire can be considered an infinite set of point charges, so the region of the Čerenkov radiation will be a rear wedge with the wire forming its thin edge and the inclined planes making an angle 2φ. At any point in the wedge the intensity $\mathbf{E}$ of the radiation field is the superposition of the intensities of the Čerenkov radiation field at that point due to all the point charges.

Consider a point P in the midplane as shown in Fig. 5.16. Obviously P has to be at the rear of the line of charges represented by the y-axis. Let the line of charges pass through P at $t = 0$, then at time t the line is at

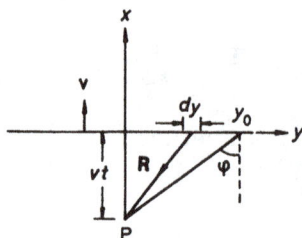

Fig. 5.16

a distance vt from P. The radius vector from a line charge element λdy to point P is

$$\mathbf{R} = -vt\mathbf{e}_x - y\mathbf{e}_y \,.$$

As $\mathbf{v} = v\mathbf{e}_x$, $\mathbf{R} \cdot \mathbf{v} = -v^2 t$. The intensity of the field at P caused by λdy is

$$d\mathbf{E} = \frac{\lambda dy \left(1 - n^2 \frac{v^2}{c^2}\right)(-vt\mathbf{e}_x - y\mathbf{e}_y)}{n^2 \left[\left(1 - \frac{v^2 n^2}{c^2}\right)(y^2 + v^2 t^2) + \frac{v^4 n^2 t^2}{c^2}\right]^{3/2}} \,.$$

The total Čerenkov field intensity at P at time t is the vector sum of the intensities contributed by all charge elements on the line. By symmetry the contributions of two charge elements located at y and $-y$ to E_y will cancel out and the total contribution is the sum of their x components. Hence the total electric field at P is in the x direction and has magnitude

$$E(t) = 2 \int_0^{y_0} dE_x \,,$$

where the upper limit of the integral is given by the requirement that P should fall within the Čerenkov cone of the charge element λdy at y_0. Thus

$$E(t) = \frac{2\lambda}{n^2} \left(n^2 \frac{v^2}{c^2} - 1\right) vt \int_0^{vt \tan \varphi} \frac{dy}{[v^2 t^2 - (\frac{n^2 v^2}{c^2} - 1)y^2]^{3/2}}$$

$$= \frac{2\lambda(\frac{n^2 v^2}{c^2} - 1) \tan \varphi}{n^2 vt \sqrt{1 - (\frac{n^2 v^2}{c^2} - 1) \tan \varphi}} \propto \frac{1}{t} \,.$$

This can be written as

$$E(t) = \frac{A}{t} \,,$$

where A is a constant.
By Fourier transform

$$E(t) = \int_{-\infty}^{\infty} E(\omega) e^{-i\omega t} d\omega$$

with

$$E(\omega) = \frac{1}{2\pi} \int_{-\infty}^{\infty} E(t) e^{i\omega t} dt = \frac{A}{2\pi} \int_{-\infty}^{\infty} \frac{e^{i\omega t}}{t} dt \,.$$

As

$$\int_{-\infty}^{\infty} \frac{e^{ix}}{x} dx = \pi i \,,$$

$$E(\omega) = \frac{Ai}{2},$$

i.e., $|E(\infty)|$ is a constant, independent of frequency. This means that the Čerenkov radiation has a "white spectrum", i.e., each of its monochromatic components has the same intensity and there is no principal frequency.

(c) As shown in Fig. 5.15, let a unit vector **S** be normal to the upper plane of the wedge forming the surface of the Čerenkov radiation. **S** is just along the direction **k** ($|\mathbf{k}| = \frac{\omega}{c} n$) of the Čerenkov radiation. Then

$$k_x = \frac{\omega}{c} n \, \sin \varphi = \frac{\omega}{c} n \left(\frac{c}{nv} \right) = \frac{\omega}{v}.$$

However not all the frequencies in the "white spectrum" of the Čerenkov radiation can propagate in the waveguide, only those that satisfy

$$\omega = \frac{c}{n} \sqrt{k_x^2 + \left(\frac{m\pi}{a} \right)^2} = \frac{c}{n} \sqrt{\left(\frac{\omega}{v} \right)^2 + \left(\frac{m\pi}{a} \right)^2},$$

or

$$\omega = \frac{m\pi}{a} \left(\frac{n^2}{c^2} - \frac{1}{v^2} \right)^{-1/2} = \omega_m. \quad (m = 1, 2, \ldots)$$

The frequencies ω_m which are allowed by the waveguide may be considered the principal modes of the Čerenkov radiation in the waveguide.

5024

A particle with mass m and electric charge q is bound by the Coulomb interaction to an infinitely massive particle with electric charge $-q$. At $t = 0$ its orbit is (approximately) a circle of radius R. At what time will it have spiraled into $R/2$? (Assume that R is large enough so that you can use the classical radiation theory rather than quantum mechanics.)

(*Columbia*)

Solution:

The massive particle can be considered stationary during the motion. The total energy of the particle of mass m is

$$E = \frac{1}{2} mv^2 + V,$$

where the potential energy of the particle is that due to the Coulomb interaction,

$$V = -\frac{q^2}{4\pi\varepsilon_0 r},$$

r being the distance of the particle from the massive particle.

As the particle moves in a circle of radius r, we have

$$m|\dot{\mathbf{v}}| = \frac{mv^2}{r} = \frac{q^2}{4\pi\varepsilon_0 r^2},$$

or

$$\frac{1}{2}mv^2 = \frac{q^2}{8\pi\varepsilon_0 r}, \quad |\dot{\mathbf{v}}| = \frac{q^2}{4\pi\varepsilon_0 m r^2}.$$

Hence

$$E = -\frac{q^2}{8\pi\varepsilon_0 r}.$$

As the particle undergoes centripetal acceleration $\dot{\mathbf{v}}$ it loses energy by radiation:

$$\frac{dE}{dt} = -\frac{q^2|\dot{\mathbf{v}}|^2}{6\pi\varepsilon_0 c^3}.$$

On the other hand, we have for the above

$$\frac{dE}{dt} = \frac{q^2}{8\pi\varepsilon_0 r^2}\frac{dr}{dt}.$$

Hence

$$\frac{dr}{dt} = -\frac{4}{3}\frac{r^2}{c^3}|\dot{\mathbf{v}}|^2 = -\frac{q^4}{12\pi^2\varepsilon_0^2 c^3 m^2 r^2}.$$

As $r = R$ at $t = 0$, the time at which $r = \frac{R}{2}$ is

$$\tau = -\frac{12\pi^2\varepsilon_0^2 c^3 m^2}{q^4}\int_R^{\frac{R}{2}} r^2 dr = \frac{7\pi^2\varepsilon_0^2 c^3 m^2 R^3}{2q^4}.$$

5025

A classical hydrogen atom has the electron at a radius equal to the first Bohr radius at time $t = 0$. Derive an expression for the time it takes the radius to decrease to zero due to radiation. Assume that the energy loss per revolution is small compared with the remaining total energy of the atom.

(Princeton)

Solution:

As the energy loss per revolution is small we may assume the motion to be nonrelativistic. Then in Gaussian units the rate of radiation loss of the electron is

$$\frac{dE}{dt} = -\frac{2e^2}{3c^3} a^2 ,$$

where a is the magnitude of the acceleration. In the Coulomb field of the hydrogen nucleus the total energy and acceleration of the electron are respectively

$$E = \frac{1}{2} mv^2 - \frac{e^2}{r} = -\frac{e^2}{2r} , \qquad a = \frac{e^2}{mr^2} ,$$

where we have used the expression for the centripetal acceleration $a = \frac{v^2}{r}$. Hence

$$\frac{dE}{dt} = \frac{dE}{dr} \frac{dr}{dt} = \frac{e^2}{2r^2} \frac{dr}{dt} = -\frac{2e^2}{3c^3} \left(\frac{e^2}{mr^2} \right)^2 ,$$

or

$$dt = -\frac{3m^2 c^3}{4e^4} r^2 dr .$$

Therefore, the time taken for the Bohr orbit to collapse completely is

$$\tau = \int_0^t dt = -\frac{3m^2 c^3}{4e^4} \int_{a_0}^0 r^2 dr = \frac{m^2 c^3 a_0^3}{4e^4} ,$$

where $a_0 = \frac{\hbar^2}{me^2}$ is the first Bohr radius.

3. MOTION OF A CHARGED PARTICLE IN ELECTROMAGNETIC FIELD (5026–5039)

5026

A particle of mass m and charge e is accelerated for a time by a uniform electric field to a velocity not necessarily small compared with c.

(a) What is the momentum of the particle at the end of the acceleration time?

(b) What is the velocity of the particle at that time?

(c) The particle is unstable and decays with a lifetime τ in its rest frame. What lifetime would be measured by a stationary observer who observed the decay of the particle moving uniformly with the above velocity?

(Wisconsin)

Solution:

(a) As

$$\frac{d(m\gamma v)}{dt} = eE, \quad m\gamma v = \int_0^t eE dt = eEt,$$

where E is the intensity of the uniform electric field,

$$\gamma = (1 - \beta^2)^{-\frac{1}{2}} \quad \text{with} \quad \beta = \frac{v}{c}.$$

(b) As

$$m\gamma\beta c = eEt,$$

or

$$\gamma\beta = (\gamma^2 - 1)^{\frac{1}{2}} = \frac{eEt}{mc},$$

we have

$$\gamma^2 = \frac{1}{1 - \beta^2} = \left(\frac{eEt}{mc}\right)^2 + 1,$$

or

$$\beta^2 = \frac{(eEt)^2}{(eEt)^2 + (mc)^2},$$

giving

$$v = \beta c = \frac{eEct}{\sqrt{(eEt)^2 + (mc)^2}}.$$

(c) On account of time dilation, the particle's lifetime in the observer's frame is

$$T = \gamma\tau = \tau\sqrt{1 + \left(\frac{eEt}{mc}\right)^2}.$$

5027

The Lagrangian of a relativistic charged particle of mass m, charge e and velocity $\mathbf{v}$ moving in an electromagnetic field with vector potential $\mathbf{A}$ is

$$L = -mc^2\sqrt{1 - \beta^2} + \frac{e}{c}\mathbf{A}\cdot\mathbf{v}.$$

The field of a dipole of magnetic moment μ along the polar axis is described by the vector potential $\mathbf{A} = \frac{\mu \sin\theta}{\gamma^2}\mathbf{e}_\phi$ where θ is the polar angle and ϕ is the azimuthal angle.

(a) Express the canonical momentum p_ϕ conjugate to ϕ in terms of the coordinates and their derivatives.

(b) Show that this momentum p_ϕ is a constant of the motion.

(c) If the vector potential **A** given above is replaced by

$$\mathbf{A}' = \mathbf{A} + \nabla\chi(r, \theta, \phi),$$

where χ is an arbitrary function of coordinates, how is the expression for the canonical momentum p_ϕ changed? Is the expression obtained in part (a) still a constant of the motion? Explain.

(Wisconsin)

Solution:

We first use Cartesian coordinates to derive an expression for the Hamiltonian.

Let $\gamma = \dfrac{1}{\sqrt{1-\beta^2}}$. The canonical momentum is

$$p_i = \frac{\partial L}{\partial v_i} = m\gamma v_i + \frac{e}{c}A_i,$$

or, in vector form,

$$\mathbf{p} = m\gamma\mathbf{v} + \frac{e}{c}\mathbf{A}.$$

The Hamiltonian is then

$$H = \mathbf{p}\cdot\mathbf{v} - L = m\gamma v^2 + \frac{mc^2}{\gamma} = \frac{mc^2}{\gamma}(\gamma^2\beta^2 + 1) = m\gamma c^2,$$

as

$$\gamma^2\beta^2 = \gamma^2 - 1.$$

(a) In spherical coordinates the velocity is

$$\mathbf{v} = \dot{r}\mathbf{e}_r + r\dot{\theta}\mathbf{e}_\theta + r\sin\theta\,\dot{\phi}\mathbf{e}_\phi.$$

The Lagrangian of the magnetic dipole in the field of vector potential **A** is therefore

$$L = -\frac{mc^2}{\gamma} + \frac{e}{c}\frac{\mu\sin^2\theta}{r}\dot{\phi}.$$

The momentum conjugates to ϕ is

$$p_\phi = \frac{\partial L}{\partial\dot{\phi}} = -mc^2\frac{\partial}{\partial\gamma}\left(\frac{1}{\gamma}\right)\frac{\partial\gamma}{\partial\beta}\frac{\partial\beta}{\partial\dot{\phi}} + \frac{e}{c}\frac{\mu\sin^2\theta}{r}.$$

As

$$\beta^2 = \frac{1}{c^2}\left(\dot{r}^2 + r^2\dot{\theta}^2 + r^2\sin^2\theta\,\dot{\phi}^2\right),$$

$$\frac{\partial\beta}{\partial\dot{\phi}} = \frac{r^2}{\beta c^2}\sin^2\theta\,\dot{\phi},$$

and as $\gamma^{-2} = 1 - \beta^2$,

$$\frac{\partial\gamma}{\partial\beta} = \gamma^3\beta.$$

Hence

$$p_\phi = m\gamma r^2\sin^2\theta\,\dot{\phi} + \frac{e}{c}\frac{\mu\sin^2\theta}{r}.$$

(b) As the Hamiltonian does not depend on ϕ,

$$\dot{p}_\phi = -\frac{\partial H}{\partial\phi} = 0.$$

Hence p_ϕ is a constant of the motion.

(c) If the vector potential is replaced by $\mathbf{A}' = \mathbf{A} + \nabla\chi(r, \theta, \phi)$, the new Lagrangian is

$$L' = \frac{-mc^2}{\gamma} + \frac{e}{c}\mathbf{A}\cdot\mathbf{v} + \frac{e}{c}\nabla\chi\cdot\mathbf{v}.$$

The canonical momentum is now

$$\mathbf{p}' = m\gamma\mathbf{v} + \frac{e}{c}\mathbf{A} + \frac{e}{c}\nabla\chi.$$

But the Hamiltonian

$$H' = \mathbf{p}'\cdot\mathbf{v} - L = m\gamma c^2$$

is the same as before.

For an arbitrary scalar function χ,

$$\nabla\chi = \frac{\partial\chi}{\partial r}\mathbf{e}_r + \frac{1}{r}\frac{\partial\chi}{\partial\theta}\mathbf{e}_\theta + \frac{1}{r\sin\theta}\frac{\partial\chi}{\partial\phi}\mathbf{e}_\phi,$$

so that

$$\nabla\chi\cdot\mathbf{v} = \dot{r}\frac{\partial\chi}{\partial r} + \dot{\theta}\frac{\partial\chi}{\partial\theta} + \dot{\phi}\frac{\partial\chi}{\partial\phi}.$$

Thus the momentum conjugate to ϕ is now

$$p'_\phi = \frac{\partial L'}{\partial \dot\phi} = m\gamma r^2 \sin^2\theta \, \dot\phi + \frac{e}{c}\left(\frac{\mu \sin^2\theta}{r} + \frac{\partial\chi}{\partial\phi}\right),$$

i.e., p'_ϕ is modified by the addition of the term $\frac{e}{c}\frac{\partial\chi}{\partial\phi}$.

As H' is still independent of ϕ, the canonical momentum p'_ϕ is a constant of the motion. However, as

$$p_\phi = p'_\phi - \frac{e}{c}\frac{\partial\chi}{\partial\phi}$$

and χ is an arbitrary scalar function, the part p_ϕ is not a constant of the motion.

5028

An electron (mass m, charge e) moves in a plane perpendicular to a uniform magnetic field. If energy loss by radiation is neglected the orbit is a circle of some radius R. Let E be the total electron energy, allowing for relativistic kinematics so that $E \gg mc^2$.

(a) Express the needed field induction B analytically in terms of the above parameters. Compute numerically, in Gauss, for the case where $R = 30$ meters, $E = 2.5 \times 10^9$ electron-volts. For this part of the problem you will have to recall some universal constants.

(b) Actually, the electron radiates electromagnetic energy because it is being accelerated by the B field. However, suppose that the energy loss per revolution, ΔE, is small compared with E. Express the ratio $\Delta E/E$ analytically in terms of the parameters. Then evaluate this ratio numerically for the particular values of R and E given above.

(*CUSPEA*)

Solution:

(a) In uniform magnetic field B the motion of an electron is described in Gaussian units by

$$\frac{d\mathbf{p}}{dt} = \frac{e}{c}\mathbf{v} \times \mathbf{B},$$

where $\mathbf{p}$ is the momentum of the electron,

$$\mathbf{p} = m\gamma\mathbf{v},$$

with $\gamma = (1 - \beta^2)^{-\frac{1}{2}}$, $\beta = \frac{v}{c}$. Since $\frac{e}{c} \mathbf{v} \times \mathbf{B} \cdot \mathbf{v} = 0$, the magnetic force does no work and the magnitude of the velocity does not change, i.e., v, and hence γ, are constant. For circular motion,

$$\left| \frac{dv}{dt} \right| = \frac{v^2}{R}.$$

Then

$$m\gamma \left| \frac{dv}{dt} \right| = \frac{e}{c} |\mathbf{v} \times \mathbf{B}|.$$

As $\mathbf{v}$ is normal to $\mathbf{B}$, we have

$$m\gamma \frac{v^2}{R} = \frac{e}{c} vB$$

or

$$B = \frac{pc}{eR}.$$

With $E \gg mc^2$, $pc = \sqrt{E^2 - m^2c^4} \approx E$ and

$$B \approx \frac{E}{eR} \approx 0.28 \times 10^4 \text{ Gs}.$$

(b) The rate of radiation of an accelerated non-relativistic electron is

$$P = \frac{2}{3} \frac{e^2}{c^3} |\dot{\mathbf{v}}|^2 = \frac{2}{3} \frac{e^2}{m^2c^3} \left(\frac{d\mathbf{p}}{dt} \cdot \frac{d\mathbf{p}}{dt} \right),$$

where $\mathbf{v}$ and $\mathbf{p}$ are respectively the velocity and momentum of the electron. For a relativistic electron, the formula is modified to

$$P = \frac{2}{3} \frac{e^2}{m^2c^5} \left(\frac{dp_\mu}{d\tau} \frac{dp^\mu}{d\tau} \right),$$

where $d\tau = \frac{dt}{\gamma}$, p_μ and p^μ are respectively the covariant and contravariant momentum-energy four-vector of the electron:

$$p_\mu = (\mathbf{p}c, -E), \quad p^\mu = (\mathbf{p}c, E).$$

Thus

$$\frac{dp_\mu}{d\tau} \frac{dp^\mu}{d\tau} = \left(\frac{d\mathbf{p}}{d\tau} \cdot \frac{d\mathbf{p}}{d\tau} \right) c^2 - \left(\frac{dE}{d\tau} \right)^2.$$

Since the energy loss of the electron per revolution is very small, we can take approximations $\frac{dE}{d\tau} \approx 0$ and $\gamma \approx$ constant. Then

$$\frac{d\mathbf{p}}{d\tau} = \gamma \frac{d\mathbf{p}}{dt} = m\gamma^2 \frac{d\mathbf{v}}{dt}.$$

Substitution in the expression for τ gives

$$P = \frac{2}{3} \frac{e^2}{m^2 c^5} m^2 \gamma^4 c^2 \left| \frac{d\mathbf{v}}{dt} \right|^2 = \frac{2}{3} \frac{e^2 c}{R^2} \left(\frac{v}{c} \right)^4 \gamma^4.$$

The energy loss per revolution is

$$\Delta E = \frac{2\pi R}{v} P = \frac{4\pi}{3} \frac{e^2}{R} \left(\frac{v}{c} \right)^3 \gamma^4.$$

As $\gamma = \frac{E}{mc^2}$,

$$\frac{\Delta E}{E} = \frac{4\pi}{3} \left(\frac{v}{c} \right)^3 \left(\frac{e^2}{mc^2 R} \right) \left(\frac{E}{mc^2} \right)^3 \approx 5 \times 10^{-4}.$$

5029

Consider the static magnetic field given in rectangular coordinates by

$$\mathbf{B} = B_0 (x\,\hat{x} - y\,\hat{y})/a.$$

(a) Show that this field obeys Maxwell's equations in free space.

(b) Sketch the field lines and indicate where filamentary currents would be placed to approximate such a field.

(c) Calculate the magnetic flux per unit length in the $\hat{z}$-direction between the origin and the field line whose minimum distance from the origin is R.

(d) If an observer is moving with a non-relativistic velocity $\mathbf{v} = v\hat{z}$ at some location (x, y), what electric potential would he measure relative to the origin?

(e) If the magnetic field $B_0(t)$ is slowly varying in time, what electric field would a stationary observer at location (x, y) measure?

(*Wisconsin*)

Solution:

(a)

$$\nabla \cdot \mathbf{B} = \left(\hat{x} \frac{\partial}{\partial x} + \hat{y} \frac{\partial}{\partial y} + \hat{z} \frac{\partial}{\partial z} \right) \cdot \left[\frac{B_0}{a} (x\hat{x} - y\hat{y}) \right]$$

$$= \frac{B_0}{a} (\hat{x} \cdot \hat{x} - \hat{y} \cdot \hat{y}) = 0 .$$

$$\nabla \times \mathbf{B} = \left(\hat{x} \frac{\partial}{\partial x} + \hat{y} \frac{\partial}{\partial y} + \hat{z} \frac{\partial}{\partial z} \right) \times \left[\frac{B_0}{a} (x\hat{x} - y\hat{y}) \right]$$

$$= \frac{B_0}{a} (\hat{x} \times \hat{x} - \hat{y} \times \hat{y}) = 0 .$$

(b) The magnetic field lines are given by the differential equation

$$\frac{dy}{dx} = \frac{B_y}{B_x} = -\frac{y}{x} ,$$

i.e.,

$$x\,dy + y\,dx = 0 ,$$

or

$$d(xy) = 0 .$$

Hence

$$xy = \text{const} .$$

The field lines are shown in Fig. 5.17. In order to create such a field, four infinitely long straight currents parallel to the z direction are symmetrically placed on the four quadrants with flow directions as shown in Fig. 5.17.

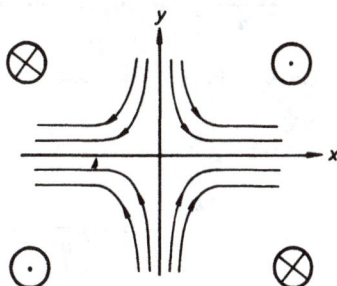

Fig. 5.17

(c) Consider a rectangle of height $z = 1$ and length R along the bisector of the right angle between the x- and y-axes in the first quadrant (i.e., along the line $x = y$). Then the unit normal to this rectangle is $\mathbf{n} = \frac{1}{\sqrt{2}}(\hat{x} - \hat{y})$. Along the length R, $\mathbf{B}(x, y) = \frac{B_0}{a}(x\hat{x} - y\hat{y}) = \frac{B_0}{a}x(\hat{x} - \hat{y})$. Taking as the area element of the rectangle $d\sigma = \sqrt{2}\,dx$, one has for the magnetic flux through the rectangle

$$\phi_B = \int \mathbf{B}(x, y) \cdot \mathbf{n}\,d\sigma = \frac{2B_0}{a}\int_0^{\frac{R}{\sqrt{2}}} x\,dx = \frac{B_0 R^2}{2a}.$$

(d) Transforming to the observer's frame, we find

$$\mathbf{E}'_\perp = \gamma(\mathbf{E}_\perp + \mathbf{v} \times \mathbf{B}_\perp) = \gamma\mathbf{v} \times \mathbf{B}, \quad E'_\parallel = E_\parallel = 0,$$

or

$$\mathbf{E}' = \mathbf{v} \times \mathbf{B},$$

as for small velocities, $\beta = \frac{v}{c} \approx 0$, $\gamma = (1 - \beta^2)^{-\frac{1}{2}} \approx 1$.

Hence

$$\mathbf{E}' = v\hat{z} \times \left[\frac{B_0}{a}(x\hat{x} - y\hat{y})\right] = \frac{B_0}{a}v(x\hat{y} + y\hat{x}).$$

The potential $\phi(x, y)$ relative to the origin $(0, 0)$ as measured by the observer is given by

$$\phi(x, y) = -\int_0^{\mathbf{r}} \mathbf{E}' \cdot d\mathbf{r},$$

where $\mathbf{r} = x\hat{x} + y\hat{y}$.

Thus

$$\phi(x, y) = \frac{B_0 v}{a}\int_0^{\mathbf{r}}(x\,dy + y\,dx)$$

$$= -\frac{B_0 v}{a}\int_0^{\mathbf{r}} d(xy) = -\frac{B_0}{a}vxy.$$

(e) Maxwell's equation $\nabla \times \mathbf{E} = -\frac{\partial \mathbf{B}}{\partial t}$ gives

$$\frac{\partial E_z}{\partial y} - \frac{\partial E_y}{\partial z} = -\dot{B}_0(t)\frac{x}{a},$$

$$\frac{\partial E_x}{\partial z} - \frac{\partial E_z}{\partial x} = \dot{B}_0(t)\frac{y}{a},$$

$$\frac{\partial E_y}{\partial x} - \frac{\partial E_x}{\partial y} = 0.$$

As **B** is only slowly varying in time, $\dot{\mathbf{B}}$ can be taken to be independent of the spatial coordinates. The solution of this set of equations is $E_x =$ constant, $E_y =$ constant, and

$$E_z = -\frac{\dot{B}_0(t)x}{a} \int dy = -\frac{\dot{B}_0(t)xy}{a} + f_1(x)$$

or

$$E_z = -\frac{\dot{B}_0(t)xy}{a} + f_2(y).$$

Hence $f_1(x) = f_2(y) =$ constant, which as well as the other constants can be taken to be zero as we are not interested in any uniform field. Therefore

$$\mathbf{E} = -\dot{B}_0(t)\frac{xy}{a}\,\mathbf{e}_z\,.$$

5030

Consider the motion of electrons in an axially symmetric magnetic field. Suppose that at $z = 0$ (the "median plane") the radial component of the magnetic field is 0 so $\mathbf{B}(z = 0) = B(r)\,\mathbf{e}_z$. Electrons at $z = 0$ then follow a circular path of radius R, as shown in Fig. 5.18.

(a) What is the relationship between the electron momentum p and the orbit radius R? In a betatron, electrons are accelerated by a magnetic field which changes with time. Let B_{av} be the average value of the magnetic field over the plane of the orbit (within the orbit), i.e.,

$$B_{\mathrm{av}} = \frac{\psi_B}{\pi R^2}\,,$$

where ψ_B is the magnetic flux through the orbit. Let $\mathbf{B}_0$ equal $\mathbf{B}(r = R, z = 0)$.

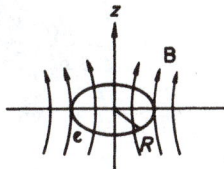

Fig. 5.18

(b) Suppose B_{av} is changed by an amount ΔB_{av} and B_0 is changed by ΔB_0. How must ΔB_{av} be related to ΔB_0 if the electrons are to remain at radius R as their momentum is increased?

(c) Suppose the z component of the magnetic field near $r = R$ and $z = 0$ varies with r as $B_z(r) = B_0(R)(\frac{R}{r})^n$. Find the equations of motion for small departures from the equilibrium orbit in the median plane. There are two equations, one for small vertical changes and one for small radial changes. Neglect any coupling between radial and vertical motion.

(d) For what range of n is the orbit stable against both vertical and radial perturbations?

(*Princeton*)

Solution:

For simplicity, we shall assume nonrelativistic motions.

(a) The equation of motion is $m \left| \frac{d\mathbf{v}}{dt} \right| = -e \left| \mathbf{v} \times \mathbf{B} \right|$, or $\frac{mv^2}{R} = -evB$. Hence $P = mv = -eBR$, where $-e$ is the electronic charge.

(b) It is required that R remains unchanged as B_0 increases by ΔB_0 and v changes by Δv. Thus

$$ m \frac{(v + \Delta v)^2}{R} = -e(v + \Delta v)(B_0 + \Delta B_0), $$

or

$$ \Delta B_0 \approx -\frac{m\Delta v}{eR}, $$

as

$$ \frac{mv^2}{R} = -evB_0. $$

The change of v arises from $\mathbf{B}$ changing with time. Faraday's law

$$ \oint_c \mathbf{E} \cdot d\mathbf{r} = -\int_S \mathbf{B} \cdot d\mathbf{S} $$

indicates that a tangential electric field

$$ E = \frac{1}{2\pi R} \frac{d\phi}{dt} $$

is induced on the orbit. Thus the resultant change of momentum is

$$ m\Delta v = \int_0^{\Delta t} \frac{-e}{2\pi R} \frac{d\phi}{dt} dt = \frac{-e\Delta\phi}{2\pi R} = \frac{-eR\Delta B_{av}}{2}, $$

as $\Delta\phi = \Delta B_{\mathrm{av}}\pi R^2$. Hence

$$\Delta B_0 = \frac{1}{2}\Delta B_{\mathrm{av}}.$$

(c) Suppose the electron suffers a radial perturbation, so that the equilibrium radius and angular velocity change by small quantities:

$$r = R + r_1, \quad \omega = \omega_0 + \omega_1,$$

where $\omega_0 = \frac{v}{R} = -\frac{eB_0}{m}$.

In cylindrical coordinates r, θ, z, the electron has velocity

$$\mathbf{v} = \dot{r}\,\mathbf{e}_r + r\dot{\theta}\,\mathbf{e}_\theta.$$

As

$$\dot{\mathbf{e}}_r = \dot{\theta}\,\mathbf{e}_\theta, \quad \dot{\mathbf{e}}_\theta = -\dot{\theta}\,\mathbf{e}_r,$$

the acceleration is

$$\mathbf{a} = \dot{\mathbf{v}} = (\ddot{r} - r\dot{\theta}^2)\,\mathbf{e}_r + (2\dot{r}\dot{\theta} + r\ddot{\theta})\,\mathbf{e}_\theta.$$

Newton's second law

$$\mathbf{F} = m\mathbf{a} = -e\mathbf{v}\times\mathbf{B}$$

then gives

$$-er\dot{\theta}B_z = m(\ddot{r} - r\dot{\theta}^2),$$

$$e\dot{r}B_z = m(r\ddot{\theta} + 2\dot{r}\dot{\theta}).$$

As $\mathbf{B} = B_z\,\mathbf{e}_z$,

$$\mathbf{v}\times\mathbf{B} = r\dot{\theta}B_z\mathbf{e}_r - \dot{r}B_z\mathbf{e}_\theta.$$

In terms of the perturbations,

$$\dot{r} = \dot{r}_1, \quad \dot{\theta} = \omega_0 + \omega_1, \quad \ddot{\theta} = \dot{\omega}_1$$

and to first approximation, the above equations respectively become

$$m(\ddot{r}_1 - R\omega_0^2 - r_1\omega_0^2 - 2R\omega_0\omega_1) \approx -e(R\omega_0 + R\omega_1 + r_1\omega_0)B_z,$$

$$e\dot{r}_1 B_z \approx mR\dot{\omega}_1 + 2m\dot{r}_1\omega_0.$$

Using $B_z(R + r_1) \approx B_z(R) + \left(\frac{\partial B_z}{\partial r}\right)_R r_1$, $eB_z(R) = m\omega_0$, these equations

become, again to first approximation,

$$-eR\omega_0 B_z'(R)r_1 - eB_z(R\omega_1 + r_1\omega_0) = m(\ddot{r}_1 - 2R\omega_0\omega_1 - r_1\omega_0^2),$$

and

$$R\dot{\omega}_1 + \dot{r}_1\omega_0 = 0,$$

where $B_z'(R) = \left(\frac{\partial B_z}{\partial r}\right)_R$. Integrating the second equation and using it in the first give

$$-eR\omega_0 B_z'(R)r_1 = m\ddot{r}_1 + m\omega_0^2 r_1.$$

Now as

$$B_z'(R) = B_0(R)n \left(\frac{R}{r}\right)^{n-1} \cdot \left(-\frac{R}{r^2}\right)\bigg|_{r=R} = -\frac{n}{R} B_z(R),$$

we have, again using $eB_z(R) = m\omega_0$, the radial equation of motion

$$\ddot{r}_1 + (1-n)\omega_0^2 r_1 = 0. \tag{1}$$

The vertical motion is given by Newton's second law $F_z = m\ddot{z}$. Now

$$F_z = -e(\mathbf{v} \times \mathbf{B}) \cdot \mathbf{e}_z = -e(\dot{r}B_\theta - r\dot{\theta}B_r)$$
$$= -e\dot{r}_1 B_\theta + e(R + r_1)(\omega_0 + \omega_1)B_r \approx eR\omega_0 B_r,$$

as B_θ and B_r are first order small quantities. Hence

$$m\ddot{z} = e\omega_0 R B_r.$$

To find B_r, consider a small loop C in a plane containing the z axis as shown in Fig. 5.19. Using Ampère's circuital law $\oint_c \mathbf{B} \cdot d\mathbf{l} = 0$ and noting that there is no radial component of $\mathbf{B}$ in the plane $z = 0$, we find

$$B_z(R)z + B_r(z)dr - B_z(R + dr)z = 0,$$

or

$$B_r(z) = \frac{B_z(R + dr) - B_z(R)}{dr} z = B_z'(R)z.$$

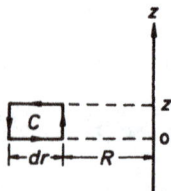

Fig. 5.19

As $B'_z(R) = -\frac{n}{R} B_z(R)$, $eB_z(R) = m\omega_0$, we have

$$m\ddot{z} + m\omega_0^2 nz = 0 . \qquad (2)$$

Equations (1) and (2) describe small departures from the equilibrium orbit.

(d) For the orbit to be stable both the vertical and radial perturbations must be sinusoidal. Then Eq. (1) requires $n < 1$ and Eq. (2) requires $n > 0$. Hence we must have $0 < n < 1$.

5031

An electron moves in a one-dimensional potential well of harmonic oscillator with frequency $\omega = 10^5$ rad/s, and amplitude $x_0 = 10^{-8}$ cm.

(a) Calculate the radiated energy per revolution.

(b) What is the ratio of the energy loss per revolution to the average mechanical energy?

(c) How much time must it take to lose half of its energy?

(*Columbia*)

Solution:

(a) The radiation reaction which acts as a damping force to the motion of a nonrelativistic electron is, in Gaussian units, (Problem **5032(a)**)

$$f = \frac{2e^2}{3c^3} \dddot{x} .$$

Thus the equation of motion for the electron is

$$m\ddot{x} = -kx + \frac{2e^2}{3c^3} \dddot{x} ,$$

or

$$\ddot{x} = -\omega_0^2 x + \frac{2e^2}{3mc^3} \dddot{x} ,$$

where $\omega_0^2 = \frac{k}{m}$. We consider the radiation damping to be small and first neglect the radiation term so that $\ddot{x} + \omega_0^2 x = 0$, or $x = x_0 e^{-i\omega_0 t}$. Then

$$\frac{2e^2}{3mc^3} \dddot{x} = \frac{i2e^2\omega_0^3}{3mc^3} x = i2\omega_0\alpha x$$

with $\alpha = \frac{e^2\omega_0^2}{3mc^3}$.

The equation of motion now becomes

$$\ddot{x} = -(\omega_0^2 - i2\omega_0\alpha)\,x\,.$$

The solution is

$$x = x_0 e^{-i\sqrt{\omega_0^2 - i2\omega_0\alpha}\,t}$$
$$\approx x_0 e^{-\alpha t} e^{-i\omega_0 t}\,.$$

Note that as $\frac{2\alpha}{\omega_0} = \frac{2}{3}r_0\frac{\omega_0}{c}$, where $r_0 = \frac{e^2}{mc^2} = 2.82\times10^{-13}$ cm is the classical radius of electron, is much smaller than unity, the above approximation holds. Furthermore, we can take

$$\ddot{x} \approx -\omega_0^2 x\,.$$

The average mechanical energy of the electron is

$$\langle E \rangle = \frac{1}{2}m\langle\dot{x}^2\rangle + \frac{1}{2}k\langle x^2\rangle$$
$$= \frac{1}{4}m\omega_0^2 x_0^2 e^{-2\alpha t} + \frac{1}{4}kx_0^2 e^{-2\alpha t}$$
$$= \frac{1}{2}m\omega_0^2 x_0^2 e^{-2\alpha t}\,.$$

The average rate of energy loss by radiation is

$$\left\langle \frac{dE}{dt} \right\rangle = \frac{1}{T}\int_0^T f\dot{x}\,dt = \frac{2e^2}{3c^3}\frac{1}{T}\int_0^T \dddot{x}\dot{x}\,dt = \frac{2e^2}{3c^3}\frac{1}{T}[\ddot{x}\,\dot{x}]_0^T - \frac{2e^2}{3c^3}\frac{1}{T}[\ddot{x}^2]_0^T$$
$$= -\frac{2e^2}{3c^3}\frac{\omega_0^4 x_0^2}{2}\,,$$

so the energy loss per revolution is

$$\Delta E = \frac{2e^2}{3c^3}\frac{\omega_0^4 x_0^2}{2}\frac{2\pi}{\omega_0} = \frac{2\pi}{3}\frac{e^2\omega_0^3 x_0^2}{c^3} = 1.8\times10^{-51}\ \text{erg} = 1.1\times10^{-39}\ \text{eV}\,.$$

(b) The ratio of the energy loss per revolution to the total mechanical energy is

$$\frac{\Delta E}{\langle E \rangle} = \frac{4\pi}{3}\cdot\frac{e^2\omega_0}{mc^3} = \frac{4\pi}{3}r_0\frac{\omega_0}{c} = 3.9\times10^{-18}\,.$$

(c)

$$E(t) = \frac{1}{2} m\omega_0^2 x_0^2 e^{-2\alpha t}.$$

Let $E(t + \tau) = \frac{1}{2} E(t)$. Then

$$\tau = \frac{\ln 2}{2\alpha} = \frac{3mc^3}{2e^2\omega_0^2} \ln 2$$

$$= \frac{3c}{2r_0\omega_0^2} \ln 2$$

$$= 1.1 \times 10^{13} \text{ s}.$$

5032

An electron of charge e and mass m is bound by a linear restoring force with spring constant $k = m\omega_0^2$. When the electron oscillates, the radiated power is expressed by

$$P = \frac{2e^2\dot{v}^2}{3c^3}$$

where $\dot{v}$ is the acceleration of the electron and c is the speed of light.

(a) Consider the radiative energy loss to be due to the action of a damping force $\mathbf{F}_s$. Assume that the energy loss per cycle is small compared with the total energy of the electron. Using the work-energy relationship over a long time period, obtain an expression for $\mathbf{F}_s$ in terms of $\ddot{v}$. Under what conditions is F_s approximately proportional to v?

(b) Write down the equation of motion for the oscillating charge, assuming that F_s is proportional to v. Solve for the position of the charge as a function of time.

(c) Is the assumption of part (a) that the energy loss per cycle is small satisfied for a natural frequency $\frac{\omega_0}{2\pi} = 10^{15}$ Hz?

(d) Now assume that the electron oscillator is also driven by an external electric field $E = E_0 \cos(\omega t)$. Find the relative time-averaged intensity $I/I_{\max}$ of the radiated power as a function of angular frequncy ω for $|\omega - \omega_0| \ll \omega_0$ (near resonance). Find the frequency ω_1 for which I is a maximum, find the fractional "level shift" $(\omega_1 - \omega_0)/\omega_0$ and the fractional full width at half-maximum $\Delta\omega_{\text{FWHM}}/\omega_0$.

(*MIT*)

Solution:

(a) The damping force is defined such that the work done against it per unit time by the electron just equals the power radiated. In Gaussian units we thus have

$$-\int_{t_1}^{t_2} \mathbf{F}_s \cdot \mathbf{v}\, dt = \int_{t_1}^{t_2} \frac{2e^2}{3c^3}\dot{v}^2 dt = \int_{t_1}^{t_2} \frac{2e^2}{3c^3}\dot{\mathbf{v}} \cdot d\mathbf{v}$$

$$= \frac{2e^2}{3c^3}\dot{\mathbf{v}} \cdot \mathbf{v}\Big|_{t_1}^{t_2} - \frac{2e^2}{3c^3}\int_{t_1}^{t_2} \ddot{\mathbf{v}} \cdot \mathbf{v}\, dt.$$

Letting $t_2 - t_1 = T$ be one period of oscillation and assuming that the energy loss per cycle is small compared with the total energy, we can treat the motion of the electron as quasi-periodic. Then $\mathbf{v}|_{t_1} = \mathbf{v}|_{t_2}$, $\dot{\mathbf{v}}|_{t_1} = \dot{\mathbf{v}}|_{t_2}$ and the first term on the right-hand side cancels out. So, the above gives

$$\mathbf{F}_s = \frac{2e^2}{3c^3}\ddot{\mathbf{v}}.$$

If the damping force is very weak compared with the restoring force on the electron, the displacement of the electron can be taken to be still $\mathbf{x} = \mathbf{x}_0 e^{-i\omega_0 t}$, and, furthermore, we can take $\ddot{\mathbf{v}} = -\omega_0^2 \mathbf{v}$. Then $\mathbf{F}_s$ would be proportional to v,

$$\mathbf{F}_s = -\frac{2e^2\omega_0^2}{3c^3}\mathbf{v}.$$

(b) The equation of motion for the electron is

$$m\ddot{x} = -m\omega_0^2 x - \frac{2e^2}{3c^3}\omega_0^2 \dot{x}.$$

By setting $\gamma = \frac{2e^2\omega_0^2}{3mc^3}$, the above equation can be written as

$$\ddot{x} + \gamma\dot{x} + \omega_0^2 x = 0.$$

If F_s is much smaller than the restoring force, i.e., $\gamma \ll \omega_0$, the above has the solution (Problem 5031)

$$x = x_0 e^{-\frac{\gamma}{2}t} e^{-i\omega_0 t}.$$

(c) For a natural frequency $f = \frac{\omega_0}{2\pi} = 10^{15}$ Hz,

$$\gamma = \frac{2}{3}r_0 \frac{\omega_0^2}{c} = 2.5 \times 10^8 \text{ s}^{-1},$$

where $r_0 = \frac{e^2}{mc^2} = 2.82 \times 10^{-13}$ cm is the classical radius of electron. The condition $\gamma \ll \omega_0$ is obviously satisfied.

The potential energy of the electron is $\frac{1}{2} m\omega_0^2 x^2$. After each cycle, x is reduced by a factor $e^{-\frac{\gamma}{2} T}$, where T is the period $\frac{2\pi}{\omega_0}$. Thus the ratio of the energy loss per cycle to the total mechanical energy can be estimated as

$$1 - e^{-\gamma T} = 1 - \exp\left(-2.5 \times 10^8 \times 10^{-15}\right)$$
$$= 2.5 \times 10^{-7} \ll 1.$$

The same goes for the kinetic part of its energy.
Thus the assumption of (a) is valid.

(d) After adding the external field, the equation of motion becomes

$$m\ddot{x} = -eE_0 e^{-i\omega t} - m\omega_0^2 x - \frac{2e^2}{3c^3} \omega_0^2 \dot{x}.$$

Putting $\gamma = \frac{2e^2\omega_0^2}{3mc^3}$, it can be written as

$$\ddot{x} + \gamma\dot{x} + \omega_0^2 x = -\frac{e}{m} E_0 e^{-i\omega t}.$$

By substituting $x = x_0 e^{-i\omega t}$ in the equation, we get the steady state solution

$$x = \frac{e}{m} \frac{1}{\omega^2 - \omega_0^2 + i\omega\gamma} E_0 e^{-i\omega t},$$

which gives

$$\ddot{x} = -\frac{e}{m} \frac{\omega^2}{\omega^2 - \omega_0^2 + i\omega\gamma} E_0 e^{-i\omega t}.$$

The time-averaged radiated power is now

$$I(\omega) = \frac{2e^2}{3c^3} \langle \ddot{x}^2 \rangle = \frac{2e^2}{3c^3} \cdot \frac{1}{2} \text{Re}\left(\ddot{x}^* \ddot{x}\right)$$
$$= \frac{e^4 E_0^2}{3m^2 c^3} \cdot \frac{\omega^4}{(\omega^2 - \omega_0^2)^2 + \omega^2\gamma^2}.$$

I_{max} occurs near the natural frequency ω_0.

Let $\Delta\omega = \omega - \omega_0$ and $u = \frac{\Delta\omega}{\omega_0}$. As $u = \frac{\Delta\omega}{\omega_0} \ll 1$, $\frac{\gamma}{\omega_0} \ll 1$, we have, correct to second order of small quantities,

$$I(u) = \frac{e^4 E_0^2}{3m^2 c^3} \frac{1 + 4u + 6u^2}{4u^2 + \frac{\gamma^2}{\omega_0^2}}.$$

From $\frac{dI(u)}{du} = 0$, we find

$$u_1 = \frac{\omega_1 - \omega_0}{\omega_0} = \frac{1}{2}\frac{\gamma^2}{\omega_0^2},$$

so the frequency corresponding to I_{max} is

$$\omega_1 = \omega_0 + \frac{1}{2}\frac{\gamma^2}{\omega_0},$$

and the maximum radiated power is

$$I_{max} = I(u_1) \approx \frac{e^4 E_0^2}{3m^2c^3}\frac{\omega_0^2}{\gamma^2}.$$

Hence

$$\frac{I(\omega)}{I_{max}} = \frac{\omega_0^2}{4(\omega - \omega_0)^2 + \gamma^2} \cdot \frac{\gamma^2}{\omega_0^2} = \frac{\gamma^2}{4(\omega - \omega_0)^2 + \gamma^2}.$$

For $I(\omega)/I_{max} = \frac{1}{2}$, $\omega = \omega_\pm = \omega_0 \pm \frac{\gamma}{2}$. The full width at half maximum is therefore

$$\frac{\Delta\omega_{FWHM}}{\omega_0} = \frac{\omega_+ - \omega_-}{\omega_0} = \frac{\gamma}{\omega_0}.$$

5033

To account for the effects of energy radiation by an accelerating charged particle, we must modify Newton's equation of motion by adding a radiative reaction force F_R.

(a) Deduce the classical result for F_R:

$$\mathbf{F_R} = \frac{2}{3}\frac{e^2}{c^3}\ddot{\mathbf{v}}$$

by using conservation of energy. Assume for simplicity that the orbit is circular so that $\dot{\mathbf{v}} \cdot \mathbf{v} = 0$, where $\mathbf{v}$ is the particle's velocity.

Now consider a free electron. Let a plane wave with electric field $\mathbf{E} = \mathbf{E}_0 e^{-i\omega t}$ be incident on the electron. Again assume $v \ll c$.

(b) What is the time-averaged force $\langle F \rangle$ on the electron due to the electromagnetic wave?

(c) Use the radiation pressure p of this wave to deduce the effective cross section for the scattering of radiation

$$\sigma = \langle F \rangle / p.$$

(*Chicago*)

Solution:

(a) See Problem 5032.

(b) The equation of motion for an electron under the action of a plane electromagnetic wave is

$$m\ddot{\mathbf{r}} = -e\mathbf{E}_0 e^{-i\omega t} + \frac{2e^2}{3c^3}\dddot{\mathbf{r}}.$$

In the steady state $\mathbf{r} = \mathbf{r}_0 e^{-i\omega t}$. Substitution in the above gives

$$\mathbf{r}_0 = \frac{e\mathbf{E}_0}{m\omega^2 + i\frac{2e^2\omega^3}{3c^3}}.$$

The force on the electron averaged over one period is

$$\langle F \rangle = \langle -e\mathbf{E}_0 e^{-i\omega t} - \frac{e}{c}\mathbf{v} \times \mathbf{B} \rangle = -\frac{e}{c}\langle \mathbf{v} \times \mathbf{B} \rangle$$

with

$$\mathbf{v} = \dot{\mathbf{r}} = -i\omega\mathbf{r}_0 e^{-i\omega t},$$

as $\langle e^{-i\omega t} \rangle = 0$.

$$\begin{aligned}
\langle \mathbf{F} \rangle &= -\frac{e}{2c}\,\text{Re}\,(\mathbf{v}^* \times \mathbf{B}) \\
&= -\frac{e}{2c}\,\text{Re}\left[\frac{i\omega e\mathbf{E}_0}{m\omega^2 - i\frac{2e^2\omega^3}{3c^3}} \times \mathbf{B}_0\right] \\
&= -\frac{e^2}{2c}\cdot\frac{8\pi}{c}\langle S \rangle\,\text{Re}\left[\frac{i\omega}{m\omega^2 - i\frac{2e^2\omega^3}{3c^3}}\right],
\end{aligned}$$

where

$$\begin{aligned}
\langle S \rangle &= \langle \frac{c}{4\pi}\mathbf{E} \times \mathbf{B} \rangle \\
&= \frac{c}{4\pi}\cdot\frac{1}{2}\,\text{Re}\,(\mathbf{E}^* \times \mathbf{B}) \\
&= \frac{c}{8\pi}\mathbf{E}_0 \times \mathbf{B}_0.
\end{aligned}$$

is the average Poynting vector.

Now

$$\text{Re}\left[\frac{i\omega}{m\omega^2 - i\frac{2e^2\omega^3}{3c^3}}\right] = -\frac{2e^2\omega^4}{3c^3}\left[m^2\omega^4 + \left(\frac{2e^2\omega^3}{3c^3}\right)^2\right]^{-1} \approx -\frac{2e^2}{3m^2c^3},$$

since the assumption $v \ll c$ means that $\omega r \ll c$, or $\frac{e^2\omega^3}{c^3} = mr_0\frac{\omega^3}{c} \ll m\omega^2$, where $r_0 = \frac{e^2}{mc^2}$, the classical radius of electron, $\sim r$. Hence

$$\langle \mathbf{F} \rangle \simeq \frac{8\pi e^4}{3m^2c^5}\langle \mathbf{S} \rangle.$$

(c) The average radiation pressure is $\langle p \rangle = \frac{\langle |S| \rangle}{c}$. It is related to $\langle \mathbf{F} \rangle$ through the effective cross section σ by

$$\sigma = \frac{\langle F \rangle}{\langle p \rangle} = \frac{8\pi e^4}{3m^2c^4} = \frac{8\pi}{3}r_0^2,$$

where r_0 is the classical radius of the electron.

5034

Consider the classical theory of the width of an atomic spectral line. The "atom" consists of an electron of mass m and charge e in a harmonic oscillator potential. There is also a frictional damping force, so the equation of motion for the electron is

$$m\ddot{\mathbf{x}} + m\omega_0^2\mathbf{x} + \gamma\dot{\mathbf{x}} = 0.$$

(a) Suppose at time $t = 0$, $\mathbf{x} = \mathbf{x}_0$ and $\dot{\mathbf{x}} = 0$. What is the subsequent motion of the electron? A classical electron executing this motion would emit electromagnetic radiation. Determine the intensity $I(\omega)$ of this radiation as a function of frequency. (You need not calculate the absolute normalization of $I(\omega)$, only the form of the ω dependence of $I(\omega)$. In other words, it is enough to calculate $I(\omega)$ up to a cosntant of proportionality.) Assume $\gamma/m \ll \omega_0$.

(b) Now suppose the damping force $\gamma\dot{x}$ is absent from the equation in (a) and that the oscillation is damped only by the loss of energy to radiation (an effect which has been ignored above). The energy U of the oscillator will decay as $U_0e^{-\Gamma t}$. What, under the above assumptions, is Γ? (You may assume that in any one oscillation the electron loses only a very small fraction of its energy.)

(c) For an atomic spectral line of 5000 Å, what is the width of the spectral line, in Angstroms, as determined from the calculation of part (b)? About how many oscillations does the electron make while losing half its energy? Rough estimates are enough.

(Princeton)

Solution:

(a) The equation of motion for the electron is

$$m\ddot{x} + m\omega_0^2 x + \gamma\dot{x} = 0,$$

with the initial conditions

$$x\big|_{t=0} = x_0,$$

$$\dot{x}\big|_{t=0} = 0.$$

Its solution is

$$x = x_0 e^{-\frac{\gamma}{2m}t}\, e^{-i\omega t},$$

where

$$\omega = \sqrt{\omega_0^2 - \frac{\gamma^2}{4m^2}}.$$

As $\frac{\gamma}{2m} \ll \omega_0$, $\omega \approx \omega_0$, and $x = x_0 e^{-\frac{\gamma}{2m}t}e^{-i\omega_0 t}$. The oscillation of the electron about the positive nucleus is equivalent to an oscillating dipole of moment $p = p_0 e^{-\frac{\gamma}{2m}t}e^{-i\omega_0 t}$, i.e., a dipole oscillator with attenuating amplitude, where $p_0 = ex_0$. Its radiation field at a large distance away is given by

$$E(r, t) = E_0(r)\, e^{-\frac{\gamma}{2m}\left(t-\frac{r}{c}\right)}\, e^{-i\omega_0\left(t-\frac{r}{c}\right)}.$$

For simplicty we shall put $t - \frac{r}{c} = t'$ and write

$$E(t) = E_0 e^{-\frac{\gamma}{2m}t'}e^{-i\omega_0 t'}.$$

Note that t' is the retarded time. By Fourier transform the oscillations are a superposition of oscillations of a spread of frequencies:

$$E(t) = \int_{-\infty}^{+\infty} E(\omega)\, e^{-i\omega t'}\, d\omega,$$

where, as $\mathbf{E}(t) = 0$ for $t < 0$,

$$\mathbf{E}(\omega) = \frac{1}{2\pi} \int_0^{+\infty} \mathbf{E}(t)e^{i\omega t'}\,dt'$$

$$= \frac{1}{2\pi} \int_0^{\infty} \left(\mathbf{E}_0 e^{-\frac{\gamma}{2m}t'} e^{-i\omega_0 t'} \right) e^{i\omega t'}\,dt'$$

$$= \frac{\mathbf{E}_0}{2\pi} \frac{1}{i(\omega - \omega_0) - \frac{\gamma}{2m}} .$$

The rate of radiation is then

$$I(\omega) \propto |\mathbf{E}(\omega)|^2 \propto \frac{1}{(\omega - \omega_0)^2 + \frac{\gamma^2}{4m^2}} . \tag{1}$$

This is a Lorentz spectrum.

(b) For $\gamma = 0$, $\mathbf{p} = \mathbf{p}_0 e^{-i\omega_0 t'}$ and the rate of dipole radiation is

$$\langle P \rangle = \frac{2}{3c^3} \langle |\ddot{\mathbf{p}}|^2 \rangle = \frac{2}{3c^3} \frac{1}{2} \mathrm{Re}\,(\ddot{p}^*\ddot{p}) = \frac{e^2 \omega_0^4 x_0^2}{3c^3} .$$

The total energy of the dipole is $U = \frac{1}{2} m\omega_0^2 x_0^2$, so $\langle P \rangle = \frac{2e^2\omega_0^2}{3c^3 m} U$. As the loss of energy is due solely to radiation, we have

$$\frac{dU}{dt'} + \frac{2e^2\omega_0^2}{3mc^3} U = 0 ,$$

which has solution $U = U_0 e^{-\Gamma t'}$, with $\Gamma = \frac{2e^2\omega_0^2}{3mc^3}$.

(c) To find the width of the spectral line, we see that, for $\gamma = 0$, $\frac{\gamma}{2m}$ in Eq. (1) is to be replaced by $\frac{\Gamma}{2}$. Then if we define $\Delta\omega = \omega_+ - \omega_-$, where $\omega_\pm$ are the frequencies at which the intensity is half the maximum intensity, we have

$$\frac{\Delta\omega}{2} = \frac{\Gamma}{2}$$

or

$$\Delta\omega = \Gamma .$$

Hence

$$\Delta\lambda = \lambda_0 \frac{\Delta\omega}{\omega_0} = \lambda_0 \frac{\Gamma}{\omega_0} = \lambda_0 \frac{2e^2\omega_0}{3mc^3} = \frac{2e^2}{3mc^3} 2\pi c = \frac{4\pi}{3} r_0$$

$$= \frac{4\pi}{3} \times 2.82 \times 10^{-5} = 1.2 \times 10^{-4} \text{ Å} ,$$

where $r_0 = \frac{e^2}{mc^2} = 2.82 \times 10^{-5}$ Å is the classical radius of electron.

The time needed for losing half the energy is $T = \frac{\ln 2}{\Gamma}$ while the time for one oscillation is $\tau = \frac{2\pi}{\omega_0}$. Hence to lose half the energy the number of oscillations required is

$$N = \frac{T}{\tau} = \frac{\ln 2}{\Gamma} \cdot \frac{\omega_0}{2\pi} = \frac{3c}{4\pi\omega_0 r_0} \ln 2 = \frac{3}{8\pi^2} \frac{\lambda_0}{r_0} \ln 2$$

$$= \frac{3\ln 2}{8\pi^2} \times \frac{5000}{2.82 \times 10^{-5}} = 4.7 \times 10^6 .$$

5035

Energy loss due to radiation is supposed to be insignificant for a non-relativistic charged particle in a cyclotron. To illustrate this fact, consider a particle of given charge, mass and kinetic energy which starts out in a circular path of given radius in a cyclotron with a uniform axial magnetic field.

(a) Determine the kinetic energy of the particle as a function of time.

(b) If the particle is a proton with the initial kinetic energy of 100 million electron volts, find how long it takes, in seconds, for it to lose 10 percent of its energy, if it starts at a radius of 10 meters.

(*UC, Berkeley*)

Solution:

(a) Let the particle's mass, charge, and kinetic energy (at time t) be m, q and T respectively. As the particle is non-relativistic, the radiation energy loss per revolution is very much smaller than the kinetic energy, so that we may consider the particle as moving along a circle of radius R at time t. Its rate of radiation is

$$P = \frac{q^2}{6\pi\varepsilon_0 c^3} \dot{v}^2 .$$

The equation of motion for the charge as it moves along a circular path in an axial uniform magnetic field $\mathbf{B}$ is

$$m|\dot{\mathbf{v}}| = \frac{mv^2}{R} = qvB .$$

The non-relativistic kinetic energy of the particle is $T = \frac{1}{2}mv^2$. Thus its rate of radiation is

$$P = \frac{q^2}{6\pi\varepsilon_0 c^3} \dot{v}^2 = \frac{q^4 B^2 T}{3\pi\varepsilon_0 m^3 c^3} .$$

The magnetic force does no work on the charge since $\mathbf{v} \times \mathbf{B} \cdot \mathbf{v} = 0$. P is therefore equal to the loss of kinetic energy per unit time:

$$P = -\frac{dT}{dt} = \frac{q^4 B^2 T}{3\pi\varepsilon_0 m^3 c^3},$$

which gives

$$T = T_0 e^{-\frac{q^4 B^2}{3\pi\varepsilon_0 m^3 c^3}},$$

where T_0 is the initial kinetic energy of the charge.

(b) For a proton, $q = 1.6 \times 10^{-19}$ C, $m = 1.67 \times 10^{-27}$ kg. The time it takes to lose 10 percent of its initial energy is

$$\tau = -\frac{3\pi\varepsilon_0 m^3 c^3}{q^4 B^2} \ln(0.9).$$

As $T = \frac{1}{2} mv^2 = \frac{1}{2m} \cdot R^2 q^2 B^2$, with $T_0 = 100$ MeV, $R = 10$ m, the magnetic field is given by

$$B^2 = \frac{2mT_0}{R^2 q^2} \approx 2.09 \times 10^{-2} \text{ Wb}^2/\text{m}^2.$$

Substituting it in the expression for τ, we find

$$\tau \approx 8.07 \times 10^{10} \text{ s}.$$

5036

A non-relativistic positron of charge e and velocity v_1 ($v_1 \ll c$) impinges head-on on a fixed nucleus of charge Ze. The positron, which is coming from far away, is decelerated unitil it comes to rest and then is accelerated again in the opposite direction until it reaches a terminal velocity v_2. Taking radiation loss into account (but assuming it is small) find v_2 as a function of v_1 and Z. What are the angular distribution and polarization of the radiation?

(Princeton)

Solution:

As the radiation loss of the positron is much smaller than its kinetic energy, it can be considered as a small perturbation. We therefore first neglect the effect of radiation. By the conservation of energy, when the distance between the positron and the fixed nucleus is r and its velocity is v we have

$$\frac{1}{2}mv^2 + \frac{1}{4\pi\varepsilon_0}\frac{Ze^2}{r} = \frac{1}{2}mv_1^2 .$$

When $v = 0$, r reaches its minimum r_0. Thus

$$\frac{1}{4\pi\varepsilon_0}\frac{Ze^2}{r_0} = \frac{1}{2}mv_1^2 ,$$

or

$$r_0 = \frac{Ze^2}{2\pi\varepsilon_0 mv_1^2} ,$$

whence

$$v^2 = v_1^2\left(1 - \frac{r_0}{r}\right) .$$

Differentiating the last equation we have

$$2\dot{r}\ddot{r} = \frac{v_1^2 r_0}{r^2}\dot{r} ,$$

or

$$\ddot{r} = \frac{v_1^2 r_0}{2r^2} .$$

The rate of radiation loss is given by

$$P = \frac{dW}{dt} = \frac{dW}{dr}\dot{r} = \frac{dW}{dr}v = \frac{2e^2}{3c^3}\ddot{r}^2 ,$$

so that

$$dW = \frac{e^2}{6c^3}\frac{v_1^3 r_0^2}{r^4}\frac{1}{\sqrt{1 - \frac{r_0}{r}}}\,dr .$$

Hence

$$\Delta W = 2\int_{r_0}^{\infty}dW = \frac{e^2 v_1^3 r_0^2}{3c^3}\int_{r_0}^{\infty}\frac{dr}{r^3\sqrt{r(r - r_0)}} .$$

By putting $r = r_0\sec^2\alpha$, we can carry out the integration and find

$$\Delta W = -\frac{8}{45}\cdot\frac{v_1^3}{Zc^3}mv_1^2 .$$

As $\frac{1}{2}mv_2^2 = \frac{1}{2}mv_1^2 - \Delta W$, we have

$$v_2^2 = v_1^2\left(1 - \frac{16}{45}\cdot\frac{v_1^3}{c^3}\cdot\frac{1}{Z}\right).$$

Hence

$$v_2 \approx v_1\left(1 - \frac{8}{45}\cdot\frac{v_1^3}{Zc^3}\right),$$

as $v_1 \ll c$.

Because $v \ll c$, the radiation is dipole in nature so that the angular distribution of its radiated power is given by

$$\frac{dP}{d\Omega} \propto \sin^2\theta,$$

θ being the angle between the directions of the radiation and the particle velocity. The radiation is plane polarized with the electric field vector in the plane containing the directions of the radiation and the acceleration (which is the same as that of the velocity in this case).

5037

A charged particle moves near the horizontal symmetry plane of a cyclotron in an almost circular orbit of radius R. Show that the small vertical motions are simple harmonic with frequency

$$\omega_v = \omega_c\left(-\frac{R}{B_z}\cdot\frac{\partial B_z}{\partial r}\right)^{1/2}, \quad \omega_c = \frac{qB_z}{m}.$$

(Wisconsin)

Solution:

As shown in Fig. 5.20, we choose a loop C for Ampère's circuital law $\oint_C \mathbf{B}\cdot d\mathbf{l} = \mu_0 I = 0$. Thus

$$B_z(r)z - B_z(r + dr)z + B_r(z)dr - B_r(z = 0)dr = 0.$$

As $B_r(z = 0) = 0$, we have

$$B_r(z) = \frac{\partial B_z(r)}{\partial r}\cdot z.$$

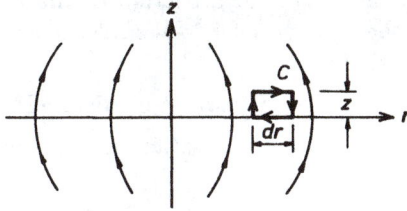

Fig. 5.20

The vertical motion of the particle is described by

$$m\ddot{z} = q(\mathbf{v} \times \mathbf{B})_z = q(v_r B_\theta - v_\theta B_r).$$

As $\mathbf{B}_\theta = 0$, $v_\theta = v$, this gives

$$m\ddot{z} = -qv \cdot \frac{\partial B_z(r)}{\partial r} \cdot z,$$

or

$$\ddot{z} = -\omega_v^2 z,$$

where

$$\omega_v^2 = \frac{qv}{m} \frac{\partial B_z}{\partial r} = \frac{v^2}{R B_z} \frac{\partial B_z}{\partial r}$$

as $\frac{mv^2}{R} = qvB_z$. The angular velocity of circular motion is $\omega_c = \frac{v}{R}$. Furthermore, $\frac{\partial B_z}{\partial r}$ is negative for $z \neq 0$ as shown in Fig. 5.20. Hence using $\frac{\partial B_z}{\partial r}$ to denote the absolute value we can write

$$\omega_v = \omega_c \left(-\frac{R}{B_z} \frac{\partial B_z}{\partial r} \right)^{\frac{1}{2}}.$$

5038

A high-current neutral plasma discharge is intended to focus a weak beam of antiprotons. The relativistic antiprotons are incident parallel to the axis of the discharge, travel a distance L through the arc, and leave the axis.

(a) Calculate the magnetic field distribution produced by a current I in the discharge, assumed to be a cylinder of uniform current density of radius R.

(b) Show that the magnetic deflection of the particles is such that the beam entering the field parallel to the axis can be focused to a point down-stream of the discharge.

(c) Which way must the arc current be directed?

(d) Using the thin lens approximation, find the focal length of such a lens.

(e) If the plasma were replaced by an electron beam with the same current, would the focal length be the same? Explain your answer.

(*UC, Berkeley*)

Solution:

(a) The magnetic field at a point distance $r \leq R$ from the axis of the current cylinder is given by Ampère's circuital law $\oint_c \mathbf{B} \cdot d\mathbf{l} = \mu_0 I$ to be

$$B = \frac{\mu_0}{2\pi r} \left(\frac{r^2}{R^2} \right) I = \frac{\mu_0 I r}{2\pi R^2} .$$

Note that the relative directions of I and $\mathbf{B}$ are given by the right-handed screw rule.

(b) (c) The antiproton carries charge $-e$. Its motion must be opposite in direction to the current for it to experience a force $-e\mathbf{v} \times \mathbf{B}$ pointing towards the axis of the discharge for focusing.

(d) From the above we see that an antiproton has velocity $\mathbf{v} = -v_z\, \mathbf{e}_z - v_r \mathbf{e}_r$. As $\mathbf{B} = B\mathbf{e}_\theta$, its equation of radial motion is

$$m \frac{dv_r}{dt} = -e(\mathbf{v} \times \mathbf{B})_r = -ev_z B \approx -evB = -\frac{\mu_0 evI}{2\pi R^2}\, r .$$

Note that

$$dt = \frac{dz}{v_z} \approx \frac{dz}{v} ,$$

and

$$v \approx \text{const.}$$

Furthermore, $m = \frac{m_0}{\sqrt{1 - v^2/c^2}}$ can also be taken to be approximately constant. Thus after traveling an arc of length L the radial velocity is

$$v_r = \int_0^L \frac{dv_r}{dt} \frac{dz}{v} = -\frac{\mu_0 eIrL}{2\pi m R^2}$$

toward the axis. In the thin lens approximation the focal length is then

$$h \approx vt = v \, \frac{r}{|v_r|} = \frac{2\pi m R^2 v}{\mu_0 e I L} \, .$$

(e) If the plasma were replaced by an electron beam of the same current, the antiprotons would experience an electric force whose direction deviates from the axis of the discharge. Under the assumption of uniform current distribution the electron number density n is constant. Applying Gauss' flux theorem to a unit length of the electron beam we find

$$2\pi r \varepsilon_0 E = -ne\pi r^2 \, ,$$

or

$$E = -\frac{ner}{2\varepsilon_0} \, .$$

As $I = -nev_e \pi R^2$, where v_e is the velocity of the electrons, the electric force on an antiproton is

$$f_e = -eE = \frac{ne^2 r}{2\varepsilon_0} = \frac{-eIr}{2\varepsilon_0 \pi R^2 v_e} \, ,$$

while the magnetic force on the antiproton is

$$f_m = -evB = -\frac{\mu_0 e v I r}{2\pi R^2} \, ,$$

where v is the velocity of the antiprotons. Hence

$$\frac{f_e}{f_m} = \frac{1}{\varepsilon_0 \mu_0 v v_e} = \frac{c^2}{v v_e} \gg 1 \, .$$

The magnetic force can therefore be neglected and the antiprotons, which come mainly under the action of electric repulsion, can no longer be focused.

5039

A beam of relativistic particles with charge $e > 0$ is passed successively through two regions, each of length l which contain uniform magnetic and electric fields **B** and **E** as shown in Fig. 5.21. The fields are adjusted so that the beam suffers fixed small deflections θ_B and θ_E ($\theta_B \ll 1, \theta_E \ll 1$) in the respective fields.

(a) Show that the momentum p of the particle can be determined in terms of B, θ_B, and l.

(b) Show that by using both the **B** and **E** fields, one can determine the velocity and mass of the particles in the beam.

(*Wisconsin*)

Solution:

The equation of the motion of a particle of charge e and rest mass m_0 in a magnetic field **B** is

$$\frac{d}{dt}(m\mathbf{v}) = e\mathbf{v} \times \mathbf{B}$$

where

$$m = \gamma m_0, \quad \gamma = \frac{1}{\sqrt{1 - v^2/c^2}}.$$

Differentiating $(\gamma v)^2 = c^2(\gamma^2 - 1)$, we have

$$2\gamma\mathbf{v} \cdot \left(\frac{d\gamma}{dt}\mathbf{v} + \gamma\frac{d\mathbf{v}}{dt}\right) = 2\gamma c^2 \frac{d\gamma}{dt}$$

or

$$\frac{d\gamma}{dt} = \frac{\gamma^3}{c^2}\mathbf{v} \cdot \frac{d\mathbf{v}}{dt}.$$

In the magnetic field $\dot{\mathbf{v}} \perp \mathbf{v}$, so $\frac{d\gamma}{dt} = 0$, i.e., $\gamma = \text{const}$. Using Cartesian coordinates such that $\mathbf{B} = B\mathbf{e}_z$, we can write the equation of motion as

$$\ddot{x} = \frac{eB}{\gamma m_0}\dot{y}, \quad \ddot{y} = -\frac{eB}{\gamma m_0}\dot{x}, \quad \ddot{z} = 0.$$

The z equation shows $\dot{z} = \text{const}$. As

$$\dot{z} = 0, \quad z = 0$$

initially, there is no z motion.
Putting $\omega_0 = \frac{eB}{\gamma m_0}$, we have

$$\begin{cases} \ddot{x} - \omega_0\dot{y} = 0, \\ \ddot{y} + \omega_0\dot{x} = 0, \end{cases}$$

and, by differentiation,

$$\begin{cases} \dddot{x} - \omega_0 \ddot{y} = 0, \\ \dddot{y} + \omega_0 \ddot{x} = 0. \end{cases}$$

Combining the above we obtain

$$\begin{cases} \dddot{x} + \omega_0^2 \dot{x} = 0, \\ \dddot{y} + \omega_0^2 \dot{y} = 0. \end{cases}$$

This set of equations shows that the particle executes circular motion with angular velocity ω_0 and radius

$$R = \frac{v}{\omega_0} = \frac{p}{m\omega_0}.$$

Note that $m = \gamma m_0$ is constant in the magnetic field. As shown in Fig. 5.21,

$$\theta_B \approx \frac{l}{R} = \frac{m\omega_0 l}{m\omega_0 R} = \frac{eBl}{p},$$

or $p = \frac{eBl}{\theta_B}$.

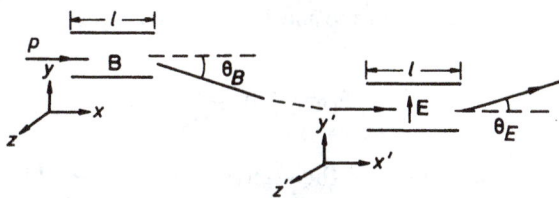

Fig. 5.21

(b) In the electric field, $\frac{d}{dt}(m\mathbf{v}) = e\mathbf{E}$. Taking Cartesian coordinates such that $\mathbf{E} = E\mathbf{e}'_y$, we have

$$mv'_y = eEt \approx eE\frac{l}{v},$$

i.e.,

$$v'_y \approx \frac{eEl}{mv}.$$

Then

$$\theta_E \approx \frac{v'_y}{v} = \frac{eEl}{pv},$$

from which v can be calculated as p can be determined from θ_B.

As $m = \gamma m_0 = \frac{p}{v} = \frac{p^2 \theta_E}{eEl}$, m_0 can also be calculated.

4. SCATTERING AND DISPERSION OF ELECTROMAGNETIC WAVES (5040–5056)

5040

Calculate the scattering cross section of a classical electron for high-frequency electromagnetic waves.

(*Columbia*)

Solution:

Let the fields of the high frequency electromagnetic waves be $\mathbf{E}_0(\mathbf{r}, t)$ and $\mathbf{B}_0(\mathbf{r}, t)$. For plane electromagnetic waves, $\sqrt{\varepsilon_0}\,|\mathbf{E}| = \sqrt{\mu_0}\,|\mathbf{H}|$, or $|\mathbf{B}| = \frac{1}{c}|\mathbf{E}|$, so that for a classical electron with $v \ll c$ the magnetic force $e\mathbf{v} \times \mathbf{B}_0$ can be neglected when compared with the electric force $e\mathbf{E}_0$. We let $\mathbf{E}_0(\mathbf{r}, t) = \mathbf{E}_0 e^{-i\omega t}$ at a fixed point $\mathbf{r}$. As the frequency of the incident waves is high, we must take into account the radiation damping (see Problem 5032). Then the equation of motion for the electron is

$$m\ddot{\mathbf{x}} = e\mathbf{E}_0 e^{-i\omega t} + \frac{e^2}{6\pi\varepsilon_0 c^3}\,\dddot{\mathbf{x}}\,,$$

where $\mathbf{x}$ is the displacement of the electron from the equilibrium position, the point $\mathbf{r}$ above. Consider small damping so that $\ddot{\mathbf{x}} = -\omega^2 \mathbf{x}$ and let $\gamma = \frac{e^2 \omega^2}{6\pi\varepsilon_0 mc^3}$. We then have

$$\ddot{\mathbf{x}} + \gamma\dot{\mathbf{x}} = \frac{e}{m}\,\mathbf{E}_0 e^{-i\omega t}\,.$$

Letting $\mathbf{x} = \mathbf{x}_0 e^{-i\omega t}$ we have

$$\mathbf{x}_0 = -\frac{e\mathbf{E}_0}{m\omega(\omega + i\gamma)}\,.$$

The radiation field of the electron at a point of radius vector $\mathbf{r}$ from it is

$$\mathbf{E}(\mathbf{x}, t) = \frac{e}{4\pi\varepsilon_0 c^2 r}\,\mathbf{n} \times (\mathbf{n} \times \ddot{\mathbf{x}})\,,$$

where

$$\mathbf{n} = \frac{\mathbf{r}}{r} \, .$$

Let α be the angle between $\mathbf{n}$ and $\mathbf{E}_0$. We then have

$$E(\mathbf{x}, t) = -\frac{e^2 \omega E_0 \sin \alpha}{4\pi \varepsilon_0 mc^2 (\omega + i\gamma) r} e^{-i\omega t} \, ,$$

whose amplitude is

$$E(\mathbf{x}) = \frac{e^2 \omega E_0 \sin \alpha}{4\pi \varepsilon_0 mc^2 (\omega^2 + \gamma^2)^{\frac{1}{2}} r} \, .$$

The intensity of the incident waves, averaged over one cycle, is

$$I_0 = \langle |\mathbf{E} \times \mathbf{H}| \rangle = \frac{1}{\mu_0} \langle |\mathbf{E} \times \mathbf{B}| \rangle = \frac{1}{c\mu_0} \langle E^2 \rangle = \frac{c\varepsilon_0}{2} \mathrm{Re}\,(E^* E) = \frac{\varepsilon_0 c}{2} E_0^2 \, .$$

Similarly, the intensity of the scattered waves in the direction α is

$$I = \frac{\varepsilon_0 c}{2} E^2 \, ,$$

or

$$I = \frac{\omega^2}{\omega^2 + \gamma^2} \frac{r_0^2}{r^2} I_0 \sin^2 \alpha \, ,$$

where $r_0 = \frac{e^2}{4\pi\varepsilon_0 mc^2}$ is the classical radius of electron.

Take coordinate axes as shown in Fig. 5.22 such that the origin O is at the equilibrium position of the electron, the z-axis is along the direction of the incident waves, and the x-axis is in the plane containing the z-axis and $\mathbf{r}$, the direction of the secondary waves. With the angles defined as shown, we have, since $\mathbf{E}_0$ is in the xy plane,

$$\cos \alpha = \sin \theta \, \cos \phi \, .$$

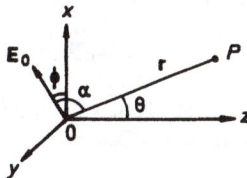

Fig. 5.22

If the incident waves are not polarized, ϕ is random and the secondary intensity $I(\theta)$ for a given scattering angle θ must be averaged over ϕ:

$$\langle I(\theta) \rangle = \frac{\omega^2}{\omega^2 + \gamma^2} \cdot \frac{r_0^2}{r^2} I_0 \frac{1}{2\pi} \int_0^{2\pi} (1 - \sin^2 \theta \, \cos^2 \phi) \, d\phi$$

$$= \frac{1}{2} \frac{\omega^2}{\omega^2 + \gamma^2} \frac{r_0^2}{r^2} (1 + \cos^2 \theta) I_0 \,.$$

The total radiated power is then

$$P = \int_0^\pi \langle I(\theta) \rangle \, 2\pi r^2 \sin \theta d\theta = \frac{8\pi}{3} \frac{\omega^2}{\omega^2 + \gamma^2} r_0^2 I_0 \,.$$

Hence the scattering cross section is

$$\sigma = \frac{P}{I_0} = \frac{8\pi}{3} \frac{\omega^2}{\omega^2 + \gamma^2} r_0^2 \,.$$

5041

A linearly polarized plane electromagnetic wave of frequency ω, intensity I_0 is scattered by a free electron. Starting with a general formula for the rate of radiation of an accelerated charge, derive the differential cross-section for scattering in the non-relativistic limit (Thompson scattering). Discuss the angular distribution and polarization of the scattered radiation.

(*UC, Berkeley*)

Solution:

Consider the forced oscillation of the electron by the incident wave. As $v \ll c$ the magnetic force could be ignored in comparison with the electric force, and we can think of the electron as in a uniform electric field since the incident wavelength is much greater than the amplitude of electron's motion. The electric intensity of the incident plane wave at the electron is $\mathbf{E} = \mathbf{E}_0 e^{-i\omega t}$ and the equation of motion is

$$m\dot{\mathbf{v}} = -e\mathbf{E} \,.$$

The rate of the radiation emitted by the electron at angle α with the direction of acceleration is, in Gaussian units,

$$\frac{dP}{d\Omega} = \frac{e^2 \dot{\mathbf{v}}^2}{4\pi c^3} \sin^2 \alpha \,.$$

As the average intensity of the incident wave is $I_0 = \langle |\mathbf{E} \times \mathbf{H}| \rangle = \frac{cE_0^2}{8\pi}$, we have

$$\frac{dP}{d\Omega} = I_0 r_e^2 \sin^2 \alpha \,,$$

where $r_e = \frac{e^2}{mc^2}$ is the classical radius of electron.

Let θ be the scattering angle and define ϕ as in Problem **5040**. We have

$$\sin^2 \alpha = 1 - \sin^2 \theta \cos^2 \phi \,.$$

So the differential cross section for scattering is

$$\frac{d\sigma}{d\Omega} = \frac{dP}{I_0 d\Omega} = r_e^2 \left(1 - \sin^2 \theta \cos^2 \varphi\right),$$

which shows that the angular distribution of the secondary radiation depends on both the scattering angle θ and the polarization angle ϕ. In the forward and backward directions of the primary radiation, the scattered radiation is maximum regardless of the polarization of the primary waves. In the transverse directions, $\theta = \frac{\pi}{2}$, the scattered radiation is minimum; it is zero for $\phi = 0, \pi$. For any other scattering angle θ, the scattered radiation intensity depends on ϕ, being maximum for $\phi = \frac{\pi}{2}, \frac{3\pi}{2}$ and minimum for $\phi = 0, \pi$.

The electric intensity of the secondary waves is

$$\mathbf{E} = -\frac{e}{c^2 r^3} \mathbf{r} \times (\mathbf{r} \times \dot{\mathbf{v}}),$$

where $\mathbf{r}$ is the radius vector of the field point from the location of the electron. This shows that $\mathbf{E}$ is in the plane containing $\mathbf{r}$ and $\dot{\mathbf{v}}$. As the incident waves are linearly polarized, $\dot{\mathbf{v}}$ has a fixed direction and the secondary radiation is linearly polarized also.

5042

A linearly polarized electromagnetic wave, wavelength λ, is scattered by a small dielectric cylinder of radius b, height h, and dielectric constant K ($b \ll h \ll \lambda$). The axis of the cylinder is normal to the incident wave vector and parallel to the electric field of the incident wave. Find the total scattering cross section.

(*UC, Berkeley*)

Solution:

As $b \ll h \ll \lambda$, the small dielectric cylinder can be considered as an elctric dipole of moment p for scattering of the electromagnetic wave. The electric field generated by p is much smaller than the electric field of the incident electromagnetic wave. Since the tangential component of the electric field intensity across the surface of the cylinder is continuous, the electric field inside the cylinder is equal to the electric field $\mathbf{E} = E_0 e^{i(\mathbf{k}\cdot\mathbf{r}-\omega t)}\mathbf{e}_z$ of the incident wave. Take the origin at the location of the dipole, then $\mathbf{r} = 0$ and the electric dipole moment of the small cylinder is

$$\mathbf{p} = \pi b^2 h \varepsilon_0 (K-1) E_0 e^{-i\omega t}\mathbf{e}_z \,,$$

the z-axis being taken along the axis of the cylinder.

The total power radiated by the oscillating electric dipole, averaged over one cycle, is

$$P = \frac{|\ddot{\mathbf{p}}|^2}{12\pi\varepsilon_0 c^3} = \frac{\pi b^4 h^2 \varepsilon_0 \omega^4 (K-1)^2 E_0^2}{12 c^3} \,.$$

The intensity of the incident wave is $I_0 = \frac{\varepsilon_0 c}{2} E_0^2$, so the total scattering cross section of the cylinder is

$$\sigma = \frac{P}{I_0} = \frac{\pi}{6} b^4 h^2 (K-1)^2 \frac{\omega^4}{c^4} \,.$$

5043

A plane electromagnetic wave of wavelength λ is incident on an insulating sphere which has dielectric constant ε and radius a. The sphere is small compared with the wavelength ($a \ll \lambda$). Compute the scattering cross section as a function of scattering angle. Comment on the polarization of the scattered wave as a function of the scattering direction.

(Princeton)

Solution:

Assume the incident electromagnetic wave to be linearly polarized and let its electric intensity be $\mathbf{E} = \mathbf{E}_0 e^{i(\mathbf{k}\cdot\mathbf{x}-\omega t)}$. In this field the insulating sphere is polarized so that it is equivalent to an electric dipole at the center of dipole moment (Problem 1064)

$$\mathbf{p} = 4\pi\varepsilon_0 a^3 \left(\frac{\varepsilon-\varepsilon_0}{\varepsilon+2\varepsilon_0}\right) \mathbf{E}_0 e^{-i\omega t'} \,.$$

Take coordinates with the origin at the center of the sphere and the z-axis parallel to $\mathbf{E}_0$ as shown in Fig. 5.23. Then

$$\ddot{\mathbf{p}} = -4\pi\varepsilon_0 a^3\omega^2 \left(\frac{\varepsilon - \varepsilon_0}{\varepsilon + 2\varepsilon_0}\right) E_0 e^{-i\omega t'}\, \mathbf{e}_z \,.$$

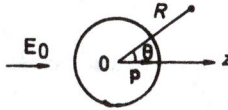

Fig. 5.23

The radiation field of the dipole under the condition $a \ll \lambda$ at a point of radius vector $\mathbf{R}$ is given by

$$\mathbf{B} = \frac{1}{4\pi\varepsilon_0 c^3 R}\ddot{\mathbf{p}} \times \mathbf{e}_R = \left(\frac{\varepsilon - \varepsilon_0}{\varepsilon + 2\varepsilon_0}\right)\frac{a^3\omega^2}{c^3 R}\, E_0\, \sin\theta e^{i(kR-\omega t)}\, \mathbf{e}_\varphi \,,$$

$$\mathbf{E} = c\mathbf{B} \times \mathbf{e}_R = \left(\frac{\varepsilon - \varepsilon_0}{\varepsilon + 2\varepsilon_0}\right)\frac{a^3\omega^2}{c^2 R}\, E_0 \sin\theta e^{i(kR-\omega t)}\, \mathbf{e}_\theta \,,$$

where t is given by the retardation condition $t' = t - \frac{R}{c}$, and $k = \frac{\omega}{c}$. The averaged Poynting vector (Problem 4049) is

$$\langle \mathbf{S} \rangle = \frac{1}{2\mu_0} R_e\left(\mathbf{E}^* \times \mathbf{B}\right) = \left(\frac{\varepsilon - \varepsilon_0}{\varepsilon + 2\varepsilon_0}\right)^2 \frac{a^6\omega^4}{2\mu_0 c^5}\frac{\sin^2\theta}{R^2}\, E_0^2 \mathbf{e}_R$$

$$= I_0 \left(\frac{\varepsilon - \varepsilon_0}{\varepsilon + 2\varepsilon_0}\right)^2 \frac{a^6\omega^4}{c^4}\frac{\sin^2\theta}{R^2}\, \mathbf{e}_R \,,$$

where $I_0 = \frac{1}{2}\varepsilon_0 c E_0^2$ is the (average) intensity of the incident wave. As the average power scattered into a solid angle $d\Omega$ in the radial direction at angle θ to the z-axis is $\langle \mathbf{S} \rangle R^2 d\Omega$, the differential scattering cross section is

$$\frac{d\sigma}{d\Omega} = \frac{\langle \mathbf{S} \rangle R^2}{I_0} = \left(\frac{\varepsilon - \varepsilon_0}{\varepsilon + 2\varepsilon_0}\right)^2 \frac{a^6\omega^4}{c^4}\sin^2\theta \,.$$

The scattered wave is polarized with the electric vectors in the plane containing the scattered direction and the direction of the electric vector of the primary wave at the dipole.

5044

A beam of plane polarized electromagnetic radiation of frequency ω, electric field amplitude E_0, and polarization x is normally incident on a region of space containing a low density plasma ($\rho = 0$, n_0 electrons/vol).

(a) Calculate the conductivity as a function of frequency.

(b) Using the Maxwell equations determine the index of refraction inside the plasma.

(c) Calculate and plot the magnitude of $\mathbf{E}$ as a function of position in the region of the edge of the plasma.

(*Wisconsin*)

Solution:

(a) As the plasma is of low density, the space is essentially free space with permittivity ε_0 and permeability μ_0. Maxwell's equations are then

$$\nabla \cdot \mathbf{E}' = \frac{\rho}{\varepsilon_0} = 0 \,,$$

$$\nabla \times \mathbf{E}' = -\frac{\partial \mathbf{B}'}{\partial t} \,,$$

$$\nabla \cdot \mathbf{B}' = 0 \,,$$

$$\nabla \times \mathbf{B}' = \mu_0 \mathbf{j} + \frac{1}{c^2} \frac{\partial \mathbf{E}'}{\partial t} \,.$$

We also have Ohm's law

$$\mathbf{j} = -n_0 e \mathbf{v} = \sigma \mathbf{E}' \,,$$

where $\mathbf{v}$ is the average velocity of the electrons inside the plasma. For $v \ll c$, the magnetic force on an electron is much smaller than the electric force and can be neglected. The equation of motion for an electron is therefore

$$\frac{d\mathbf{v}}{dt} = -\frac{e}{m} \mathbf{E}' \,.$$

As the traversing radiation has electric intensity $\mathbf{E}' = \mathbf{E}'_0(\mathbf{x}) e^{-i\omega t}$, the displacement of the electron from the equilibrium position is $\mathbf{r} = \mathbf{r}_0 e^{-i\omega t}$ in the steady state. The equation of motion then gives

$$\mathbf{r} = \frac{e}{m\omega^2} \mathbf{E}'$$

and

$$\mathbf{v} = \dot{\mathbf{r}} = -i \frac{e}{\omega m} \mathbf{E}' \,,$$

Hence

$$\mathbf{j} = i\frac{n_0 e^2}{m\omega}\mathbf{E}'$$

and the conductivity is

$$\sigma = i\frac{n_0 e^2}{m\omega}.$$

(b) The polarization vector of the plasma is by definition

$$\mathbf{P} = -n_0 e\mathbf{r} = \frac{-n_0 e^2}{m\omega^2}\mathbf{E}',$$

so that the electric displacement is

$$\mathbf{D} = \varepsilon\mathbf{E}' = \varepsilon_0\mathbf{E}' + \mathbf{P}.$$

Hence the effective dielectric constant of the plasma is given by

$$\varepsilon = \varepsilon_0 + \frac{P}{E'} = \varepsilon_0 - \frac{n_0 e^2}{m\omega^2},$$

or

$$\frac{\varepsilon}{\varepsilon_0} = 1 - \left(\frac{\omega_p}{\omega}\right)^2,$$

where

$$\omega_p = \sqrt{\frac{n_0 e^2}{\varepsilon_0 m}}$$

is called the plasma (angular) frequency.

The index of refraction of the plasma is therefore

$$n = \sqrt{\frac{\mu_0\varepsilon}{\mu_0\varepsilon_0}} = \sqrt{\frac{\varepsilon}{\varepsilon_0}} = \sqrt{1 - \left(\frac{\omega_p}{\omega}\right)^2}.$$

(c) Consider the primary wave $\mathbf{E}_0 = E_0 e^{i(kz-\omega t)}\mathbf{e}_x$, where $k = \frac{\omega}{c}$, as incident normally on the boundary of the plasma, then the wave inside the plasma is also a plane polarized wave, with amplitude $E_0' = \frac{2E_0}{1+n}$ and wave number $k' = \frac{\omega}{c}n = kn$ (see Problem 4011). Hence the electric intensity of the wave in the region of the edge of the plasma is

$$\mathbf{E}' = \frac{2E_0}{1+n}e^{i(knz-\omega t)}\mathbf{e}_z.$$

Note that for $\omega > \omega_p$, n and kn are real and $\mathbf{E'}$ propagates as wave, but for $\omega < \omega_p$, n and kn are imaginary and $\mathbf{E'}$ attenuates exponentially as shown in Fig. 5.24.

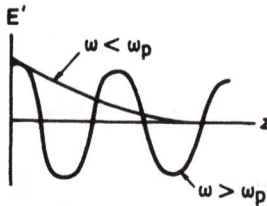

Fig. 5.24

5045

A "tenuous plasma" consists of free electric charges of mass m and charge e. There are n charges per unit volume. Assume that the density is uniform and that interactions between the charges may be neglected. Electromagnetic plane waves (frequency ω, wave number k) are incident on the plasma.

(a) Find the conductivity σ as a function of ω.

(b) Find the dispersion relation, i.e., find the relation between k and ω.

(c) Find the index of refraction as a function of ω. The plasma frequency is defined by

$$\omega_p^2 = \frac{4\pi n e^2}{m},$$

if e is expressed in e.s.u. units. What happens if $\omega < \omega_p$?

(d) Now suppose there is an external magnetic field $\mathbf{B_0}$. Consider plane waves traveling parallel to $\mathbf{B_0}$. Show that the index of refraction is different for right and left circularly polarized waves. (Assume that $\mathbf{B}$ of the traveling wave is negligible compared with $\mathbf{B_0}$.)

(*Princeton*)

Solution:

Gaussian units are to be used for this problem.

The electric vector of the incident wave at a charge is $\mathbf{E_0}e^{-i\omega t}$, while the effect of the magnetic vector can be ignored in the non-relativistic case. Thus the equation of motion for a charge e in the plasma is

$$m\ddot{\mathbf{x}} = e\mathbf{E_0}e^{-i\omega t}.$$

In the steady state $\mathbf{x} = \mathbf{x}_0 e^{-i\omega t}$. Substitution gives

$$\mathbf{x}_0 = -\frac{e\mathbf{E}_0}{m\omega^2},$$

or

$$\mathbf{x} = -\frac{e\mathbf{E}}{m\omega^2}.$$

(a) The motion of the charges gives rise to a current density

$$\mathbf{j} = ne\dot{\mathbf{x}} = i\frac{ne^2}{m\omega}\mathbf{E},$$

so that the conductivity is

$$\sigma = \frac{j}{E} = i\frac{ne^2}{m\omega}.$$

(b) The polarization of the plasma is

$$\mathbf{P} = ne\mathbf{x} = -\frac{ne^2}{m\omega^2}\mathbf{E} = \chi_e\mathbf{E},$$

where χ_e is the polarizability of the plasma. The dielectric constant is by definition

$$\varepsilon = 1 + 4\pi\chi_e = 1 - \frac{4\pi ne^2}{m\omega^2} = 1 - \frac{\omega_p^2}{\omega^2},$$

where

$$\omega_p = \sqrt{\frac{4\pi ne^2}{m}}$$

is the plasma frequency. Then the refractive index of the plasma is

$$n = \sqrt{\varepsilon} = \sqrt{1 - \frac{\omega_p^2}{\omega^2}},$$

as we may assume $\mu = \mu_0 = 1$ for the plasma. The wave number in the plasma is therefore given by

$$k^2 = \frac{\omega^2 n^2}{c^2} = \frac{1}{c^2}(\omega^2 - \omega_p^2),$$

which is the dispersion relation.

(c) The index of refraction is

$$n = \sqrt{\varepsilon\mu} = \sqrt{\varepsilon} = \left(1 - \frac{\omega_p^2}{\omega^2}\right)^{\frac{1}{2}}.$$

If $\omega < \omega_p$, n is an imaginary number, and so is k. Take the z-axis along the direction of propagation. Writing $k = i\kappa$, where κ is real, we see that $e^{ikz} = e^{-\kappa z}$, so the wave will attenuate exponentially and there is no propagation, the plasma serving only to reflect the incident wave.

(d) As $\mathbf{B} = B_0\mathbf{e}_z$, $\mathbf{k} = k\mathbf{e}_z$, the equation of motion for a charge e in the plasma is

$$m\ddot{\mathbf{x}} = e\mathbf{E} + \frac{e}{c}\mathbf{v} \times \mathbf{B}_0.$$

Since the plane wave is transverse, we have $\mathbf{E} = E_x\mathbf{e}_x + E_y\mathbf{e}_y$. In the steady state, the charge will oscillate with the same frequency ω and the solution will have the form $\mathbf{x} = \mathbf{x}_0 e^{-i\omega t}$. Thus $\mathbf{v} = \dot{\mathbf{x}} = (-i\omega)\mathbf{x}$, and the component equations are

$$m\ddot{x} = eE_x + \frac{e}{c}\dot{y}B_0, \tag{1}$$

$$m\ddot{y} = eE_y - \frac{e}{c}\dot{x}B_0, \tag{2}$$

$$m\ddot{z} = 0.$$

Suppose z and $\dot{z}$ are both zero initially. Then $z = 0$ and $\mathbf{x} = x\mathbf{e}_x + y\mathbf{e}_y$.

For the right circularly polarized wave (looking against the direction of propagation $\mathbf{E}$ rotates to the right, i.e., clockwise) the electric vector is

$$\mathbf{E}_R = \mathrm{Re}\left\{E_0(\mathbf{e}_x + e^{-i\frac{\pi}{2}}\mathbf{e}_y)e^{-i\omega t}\right\}$$
$$= E_0\cos(\omega t)\,\mathbf{e}_x - E_0\sin(\omega t)\,\mathbf{e}_y,$$

and Eqs. (1) and (2) reduce to

$$m\ddot{x} = eE_0\cos\omega t + \frac{e}{c}B_0\dot{y},$$

$$m\ddot{y} = -eE_0\sin\omega t - \frac{e}{c}B_0\dot{x}.$$

Let $u = x - iy$, $\omega_c = \frac{eB_0}{mc}$, the above equations can be combined into

$$\ddot{u} - i\omega_c\dot{u} = \frac{eE_0}{m}(\cos\omega t + i\sin\omega t) = \frac{eE_0}{m}e^{i\omega t}.$$

The steady state solution is

$$u = u_0 e^{i\omega t} .$$

Substitution gives

$$u_0 = -\frac{eE_0}{m\omega(\omega - \omega_c)} .$$

hence

$$u = -\frac{eE_0 (\cos \omega t + i \sin \omega t)}{m\omega(\omega - \omega_c)} ,$$

whose real and imaginary parts respectively give

$$x = -\frac{eE_0 \cos \omega t}{m\omega(\omega - \omega_c)} , \quad y = \frac{eE_0 \sin \omega t}{m\omega(\omega - \omega_c)}$$

or, in vector form,

$$\mathbf{x} = -\frac{e\mathbf{E_R}}{m\omega(\omega - \omega_c)} .$$

Thus for the right circularly polarized wave, the polarizability of the plasma is

$$\chi_{eR} = -\frac{ne^2}{m\omega(\omega - \omega_c)} ,$$

and the corresponding index of refraction is

$$n_R = \sqrt{\varepsilon_R} = \sqrt{1 + 4\pi\chi_{eR}}$$
$$= \left(1 - \frac{4\pi ne^2}{m\omega(\omega - \omega_c)}\right)^{\frac{1}{2}} ,$$

or, in terms of the plasma frequency ω_p,

$$n_R = \left[1 - \frac{\omega_p^2}{\omega(\omega - \omega_c)}\right]^{\frac{1}{2}} .$$

For the left circularly polarized wave, the electric vector is

$$\mathbf{E_L} = \mathrm{Re}\left\{E_0(\mathbf{e}_x + e^{i\frac{\pi}{2}}\mathbf{e}_y)e^{-i\omega t}\right\}$$
$$= E_0 \cos(\omega t)\mathbf{e}_x + E_0 \sin(\omega t)\mathbf{e}_y .$$

Then, putting $u = x + iy$ and repeating the above procedure, we obtain the index of refraction

$$n_L = \left[1 - \frac{\omega_p^2}{\omega(\omega + \omega_c)}\right]^{\frac{1}{2}}.$$

It is obvious that $n_L \neq n_R$, unless $\omega_c = 0$, i.e., $B_0 = 0$.

5046

The dispersion relation for electromagnetic waves in a plasma is given by

$$\omega^2(k) = \omega_p^2 + c^2 k^2,$$

where the plasma frequency ω_p is defined as

$$\omega_p^2 = \frac{4\pi N e^2}{m}$$

for an electron density N, charge per electron e, and mass per electron m.

(a) For $\omega > \omega_p$, find the index of refraction n of the plasma.

(b) For $\omega > \omega_p$, is n greater than or less than 1? Discuss.

(c) For $\omega > \omega_p$, calculate the velocity at which messages can be transmitted through the plasma.

(d) For $\omega < \omega_p$, describe quantitatively the behavior of an electromagnetic wave in the plasma.

(*UC, Berkeley*)

Solution:

(a) $n = (1 - \omega_p^2/\omega^2)^{\frac{1}{2}}$.

(b) For $\omega > \omega_p$, $n < 1$.

(c) For $\omega > \omega_p$, the phase velocity in the plasma is

$$v_p = \frac{c}{n} > c.$$

However, messages or signals are transmitted with the group velocity

$$v_g = \frac{d\omega}{dk} = \frac{c^2 k}{\omega} = c\left(1 - \omega_p^2/\omega^2\right)^{\frac{1}{2}}.$$

As $\omega > \omega_p$, it is clear that $v_g < c$.

(d) For $\omega < \omega_p$, n and k are imaginary and the electromagnetic waves attenuate exponentially after entering into the plasma. Hence electromagnetic waves of frequencies $\omega < \omega_p$ cannot propagate in the plasma.

5047

Discuss the propagation of electromagnetic waves of frequency ω through a region filled with free electric charges (mass m and charge e) of density N per cm^3.

(a) In particular, find an expression for the index of refraction and show that under certain conditions it may be complex.

(b) Discuss the reflection and transmission of waves at normal incidence under conditions when the index of refraction is real, and when it is complex.

(c) Show that there is a critical frequency (the plasma frequency) dividing the real and complex regions of behavior.

(d) Verify that the critical frequency lies in the radio range ($N = 10^6$) for the ionosphere and in the ultraviolet for metallic sodium ($N = 2.5 \times 10^{22}$).

(UC, Berkeley)

Solution:

(a) See Problem **5044**.

(b) For normal incidence, if the index of refraction is real, both reflection and transmission will take place. Let the amplitude of the incident wave be E_0, then the amplitude of the reflected wave and the reflectivity are respectively (Problem **4011**)

$$E_0' = \frac{1-n}{1+n} E_0, \quad R = \left(\frac{1-n}{1+n}\right)^2,$$

and the amplitude of the transmitted wave and the transmittivity are respectively

$$E_0'' = \frac{2}{1+n} E_0, \quad T = \frac{4n}{(1+n)^2}.$$

If n is complex, the transmitted wave will attenuate exponentially so that effectively only reflection occurs (see Problem **5044**).

(c) The index of refraction is $n = (1 - \frac{\omega_p^2}{\omega^2})^{\frac{1}{2}}$, where $\omega_p^2 = \frac{Ne^2}{m\varepsilon_0}$ in SI units and $\omega_p^2 = \frac{4\pi Ne^2}{m}$ in Gaussian units. n is real for $\omega > \omega_p$ and imaginary for $\omega < \omega_p$. Thus ω_p can be considered a critical frequency.

(d) For the electron

$$e = 1.6 \times 10^{-19} \text{ C}, \quad m = 9.1 \times 10^{-31} \text{ kg}.$$

With $N = 10^6/\text{cm}^3$ for the ionosphere and $\varepsilon_0 = 8.85 \times 10^{-12}$ F/m, they give

$$\omega_p = \sqrt{\frac{Ne^2}{m\varepsilon_0}} = \left(\frac{10^6 \times 10^6 \times (1.6 \times 10^{-19})^2}{9.1 \times 10^{-31} \times 8.85 \times 10^{-12}}\right)^{\frac{1}{2}} = 5.64 \times 10^7 \text{ s}^{-1},$$

within the range of radio frequencies.
For metallic sodium, $N = 2.5 \times 10^{22}/\text{cm}^3$, so that

$$\omega_p = \left(\frac{2.5 \times 10^{22} \times 10^6 \times (1.6 \times 10^{-19})^2}{9.1 \times 10^{-31} \times 8.85 \times 10^{-12}}\right)^{\frac{1}{2}}$$
$$= 8.91 \times 10^{15} \text{ s}^{-1},$$

in the ultraviolet range.

5048

Assume that the ionosphere consists of a uniform plasma of free electrons and neglect collisions.

(a) Derive an expression for the index of refraction for electromagnetic waves propagating in this medium in terms of the frequency.

(b) Now suppose that there is an external uniform static magnetic field due to the earth, parallel to the direction of propagation of the electromagnetic waves. In this case, left and right circularly polarized waves will have different indices of refraction; derive the expressions for both of them.

(c) There is a certain frequency below which the electromagnetic wave incident on the plasma is completely reflected. Calculate this frequency for both left and right polarized waves, given that the density of electrons is 10^5 cm^{-3} and $B = 0.3$ gauss.

(UC, Berkeley)

Solution:

(a), (b) See Problem **5045**.

(c) The refractive index n of the plasma is given by

$$n^2 = 1 - \frac{\omega_p^2}{\omega^2} \,,$$

where $\omega_p = \sqrt{\frac{4\pi N e^2}{m}}$ is the plasma frequency, N being the density of free electrons. When $n^2 < 0$, n is imaginary and electromagnetic waves of (angular) frequency ω cannot propagate in the plasma. Hence the cutoff frequency is that for which $n = 0$, i.e., ω_p.

For frequencies $< \omega_p$, the wave will be totally reflected by the plasma. For the right and left circularly polarized waves, the refractive indices n_R and n_L are given by (Problem **5045**)

$$n_{\mp}^2 = 1 - \frac{\omega_p^2}{\omega(\omega \mp \omega_c)} \,,$$

where $n_- = n_R$, $\omega_c = \frac{eB}{mc}$. The cutoff frequencies are given by $n_{\mp}^2 = 0$. Thus the cutoff frequencies for right and left circularly polarized waves are respectively

$$\omega_{Rc} = \frac{\omega_c + \sqrt{\omega_c^2 + 4\omega_p^2}}{2} \,, \qquad \omega_{Lc} = \frac{-\omega_c + \sqrt{\omega_c^2 + 4\omega_p^2}}{2} \,.$$

With $N = 10^5/\text{cm}^3$, $B = 0.3$ Gs, and $m = 9.1 \times 10^{-28}$ g, $e = 4.8 \times 10^{-10}$ e.s.u. for the electron, we have

$$\omega_p^2 = \frac{4\pi \times 10^5 \times (4.8 \times 10^{-10})}{9.1 \times 10^{-28}} = 3.18 \times 10^{14} \text{ s}^{-2} \,,$$

$$\omega_c = \frac{4.8 \times 10^{-10} \times 0.3}{3 \times 10^{10} \times 9.1 \times 10^{-28}} = 5.27 \times 10^6 \text{ s}^{-1} \,,$$

and hence

$$\omega_{Rc} = 2.1 \times 10^7 \text{ s}^{-1} \,, \qquad \omega_{Lc} = 1.5 \times 10^7 \text{ s}^{-1} \,.$$

5049

Derive an expression for the penetration depth of a very low frequency electromagnetic wave into a plasma in which electrons are free to move.

Express your answer in terms of the electron density n_0, electron charge e and mass m. What does "very low" mean in this context? What is the depth in cm for $n_0 = 10^{14}$ cm^{-3}?

(UC, Berkeley)

Solution:

The dispersion relation for a plasma is (Problem **5045**)

$$k^2 = \frac{\omega^2}{c^2}\left(1 - \frac{\omega_p^2}{\omega^2}\right),$$

where $\omega_p^2 = \frac{n_0 e^2}{m\epsilon_0}$ is the plasma frequency. A "very low" frequency means that the frequency satisfies $\omega \ll \omega_p$. For such frequencies k is imaginary. Let $k = i\kappa$, where $\kappa = \frac{1}{c}\sqrt{\omega_p^2 - \omega_c^2}$, then $e^{ikz} = e^{-\kappa z}$, the z-axis being taken along the direction of propagation. This means the wave amplitude attenuates exponentially in the plasma. The penetration depth δ is defined as the depth from the plasma surface where the amplitude is e^{-1} of its surface value, i.e.,

$$\kappa\delta = 1,$$

or

$$\delta = \frac{1}{\kappa} = \frac{c}{\sqrt{\omega_p^2 - \omega^2}} \approx \frac{1}{\omega_p}.$$

With $n_0 = 10^{14}$ cm^{-3}, we have

$$\omega_p = \left(\frac{10^{14} \times 10^6 \times (1.6 \times 10^{-19})^2}{9.1 \times 10^{-31} \times 8.85 \times 10^{-12}}\right)^{\frac{1}{2}} \approx 5.64 \times 10^{11}\ \text{s}^{-1},$$

and

$$\delta = \frac{3 \times 10^{10}}{5.64 \times 10^{11}} = 0.053\ \text{cm}.$$

5050

In the presence of a uniform static magnetic field **H**, a medium may become magnetized. The magnetization may be coupled self-consistently to an electromagnetic field set up in the medium.

(a) Write down an equation of motion governing the time variation of the magnetization under the influence of a (generally time-dependent) magnetic field.

(b) An electromagnetic field is, of course, in turn generated by the time-dependent magnetization as described by the appropriate Maxwell's equations. Assume the dielectric constant $\varepsilon = 1$ for the magnetic medium. Find the dispersion relation $\omega = \omega(k)$ for the propagation of a plane wave of magnetization in the medium.

(SUNY, Buffalo)

Solution:

(a) Assuming the medium to be linear, $\mathbf{M} = \chi_m \mathbf{H}$, we have, by the definition of permeability μ,

$$\mathbf{M}(t) = \chi_m(\omega)\mathbf{H}(t) = \frac{\mu(\omega) - 1}{4\pi}\mathbf{H}(t).$$

(b) Maxwell's equations for the medium are

$$\nabla \cdot \mathbf{D} = 4\pi\rho, \qquad \nabla \cdot \mathbf{B} = 0,$$

$$\nabla \times \mathbf{E} = -\frac{1}{c}\frac{\partial \mathbf{B}}{\partial t}, \qquad \nabla \times \mathbf{H} = \frac{4\pi}{c}\mathbf{j} + \frac{1}{c}\frac{\partial \mathbf{D}}{\partial t}.$$

Note that $\mathbf{B}$ in the equations is the superposition of the external field and the time-dependent field produced by the magnetization $\mathbf{M}(t)$.

Consider the medium as isolated and uncharged, then $\rho = \mathbf{j} = 0$. Also $\mathbf{D} = \mathbf{E}$ as $\varepsilon = 1$. Maxwell's equations now reduce to

$$\nabla \cdot \mathbf{E} = 0, \qquad \nabla \cdot \mathbf{B} = 0,$$

$$\nabla \times \mathbf{E} = -\frac{1}{c}\frac{\partial \mathbf{B}}{\partial t}, \qquad \nabla \times \mathbf{H} = \frac{1}{c}\frac{\partial \mathbf{E}}{\partial t},$$

with

$$\mathbf{B} = \mu\mathbf{H}.$$

Deduce from these equations the wave equation

$$\nabla^2\mathbf{H} - \frac{\mu}{c^2}\frac{\partial^2\mathbf{H}}{\partial t^2} = 0.$$

Consider a plane wave solution

$$\mathbf{H}(t) = \mathbf{H}_0 e^{i(\mathbf{k}\cdot\mathbf{r} - \omega t)}.$$

Substitution in the wave equation gives the dispersion relation

$$k^2 - \mu\frac{\omega^2}{c^2} = 0,$$

or

$$k = \frac{\omega}{v} = \frac{\omega}{c} \sqrt{\mu(\omega)}.$$

The magnetization $\mathbf{M}$ can then be represented by a plane wave

$$\mathbf{M}(t) = \chi_m(\omega)\mathbf{H}(t) = \chi_m \mathbf{H}_0 \, e^{i(\mathbf{k}\cdot\mathbf{r}-\omega t)}.$$

As $\mathbf{M}$ satisfies the same wave equation the dispersion relation above remains valid.

5051

In a classical theory of the dispersion of light in a transparent dielectric medium one can assume that the light wave interacts with atomic electrons which are bound in harmonic oscillator potentials. In the simplest case, the medium contains N electrons per unit volume with the same resonance frequency ω_0.

(a) Calculate the response of one such electron to a linearly polarized electromagnetic plane wave of electric field amplitude E_0 and frequency ω.

(b) For the medium, give expressions for the atomic polarizability, the dielectric susceptibility and the refractive index as functions of the light frequency. What happens near resonance? What happens above resonance? The phase velocity of the light wave exceeds the vacuum velocity of light if the refractive index becomes smaller than 1. Does this violate the principles of the special theory of relativity?

(*SUNY, Buffalo*)

Solution:

(a) The equation of motion for the electron is

$$m\ddot{\mathbf{x}} = -m\omega_0^2 \mathbf{x} - e\mathbf{E}_0 e^{-i\omega t}.$$

Consider the steady state solution $\mathbf{x} = \mathbf{x}_0 e^{-i\omega t}$. Substituting in the equation gives

$$\mathbf{x} = \frac{e\mathbf{E}_0}{m(\omega^2 - \omega_0^2)} e^{-i\omega t}.$$

(b) Assume that each atom contributes only one oscillating electron. The electric dipole moment of the atom in the field of the light wave is

$$\mathbf{p} = -e\mathbf{x} = -\frac{e^2\mathbf{E}}{m(\omega^2 - \omega_0^2)},$$

giving the atomic polarizability

$$\alpha = \frac{p}{E} = \frac{e^2}{m(\omega_0^2 - \omega^2)} .$$

As there are N electrons per unit volume, the polarization of the medium is

$$\mathbf{P} = n\mathbf{p} = \frac{Ne^2\mathbf{E}}{m(\omega_0^2 - \omega^2)} .$$

The electric displacement is by definition

$$\mathbf{D} = \varepsilon\mathbf{E} = \varepsilon_0\mathbf{E} + \mathbf{P} .$$

Hence the dielectric constant is

$$\varepsilon = \varepsilon_0 + \frac{Ne^2}{m(\omega_0^2 - \omega^2)} .$$

Assume the medium to be non-ferromagnetic, then $\mu = \mu_0$, and the refractive index is given by

$$n = \sqrt{\frac{\mu\varepsilon}{\mu_0\varepsilon_0}} = \sqrt{\frac{\varepsilon}{\varepsilon_0}} .$$

Putting

$$\omega_p^2 = \frac{Ne^2}{m\varepsilon_0} ,$$

we have

$$n = \left(1 + \frac{\omega_p^2}{\omega_0^2 - \omega^2}\right)^{\frac{1}{2}} . \tag{1}$$

For $\omega < \omega_0$, we have $\varepsilon > \varepsilon_0$, $n > 1$, and the phase velocity of the wave in the medium is $v_p = \frac{c}{n} < c$.

For $\omega_0 < \omega < \sqrt{\omega_p^2 + \omega_0^2}$, n is real but smaller than unity. This means that $v_p > c$ and the wave propagates with a velocity greater than the velocity of light in free space. However, the energy or signal carried by the wave travels with the group velocity v_g given by

$$\frac{1}{v_g} = \frac{dk}{d\omega} = \frac{n}{c} + \frac{\omega}{c}\frac{dn}{d\omega}$$

as $k = \frac{\omega n}{c}$. Equation (1) gives

$$\frac{1}{v_g} = \frac{1}{cn}\left(\frac{\omega^2 - n^2\omega_0^2}{\omega^2 - \omega_0^2}\right) > \frac{1}{cn} = \frac{v_p}{c^2}$$

as $n > 1$. Then as $v_p > c$, $v_g < c$. Hence there is no violation of the principles of special relativity. For both the above cases, n increases with increasing ω and the dispersion is said to be normal.

For $\omega \approx \omega_0$, Eq. (1) does not hold but damping (collision and radiation) must be taken into account. Equation (1) is modified to

$$n \approx \left[1 + \frac{\omega_p^2(\omega_0^2 - \omega^2)}{(\omega_0^2 - \omega^2)^2 + \omega^2\gamma^2}\right]^{\frac{1}{2}}$$

$$\approx 1 + \frac{\omega_p^2(\omega_0 - \omega)}{\gamma^2\omega_0}.$$

Thus $n \approx 1$ for $\omega = \omega_0$. As ω increases from a value smaller than ω_0 to one larger than ω_0, n decreases from a value greater than unity to one smaller than unity. In this region n decreases with increasing ω and the dispersion is said to be anomalous.

For $\omega > \sqrt{\omega_p^2 + \omega_0^2}$, n is imaginary so that $k = \frac{\omega}{c}n$ is also imaginary. Let it be $i\kappa$. Then the wave amplitude at a point distance r from the surface of the medium simply attenuates according to $e^{-\kappa r}$ and propagation is not possible in the medium.

Near resonance $\omega \approx \omega_0$, the absorption coefficient becomes very large. Thus the medium is essentially opaque to the wave at $\omega \approx \omega_0$ and for $\omega > \sqrt{\omega_p^2 + \omega_0^2}$.

5052

Consider a model of an isotropic medium composed of N harmonically bound particles of charge e, mass m and natural frequency ω_0, per unit volume.

(a) Show that, for a zero magnetic field, the dielectric function of the medium is given by

$$\varepsilon(\omega) = 1 + \frac{4\pi Ne^2/m}{\omega_0^2 - \omega^2}.$$

(b) (The Faraday effect) Now a static magnetic field $\mathbf{B}$ in the direction of propagation of the electromagnetic wave is added. Show that the left and right circularly polarized electromagnetic waves have different dielectric functions, with the difference equals to

$$\delta\varepsilon(\omega) = \frac{4\pi N e^2}{m} \frac{2eB\omega/mc}{(\omega_0^2 - \omega^2)^2 - (eB\omega/mc)^2} .$$

(*Chicago*)

Solution:

(a) See Problem **5051**.

(b) Take the z-axis along the direction of propagation, then $\mathbf{k} = k\mathbf{e}_z$. When a static magnetic field $\mathbf{B} = B\mathbf{e}_z$ is added, the equation of motion for a harmonically bounded particle is

$$m\ddot{\mathbf{x}} = -m\omega_0^2\mathbf{x} + e\mathbf{E} + \frac{e}{c}\dot{\mathbf{x}} \times \mathbf{B} .$$

As plane electromagnetic waves are transverse, $\mathbf{E}$ has only x and y components. The component equations are

$$m\ddot{x} = -m\omega_0^2 x + eE_x + \frac{e}{c}B\dot{y} ,$$

$$m\ddot{y} = -m\omega_0^2 y + eE_y - \frac{e}{c}B\dot{x} ,$$

$$m\ddot{z} = -m\omega_0^2 z .$$

The last equation shows that motion along the z direction is harmonic but not affected by the applied fields and can thus be neglected. For the right circularly polarized wave (Problem **5045**)

$$\mathbf{E}_R = E_0 \cos(\omega t)\mathbf{e}_x - E_0 \sin(\omega t)\mathbf{e}_y ,$$

so the remaining equations of motion are

$$m\ddot{x} = -m\omega_0^2 x + eE_0 \cos\omega t + \frac{e}{c} B\dot{y} , \tag{1}$$

$$m\ddot{y} = -m\omega_0^2 y - eE_0 \sin\omega t - \frac{e}{c} B\dot{x} . \tag{2}$$

Putting

$$u = x + iy , \quad \omega_c = \frac{eB}{mc} ,$$

(1) $- i \times$ (2) gives

$$\ddot{u} - i\omega_c \dot{u} + \omega_0^2 u = \frac{eE_0}{m}(\cos \omega t + i \sin \omega t) = \frac{eE_0}{m} e^{i\omega t}.$$

In the steady state, $u \sim e^{i\omega t}$. Substitution in the above gives

$$u = \frac{eE_0(\cos \omega t + i \sin \omega t)}{m(\omega_0^2 - \omega^2 + \omega\omega_c)}.$$

Separating the real and imaginary parts we have

$$x = \frac{eE_0 \cos \omega t}{m(\omega_0^2 - \omega^2 + \omega\omega_c)}, \qquad y = -\frac{eE_0 \sin \omega t}{m(\omega_0^2 - \omega^2 + \omega\omega_c)}.$$

Combining the above in vector form gives

$$\mathbf{x} = \frac{e\mathbf{E_R}}{m(\omega_0^2 - \omega^2 + \omega\omega_c)}.$$

Hence the polarization of the medium due to the right circularly polarized wave is

$$\mathbf{P} = Ne\mathbf{x} = \frac{Ne^2\mathbf{E_R}}{m(\omega_0^2 - \omega^2 + \omega\omega_c)}.$$

As $\varepsilon = 1 + 4\pi \frac{P}{E}$, the above gives

$$\varepsilon_R = 1 + \frac{4\pi Ne^2}{m(\omega_0^2 - \omega^2 + \omega\omega_c)}.$$

Similarly for the left circularly polarized wave

$$\mathbf{E_L} = E_0 \cos(\omega t)\mathbf{e}_x + E_0 \sin(\omega t)\mathbf{e}_y,$$

we find

$$\varepsilon_L = 1 + \frac{4\pi Ne^2}{m(\omega_0^2 - \omega^2 - \omega\omega_c)}.$$

The difference between ε_L and ε_R is therefore

$$\delta\varepsilon(\omega) = \varepsilon_L - \varepsilon_R$$
$$= \frac{4\pi Ne^2}{m}\left[\frac{1}{\omega_0^2 - \omega^2 - \omega\omega_c} - \frac{1}{\omega_0^2 - \omega^2 + \omega\omega_c}\right]$$
$$= \frac{4\pi Ne^2}{m} \cdot \frac{2eB\omega/mc}{(\omega_0^2 - \omega^2)^2 - (eB\omega/mc)^2}.$$

5053

An electrically neutral collisionless plasma of uniform density n_0 is at rest and is permeated by a uniform magnetic field $(0, 0, B_0)$. Consider an electromagnetic wave of frequency ω propagating parallel to the magnetic field. Show that the wave splits into two waves for which the refractive indices are

$$n_R^2 = 1 - \frac{\omega_p^2/\omega^2}{1 - (\omega_c/\omega)}, \quad n_L^2 = 1 - \frac{\omega_p^2/\omega^2}{1 + (\omega_c/\omega)},$$

where the plasma frequency is $\omega_p = (4\pi n_0 e^2/m_e)^{\frac{1}{2}}$ and the cyclotron frequency is $\omega_c = eB_0/m_e c$. Show that these waves are, respectively, right-hand and left-hand circularly polarized. Explain physically why the refractive index can be less than one. What happens when it vanishes? What happens when it becomes infinite? (You may assume that only the electrons respond to the wave and that the positive charges remain uniformly distributed.)

(*UC, Berkeley*)

Solution:

Suppose the neutral plasma consists of free electrons and an equal number of positive charges. Only the free electrons, for which $\omega_0 = 0$, take part significantly in the oscillations. Using the results of problem 5045 we have, as $\mu \approx \mu_0 = 1$,

$$n_R^2 = \varepsilon_R = 1 - \frac{4\pi n_0 e^2}{m_e \omega^2 (1 - \omega_c/\omega)} = 1 - \frac{\omega_p^2}{\omega(\omega - \omega_c)},$$

$$n_L^2 = \varepsilon_L = 1 - \frac{4\pi n_0 e^2}{m_e \omega^2 (1 + \omega_c/\omega)} = 1 - \frac{\omega_p^2}{\omega(\omega + \omega_c)}.$$

Since the phase velocity of electromagnetic waves in the medium, $c(\mu\varepsilon)^{-\frac{1}{2}} \approx c\varepsilon^{-\frac{1}{2}}$, may exceed the velocity of light speed c in vacuum, the refractive index $n = \frac{c}{v}$ may be less than one. Physically, as

$$n \approx \sqrt{\varepsilon} = \sqrt{1 + 4\pi\chi_e},$$

where χ_e is the polarizability of the medium, $n < 1$ means that $\chi_e < 0$. As the electric dipole moment per electron is $\mathbf{p} = \chi_e \mathbf{E}$, this means that the polarization vector $\mathbf{P}$ of the medium caused by the external waves is antiparallel to $\mathbf{E}$.

With $\omega_0 = 0$, the group velocity $v_g = cn$ (Problem **5051**). Thus $v_g = 0$ when $n = 0$. This means that a signal consisting of such waves will be turned back at that point. Consider the case $n_L^2 = 0$. We have

$$\omega^2 + \omega_c \omega - \omega_p^2 = 0,$$

or

$$\omega = \omega_{Lc} = \frac{-\omega_c + \sqrt{\omega_c^2 + 4\omega_p^2}}{2}.$$

Similarly we have for $n_R^2 = 0$,

$$\omega = \omega_{Rc} = \frac{\omega_c + \sqrt{\omega_c^2 + 4\omega_p^2}}{2}.$$

Thus $\omega_{Rc} > \omega_{Lc}$. If $\omega > \omega_{Rc}$, then both n_R and n_L are real, and propagation is possible for both polarizations. A plane electromagnetic wave in the medium will split into two circularly polarized waves with different refraction properties. If $\omega_{Rc} > \omega > \omega_{Lc}$, n_R is imaginary and propagation is possible for only the left circularly polarized wave. A plane electromagnetic wave will become left circularly polarized in the medium. If $\omega < \omega_{Lc}$, then propagation is not possible for both circular polarizations. Note that for n imaginary, say $n = i\eta$, $e^{ikr} = e^{-\eta r}$ and the amplitude attenuates exponentially.

If $\omega = \omega_c$, $n_R = i\infty$, $n_L = (1 - \frac{\omega_p^2}{2\omega_c^2})^{\frac{1}{2}}$. As $\omega_p \gg \omega_c$ generally, n_L is also a large imaginary number. No propagation is possible. Both n_R, $n_L = i\infty$ if $\omega = 0$. In such a case, there is no wave but only an electrostatic field which separates the positive and negative charges at the boundary of the plasma. Then the plasma surface acts as a shield to external electrostatic fields.

5054

A radio source in space emits a pulse of "noise" containing a wide band of frequencies. Because of dispersion in the interstellar medium the pulse arrives at the earth as a whistle whose frequency changes with time. If this rate of change (frequency versus time) is measured and the distance D to the source is known, show that it is possible to deduce the average electron density in the interstellar medium (assumed fully ionized). (Hint: Look at the response of a free electron to a high frequency electric field to deduce the relation between frequency and the wave number $2\pi/\lambda$).

(*CUSPEA*)

Solution:

Considering the interstellar medium as a tenuous plasma, we have from Problem **5044**

$$n = (1 - \omega_p^2/\omega^2)^{\frac{1}{2}}, \tag{1}$$

where ω is the frequency of the transversing radio wave, $\omega_p = \sqrt{\frac{Ne^2}{m\varepsilon_0}}$ is the plasma frequency of the medium, N being the average number density of the electrons in the medium. With the wave number $k = \frac{\omega}{c} n$, the group velocity v_g of the electromagnetic wave is given by

$$v_g^{-1} = \frac{dk}{d\omega} = \frac{n}{c} + \frac{\omega}{c}\frac{dn}{d\omega}.$$

Equation (1) gives $\frac{dn}{d\omega} = \frac{\omega_p^2}{\omega^3} \cdot \frac{1}{n}$, so that

$$v_g = nc\left(n^2 + \frac{\omega_p^2}{\omega^2}\right)^{-1} = nc.$$

Since a pulse propagates with the group velocity, the propagating time from the source to the earth is approximately

$$t = \frac{D}{v_g} = \frac{D}{nc} = \frac{D}{c}\left(1 - \frac{\omega_p^2}{\omega^2}\right)^{-\frac{1}{2}}.$$

Thus

$$dt = \frac{D}{c} \cdot \left(-\frac{1}{2}\right)\left(1 - \frac{\omega_p^2}{\omega^2}\right)^{-\frac{3}{2}}\left(2\frac{\omega_p^2}{\omega^3}\right)d\omega,$$

or

$$\frac{d\omega}{dt} = -\frac{c}{D}\frac{\omega^3}{\omega_p^2}\left(1 - \frac{\omega_p^2}{\omega^2}\right)^{\frac{3}{2}}.$$

Thus if D and $\frac{d\omega}{dt}$ are known, we can calculate $\omega_p^2 = \frac{Ne^2}{m\varepsilon_0}$ and hence the average electron density N.

5055

A pulsar emits a pulse of broadband electromagnetic radiation which is 1 millisecond in duration. This pulse then propagates 1000 light years (10^{21} cm) through interstellar space to reach radio astronomers on earth.

(a) What must be the minimum bandwidth of a radio telescope receiver in order that the observed pulse shape be not distorted greatly?

(b) Now consider that the interstellar medium contains a low density plasma (plasma frequency $\omega_p = 5000$ radians/sec). Estimate the difference in measured pulse arrival times for radio telescopes operating at 400 MHz and 1000 MHz. Recall that the dispersion relation for a plasma is $\omega^2 = k^2 c^2 + \omega_p^2$.

(MIT)

Solution:

(a) The uncertainty principle $\Delta\nu\Delta t \approx 1$ gives the minimum bandwidth of the radio telescope receiver as

$$\Delta\nu \sim \frac{1}{\Delta t} = 10^3 \text{ Hz}.$$

(b) The group velocity of electromagnetic waves in the interstellar medium is

$$v_g = \frac{\partial\omega}{\partial k} = c\sqrt{1 - \omega_p^2/\omega^2}.$$

For operating frequencies $\omega_1 = 4 \times 10^8 \text{ s}^{-1}$, $\omega_2 = 10^9 \text{ s}^{-1}$ we have

$$\frac{\omega_p}{\omega_1} = 1.25 \times 10^{-5}, \quad \frac{\omega_p}{\omega_2} = 10^{-6}.$$

With an interstellar distance $L = 10^{19}$ m, the difference in the measured pulse arrival times is

$$\Delta t = \frac{L}{v_{g1}} - \frac{L}{v_{g2}} = \frac{L}{c}\left[\left(1 - \frac{\omega_p^2}{\omega_1^2}\right)^{-\frac{1}{2}} - \left(1 - \frac{\omega_p^2}{\omega_2^2}\right)^{-\frac{1}{2}}\right]$$

$$\approx \frac{L}{2c}\left[\left(\frac{\omega_p}{\omega_1}\right)^2 - \left(\frac{\omega_p}{\omega_2}\right)^2\right] = 0.052 \text{ s}.$$

5056

A pulsar emits short regularly spaced burst of radio waves which have been observed, e.g., at the frequencies $\omega_1 = 2\pi f_1 = 2563$ MHz and $\omega_2 = $

$2\pi f_2 = 3833$ MHz. It is noted that the arrival times of these bursts are delayed at the lower frequencies: the pulse at f_1 arrives 0.367 sec after the pulse at f_2. Attributing this delay to dispersion in the interstellar medium which is assumed to consist of ionized hydrogen with 10^5 electrons per m^3, give an estimate for the distance of this pulsar from earth.

(a) Show that the electron plasma frequency for a tenuous neutral gas consisting of heavy ions and free electrons is given in Gaussian units by

$$\omega_{\mathrm{p}} = \left(\frac{4\pi N e^2}{m_{\mathrm{e}}} \right)^{\frac{1}{2}},$$

where N is the electron density.

(b) Using that result and remembering that the index of refraction of a tenuous plasma is given by $n = \sqrt{\varepsilon} = (1 - \frac{\omega_{\mathrm{p}}^2}{\omega^2})^{\frac{1}{2}}$, calculate the distance of the pulsar.

(*Chicago*)

Solution:

(a) In a neutral plasma, when the distribution of the electrons is perturbed and undulates non-uniformly, an electric field will be produced which causes the electrons to move in a way to tend to return the plasma to the neutral state. The characteristic (angular) frequency of the undulation can be estimated as follows. Consider an electron of the plasma in an electric field $\mathbf{E}$, the equation of motion is

$$m_{\mathrm{e}} \frac{d\mathbf{v}}{dt} = -e\mathbf{E}.$$

The motion of electrons produces a current of density

$$\mathbf{j} = -N e \mathbf{v}.$$

Combining the above we have

$$\frac{\partial \mathbf{j}}{\partial t} = \frac{N e^2}{m_{\mathrm{e}}} \mathbf{E},$$

or, taking divergence of the two sides,

$$\frac{\partial}{\partial t} (\nabla \cdot \mathbf{j}) = \frac{N e^2}{m} \nabla \cdot \mathbf{E}.$$

Using the continuity equation $\nabla \cdot \mathbf{j} = -\frac{\partial \rho}{\partial t}$ and Maxwell's equation $\nabla \cdot \mathbf{E} = 4\pi\rho$, we obtain

$$\frac{\partial^2 \rho}{\partial t^2} + \frac{4\pi N e^2}{m_e}\rho = 0 ,$$

or

$$\frac{\partial^2 \rho}{\partial t^2} + \omega_p^2 \rho = 0$$

with

$$\omega_p = \sqrt{\frac{4\pi N e^2}{m_e}} .$$

This equation shows that the charge density at a point oscillates simple harmonically with characteristic (angular) frequency ω_p.

(b) Using the results of Problem **5055** we find the distance from the pulsar to the earth:

$$L = c\Delta t \left[\left(1 - \frac{\omega_p^2}{\omega_1^2}\right)^{-\frac{1}{2}} - \left(1 - \frac{\omega_p^2}{\omega_2^2}\right)^{-\frac{1}{2}} \right]^{-1} .$$

The electron plasma frequency

$$\omega_p = \sqrt{\frac{4\pi N e^2}{m_e}} = \sqrt{4\pi N r_0 c^2} = 1.79 \times 10^4 \text{ s}^{-1} ,$$

$r_0 = 2.82 \times 10^{-13}$ cm being the classical radius of electron.

The observed (angular) frequencies are $\omega_1 = 2.563 \times 10^9$ s^{-1}, $\omega_2 = 3.833 \times 10^9$ s^{-1}. As $\omega_1, \omega_2 \gg \omega_p$ we have approximately

$$L \approx \frac{2c\Delta t}{\omega_p^2 \left(\frac{1}{\omega_1^2} - \frac{1}{\omega_2^2}\right)} .$$

With $\Delta t = 0.367$ s,

$$L = 8.16 \times 10^{20} \text{ cm} = 8.5 \times 10^2 \text{ light years} .$$

INDEX TO PROBLEMS

(Classified according to specific character of a problem.)

Printed in the USA
CPSIA information can be obtained
at www.ICGtesting.com
JSHW010425231023
50599JS00003B/6

9 789810 206260